Entomology Monographs

Series Editor

Hideharu Numata, Kyoto University
Kyoto, Japan

Insects are the most diverse group of organisms, and many biological advances have been derived from this group. Although entomology is considered to be a classic science, recent developments in molecular methods, application of modern theoretical concepts and collaboration with related sciences have opened new directions in entomology. Japanese scientists play a significant role in these fields, and this book series will focus on such developments. The book series Entomology Monographs publishes refereed volumes on all aspects of entomology, including ecology, ethology, physiology, taxonomy, systematics, morphology, evolutionary developmental biology, genetics, biochemistry, and molecular biology in insects and related arthropods. Authors are not restricted to Japanese entomologists, and other international experts will also be considered on the basis of their recent contribution to these fields.

Tetsuro Shinoda • Toyomi Kotaki
Hideharu Numata
Editors

Juvenile Hormone

Structure, Synthesis and Function

Springer

Editors
Tetsuro Shinoda
Faculty of Food and Agricultural Sciences
Fukushima University
Fukushima, Fukushima, Japan

Toyomi Kotaki
Institute of Agrobiological Sciences
National Agriculture and Food Research Organization
Tsukuba, Japan

Hideharu Numata
Kyoto University
Kyoto, Kyoto, Japan

ISSN 2522-526X ISSN 2522-5278 (electronic)
Entomology Monographs
ISBN 978-981-95-8604-2 ISBN 978-981-95-8605-9 (eBook)
https://doi.org/10.1007/978-981-95-8605-9

This Springer imprint is published by the registered company Springer Nature Singapore Pte Ltd.
The registered company address is: 152 Beach Road, #21-01/04 Gateway East, Singapore 189721, Singapore

Preface

Photo A. Sir Vincent B. Wigglesworth (1899–1994), who first identified juvenile hormone as an inhibitor of insect metamorphosis. (Source: Sláma K (1994) *Eur J Entomol* 91:255–256 under open-access license, CC BY 4.0)

Photo B. Carroll M. Williams (1916–1991), pioneer of juvenile hormone research and proponent of "third-generation insecticides." (Courtesy of the National Academies Archive)

Juvenile hormone (JH) was first discovered as an inhibitor of insect metamorphosis by Sir Vincent B. Wigglesworth in 1934 (photo A). This hormone possesses a unique sesquiterpenoid structure that is almost unparalleled among known hormones. In addition to maintaining larval traits as initially described, subsequent research has revealed that JH regulates a broad spectrum of physiological processes, including reproduction, diapause, and phenotypic polymorphism.

In the 1960s, Carroll M. Williams (photo B) proposed the use of JH analogs as "third-generation insecticides," a concept that directly led to the development of methoprene, pyriproxyfen, and other insect growth regulators that remain essential tools in pest management. Because of its unusual chemical structure, the biosynthesis and mode of action of JH remained largely unresolved for many years; however, recent advances in molecular biology, genomics, and analytical chemistry have propelled significant progress. This book summarizes these developments and provides an updated overview of current research on JH.

The volume is organized into four parts, covering the history, structure, synthesis, function, and application of JH. Each chapter is self-contained yet interconnected, enabling readers to approach the volume from any topic of interest.

Part I (Introduction) outlines the historical development of JH research, from early physiological observations to the emergence of molecular mechanisms. Chapter 1 reviews major milestones, unresolved issues, and future perspectives.

Part II (Structure and Synthesis) details the molecular structures, biosynthetic pathways, key enzymes, and regulatory mechanisms of JH. Topics include the history of JH structural elucidation and the identification of $JHSB_3$ (Chap. 2), the central role of JHAMT in JH synthesis (Chap. 3), mechanisms of transport and degradation (Chap. 4), unresolved late steps in the biosynthetic pathway (Chap. 5), endocrine and neuronal regulation of synthesis (Chap. 6), and genetic analysis of JH production in *Drosophila* (Chap. 7).

Part III (Function) examines how JH regulates metamorphosis, reproduction, diapause, sexual differentiation, phenotypic plasticity, and environmental responses. Chapters cover transcriptional networks underlying metamorphosis (Chap. 8), comparisons between hemimetabolous and holometabolous insects (Chap. 9), molecular action of the JH receptor Met/Tai (Chap. 10), regulation of oogenesis and vitellogenesis (Chap. 11), endocrine and neural control of diapause (Chap. 12), JH-mediated polyphenism (Chap. 13), and methyl farnesoate–dependent signaling in crustaceans (Chap. 14).

Part IV (Application) focuses on applied aspects of JH biology. Topics include the role of JH in the mass production of edible crickets (Chap. 15), the development and application of JH analogs in pest management (Chap. 16), and strategies for discovering new hormone-targeting insecticides through high-throughput screening (Chap. 17).

Most contributors to this book have long-standing scientific relationships with the editors, including direct collaborations, shared research programs, and academic connections. Numata's work on seasonal adaptation and diapause naturally intersects with several authors investigating JH-mediated developmental timing. Shinoda's interest in JH began with studies on the bean bug, *Riptortus pedestris*, under Yasuo Chinzei, whose group was closely associated with Numata's laboratory. His subsequent research in the laboratory of Lynn M. Riddiford and James W. Truman at the University of Washington (1994–1996) deepened his involvement in the field and established a long-term collaboration with Marek Jindra, a leading researcher on JH receptors. During Shinoda's tenure at the National Institute of Agrobiological Sciences (NIAS) in Tsukuba, collaborative projects with many authors in this volume advanced research on JH biosynthesis, transport, degradation, and signaling. Kotaki, a colleague at NIAS, identified the heteropteran-specific $JHSB_3$ through collaboration with Numata and chemists at Osaka City University (currently Osaka Metropolitan University). This achievement expanded our understanding of JH structural diversity and complementing the biochemical and functional studies presented here. All the contributors to this volume collectively add depth and breadth to the book, and their diverse expertise has greatly enriched the scope and quality of this work.

We extend our sincere thanks to all the authors for their contributions and collaboration throughout the preparation of this volume. We also appreciate the guidance and support of Fumiko Yamaguchi, Springer Nature, Japan, which greatly facilitated the organization and completion of this book. We hope that this collection serves not only as a reference for researchers in insect endocrinology, physiology, molecular biology, and applied entomology but also as a valuable resource for students and younger scientists entering this field.

Fukushima, Japan — Tetsuro Shinoda
Tsukuba, Japan — Toyomi Kotaki
Kyoto, Japan — Hideharu Numata
January 2026

Contents

Part I
Introduction

Chapter 1
Juvenile Hormone: How Did We Get Here from There and Where Are We Going?

Lynn M. Riddiford and James W. Truman

Abstract Juvenile hormone is a key hormone in the control of insect development and reproduction. This chapter provides a brief overview of its discovery and subsequent identification as a sesquiterpene, the first example of this type of lipid serving as a circulating hormone. We then describe its biosynthesis and the relationship of its circulating titer to the diverse life history strategies that resulted in the great array of insect species that are evident today. We then turn to the cellular and molecular actions of juvenile hormone. Its receptor is a basic-helix-loop-helix transcription factor. The identification of the juvenile hormone receptor has led to a better understanding of the relationship of this hormone to the molting hormone, ecdysone, and an understanding of its ligand-binding properties has opened the door to the long-standing promise of using juvenile hormone-like molecules for selective insecticides. Some aspects of juvenile hormone action, though, remain poorly understood. The morphostatic actions of juvenile hormone likely involve interactions with both local and circulating morphogen systems, and details of these interactions are yet to be revealed.

Keywords Juvenile hormone · Krüppel homolog 1 · Methoprene-tolerant · Germ cell-expressed

Dedication: We dedicate this article to Carroll M. Williams, who guided us in our first studies with juvenile hormone and alerted us to its fascinating roles in the life of insects.

L. M. Riddiford · J. W. Truman (✉)
Friday Harbor Laboratories and Department of Biology, University of Washington, Friday Harbor, WA, USA
e-mail: jwt@uw.edu

T. Shinoda et al. (eds.), *Juvenile Hormone*, Entomology Monographs,
https://doi.org/10.1007/978-981-95-8605-9_1

1.1 Introduction

From the very beginning, there has been nothing "usual" about juvenile hormone (JH). In amphibians, the other group of terrestrial animals with a pronounced metamorphosis, their metamorphic transition is stimulated by the appearance of thyroxine and its metabolites. Metamorphic control in insects is quite different. The ecdysteroid molting hormones work through a molecular pathway similar to that of thyroxine, and ecdysteroids drive the metamorphic process. However, ecdysteroids drive all of the molts, including those from one juvenile stage to the next. The ability of ecdysteroids to initiate a metamorphic transition is controlled by JH, which is a negative controller of metamorphosis. Its presence maintains the stability of the larval form, and metamorphosis can only begin when JH disappears.

The structure of JH is as unusual as its actions. It was the first sesquiterpene hormone to be discovered, and its immediate precursor, methyl farnesoate, which is the JH forerunner from crustaceans, remains as the only other hormone in this class of molecules. The very hydrophobic nature of JH presents problems for a hormone that must circulate in the blood. Consequently, special binding proteins are present in the blood to carry JH to its targets and to protect it from metabolism by blood esterases.

The lipid nature of JH suggested that it likely had an intracellular receptor. The most likely possibility was that the receptor would be from the nuclear receptor family since these bind a variety of small ligands ranging from 9-*cis*-retinoic acid and thyroxine to a wide variety of steroids. The eventual discovery that the JH receptor was derived from the family of basic-helix-loop-helix (bHLH) transcription factors was a surprise since very few of the family members bind small ligands, and these are typically xenobiotics or tissue metabolites, as seen in the case of the aryl hydrocarbon receptor.

This chapter provides a brief introduction to research on JH and its actions. We also endeavor to point out interesting aspects of JH research that may have been neglected as the field has moved forward.

1.2 Discovery of JH and Determination of Its Structure

The history of juvenile hormone (JH) research has been recently reviewed (Riddiford 2020), so we will focus here on the high points. The key to the control of insect metamorphosis began about 90 years ago with the studies of V.B. Wigglesworth. His favorite experimental animal, the blood-sucking bug *Rhodnius prolixus*, was specialized so that a single, massive blood meal would allow the nymph to grow and molt to the next stage or, if a female adult, to produce a complete egg batch. Using *R. prolixus* in a series of decapitation and parabiosis experiments, Wigglesworth (1934) showed that there were two hormones released from the head region—one that promoted molting after a blood meal and one that inhibited metamorphosis. He found a pair of glands named the corpora allata (CA) in the head region behind the brain that showed cyclical changes during nymphal molting and metamorphosis. At the time, he thought that the CA contained both hormones but later found that the CA only produced the inhibitory hormone, which he then named juvenile hormone (Wigglesworth 1940). Wigglesworth (1936) also showed that the CA were necessary for ovarian maturation. The early experiments on JH function in either reproduction or metamorphosis were largely confined to manipulations involving the CA, however. The extirpation of the CA from young larvae or nymphs resulted in their premature metamorphosis or, from the adult female, prevented egg production. Conversely, the implantation of "active" CA into older larvae or nymphs prevented their metamorphosis and caused supernumerary larval molts. Such implants also stimulated reproduction in adults.

The determination of the structure of JH was hampered by the problem of obtaining enough hormone for structural purification. The solution to the problem came serendipitously from the work of the pioneering insect endocrinologist, Carroll Williams, at Harvard University. Williams had a long fascination with wild silk moths, such as the cecropia moth, *Hyalophora cecropia.* During experiments comparing the short lifespan of the adult moth with the extended longevity of the pupa, which undergoes an obligatory 10-month diapause arrest, Williams parabiosed adult abdomens of *H. cecropia* to diapausing pupae. As expected, this procedure extended the lifespan of the adult abdomen, but when the pupae finished their metamorphic development, some repeated the pupal stage, rather than becoming an adult (Williams 1956, 1961). Second pupae were produced, though, only when male abdomens were parabiosed, and only when using abdomens of male *H. cecropia,* but not other silkmoths such as *Antheraea polyphemus*. Williams showed that the source of the abdominal JH was the male's CA. Later work showed that the male CA secretes JH acid, rather than JH, but the former is methylated to JH when it is stored in the accessory glands (Shirk et al. 1976). Why this should occur in *H. cecropia* and not other wild silk moths, such as *A. polyphemus,* is still unknown.

Williams' discovery provided a source for extractable JH. The "golden oil" obtained by ether extracts of male abdomens of *H. cecropia* was rich in JH activity, and application of this material mimicked the effects of implantation of active corpora allata. Cecropia extracts provided the raw material for the eventual purification and structural determination of JH. Applications of farnesol had previously been shown to have JH-like effects (Wigglesworth 1963), and the insect JH proved to be a modified version of methyl farnesoate. Röller et al. (1967) first chemically identified the hormone from cecropia as methyl-*dl-trans, trans, cis*-10-epoxy-7-ethyl-3,11-dimethyl-2,6-tridecadienoate (JH I). Subsequently, a second JH (JH II) was isolated from the JH mixture and had a methyl group instead of one of the ethyl groups. Later, Judy et al. (1973) identified an all-methyl form of JH (JH III) from cultures of the corpora allata of adult female *Manduca sexta* (the tobacco hornworm). Interestingly, JH I and JH II are specific to Lepidoptera, and JH III is most widely found among the insects, including very basal insects, such as the firebrat, *Thermobia domestica* (Baker et al. 1984). Features of the hormone structure that are essential for activity are the terminal epoxide and methyl ester. At least seven different JHs are now known that differ in methyl versus ethyl side chains and the presence or absence of internal epoxide groups, but they all possess the terminal epoxide and methyl ester.

1.3 The JH Titer and Its Control

As with other hormones, the circulating titer of JH is a product of its release into the blood and its degradation and uptake by tissues. Early studies of the titer of JH in the blood depended on bioassays of extracts either of whole animals or of blood. Bioassays dependent on the response of whole insects typically required substantial amounts of material, so favored assays were those that depended on local responses of the test insect, such as the wax-wound test in *Galleria mellonella* (Gilbert and Schneiderman 1960) or the black larval assay in *M. sexta* (Fain and Riddiford 1975). These assays provided semiquantitative estimates of the amount of JH in extracts and were the basis of a number of blood titers. With the eventual determination of the structure of JH, it was possible to devise chemical assays, with the current "gold standard" being liquid chromatography-mass spectrometry, such as that carried out by Fernando Noriega and coworkers (Ramirez et al. 2020). As detailed below, an important target gene for JH action is the transcription factor *Krüppel homolog 1* (*Kr-h1*). In several cases, the levels of Kr-h1 transcripts provide a proxy for the JH titer (e.g., Minakuchi et al. 2009). Although not a definitive determination of JH levels, as would be given by chemical determination, most cases show that *Kr-h1* transcripts provide a reasonable estimate of the JH titer.

As shown in Fig. 1.1, there is a strong phylogenetic influence on the JH titer. In the ametabolous insect *Thermobia domestica*, JH levels peak at the end of embryogenesis, but JH determinations (Truman et al. 2024) and Kr-h1 expression (Fernandez-Nicolas et al. 2023) indicate that levels then remain low through juvenile growth without an obvious decline prior to the molt to the adult stage. Hemimetabolous insects, by contrast, also show a prominent JH peak during late embryogenesis, but JH levels then become elevated through much of the nymphal series with a prominent drop on entry into the last nymphal instar to allow the transformation of the nymph into the adult. The holometabolous pattern likewise shows a late embryonic peak followed by the reappearance of JH during larval growth and a decline in JH after entry into the last larval stage. JH reappears transiently during the formation of the pupa and then is absent until after the emergence of the adult. There are both tissue- and species-level differences in the need for this prepupal JH, and it remains as one of the poorly understood features of metamorphic physiology.

The titer of circulating JH is a product of two factors: (1) the biosynthesis of JH by the cells of the CA and (2) its subsequent degradation by JH esterase and/or by JH epoxide hydrolase. The first insights into the control of JH biosynthesis came from studies on adult cockroaches, *Leucophaea maderae,* which showed that

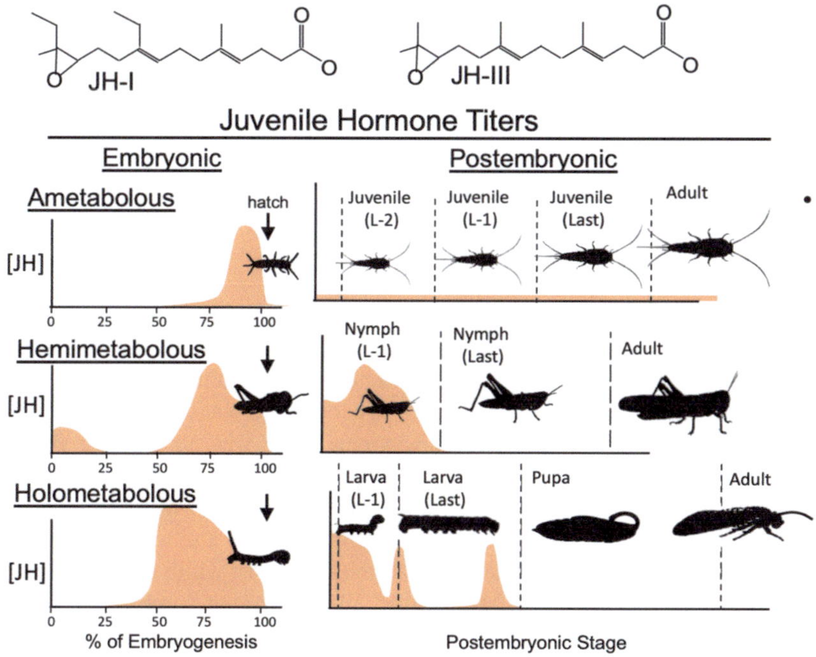

Fig. 1.1 Structure and titers of juvenile hormones in insects. The chemical structures show the first juvenile hormone (JH I), whose structure was determined, and the most widely occurring juvenile hormone (JH III) in insects. The juvenile hormone titers show representative embryonic and postembryonic titers from ametabolous (the firebrat, *Thermobia domestica*), hemimetabolous (the locusts *Locusta migratoria* [embryonic] and *Schistocerca gregaria* [postembryonic]), and holometabolous (the tobacco hornworm, *Manduca sexta*) insects. Thermobia titers are based on JH measurements (Truman et al. 2024) and *Krüppel homolog 1* (*Kr-h1*) expression profiles (Fernandez-Nicolas et al. 2023). Hemimetabolous and holometabolous patterns based on the review by Truman and Riddiford (1999)

transection of the connections between the brain and CA resulted in the hypertrophy of the gland and stimulation of oogenesis (Scharrer 1952). Work on another cockroach, *Diploptera punctata*, led to the isolation of the allatostatin peptide (see review by Stay and Tobe 2007). The factors that control JH biosynthesis in juvenile stages, though, are both complex and phylogenetically varied. Regulation of JH biosynthesis in larvae and nymphs cannot be explained solely at the level of suppression and is still an open question.

In insects that use JH III, JH is synthesized from mevalonate via farnesol and farnesoic acid, with the final two steps typically being epoxidation followed by methylation. For JH I and II in Lepidoptera, homomevalonate is the starting point. Primary inactivation of JH occurs in the hemolymph by a JH-specific esterase, which appears after the larva has reached the critical weight for metamorphosis (see review by Goodman and Cusson 2012).

1.4 JH Action—The JH Receptor

JH has diverse actions on reproduction and on the progression of life histories (Wyatt and Davey 1996; Goodman and Granger 2005). The requirement of JH in reproduction is most pronounced in females, but insects show a bewildering array of reproductive strategies, which may or may not include JH. When JH is involved, it is often in the early progression of the primary oocyte and in the production and uptake of yolk. The "flag ship" action of JH and the reason for its name is its involvement in maintaining the juvenile state. Beyond this anti-metamorphic action, JH is also involved in color polyphenisms of larvae and nymphs, in caste determination in some social insects, in seasonal wing polyphenisms in aphids and other insects, and in regulating behavioral shifts in worker honeybees (Nijhout and Wheeler 1982; Robinson and Vargo 1997). The search for JH receptors was complicated because, unlike for insect steroid and peptide hormones, there was no existing vertebrate precedent for a sesquiterpene hormone receptor. Also, some of the actions of JH, such as its induction of shape change in the ovarian follicle cells of *R. prolixus* to allow yolk uptake by the oocyte, occurred within seconds and were thought to act via a membrane receptor (e.g., Davey and Huebner 1974), while the developmental actions of JH were assumed to be genomic, likely through some kind of nuclear receptor. In 1986, Tom Wilson isolated a *Drosophila* mutant that could feed on a diet containing high amounts of the JH mimic methoprene. He named the corresponding gene *Methoprene-tolerant* (*Met*) and suggested that *Met* was involved in JH action (Wilson and Fabian 1986). There was resistance to the idea that Met was the , though, because the lack of Met did not produce a phenotype that mimicked the lack of JH. Also, Met was a member of the bHLH family of transcription factors, rather than being a nuclear receptor. Its closest relative was the vertebrate aryl hydrocarbon receptor, which links xenobiotics and cellular metabolites to cell function.

Wilson's group later found a second gene, *germ cell-expressed* (*gce*), that was structurally similar to *Met* and thought to possibly be an alternative JH receptor. The difficulties posed by *Drosophila* potentially having redundant JH receptor genes could be circumvented in the flour beetle *Tribolium castaneum*, which had only a single such gene. Konopova and Jindra (2007) showed that in *T. castaneum*, suppression of *Met* by RNA interference (RNAi) in the third or fourth larval instar caused precocious metamorphosis after the fifth or sixth instar. This precocious metamorphosis could not be prevented by the application of JH III or methoprene. Moreover, *Met*-RNAi in the pupa prevented the formation of a "second-pupa" after exposure of the pupa to JH III or methoprene. These findings strongly suggested that Met was the JH receptor. Later studies on *Drosophila* showed that genetic removal of both *gce* and *Met* robustly mimicked the effect of removing JH in fly larvae (Abdou et al. 2011). Although the use of the name Met for the JH receptor is firmly entrenched in the literature, *gce* is thought to be the ancestral gene, with the *Met* gene arising as a duplication of *gce* (Baumann et al. 2010). Met was shown to be a basic helix-loop-helix (bHLH) protein with two PAS domains, one of which binds JH (Charles et al. 2011). After binding JH, Met also binds another bHLH-PAS domain protein, Taiman, and the heterodimer can move into the nucleus, where they bind to JH response element(s) on JH target genes (Jindra et al. 2024).

The resolution of the structure of the JH binding pocket of Gce has provided important insights into how a diverse set of agonist structures can exhibit JH-like actions (Jindra et al. 2015; Jindra et al. 2024). Understanding the principles by which ligands interact with the binding pocket may finally allow the fulfillment of the promise of finding JH mimics that are selectively active on particular insect groups (Jindra et al. 2024).

1.5 The Cellular and Molecular Actions of JH

The reproductive and developmental actions of JH were originally thought to have little or nothing in common. Some reproductive actions of JH occur within seconds, such as the JH-induction of yolk uptake by the ovarioles of *R. prolixus* (Davey and Huebner 1974), while the developmental actions of JH usually require hours to become manifest. Also, the reproductive actions of JH often involved prominent gene activation, as in stimulating yolk protein synthesis, while the developmental actions primarily involved the repression of new gene activity to maintain the juvenile form. However, over the years, there has been a convergence of the molecular details of JH action during juvenile growth versus adult reproduction. The genomic actions for both phenomena involve Met/Gce. The functioning of Taiman as a binding partner for Met/Gce was first discovered for mosquito reproduction (Li et al. 2011), but later was also shown to be the case for the anti-metamorphic actions of JH.

The major target gene mediating both the developmental and the reproductive actions of JH is Kr-h1 (Jindra et al. 2015; Song et al. 2014). In the silkworm *Bombyx mori,* Kr-h1 directly represses both the pupal specifier *broad* and the adult specifier *E93* genes (Kayukawa et al. 2016, 2017) and thus prevents metamorphosis. In mosquitoes, *Kr-h1* is a key gene in mediating the reproductive effects of JH (Ojani et al. 2018). Recent studies on the locust suggest that developmental and reproductive actions of JH may involve the recruitment of different protein cofactors (Wu et al. 2021). In the nymph, Kr-h1 recruits a corepressor, C-terminal binding protein (CtBP), to suppress transcription of *E93*, whereas in the adult, it recruits the coactivator, CREB-binding protein, to drive expression of ribosomal protein L36 to support fat body growth to make vitellogenin.

The signifying character for the different life stages in insects is enshrined in the cuticle that is deposited by the epidermal cells. Not only does this cuticle make the exoskeleton that gives each stage its characteristic form, but it also has a complex set of proteins that are characteristic of each stage. Growth and changes in form require the making of a new cuticle and the shedding of the old one. These molts are orchestrated by ecdysteroids, but the nature and trajectory of each molt are set at the start of the ecdysteroid surge by the presence or absence of JH. Cellular aspects of the metamorphic actions of JH have been studied using the epidermis of *M. sexta.*

Normally, the larval epidermis becomes irreversibly committed to making pupal cuticle at the time the larva stops feeding and begins "wandering" in search of a pupation site. This is the first time the epidermis experiences ecdysteroids acting in the absence of JH. Culturing epidermis from feeding larvae in vitro showed that exposure to 20-hydroxyecdysone (20E) in the absence of JH caused this pupal commitment, whereas exposure to 20E in the presence of JH I prevented this change (Riddiford 1976). Further studies showed that at the time of pupal commitment, a new transcription factor, Broad (Broad-complex), appeared across the segment in the same pattern as the change to pupal commitment (Zhou and Riddiford 2001). In the presence of JH, Broad did not appear. In *Drosophila*, the *broad* null mutant cannot metamorphose (DiBello et al. 1991), and Broad was subsequently found to be the pupal specification factor in *Drosophila* (Zhou and Riddiford 2002). Indeed, the expression of *broad* is associated with the formation of the pupal stage in all holometabolous insects.

The critical time for JH to influence the nature of a juvenile molt—e.g., from larva to pupa—is at the very outset of the molt. The end of this critical period, determining the nature of a molt, though, does not end the role for JH. A number of insects have juvenile forms that display color polyphenisms that may depend on environmental conditions. For *M. sexta*, for instance, larvae can show a black versus green polyphenism. The window during which JH can affect the color of the larva occurs about 18 h after the start of the molt (Safranek and Riddiford 1975), well after the larval nature of the molt has been established. Little is known about the molecular basis for this JH effect beyond the fact that the lack of JH results in the induction of enzymes required for melanin formation.

Another molt-related function of JH that has been understudied is its role during the formation of the prepupa. Early studies of the role of JH in determining life stages in the Holometabola involved a "high-low-no" relationship. High levels of JH resulted in a larva repeating the larval stage, intermediate levels of JH resulted in the formation of the pupa, while no JH resulted in adult differentiation. This idea made for simple summary diagrams that persist in developmental and entomological texts. While being easy to denote pictorially, this representation misinterprets the complexity of JH action during the larval-pupal transition. As noted above, the absence of JH is essential for ecdysone to commit larval tissues to pupal differentiation. The imaginal discs and primordia that are carried in the larva, though, undergo their commitment (and induction of *broad*) earlier in the last larval instar at the critical weight checkpoint (Mirth et al. 2009). Once the imaginal primordia begin their morphogenesis, JH treatment can arrest their further progression, but it cannot cause them to revert back to a premetamorphic condition. The result of JH treatment at these times is an additional larval instar with various imaginal primordia that produce pupal cuticle (Nijhout and Kremen 1998).

Different tissues in the larva may vary in their requirements for JH. The larval epidermis, for example, requires the presence of JH during a narrow window at the onset of an ecdysteroid-induced molt to determine the nature of the cuticle for the ensuing instar. The imaginal discs and imaginal primordia, though, have a tonic requirement for JH. Studies on *M. sexta* show that these imaginal tissues require the continuous presence of JH through the intermolt period to link their growth to nutrient intake and to prevent the discs and primordia from initiating premature adult morphogenesis (Truman et al. 2006). This morphostatic action of JH does not require ecdysteroids, and it ensures that adult morphogenesis is delayed until the critical weight checkpoint in the last larval stage. A similar morphostatic action of JH is observed in some insect embryos. For example, in the firebrat, *Thermobia domestica,* JH normally appears late in embryogenesis after morphogenesis is completed and as the embryo is undergoing tissue differentiation and maturation. Application of JH mimics earlier in embryogenesis, though, has potent morphostatic effects (Truman et al. 2024; Lv et al. 2024). It induces premature expression of Kr-h1 and arrests morphogenesis while causing premature tissue differentiation. This morphostatic function of JH is extremely pronounced in the embryos of insects derived from groups at the base of the insect radiation and may represent the ancestral developmental function of JH.

Developmental actions of farnesol derivatives extend well beyond the arthropods. In the yeast *Candida albicans*, farnesoic acid is secreted to induce the transition from a budding form to a hyphal form of growth (Hornby et al. 2001). In the annelid *Platynereis dumerilii,* methyl farnesoate has been identified as the brain hormone that maintains the juvenile stage and allows this stage to undergo posterior regeneration after injury (Álvarez-Campos et al. 2023). Methyl farnesoate in Crustacea affects reproduction and also various aspects of development, although the latter effects are subtle and diverse (Laufer and Biggers 2001). The developmental actions of farnesol derivatives outside of the arthropods indicate that these molecules have had a long involvement in development prior to their capture by insects to control the nature of molts.

1.6 Relationship of JH to Morphogens

The idea that JH was required for every molt from one juvenile stage to the next was established early in the field. Experiments on *R. prolixus* by Wigglesworth showed that when a headless first-stage nymph was parabiosed to a last-stage nymph, it would develop precocious adult-like characters when its partner molted to the adult stage (Wigglesworth 1934). Similarly, Piepho (1938) showed that integument from first instar *G. mellonella* larvae implanted into the body cavity of a last larval stage would form pupal cuticle when its host also molted to the pupal stage. Results like these led to the conclusion that JH was required for every juvenile molt.

First-stage juveniles were typically too small for the mechanical removal of their corpora allata. The application of allatocidal agents such as precocenes (see below) failed to induce a first-stage juvenile to precociously start metamorphosis. Rather, they molted to a second-stage juvenile and then often initiated metamorphosis at the next molt (Bowers and Martinez-Pardo 1977). These results were typically interpreted as a delay in the ability of the drugs to completely suppress the production of JH. Conclusive results came from Daimon et al. (2015) on *B. mori* and Smykal et al.

(2014) on the linden bug, *Pyrrhocoris apterus*. The embryonic silencing of genes involved in JH biosynthesis (genes) or with JH action (other genes) had no effect on the production of first-stage juveniles or on the ability of these juveniles to progress to the second instar. Metamorphosis was initiated, though, when these juveniles progressed to the next stage. The earliest juvenile instars of both *B. mori* and *P. apterus*, then, do not require JH for their production. Nor do they require Kr-h1, the main effector gene involved in JH action. In the absence of JH biosynthesis or JH action, *B. mori* larvae are maintained despite Kr-h1 transcripts remaining at basal levels throughout (Daimon et al. 2015).

How, then, does one reconcile the results of the *Bombyx* experiments with the classic results from Wigglesworth and Piepho? A simple explanation is that the blood of the last-stage larva or nymph contains a metamorphosis-promoting factor in addition to lacking JH (Truman et al. 2006). The best candidate for such a factor is myoglianin, a member of the Transforming Growth Factor-β family of ligands. Studies on *Gryllus bimaculatus* (Ishimaru et al. 2016) and on *Blattella germanica* (Kamsoi and Belles 2019) show that high levels of myoglianin in the nymph suppress the expression of JH acid methyltransferase in the CA, thereby promoting the JH decline needed for metamorphosis. The later study by He et al. (2020) showed that in *M. sexta*, relative muscle growth and the resulting elevation of myoglianin levels provided the mechanism for threshold size signaling. Also, the suppression of myoglianin production in *T. castaneum* resulted in maintained larval molting well past the critical weight. The role of myoglianin likely goes beyond actions on the CA and JH production. For example, in *Drosophila* larvae myoglianin is necessary at a critical weight to induce EcR-B1 in larval neurons (Awasaki et al. 2011). In the absence of this ecdysone receptor isoform, larval neurons remain locked in their larval form and cannot undergo pruning or death in response to the metamorphic surges of ecdysteroids. Importantly, there are likely other metamorphosis-promoting factors that are called into play before the threshold size for metamorphosis. For example, the studies by Chafino et al. (2019) on *T. castaneum* show that prior to a couple of instars before attaining threshold size, larvae become competent to express E93, the adult-specification gene, and that JH is then needed to suppress its expression.

1.7 Applied Aspects of JH Action

A breakthrough in JH research came when a Czech entomologist, Karel Sláma, came to work at Harvard and found that *P. apte*rus that he brought from Czechoslovakia would not undergo metamorphosis at Harvard, acting like they had been exposed to JH. A search through possible suspects finally revealed that American paper products made from the balsam fir had this juvenilizing effect while European paper, made from pine, was inactive (Sláma and Williams 1966). The active JH mimic in the American paper was purified, its structure determined, and it was called juvabione (Bowers et al. 1966). Intriguingly, it was extremely active against *P. apterus* and other bugs in the family Pyrrhocoridae but not against bugs in the closely related family Lygaeidae. This finding started the search for insect growth regulators that would selectively prevent metamorphosis in a particular group of insects. The discovery of the "paper factor" showed that compounds with high JH activity could be found in natural products and that it might be possible to find JH mimics that were very selective in the species that they targeted. The goal of finding extreme selectivity of JH agonists has been slow to be achieved, though. New approaches that are based on the JH receptor ligand binding pocket (Jindra et al. 2024) may eventually bring this promise to fruition.

The work on the paper factor ushered in the search for Insect Growth Regulators (IGRs)—insect control agents that targeted specific features of insect physiology—typically molting and metamorphosis—with the idea that such agents would pose less environmental risk to vertebrates as compared to insecticides that target mechanisms that are found in birds and mammals as well as insects. Also, among the insects, one might be able to tailor agents to act against pest groups of insects while leaving beneficial insects, such as bees, unharmed.

The search for IGRs spurred work on both JH and ecdysone. It was clear from the outset that the appropriateness of JH for insect control depended on whether or not the juvenile form of the pest was the destructive phase. While JH application to a lepidopteran crop pest would eventually cause the pest population to crash because the larvae could not enter metamorphosis, it would be little solace to the farmer who had had his/her crop devastated by giant caterpillars. A positive side effect is the ability of JH treatment to generate giant larvae, which have been used commercially to produce giant mealworms for fish bait and the pet food trade.

JH mimics have been a commercially viable option for control of pest insects whose adult stage is the damaging stage, such as mosquitoes and fleas. While the larvae are not killed, the pestiferous adult stage is not formed. JH mimics are not cheap to make but are used in niche markets, such as for flea and mosquito control, where resistance excludes the use of cheaper pesticides.

The impact of insects on agriculture generally involves the immature stages, especially in the case of holometabolous pests. Consequently, there has been a search for compounds that could block JH action. While a potent anti-JH has yet to be found, in their search for plant compounds that would inhibit JH action, Bowers and Martinez-Pardo (1977) found that chromene derivatives would cause precocious metamorphosis of the milkweed bug, so they called them precocenes. It turned out that precocenes were allatocidal since they are converted to cytotoxic agents by the epoxidase that is used in JH biosynthesis, which then destroys the corpora allata. Thus, they can be used for chemical allatectomy experiments. For some unknown reason, precocenes are less effective in holometabolous insects (Bowers and Feldlaufer 1982). Hopefully, the detailed knowledge of the binding pocket for JH may allow new strategies for designing and testing new compounds for anti-JH activity.

References

Abdou MA, Heb Q, Wen D, Zyaan O, Wang J, Xu J et al (2011) *Drosophila* Met and Gce are partially redundant in transducing juvenile hormone action. Insect Biochem Mol Biol 41:938–945. https://doi.org/10.1016/j.ibmb.2011.09.003

Álvarez-Campos P, Planques A, Bideau L, Vervoort M, Gazave E (2023) On the hormonal control of posterior regeneration in the annelid *Platynereis dumerilii.* J Exp Zool B Mol Dev Evol 340:298–315. https://doi.org/10.1002/jez.b.23182

Awasaki T, Huang Y, O'Connor MB, Lee T (2011) Glia instruct developmental neuronal remodeling through TGF-β signaling. Nat Neurosci 14:821–823. https://doi.org/10.1038/nn.2833

Baker FC, Lanzrein B, Miller CA, Tsai LW, Jamieson GC, Schooley DA (1984) Detection of only JH III in several life-stages of *Nauphoeta cinerea* and *Thermobia domestica.* Life Sci 35:1553–1560. https://doi.org/10.1016/0024-3205(84)90353-9

Baumann A, Barry J, Wang S, Fujiwara Y, Wilson TG (2010) Paralogous genes involved in juvenile hormone action in *Drosophila melanogaster.* Genetics 185:1327–1336. https://doi.org/10.1534/genetics.110.116962

Bowers WS, Feldlaufer MF (1982) *In vitro* inactivation of *Tenebrio molitor* corpora allata by a synthetic precocene analog. Gen Compar Endocrin 47:120–124. https://doi.org/10.1016/0016-6480(82)90093-4

Bowers WS, Martinez-Pardo R (1977) Antiallatotropins: inhibition of corpus allatum development. Science 197:1369–1371. https://doi.org/10.1126/science.197.4311.1369

Bowers WS, Fales HM, Thompson MJ, Uebel EC (1966) Juvenile hormone: identification of an active compound from balsam fir. Science 154:1020–1021. https://doi.org/10.1126/science.154.3752.1020

Chafino S, Ureña E, Casanova J, Casacuberta E, Franch-Marro X, Martín D (2019) Upregulation of *E93* gene expression acts as the trigger for metamorphosis independently of the threshold size in the beetle *Tribolium castaneum.* Cell Rep 27:1039–1049. https://doi.org/10.1016/j.celrep.2019.03.094

Charles JP, Iwema T, Epa VC, Takaki K, Rynes J, Jindra M (2011) Ligand-binding properties of a juvenile hormone receptor, *Methoprene-tolerant.* Proc Natl Acad Sci USA 108:21128–21133. https://doi.org/10.1073/pnas.1116123109

Daimon T, Uchibori M, Nakao H, Sezutsu H, Shinoda T (2015) Knockout silkworms reveal a dispensable role for juvenile hormones in holometabolous life cycle. Proc Natl Acad Sci USA 112:E4226–E4235. https://doi.org/10.1073/pnas.1506645112

Davey KG, Huebner E (1974) The response of the follicle cells of *Rhodnius prolixus* to juvenile hormone and antigonadotropin in vitro. Can J Zool 52:1407–1411. https://doi.org/10.1139/z74-178

DiBello PR, Withers DA, Bayer CA, Fristrom JW, Guild GM (1991) The *Drosophila* Broad-Complex encodes a family of related proteins containing zinc fingers. Genetics 129:385–397. https://doi.org/10.1093/genetics/129.2.385

Fain MJ, Riddiford LM (1975) Juvenile hormone titers in the hemolymph during late larval development of the tobacco hornworm, *Manduca sexta* (L.). Biol Bull 149:506–521. https://doi.org/10.2307/1540383

Fernandez-Nicolas A, Machaj G, Ventos-Alfonso A, Pagone V, Minemura T, Ohde T et al (2023) Reduction of embryonic *E93* expression as a hypothetical driver of the evolution of insect metamorphosis. Proc Natl Acad Sci USA 120:e2216640120. https://doi.org/10.1073/pnas.2216640120

Gilbert LI, Schneiderman HA (1960) The development of a bioassay for the juvenile hormone of insects. Trans Am Microsc Soc 79:38–67

Goodman W, Cusson M (2012) The juvenile hormones. In: Gilbert LI (ed) Insect endocrinology. Elsevier, London, pp 310–365

Goodman W, Granger NA (2005) The juvenile hormones. In: Gilbert LI, Iatrou K, Gill SS (eds) Comprehensive molecular insect science, vol 3. Elsevier/Pergamon, Oxford, pp 319–409

He LL, Shin SH, Wang Z, Yuan I, Weschler R, Chiou A et al (2020) Mechanism of threshold size assessment: metamorphosis is triggered by the TGF-beta/Activin ligand Myoglianin. Insect Biochem Mol Biol 126:103452. https://doi.org/10.1016/j.ibmb.2020.103452

Hornby JM, Jensen EC, Lisec AD, Tasto JJ, Jahnke B, Shoemaker R et al (2001) Quorum sensing in the dimorphic fungus *Candida albicans* is mediated by farnesol. Appl Environ Microbiol 67:2982–2992. https://doi.org/10.1128/AEM.67.7.2982-2992.2001

Ishimaru Y, Tomonari S, Matsuoka Y, Watanabe T, Miyawaki K, Bando T et al (2016) TGF-β signaling in insects regulates metamorphosis via juvenile hormone biosynthesis. Proc Natl Acad Sci USA 113:5634–5639. https://doi.org/10.1073/pnas.1600612113

Jindra M, Tumova S, Bittova L, Tuma R, Sedlak D (2024) Agonist-dependent action of the juvenile hormone receptor. Curr Opin Insect Sci 65:101234. https://doi.org/10.1016/j.cois.2024.101234

Jindra M, Uhlirova M, Charles JP, Smykal V, Hill RJ (2015) Genetic evidence for function of the bHLH-PAS protein Gce/Met as a juvenile hormone receptor. PLoS Genet 11:e1005394. https://doi.org/10.1371/journal.pgen.1005394

Judy KJ, Schooley DA, Dunham LL, Hall MS, Bergot J, Siddall JB (1973) Isolation, structure, and absolute configuration of a new natural juvenile hormone from *Manduca sexta*. Proc Natl Acad Sci USA 70:1509–1513. https://doi.org/10.1073/pnas.70.5.1509

Kamsoi O, Belles X (2019) Myoglianin triggers the premetamorphosis stage in hemimetabolan insects. FASEB J 33:3659–3669. https://doi.org/10.1096/fj.201801511R

Kayukawa T, Jouraku A, Ito Y, Shinoda T (2017) Molecular mechanism underlying juvenile hormone-mediated repression of precocious larval-adult metamorphosis. Proc Natl Acad Sci USA 114:1057–1062. https://doi.org/10.1073/pnas.1615423114

Kayukawa T, Nagamine K, Ito Y, Nishita Y, Ishikawa Y, Shinoda T (2016) Krüppel homolog 1 inhibits insect metamorphosis via direct transcriptional repression of *broad-complex*, a pupal specifier gene. J Biol Chem 291:1751–1762. https://doi.org/10.1074/jbc.M115.686121

Konopova B, Jindra M (2007) Juvenile hormone resistance gene *Methoprene-tolerant* controls entry into metamorphosis in the beetle *Tribolium castaneum*. Proc Natl Acad Sci USA 104:10488–10493. https://doi.org/10.1073/pnas.0703719104

Laufer H, Biggers WJ (2001) Unifying concepts learned from methyl farnesoate for invertebrate reproduction and post-embryonic development. Am Zool 41:s442–s457. https://doi.org/10.1093/icb/41.3.442

Li M, Mead EA, Zhu J (2011) Heterodimer of two bHLH-PAS proteins mediates juvenile hormone-induced gene expression. Proc Natl Acad Sci USA 108:638–643. https://doi.org/10.1073/pnas.101391410

Lv YN, Zeng M, Yan ZY, Lang PV, Ban N et al (2024) Juvenile hormone signaling is indispensable for late embryogenesis in ametabolous and hemimetabolous insects. BMC Biol 22:232. https://doi.org/10.1186/s12915-024-02029-2

Minakuchi C, Namiki T, Shinoda T (2009) *Krüppel homolog 1*, an early juvenile hormone-response gene downstream of *Methoprene-tolerant*, mediates its anti-metamorphic action in the red flour beetle *Tribolium castaneum*. Dev Biol 325:341–350. https://doi.org/10.1016/j.ydbio.2008.10.016

Mirth CK, Truman JW, Riddiford LM (2009) The ecdysone receptor controls the post-critical weight switch to nutrition-independent differentiation in *Drosophila* wing imaginal discs. Development 136:2345–2353. https://doi.org/10.1242/dev.032672

Nijhout HF, Kremen C (1998) Control of pupal commitment in the imaginal disks of *Precis coen*ia (Lepidoptera: Nymphalidae). J Insect Physiol 44:287–296. https://doi.org/10.1016/s0022-1910(97)00121-2

Nijhout HF, Wheeler DE (1982) Juvenile hormone and the physiological basis of insect polyphenisms. Q Rev Biol 57:109–133

Ojani R, Fu X, Ahmed T, Liu P, Zhu J (2018) Krüppel homologue 1 acts as a repressor and an activator in the transcriptional response to juvenile hormone in adult mosquitoes. Insect Mol Biol 27:268–278. https://doi.org/10.1111/imb.12370

Piepho H (1938) Über die experimentelle auslösbarkeit überzähliger häutungen und vorzeitiger verpuppung an hautstücken bei kleinschmetterlingen. Naturwissenschaften 52:141–142. (in German)

Ramirez CE, Nouzova M, Michalkova V, Fernandez-Lima F, Noriega F (2020) Common structural features facilitate the simultaneous identification and quantification of the five most common juvenile hormones by liquid chromatography-tandem mass spectrometry. Insect Biochem Mol Biol 116:103287. https://doi.org/10.1016/j.ibmb.2019.103287

Riddiford LM (1976) Hormonal control of insect epidermal cell commitment *in vitro*. Nature 259:115–117. https://doi.org/10.1038/259115a0

Riddiford LM (2020) *Rhodnius*, golden oil, and met: a history of juvenile hormone research. Front Cell Dev Biol 8:679. https://doi.org/10.3389/fcell.2020.00679

Robinson GE, Vargo EL (1997) Juvenile hormone in adult eusocial Hymenoptera: gonadotropin and behavioral pacemaker. Arch Insect Biochem Physiol 35:559–583. https://doi.org/10.1002/(SICI)1520-6327(1997)35:4<559::AID-ARCH13>3.0.CO;2-9

Röller H, Dahm KH, Sweeley CC, Trost BM (1967) The structure of juvenile hormone. Angew Chem 6:179–180. https://doi.org/10.1002/anie.196701792

Safranek L, Riddiford LM (1975) The biology of the black larval mutant of the tobacco hornworm, *Manduca sexta*. J Insect Physiol 21:1931–1938. https://doi.org/10.1016/0022-1910(75)90225-5

Scharrer B (1952) Neurosecretion. XI. The effects of nerve section on the intercerebralis-cardiacum-allatum system of the insect *Leucophaea maderae*. Biol Bull 102:261–272

Shirk PD, Dahm KH, Röller H (1976) The accessory sex glands as the repository for juvenile hormone in male cecropia moths. Zeitschr Naturforsch Sec C Biosci 31:199–200. https://doi.org/10.1515/znc-1976-3-421

Sláma K, Williams CM (1966) 'Paper factor' as an inhibitor of the embryonic development of the European bug, *Pyrrhocoris apterus*. Nature 210:329–330. https://doi.org/10.1038/210329a0

Smykal V, Daimon T, Kayukawa T, Takaki K, Shinoda T, Jindra M (2014) Importance of juvenile hormone signaling arises with competence of insect larvae to metamorphose. Dev Biol 390:221–230. https://doi.org/10.1016/j.ydbio.2014.03.006

Song J, Wu Z, Wang Z, Deng S, Zhou S (2014) Krüppel-homolog 1 mediates juvenile hormone action to promote vitellogenesis and oocyte maturation in the migratory locust. Insect Biochem Mol Biol 52:94–101. https://doi.org/10.1016/j.ibmb.2014.07.001

Stay B, Tobe SS (2007) The role of allatostatins in juvenile hormone synthesis in insects and crustaceans. Annu Rev Entomol 52:277–299. https://doi.org/10.1146/annurev.ento.51.110104.151050

Truman JW, Riddiford LM (1999) The origins of insect metamorphosis. Nature 401:447–452. https://doi.org/10.1038/46737

Truman JW, Hiruma K, Allee JP, Macwhinnie SG, Champlin DT, Riddiford LM (2006) Juvenile hormone is required to couple imaginal disc formation with nutrition in insects. Science 312:1385–1388. https://doi.org/10.1126/science.1123652

Truman JW, Riddiford LM, Konopova B, Nouzova M, Noriega FG, Herko M (2024) The embryonic role of juvenile hormone in the firebrat, *Thermobia domestica,* reveals its function before its involvement in metamorphosis. eLife 12:RP92643. https://doi.org/10.7554/eLife.92643

Wigglesworth VB (1934) The physiology of ecdysis in *Rhodnius prolixus* (Hemiptera) II. Factors controlling moulting and metamorphosis. Quart J Microsc Sci 77:191–222

Wigglesworth VB (1936) The function of the corpus allatum in the growth and reproduction of *Rhodnius prolixus* (Hemiptera). Quart J Microsc Sci 79:91–121. https://doi.org/10.1242/jcs.s2-77.306.191

Wigglesworth VB (1940) The determination of characters at metamorphosis in *Rhodnius prolixus* (Hemiptera). J Exp Biol 17:201–223. https://doi.org/10.1242/jeb.17.2.201

Wigglesworth VB (1963) The juvenile hormone effect of farnesol and some related compounds: quantitative experiments. J Insect Physiol 9:105–112. https://doi.org/10.1016/0022-1910(63)90087-8

Williams CM (1956) The juvenile hormone of insects. Nature 178:212–213. https://doi.org/10.1038/178212b0

Williams CM (1961) The juvenile hormone. II. Its role in the endocrine control of molting, pupation, and adult development in the Cecropia silkworm. Biol Bull 121:572–585. https://doi.org/10.2307/1539456

Wilson TG, Fabian J (1986) A *Drosophila melanogaster* mutant resistant to a chemical analog of juvenile hormone. Dev Biol 118:190–201. https://doi.org/10.1016/0012-1606(86)90087-4

Wu Z, Yang L, Li H, Zhou S (2021) Krüppel-homolog 1 exerts anti-metamorphic and vitellogenic functions in insects via phosphorylation-mediated recruitment of specific cofactors. BMC Biol 19:222. https://doi.org/10.1186/s12915-021-01157-3

Wyatt GR, Davey KG (1996) Cellular and molecular actions of juvenile hormone. II. Roles of juvenile hormone in adult insects. Adv Insect Physiol 26:2–155. https://doi.org/10.1016/S0065-2806(08)60030-2

Zhou B, Riddiford LM (2001) Hormonal regulation and patterning of the Broad-complex in the epidermis and wing discs of the tobacco hornworm, *Manduca sexta*. Dev Biol 231:125–137. https://doi.org/10.1006/dbio.2000.0143

Zhou X, Riddiford LM (2002) Broad specifies pupal development and mediates the 'status quo' action of juvenile hormone on the pupal-adult transformation in *Drosophila* and *Manduca*. Development 129:2259–2269. https://doi.org/10.1242/dev.129.9.2259

Part II
Structure and Synthesis

Chapter 2
Chemical Structure of Juvenile Hormone: A Journey of Half a Century and More

Toyomi Kotaki

Abstract The chemical structure of juvenile hormone (JH) was first determined in 1967, nearly three decades after the discovery of the "inhibitory factor" that inhibited metamorphosis of the blood-sucking bug *Rhodnius prolixus*. Since then, it took almost half a century to identify JH occurring in Hemiptera, including this important model insect. This chapter summarizes the history of JH chemistry, along with related chemicals. The finding that "Golden oil," extracts of the abdomen of Cecropia male moths, contained a large amount of JH activity was a breakthrough for the structure determination of JH I. Following this success, a group of epoxides of methyl farnesoate (MF) was described as insect JHs, and the structure of the Hemiptera-specific JH, JH III skipped bisepoxide ($JHSB_3$), was finally determined in 2009. Because MF was shown to be a JH equivalent in Decapoda, Crustacea, it was likely to be the simplest form of JH. MF was then modified to 10,11-epoxide, JH III, which was known to be most widespread among insects. This structure was further diversified in an order-specific manner. Many questions about JH chemistry are unanswered yet, and efforts to answer these questions will open new avenues for advancing JH research and its potential applications.

Keywords Golden oil · Sesquiterpenoids · Methyl esters · Epoxides · Radiochemical assay

2.1 Introduction

Juvenile hormone (JH) is one of the most important hormones in insects and some crustaceans. The name "juvenile hormone" was first given to a humoral factor that inhibits metamorphosis from nymph to adult in the blood-sucking bug *Rhodnius*

T. Kotaki (✉)
Institute of Agrobiological Sciences, National Agriculture and Food Research Organization, Tsukuba, Japan
e-mail: kotakitoyomi1@gmail.com

T. Shinoda et al. (eds.), *Juvenile Hormone*, Entomology Monographs,
https://doi.org/10.1007/978-981-95-8605-9_2

prolixus (Wigglesworth 1940). In an earlier study, Vincent B. Wigglesworth discovered an "inhibitory factor" secreted from the corpus allatum (CA), which suppressed metamorphosis (Wigglesworth 1934). Later, he found that the same factor also enhanced in adult insects (Wigglesworth 1936). A series of his studies marked the beginning of JH research. Since then, JH research has been an important topic in insect physiology and endocrinology. However, it was in 1967, three decades after the dawn of JH research, that the chemical structure of JH was determined for the first time (Röller et al. 1967), and it took about half a century thereafter to elucidate the structure of JH specific to the order Hemiptera, to which *R. prolixus* belongs. This chapter summarizes the history of JH chemistry.

2.2 Biological Function and Structure Determination of JH

2.2.1 Biological Function

Wigglesworth explored the humoral factors controlling molting and metamorphosis in *R. prolixus* using experimental morphological techniques (Wigglesworth 1934, 1936). Once the nymphs of this insect fully ingested a blood meal, they developed into the next stage. They also underwent molting even after decapitation and could survive a parabiosis experiment in which one individual was connected to another via a glass tube so that they shared the circulating hemolymph. These physiological traits were advantageous for exploring the hormones controlling metamorphosis and led Wigglesworth to conclude that a humoral factor secreted from the CA was responsible for inhibiting metamorphosis from nymphs into adults, thereby maintaining nymphal morphology upon the following ecdysis. Similarly, in the silkworm, *Bombyx mori*, it was shown that a hormone released from the CA inhibited the metamorphosis of larvae into pupae (Bounhiol 1936; Kim 1939; Fukuda 1940).

Based on this function, Wigglesworth (1940) referred to this inhibitory factor as JH for the first time. The same factor was also shown to play a role in stimulating egg production in female adults and accessory gland development in male adults (Wigglesworth 1936). It was later unequivocally established that JH exerts an inhibitory effect on metamorphosis in both hemimetabolous and holometabolous insects (Goodmann and Cusson 2012). JH also stimulates reproduction in multiple ways, such as inducing vitellogenin biosynthesis and uptake by the ovary, sex pheromone production, sperm production, accessory gland development, and mating.

2.2.2 Golden Oil: A Key to First Structure Determination of JH

In a series of studies on metamorphosis and pupal diapause in the Cecropia moth, *Hyalophora cecropia,* Carroll M. Williams found that male adults of the Cecropia accumulated JH in their abdomens (Williams 1956, 1963), as mentioned in Chap. 1. This discovery opened a door into JH chemistry. He obtained the so-called "golden oil" by extracting male moth abdomens with ether and evaporating the solvent. This oil exhibited strong JH activity, which originated from the CA. This conclusion was supported by the finding that extracts from males whose CA had been removed before pupation were no longer active (Williams 1963).

Starting with the golden oil, Röller et al. (1965) and Williams and Law (1965) succeeded in purification of JH activity up to 105,000-fold and 50,000-fold, respectively, and isolation as a single peak in gas chromatography (Williams and Law 1965; Röller et al. 1965; Röller and Bjerke 1965). Based on the results of the mass spectrum analysis, Williams and Law (1965) speculated and synthesized methyl 9,10-*cis*-epoxyhexadecanoate and methyl 9,10-*trans*-epoxyhexadecanoate as candidates of JH having the same retention time as the purified fraction in their gas chromatography (GC) analysis. However, these compounds showed no JH activity in their bioassay. On the other hand, using a different GC column, Röller and his colleagues determined the structure of JH, which nowadays is referred to as JH I (Fig. 2.1), by means of the combination of chemical modification, mass spectrometry, and nuclear magnetic resonance (NMR) (Röller et al. 1967; Röller and Dahm 1968). Meyer and his colleagues also purified JH activity from the golden oil and found two JH-active components at varying ratios from about 7:1 to 4:1, which were inferred to have a close structural relationship to each other (Meyer et al. 1965). In 1968, they successfully identified the structure of the minor component as the second JH, JH II (Fig. 2.1), by the combination of chemical modification, mass spectrometry, ^{1}H-NMR, and infrared spectrum analyses (Meyer et al. 1968). The key to Meyer's group successfully identifying JH II was the use of multiple GC columns together with a GC system equipped with a metal-free injector, which enabled the separation of the two JHs.

The JH I molecule has two asymmetric carbons at the 10- and 11-positions constituting an epoxide ring, and these carbons determine the direction of the epoxide ring relative to the plane of the carbon double bond. Accordingly, the natural JHs extracted from Cecropia male moths were assumed to be either one of two enantiomers or a racemic mixture. Meyer and Hanzmann, however, indicated that natural JH was optically active with a positive dispersion curve (Meyer and Hanzmann 1970). Furthermore, Meyer's group determined the absolute configurations of JH I and JH II to be 10*R*,11*S* (Meyer et al. 1971).

Judy et al. (1973) incubated the CA of adult females of the tobacco hornworm, *Manduca sexta*, in a medium containing L-[methyl-^{14}C]-methionine in vitro and extracted products released by the CA. Starting from these extracts, they found two components with high JH activity and radioactivity by both GC and liquid

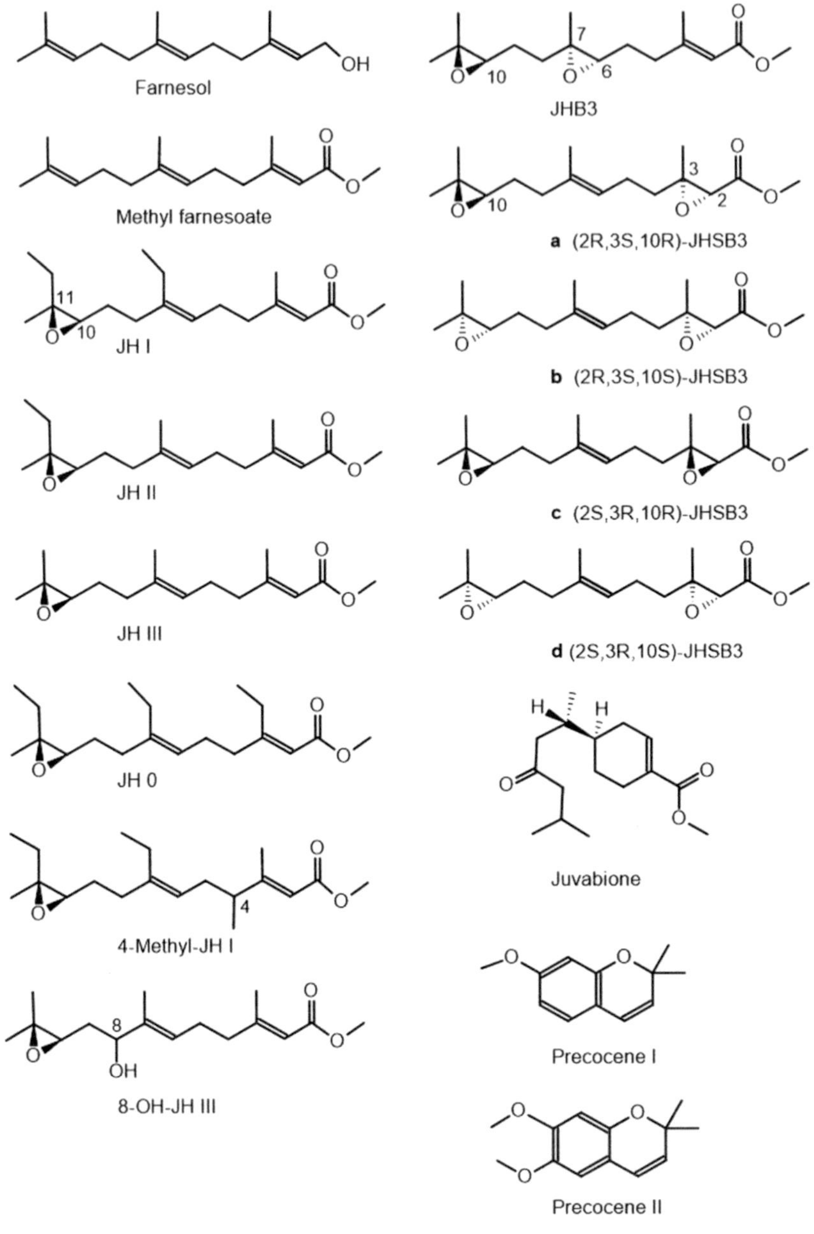

Fig. 2.1 Known juvenile hormones (JH) and related compounds. JH I, Röller et al. (1967); JH II, Meyer et al. (1968); JH III, Judy et al. (1973); JH 0 and 4-methyl-JH I, Bergot et al. (1981); 8-OH-JH III, Darrouzet et al. (1997) and Mauchamp et al. (1999); Juvenile hormone III bisepoxide (JHB_3), Richard et al. (1989); (**a**) Juvenile hormone III skipped bisepoxide ($JHSB_3$) and its stereoisomers (**b–d**), Kotaki et al. (2009); Juvabione, Sláma and Williams (1965); Precocene I and Precocene II, Bowers et al. (1976)

chromatography (LC); one was confirmed to be JH II, and the other, the third JH, was identified as methyl 10,11-epoxyfarnesoate, JH III (Fig. 2.1) (Judy et al. 1973). This compound had previously been synthesized and shown to have strong JH activity (Bowers et al. 1965). They also determined the absolute configuration of JH II and JH III as 10*R*,11*S* and 10*R*, respectively (Judy et al. 1973).

Bergot et al. (1981) examined JH titers in the eggs of *M. sexta*. In addition to JH I, they detected a peak with a slightly longer retention time than JH I in samples of eggs 6–29 h before hatching. They determined the structure of the unknown peak and named it JH 0 (Fig. 2.1). Its structure is characterized by the substitution of all three methyl branches of the farnesyl skeleton of JH III with ethyl groups. This new JH showed high JH activity in *M. sexta* larvae. 4-methyl-JH I (Fig. 2.1) was also detected in *Maduca* eggs.

In newly laid eggs, neither JH I nor JH 0 was detected, but the levels of these JHs increased 2–3 days after egg deposition and then precipitously declined at hatching, suggesting their function as hormones in *M. sexta* eggs. Even in a primitive ametabolous insect, the firebrat, *Thermobia domestica*, JH is known to exert morphostatic effects during embryogenesis (Truman et al. 2024; Lv et al. 2024), suggesting that the function during embryogenesis is the ancestral role of JH.

Until the end of the 1980s, many studies were conducted to elucidate the physiological function and identity of JH in various insects, and these studies indicated that JH III is the most widespread in insects not belonging to Lepidoptera, whereas lepidopteran insects possess ethyl-branched JHs, namely, JH 0, 4-methyl JH I, JH I, and JH II, as well as JH III.

In addition, Darrouzet and colleagues reported de novo synthesis of hydroxylated forms of JH III in the migratory locust, *Locusta migratoria*. They found that the CA of *L. migratoria* adults released unidentified products with retention times different from that of JH III in their LC analysis and determined their structures to be hydroxy-JH III at positions 4, 8, or 12, respectively (Fig. 2.1) (Darrouzet et al. 1997; Darrouzet et al. 1998; Mauchamp et al. 1999). Although the biological activity of these products in *L. migratoria* was not tested in their study, 12-OH-JH III was shown to be active in a bioassay using larvae of the yellow mealworm, *Tenebrio molitor* (*Tenebrio* test), with an activity level approximately 100 times lower than that of JH III.

Interestingly, JH III has also been detected in several plants, such as the ricefield flatsedge, *Cyperus iria* (Cyperaceae) (Toong et al. 1988); an African grass, *Euclasta condylotricha* (Poaceae) (Bohounton et al. 2023); and a tropical tree, *Cananga latifolia* (Annonaceae) (Yang et al. 2013), although the ecological significance or physiological role of its presence in these plants has not been addressed in these studies.

2.2.3 A JH with Two Epoxide Moieties in Higher Diptera

Richard et al. (1989) reported that a molecule having a JH III structure with a 6,7-double bond converted to an epoxide moiety, JH III bisepoxide (JHB_3), was the JH in the fruit fly *Drosophila melanogaster*. They incubated the ring gland—a complex endocrine organ that includes the CA, prothoracic glands, and corpora cardiaca—from *D. melanogaster* larvae in a medium containing L-[methyl-^{3}H]-methionine. They analyzed hexane extracts of the incubation medium by thin-layer chromatography (TLC) and found that most of the radiolabel migrated as a spot (Rf value = 0.55), more polar than JH III (Rf = 0.66), but less polar than JH III diol (Rf = 0.33). This product was demonstrated to be coeluted with synthetic JHB_3, methyl 6,7;10,11-bisepoxyfarnesoate in reversed-phase liquid chromatography (RPLC). When JH III was added to the incubation medium of the ring gland, only a small portion of it was metabolized to JHB_3, indicating that JHB_3 was not likely to be a metabolite of JH III after its release into the medium. In contrast, since JH III added to homogenates of the ring gland was converted to JHB_3, the former may be a precursor of the latter. Topical application of JHB_3 onto newly formed white puparia revealed that JHB_3 was biologically active, though it required 10 times more to interfere with normal adult development than JH III.

The ring glands from some other cyclorrhaphous dipterans, *Sarcophaga bullata*, *Musca domestica*, and *Calliphora vicina*, also produced JHB_3, while the CA of the mosquito *Toxorhynchites brevipalpis* only produced JH III (Richard et al. 1989). Therefore, JHB_3 is likely specific to the higher Diptera. The absolute configuration of JHB_3 biosynthesized by the ring glands of the Australian sheep blowfly, *Lucilia cuprina*, was determined as a (6*S*,7*S*,10*R*)-isomer using high-performance liquid chromatography (HPLC) and micellar electrokinetic capillary chromatography (Herlt et al. 1993).

2.2.4 Another JH with Two Epoxide Moieties in Hemiptera

In *R. prolixus*, Wigglesworth first described the presence of JH in the 1930s. However, the chemical nature of JH in the order Hemiptera, including *R. prolixus*, has been a longstanding question. Rankin and Riddiford (1977, 1978) studied the hormonal control of migration and adult diapause in the large milkweed bug, *Oncopeltus fasciatus*. They compared JH titers in adults destined to reproduce shortly after emergence and those destined to migrate to overwintering sites, followed by adult diapause induction using the *Manduca* black mutant assay, and concluded that JH plays important roles in reproduction and migration in Hemiptera.

Three decades after the first structural determination of JH I, Baker et al. (1988) reported the absence of any known JHs and related compounds in *O. fasciatus*. They attempted to detect JHs, their metabolites (JH acids and JH diols), and JH precursors, farnesol and methyl farnesoate (MF), but failed to detect significant levels of any of them. In the 1980s, some confusion occurred in attempts to determine the identity of JH in Hemiptera. In *O. fasciatus,* the cotton stainer, *Dysdercus fasciatus*, and the southern green stink bug, *Nezara viridula*, the production of JH III by the CA in vitro was reported (Bowers et al. 1983), but the same research group showed in the previous year that the CA of *D. fasciatus* produced not JH III but MF (Feldlaufer et al. 1982). In the bean bug, *Riptortus pedestris*, Numata et al. (1992)

detected JH I in hemolymph extracts using LC-MS. However, subsequent studies failed to detect JH I in this species.

In the early 1990s I tried incubation experiments of the CA of the brown-winged green bug, *Plautia stali*, in a medium containing L-[methyl-^{3}H]-methionine (radiochemical assay for JH biosynthesis, RCA. See Sect. 2.2.6 for details) to measure biosynthetic activity of JH as a factor controlling adult diapause in this species (Kotaki 1993). Since JH III is widespread among non-lepidopteran insects, the first question of my experiment was whether the CA of *P. stali* would produce JH III. The results showed, however, that the CA of this species, as well as three other stink bugs, released products that behaved differently from JH I, JH II, JH III, and JHB_3 in TLC (Kotaki 1993). A bioassay indicated that the CA products of *P. stali* were JH active, which suggested the presence of a new JH in hemipteran insects (Kotaki 1996). Since then, it took more than a decade to determine the structure of JH, $JHSB_3$ in this stink bug (Kotaki et al. 2009; Kotaki et al. 2012). Collaboration with chemists and their unique approach to the problem were the keys to the success in the discovery of the Hemiptera-specific JH. Four steps of our approach were as follows: (1) deduction of candidate structure based on what was known at that time; (2) synthesis of a mixture containing candidate compounds with the estimated molecular formula, separation of the mixture using a chiral column, and the bioassay; (3) stereospecific synthesis of candidates exhibiting high JH activity in Step 2; (4) comparison of candidates and natural product using GC-MS.

Structure deduction The CA of *P. stali* incubated in a medium containing L-[methyl-^{3H}]-methionine released a radiolabeled product into the incubation medium, which behaved differently from any known JHs (Kotaki 1993, 1996). Because the precursors of JH III, farnesol and farnesoic acid, enhanced its production, the product likely shares the same C15-skeleton as JH III. The fact that radiolabel derived from the methyl-^{3}H moiety of methionine was incorporated into the CA product suggests that the product was a methyl ester, as well as all other JHs. The molecular formula of the CA product was estimated to be $C_{16}H_{26}O_4$ based on the measurement of its molecular mass, $[M + H]^+$: 283.1885 m/z, using high-resolution mass spectrometry. The formula was the same as that of JHB_3. Altogether, it was inferred that the CA product was a structural isomer of JHB_3, namely, methyl 2,3;6,7-bisepoxyfarnesoate or methyl 2,3;10,11-bisepoxyfarnesoate.

Synthesis of candidate mixture, separation of the mixture using a chiral column and bioassay Four stereoisomers were theoretically possible for each bisepoxide mentioned above because of the two possible directions of the epoxide ring relative to the plane of the double bond. A mixture containing these stereoisomers and other geometric isomers was prepared in two steps. This mixture was subjected to HPLC equipped with a chiral column, separated into 21 fractions, and each fraction was tested for JH activity. Only two fractions exhibited pronounced metamorphosis-inhibiting activity upon topical application to the last instar nymphs of *P. stali*. NMR analysis of these fractions revealed that both active fractions were likely to be isomers of methyl 2,3;10,11-bisepoxyfarnesoate, but NMR could not assign these two fractions to any of the four possible stereoisomers.

Stereospecific synthesis of candidates To obtain chemically pure four stereoisomers (Fig. 2.1a–d), each isomer was prepared by more than 10 steps of chemical synthesis, including stereoselective Katsuki-Sharpless epoxidation reaction and Sharpless asymmetric dihydroxylation, followed by purification with the chiral column. Among the four isomers thus obtained, isomers **a** and **b** showed the same retention times as the two JH-active fractions of the isomer mixture.

Comparison of candidates and natural products These synthetic samples were subsequently subjected to the bioassay, and two isomers, **a** and **b**, were found to be highly JH active. Finally, these active isomers and the CA product were analyzed using GC-MS equipped with a chiral column, and the retention times for these samples were compared. Isomers **b** and **a** were detected by GC-MS in this order, and the CA product was detected at the retention time for **a**. Mass spectra of **a** and the CA products coincided with each other. Co-injection of **a** and CA product demonstrated that these were detected at the same time and gave a single peak in the chromatography. In this way, the structure of the CA product was determined as **a**, methyl (2*R*,3*S*,10*R*)-2,3;10,11-bisepoxyfaresoate. We named this unprecedented structure juvenile hormone III skipped bisepoxide, $JHSB_3$, based on the arrangement of two epoxides skipping the double bond in between them. When the structures of JH I, II, III, and JHB_3 were first elucidated, the absolute configuration of asymmetric carbon(s) had not been determined. In contrast to these JHs, the absolute configurations of three asymmetric carbons in the $JHSB_3$ molecule were fully determined. In this regard, the (2*R*,3*S*,10*R*)-isomer alone should be referred to as $JHSB_3$. In later studies focusing on JH in the suborder Heteroptera, the order Hemiptera indicated that $JHSB_3$ was widespread among heteropteran insects, including *R. prolixus* (Matsumoto et al. 2021; Villalobos-Sambucaro et al. 2020; Matsumoto et al. 2020), though the report by Villalobos-Sambucaro et al. was in part incomplete in that they did not specify the absolute configuration of the molecule.

2.2.5 Stereoisomers and Their JH Activity

All JHs identified so far share the 10*R*-configuration, with the only exception in the brown marmorated stink bug, *Halyomorpha halys*, as described below (Kotaki et al. 2020). In most insects, comparison of JH activity between 10*R*- and 10*S*-isomers showed that the former was much more active than the latter (Sakurai et al. 1990; Kindle et al. 1989). In *P. stali*, comparisons among the four $JHSB_3$ isomers, however, revealed a more complicated aspect. $JHSB_3$, **a** with (2*R*,3*S*,10*R*)-configuration and **b** with (2*R*,3*S*,10*S*)-configuration, exhibited comparable JH activity, whereas **c** with (2*S*,3*R*,10*R*)-configuration and **d** with (2*S*,3*R*,10*S*)-configuration were much less active than **a** and **b** (Kotaki et al. 2011). Therefore, in this stink bug, the configuration in 2,3-epoxide was more important than that in 10,11-epoxide for JH activity. Interestingly, *H. halys* is the only species where a significant amount of **b**, (2*R*,3*S*,10*S*)-isomer, was detected as one of the CA products (Kotaki et al. 2020). The CA of *H. halys* produced **a** and **b** in a ratio of about 5:1 in vitro. Bioassay revealed that ***b*** was about twice less active than ***a***. As the presence of 10*S*-JH III has recently been reported in a parasitic wasp (Ouyang et al. 2025), the possibility of the occurrence of 10*S*-JHs in any other species cannot be ruled out.

The reason for the difference in JH activity between the 10*R*- and 10*S*-isomers or among the four isomers of $JHSB_3$ is not well understood. A likely explanation may be attributed to the difference in the binding affinity of the JH receptor to the 10*R*- and 10*S*-isomers. The rate of degradation of JH circulating in the hemolymph and excretion from the insect body may also differ according to the isomers.

As mentioned above, the absolute configuration for the asymmetric carbon at the 10-position of the naturally occurring JH molecule was already determined as the *R* configuration, and that at the 11-position of JH I and JH II as the *S* configuration. In JH with two epoxides, JHB_3 and $JHSB_3$, the absolute configurations of the three asymmetric carbons were also elucidated. The biological activities of stereoisomers differ from each other, and advances in analytical techniques such as HPLC, UPLC, and GC with chiral columns have made it much easier to distinguish them. Therefore, it is essential to specify which stereoisomer is being analyzed or referenced in the experiments.

2.2.6 Radiochemical Assay (RCA): Magic Wand for JH Research

In JH research, in vitro incubation of the CA has been a crucial technique, contributing to the structural elucidation of JH III (Judy et al. 1973) and bisepoxide JHs (Richard et al. 1989; Kotaki et al. 2009). One of the reasons for this is that the CA does not accumulate JH but releases it into the hemolymph or incubation medium as soon as it is biosynthesized (Pratt and Tobe 1974). This makes it impossible to extract JH from the CA, although it is the endocrine organ responsible for JH biosynthesis. Extraction of the end products by CA from the incubation medium for JH analysis, on the other hand, is highly advantageous because samples thus prepared contain much fewer impurities than extracts from insect body homogenates or hemolymph. This simple method is also useful for elucidating the biosynthesis pathway (Jennings et al. 1975; Schooley et al. 1973) as well as for identifying neuropeptides that regulate JH biosynthesis, such as allatotropins and allatostatins.

The JH biosynthesis pathway is divided into two parts: an early process called the mevalonate pathway, which produces farnesol from acetyl-CoA, and a later JH-specific process (see Chaps. 3 and 7 for further details). The mevalonate pathway is involved in the production of the C-5 (consisting of 5 carbon atoms) isoprene unit, a building block of terpenoids. A monoterpene with 10 carbon atoms consists of two isoprene units, and a C-20 diterpene is a product of four isoprene units. The farnesyl skeleton, the main carbon chain of the JH III molecule, consists of three isoprene units with 15 carbon atoms and is called a sesquiterpene (the prefix "sesqui-" means one and a half). The ethyl-branched skeletons of JH I and JH II are generated by substituting one or two of the three isoprene units of the farnesyl skeleton with a C-6 homoisoprene unit derived from propionyl-CoA instead of acetyl-CoA (Schooley and Baker 1985; Brindle et al. 1988; Cusson et al. 1996).

The JH-specific pathway starting from farnesol (or ethyl-branched homologs, homofarnesol and dihomofarnesol) to the end products, JHs, includes two steps of oxidation from alcohol to aldehyde and then to acid, esterification of the resulting acid to methyl ester, and epoxidation of a double bond at the 10,11-position. In addition, in the case of bisepoxides, further epoxidation of the double bond at either the 6,7- or 2,3-position occurs. The conversion of farnesoic acid or JH acid to methyl ester is catalyzed by a member of *O*-methyl transferase, JH acid methyl transferase (JHAMT; see Chap. 3 for details).

In JH research, the esterification process has been important for measuring the JH biosynthetic activity in vitro. The methyl group of methyl ester in the JH molecule is derived solely from the S-methyl moiety of methionine by the action of JHAMT (Metzler et al. 1971). When ^{3}H-farnesoic acid and [methyl-^{14}C]-methionine were added to the incubation medium, these labels were incorporated into JH III at a ratio of 1:1 on a molar basis (Pratt and Tobe 1974). Assuming that the methionine pool within CA cells as a donor of methyl moiety was small enough to be depleted quickly and farnesol was synthesized in CA cells de novo, the radiolabeled methyl group derived from external methionine is taken up into JH III in a stoichiometric way. This is the principle of the RCA to measure the rate of JH biosynthesis in vitro (Pratt and Tobe 1974). This method has been widely used for analyses of JH biosynthesis and its regulatory factors in various insects because of its high sensitivity, along with a simpler procedure compared with the JH quantification using mass spectrometry or radioimmunoassay methods. Today, the use of radioisotopes to detect the CA products has largely been replaced by highly sensitive UPLC/MS-MS systems equipped with chiral columns.

2.3 JH Analogs and Anti-JH Agents

2.3.1 JH Analogs

Attempts to search for substances with JH activity began earlier than JH structure determination. Isoprenoids such as farnesol and farnesyl methyl ester were found active in *T. molitor* (Bowers and Thompson 1963) and *R. prolixus* (Wigglesworth 1963). As mentioned in 2.2.2, Bowers et al. (1965) synthesized methyl 10,11-epoxyfarnesoate and described its high JH activity. Interestingly, this compound was later found to be a natural JH, nowadays known as JH III (Judy et al.

1973). Soon after the first report of the JH I structure, Williams (1967) pointed out the possibility of the development of the "third-generation pesticides" based on JH activity. Indeed, the simplicity of the JH structure and the ease of synthesizing related compounds have led to the production of many JH analogs. Based on structure-activity analyses of JH and its derivatives, highly active JH analogs with a broad valid range, methoprene (ZR515, Altosid, Diacon, and Precor) and hydroprene (ZR512, Altozar, and Gentrol), were developed in this period and commercialized. These chemicals were characterized by the substitution of methyl ester with isopropyl or ethyl ester, respectively, and a double bond at the 4,5-position instead of the 6,7-position. These analogs omitted 10,11-epoxyide and, in addition, the former was accompanied by a methoxy group at the 11-position (Henrick et al. 1973; Henrick et al. 1976) (see Chap. 16 for details of JH analogs, in particular pyriproxyfen, and their application).

While working in Williams' laboratory at Harvard University, a Czech entomologist, Sláma, observed a strange phenomenon where nymphs of the linden bug, *Pyrrhocoris apterus,* transferred from Czechia failed to undergo normal metamorphosis but molted into sixth or seventh instar nymphs or adultoids. They explored the origin of the factors that exerted an inhibitory effect on the bugs' metamorphosis and discovered that paper of American origin contained such a factor; hence, they named it the paper factor. Interestingly, European or Japanese paper did not exhibit metamorphosis-inhibiting activity (Sláma and Williams 1965). Bowers et al. isolated this factor from extracts of balsam fir, the main ingredient of American paper, elucidated its structure as a methyl ester of a sesquiterpenoid, todomatuic acid, and named it juvabione (Bowers et al. 1966). Paper factor seemed to be particularly active in Pyrrhocoridae, to which *P. apterus* belongs, but not so much when tested on *O. fasciatus* (Lygaeidae) and *R. prolixus.* (Reduviidae) (Sláma and Williams 1965; Bowers et al. 1966). Peptides exhibiting JH activity only in the family Pyrrhocoridae have also been described (Poduška et al. 1971). These peptides act through the JH receptor, methoprene-tolerant (MET), and thus serve as JH agonists in a genus-specific manner (Tumova et al. 2022) (see Chap. 10).

It is worth noting that in *P. stali*, belonging to Pentatomidae, methoprene did not show any JH activity (Kotaki 1996), while bugs of other families, such as *R. prolixus* (Langley et al. 1990), *Cimex lectularius* (Gujar and Palli 2016), *P. apterus* (Kodrik et al. 2010), and *R. pedestris* (Numata and Hidaka 1984), responded well to methoprene. Although heteropteran insects share $JHSB_3$ in common (Matsumoto et al. 2021), the reason for the difference in responses to paper factor and JH analogs among heteropteran families remains to be elucidated.

2.3.2 *Anti-JH Agents, Precocenes*

As mentioned above, the "third-generation pesticide" based on JH activity seemed promising, but one of the difficulties in the practical use of JH analogs is that JH analogs are effective only in the developmental period when natural JH disappears or decreases, typically in the last larval and/or pupal stage before metamorphosis. In contrast to this limitation, anti-JH or substances with an activity to antagonize JH, may be practically useful because anti-JH is assumed to be effective by disrupting JH-dependent processes during developing periods when JH is present in the insect body. For this reason, Bowers et al. (1976) referred to anti-JH as the fourth-generation pesticide (see Chap. 17 for additional information).

In search for chemicals with such an activity, two active components were discovered from lipid extracts of the floss flower, *Ageratum houstonianum* (Asteraceae). These compounds induced precocious metamorphosis in nymphs of *O. fasciatus* and *Dysdercus cingulatus* and were named precocene I and precocene II, respectively (Bowers et al. 1976). In Diptera and Coleoptera, precocenes indicated not precocious metamorphosis-inducing activity in larvae but sterilizing activity in adults. Lepidopterans seemed not much sensitive to precocenes. These compounds were known to exert a cytotoxic effect on the CA cells so that they inhibit JH biosynthesis (Bowers 1981; Müller et al. 1979; Pratt and Bowers 1977; Pratt et al. 1980; Unnithan et al. 1977). Epoxidation in the late step of JH biosynthesis is catalyzed by a member of cytochrome P450s (CYPs), and which member of this large enzyme group is involved in this step may depend on insect orders (see Chap. 7 for further details). Different CYPs involved in the biosynthesis of JH may explain the difference in precocene sensitivity among insect orders and the fact that precocenes are particularly effective in hemipteran insects using $JHSB_3$.

2.4 Conclusions

Two major hormones, the molting hormone secreted from the prothoracic gland and JH, play an important role in insect metamorphosis, and the production of the former is controlled by a brain hormone, prothoracicotropic hormone (PTTH). This principle, called the classical scheme, is consistent throughout Insecta, except for primitive insects that do not undergo metamorphosis. In the adult stage, JH stimulates reproduction in various ways. It is worth emphasizing that JH is a multifunctional hormone, in contrast to hormones in vertebrates, which usually play a role in specific physiological processes.

The most widespread JH, JH III, is an epoxide of MF. In several species of Decapoda, one of the orders belonging to Crustacea, MF is known as an equivalent of JH, although in water fleas belonging to another group of crustaceans, the JH molecule is not determined yet (see Chap. 14). This indicates that the skeleton of the JH molecule, methyl ester of farnesoic acid, is conserved throughout and outside Insecta (Fig. 2.2). The farnesyl group bears three methyl side chains, while higher homologs of JH, JH 0, JH I, JH II, and 4-methyl JH I are accompanied by ethyl side chains instead. These homologs are basically specific to Lepidoptera. Another

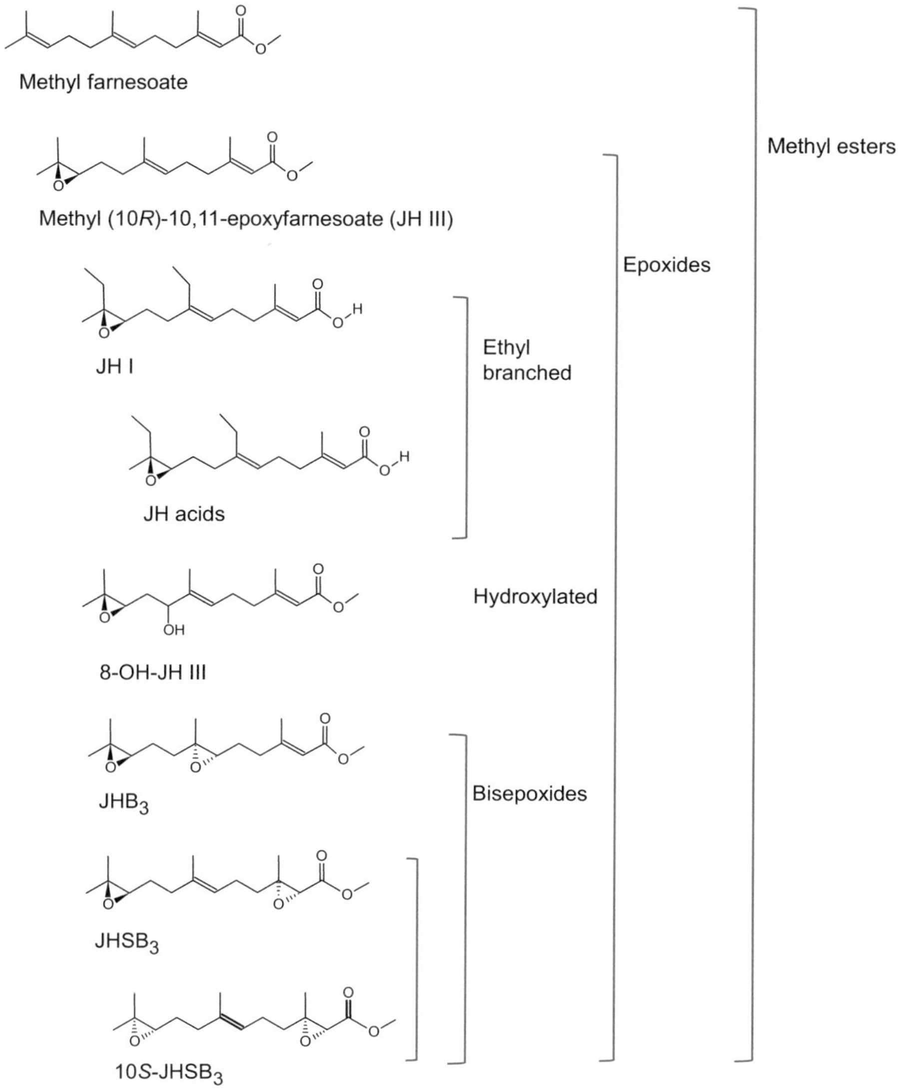

Fig. 2.2 Juvenile hormone structures: common and diversified features

characteristic feature of Lepidoptera is that the CA of male adults produce JH acids, which are subsequently sequestered in reproductive accessory glands and converted to methyl ester JHs. In the locust, hydroxylated forms of JH III were reported, but whether these molecules function as JH was not fully examined. All the above-mentioned JHs share a common characteristic of (10*R*)-10,11-epoxide. On the contrary, JHs in higher Diptera and Hemiptera are distinctive from other JHs in that they are bisepoxides of MF, though the position of the second epoxide moiety is different from one another. Hemipteran insects predominantly use $JHSB_3$. However, in only two exceptional cases, the production and occurrence of 10*S*-stereoisomers is reported in *H. halys* (Kotaki et al. 2020) and the parasitic wasp *Scleroderma guani* (Ouyang et al. 2025), respectively. In summary, the simplest form of JH equivalent in decapods is MF, which is first modified to the (10*R*)-10,11-epoxide and then diversified in an order-specific manner into ethyl-branched and hydroxylated forms and bisepoxides. Further diversification has occurred in limited species of stink bug and parasitic wasp, resulting in the biosynthesis of the 10*S*-isomers. Thus, the molecular diversity of JHs appears to form a nested pattern of conservation and innovation—unity with layers of diversification (Fig. 2.2).

More than half a century has passed between the first structure determination of JH I and the latest finding of the $JHSB_3$ structure, but this journey is far from over. The presence of other JH structures should be explored further. Why such structural variation occurs among insect orders is yet to be answered, despite the fact that the JH chemical structure itself is not likely to be subjected to natural selection. Nouzova et al. (2021) have, however, proposed a hypothesis that the epoxidated molecules improve reproductive fitness in the mosquito. Indeed, JH III is more active than MF in both the water flea *Daphnia pulex* and insects, even if in *D. pulex* the presence of JH has not been demonstrated. In both the water flea and insects, JH III induces a receptor response to a greater extent compared to MF, and this difference is attributed to the amino acid substitution in the particular position of the JH receptor (Miyakawa and Iguchi 2017).

Whether structural change in the JH molecule causes the change in the amino acid sequence of the JH receptor or vice versa, and whether this change leads to a change in reproductive fitness, are to be elucidated in the future. Phylogenetic and 3D conformation analyses of JH receptors are indispensable to address these questions. Similarly, these analyses may explain the reason why pentatomids are insensitive to methoprene, and the paper factor and peptidic JH analogs are effective only in the family Pyrrhocoridae. Answers to these questions will provide further understanding of the multifaceted function of JH and the "evolution" of JH molecules. Such efforts will embark on a new journey to an unexplored "JH world" and the practical application of JH-based insect growth regulators.

References

Baker FC, Tsai LW, Reuter CC, Schooley DA (1988) The absence of significant levels of the known juvenile hormones and related compounds in the milkweed bug, *Oncopeltus fasciatus*. Insect Biochem 18:453–462. https://doi.org/10.1016/0020-1790(88)90062-5

Bergot BJ, Baker FC, Cerf DC, Jamieson G, Schooley DA (1981) Qualitative and quantitative aspects of juvenile hormone titers in developing embryos of several insect species: discovery of a new JH-like substance extracted from eggs of *Manduca sexta*. In: Pratt GE, Brooks GT (eds) Juvenile hormone biochemistry: action, agonism, and antagonism: proceedings of the third international congress on juvenile hormone biochemistry--action, agonism, and antagonism. Elsevier/North-Holland Biomedical Press, Amsterdam, pp 33–45

Bohounton RB, Djihinto OY, Dedome OS-L, Yovo RM, Djossou L, Koba K et al (2023) *Euclasta condylotricha* flowers essential oils: a new source of juvenile hormones and its larvicidal activity against *Anopheles gambiae* s.s. (Diptera: Culicidae). PLoS One 18:e0278834. https://doi.org/10.1371/journal.pone.0278834

Bounhiol JJ (1936) Métamorphose après ablation des Corpora allata chez le Ver à soie (B. mori L.). C R Acad Sci 203:388–389

Bowers WS (1981) How anti-juvenile hormones work. Am Zool 21:737–742. https://doi.org/10.1093/icb/21.3.737

Bowers WS, Fales HM, Thompson MJ, Uebel EC (1966) Juvenile hormone: identification of an active compound from balsam fir. Science 154:1020–1021. https://doi.org/10.1126/science.154.3752.1020

Bowers WS, Marsella PA, Evans PH (1983) Identification of a hemipteran juvenile hormone: in vitro biosynthesis of JH III by *Dysdercus fasciatus*. J Exp Zool 228:555–559. https://doi.org/10.1002/jez.1402280316

Bowers WS, Ohta T, Cleere JS, Marsella PA (1976) Discovery of insect anti-juvenile hormones in plants. Science 193:542–547. https://doi.org/10.1126/science.986685

Bowers WS, Thompson MJ (1963) Juvenile hormone activity: effects of isoprenoid and straight-chain alcohols on insects. Science 142:1469–1470. https://doi.org/10.1126/science.142.3598.1469

Bowers WS, Thompson MJ, Uebel EC (1965) Juvenile and gonadotropic hormone activity of 10,11-epoxyfarnesenic acid methyl ester. Life Sci 4:2323–2331. https://doi.org/10.1016/0024-3205(65)90256-0

Brindle PA, Schooley DA, Tsai LW, Baker FC (1988) Comparative metabolism of branched-chain amino acids to precursors of juvenile hormone biogenesis in corpora allata of lepidopterous versus nonlepidopterous insects. J Biol Chem 263:10653–10657. https://doi.org/10.1016/S0021-9258(18)38021-9

Cusson M, Le Page A, McNeil JN, Tobe SS (1996) Rate of isoleucine metabolism in lepidopteran corpora allata: regulation of the proportion of juvenile hormone homologues released. Insect Biochem Mol Biol 26:195–201. https://doi.org/10.1016/0965-1748(95)00085-2

Darrouzet E, Mauchamp B, Prestwich GD, Kerhoas L, Ujary I, Couilloaud F (1997) Hydroxy juvenile hormones: new putative juvenile hormones biosynthesized by locust corpora allata *in vitro*. Biochem Biophys Res Comm 240:752–758. https://doi.org/10.1006/bbrc.1997.7739

Darrouzet E, Rossignol F, Couillaud F (1998) The release of isoprenoids by locust corpora Allata in vitro. J Insect Physiol 44:103–111. https://doi.org/10.1016/S0022-1910(97)00100-5

Feldlaufer MF, Bowers WS, Soderlund DM, Evans PH (1982) Biosynthesis of the sesquiterpenoid skeleton of juvenile hormone 3 by *Dysdercus fasciatus* corpora allata in vitro. J Exp Zool 223:295–298. https://doi.org/10.1002/jez.1402230311

Fukuda S (1940) Hormonal control of molting and pupation in the silkworm. Proc Jpn Acad 16:417–420

Goodmann W, Cusson M (2012) 8. The juvenile hormone. In: Gilbert LI (ed) Insect endocrinology. Academic Press, London, pp 310–365. https://doi.org/10.1016/B978-0-12-384749-2.10008-1

Gujar H, Palli SR (2016) Juvenile hormone regulation of female reproduction in the common bed bug, *Cimex lectularius*. Sci Rep 6:35546–35546. https://doi.org/10.1038/srep35546

Henrick CA, Staal GB, Siddall JB (1973) Alkyl 3,7,11-trimethyl-2,4-dodecadienoates, a new class of potent insect growth regulators with juvenile hormone activity. J Agric Food Chem 21:354–359. https://doi.org/10.1021/jf60187a043

Henrick CA, Willy WE, Staal GB (1976) Insect juvenile hormone activity of alkyl (2E,4E)-3,7,11-trimethyl-2,4-dodecadienoates. Variations in the ester function and in the carbon chain. J Agric Food Chem 24:207–218. https://doi.org/10.1021/jf60204a018

Herlt AJ, Rickards RW, Thomas RD, East PD (1993) The absolute configuration of juvenile hormone III bisepoxide. J Chem Soc Chem Commun 19:1497–1498. https://doi.org/10.1039/c39930001497

Jennings RC, Judy KJ, Schooley DA (1975) Biosynthesis of the homosesquiterpenoid juvenile hormone JH II [methyl (2E,6E,10Z)-10,11 -epoxy-3,7,11-trimethyltridecadienoate] from [5-3H]homomevalonate in *Manduca sexta*. J Chem Soc Chem Comm 1:21–22. https://doi.org/10.1039/c39750000021

Judy KJ, Schooley DA, Dunham LL, Hall MS, Bergot BJ, Siddall JB (1973) Isolation, structure, and absolute configuration of a new natural insect juvenile hormone from *Manduca sexta*. Proc Natl Acad Sci USA 70:1509–1513. https://doi.org/10.1073/pnas.70.5.1509

Kim J (1939) On the corpus allatum hormone in the silkworm (in Japanese). Sericult Sci Jpn 10:86–97. https://doi.org/10.11416/kontyushigen1930.10.2_86

Kindle H, Winistörfer M, Lanzrein B, Mori K (1989) Relationship between the absolute configuration and the biological activity of juvenile hormone III. Experientia 45:356–360. https://doi.org/10.1007/bf01957477

Kodrik D, Alquicer G, Socha R (2010) Methoprene modifies adipokinetic hormone characteristics in the firebug *Pyrrhocoris apterus* (Heteroptera: Pyrrhocoridae). Eur J Entomol 107:33–39. https://doi.org/10.14411/eje.2010.003

Kotaki T (1993) Biosynthetic products by heteropteran corpora allata *in vitro*. Appl Entomol Zool 28:242–245. https://doi.org/10.1303/aez.28.242

Kotaki T (1996) Evidence for a new juvenile hormone in a stink bug, *Plautia stali*. J Insect Physiol 42:279–286. https://doi.org/10.1016/0022-1910(95)00113-1

Kotaki T, Matsumoto K, Kaihara K, Ando Y, Misaki K, Shinada T (2020) A stereoisomer of JHSB3 with 10S-configuration, 10S-JHSB3, biosynthesized by the corpus allatum of the brown marmorated stink bug, *Halyomorpha halys* (Hemiptera: Pentatomidae). Appl Entomol Zool 55:223–229. https://doi.org/10.1007/s13355-020-00674-x

Kotaki T, Shinada T, Kaihara K, Ohfune Y, Numata H (2009) Structure determination of a new juvenile hormone from a heteropteran insect. Org Lett 11:5234–5237. https://doi.org/10.1021/ol902161x

Kotaki T, Shinada T, Kaihara K, Ohfune Y, Numata H (2011) Biological activities of juvenile hormone III skipped bisepoxide in last instar nymphs and adults of a stink bug, *Plautia stali*. J Insect Physiol 57:147–152. https://doi.org/10.1016/j.jinsphys.2010.10.003

Kotaki T, Shinada T, Numata H (2012) Structure determination of a natural juvenile hormone isolated from a heteropteran insect. Psyche 2012:924256. https://doi.org/10.1155/2012/924256

Langley PA, Howl V, Oouchi H (1990) Regulation of reproduction in *Rhodnius prolixus* by the juvenile hormone mimic pyriproxyfen. Entomol Exp Appl 57:271–279. https://doi.org/10.1111/j.1570-7458.1990.tb01439.x

Lv Y-N, Zeng M, Yan Z-Y, Zhang P-Y, Ban N, Yuan D-W et al (2024) Juvenile hormone signaling is indispensable for late embryogenesis in ametabolous and hemimetabolous insects. BMC Biol 22:232–232. https://doi.org/10.1186/s12915-024-02029-2

Matsumoto K, Kotaki T, Numata H, Shinada T, Goto SG (2021) Juvenile hormone III skipped bisepoxide is widespread in true bugs (Hemiptera: Heteroptera). R Soc Open Sci 8:202242. https://doi.org/10.1098/rsos.202242

Matsumoto K, Yasuno Y, Yasuda K, Hayashi T, Goto SG, Shinada T (2020) Structure determination of juvenile hormone from chagas disease vectors, *Rhodnius prolixus* and *Triatoma infestans*. Chem Lett 49:538–541. https://doi.org/10.1246/cl.200126

Mauchamp B, Darrouzet E, Malosse C, Franck C (1999) 4′-OH-JH-III: an additional hydroxylated juvenile hormone produced by locust corpora allata in vitro. Insect Biochem Mol Biol 29:475–480. https://doi.org/10.1016/S0965-1748(99)00044-2

Metzler M, Dahm KH, Meyer D, Röller H (1971) On the biosynthesis of juvenile hormone in the adult Cecropia moth. Z Naturforsch B 26:1270–1276. https://doi.org/10.1515/znb-1971-1215

Meyer AS, Hanzmann E (1970) The optical activity of the Cecropia juvenile hormones. Biochem Biophys Res Commun 41:891–893. https://doi.org/10.1016/0006-291X(70)90167-1

Meyer AS, Hanzmann E, Murphy RC (1971) Absolute configuration of cecropia juvenile hormone. Proc Natl Acad Sci USA 68(9):2312–2315. https://doi.org/10.1073/pnas.68.9.2312

Meyer AS, Schneiderman HA, Gilbert LI (1965) A highly purified preparation of juvenile hormone from the silk moth *Hyalophora cecropia* L. Nature 206:272–275. https://doi.org/10.1038/206272a0

Meyer AS, Schneiderman HA, Hanzmann E, Ko JH (1968) The two juvenile hormones from the *cecropia* silk moth. Proc Natl Acad Sci USA 60:853–860. https://doi.org/10.1073/pnas.60.3.853

Miyakawa H, Iguchi T (2017) Comparative luciferase assay for establishing reliable in vitro screening system of juvenile hormone agonists. J Appl Toxicol 37:1082–1090. https://doi.org/10.1002/jat.3459

Müller PJ, Masner P, Kälin M, Bowers WS (1979) In vitro inactivation of corpora allata of the bug *Oncopeltus fasciatus* by precocene II. Experientia 35:704–705. https://doi.org/10.1007/BF01960409

Nouzova M, Edwards MJ, Michalkova V, Ramirez CE, Ruiz M, Areiza M et al (2021) Epoxidation of juvenile hormone was a key innovation improving insect reproductive fitness. Proc Natl Acad Sci USA 118:e2109381118. https://doi.org/10.1073/pnas.2109381118

Numata H, Hidaka T (1984) Termination of adult diapause by a juvenile hormone analogue in the bean bug, *Riptortus clavatus*. Zool Sci 1:751–754. https://doi.org/10.34425/zs000080

Numata H, Numata A, Takahashi C, Nakagawa Y, Iwatani K, Takahashi S, Miura K, Chinzei Y (1992) Juvenile hormone I is the principal juvenile hormone in a hemipteran insect, *Riptortus clavatus*. Experientia 48:606–610. https://doi.org/10.1007/BF01920248

Ouyang H, Yi G, Xiang P, Wei S, Yu J, Wang X, Li H, Li X (2025) A stereoselective CYP15A1 epoxidase mediates (10S)-JH III synthesis in a parasitic wasp *Scleroderma guani*. J Agr Food Chem 73:27288–27304. https://doi.org/10.1021/acs.jafc.5c07533

Poduška K, Šorm F, Sláma K (1971) 9. Structure - juvenile activity relationships in simple peptides. Z Naturforsch B 26:719–722. https://doi.org/10.1515/znb-1971-0716

Pratt GE, Bowers WS (1977) Precocene II inhibits juvenile hormone biosynthesis by cockroach corpora allata *in vitro*. Nature 265:548–550. https://doi.org/10.1038/265548a0

Pratt GE, Jennings RC, Hamnett AF, Brooks GT (1980) Lethal metabolism of precocene-I to a reactive epoxide by locust corpora allata. Nature 284:320–323. https://doi.org/10.1038/284320a0

Pratt GE, Tobe SS (1974) Juvenile hormone radiobiosynthesised by corpora allata of adult female locusts in vitro. Life Sci 14:575–586. https://doi.org/10.1016/0024-3205(74)90372-5

Rankin MA, Riddiford LM (1977) Hormonal control of migratory flight in *Oncopeltus fasciatus*: the effects of the corpus cardiacum, corpus allatum, and starvation on migration and reproduction. Gen Comp Endocrinol 33:309–321. https://doi.org/10.1016/0016-6480(77)90045-4

Rankin MA, Riddiford LM (1978) Significance of haemolymph juvenile hormone titer changes in timing of migration and reproduction in adult *Oncopeltus fasciatus*. J Insect Physiol 24:31–38. https://doi.org/10.1016/0022-1910(78)90008-2

Richard DS, Applebaum SW, Sliter TJ, Baker FC, Schooley DA, Reuter CC et al (1989) Juvenile hormone bisepoxide biosynthesis in vitro by the ring gland of *Drosophila melanogaster*: a putative juvenile hormone in the higher Diptera. Proc Natl Acad Sci USA 86:1421–1425. https://doi.org/10.1073/pnas.86.4.1421

Röller H, Bjerke JS (1965) Purification and isolation of juvenile hormone and its action in lepidopteran larvae. Life Sci 4:1617–1624. https://doi.org/10.1016/0024-3205(65)90141-4

Röller H, Bjerke JS, McShan WH (1965) The juvenile hormone—I. Methods of purification and isolation. J Insect Physiol 11:1185–1197. https://doi.org/10.1016/0022-1910(65)90111-3

Röller H, Dahm KH (1968) The chemistry and biology of juvenile hormone. In: Astwood EB (ed) Proceedings of the 1967 Laurentian hormone conference, vol 24. Academic Press, Boston, pp 651–680. https://doi.org/10.1016/B978-1-4831-9827-9.50018-6

Röller H, Dahm KH, Sweely CC, Trost BM (1967) The structure of juvenile hormone. Angew Chem Int Ed Engl 6:179–180. https://doi.org/10.1002/anie.196701792

Sakurai S, Ohtaki T, Mori H, Fujiwhara M, Mori K (1990) Biological activity of enantiomerically pure forms of insect juvenile hormone I and III in *Bombyx mori*. Experientia 46:220–221. https://doi.org/10.1007/BF02027321

Schooley DA, Baker FC (1985) Juvenile hormone biosynthesis. In: Kerkut GA, GIlbert LI (eds) Comprehensive insect physiology, biochemistry and pharmacology, vol 7 Endocrinology. Pergamon Press, Oxford, pp 363–398

Schooley DA, Judy KJ, Bergot BJ, Hall MS, Siddall JB (1973) Biosynthesis of the juvenile hormones of *Manduca sexta*: labeling pattern from mevalonate, propionate, and acetate. Proc Natl Acad Sci USA 70:2921–2925. https://doi.org/10.1073/pnas.70.10.2921

Sláma K, Williams CM (1965) Juvenile hormone activity for the bug *Pyrrhocoris apterus*. Proc Natl Acad Sci USA 54:411–414. https://doi.org/10.1073/pnas.54.2.411

Toong YC, Schooley DA, Baker FC (1988) Isolation of insect juvenile hormone III from a plant. Nature 333:170–171. https://doi.org/10.1038/333170a0

Truman JW, Riddiford LM, Konopova B, Nouzova M, Noriega FG, Herko M (2024) The embryonic role of juvenile hormone in the firebrat, *Thermobia domestica*, reveals its function before its involvement in metamorphosis. eLife 12:RP92643. https://doi.org/10.7554/eLife.92643

Tumova S, Milacek M, Šnajdr I, Muthu M, Tuma R, Reha D et al (2022) Unique peptidic agonists of a juvenile hormone receptor with species-specific effects on insect development and reproduction. Proc Natl Acad Sci USA 119:e2215541119. https://doi.org/10.1073/pnas.2215541119

Unnithan GC, Nair KK, Bowers WS (1977) Precocene-induced degeneration of the corpus allatum of adult females of the bug *Oncopeltus fasciatus*. J Insect Physiol 23:1081–1094. https://doi.org/10.1016/0022-1910(77)90137-8

Villalobos-Sambucaro MJ, Nouzova M, Ramirez CE, Eugenia Alzugaray M, Fernandez-Lima F et al (2020) The juvenile hormone described in *Rhodnius prolixus* by Wigglesworth is juvenile hormone III skipped bisepoxide. Sci Rep 10:3091. https://doi.org/10.1038/s41598-020-59495-1

Wigglesworth VB (1934) The physiology of ecdysis in *Rhodnius prolixus* (Hemiptera). II. Factors controlling moulting and 'metamorphosis'. Q J Microsc Sci 77:191–222. https://doi.org/10.1242/jcs.s2-77.306.191

Wigglesworth VB (1936) The function of the corpus allatum in the growth and reproduction of *Rhodnius prolixus* (Hemiptera). Q J Microsc Sci 79:91–120. https://doi.org/10.1242/jcs.s2-79.313.91

Wigglesworth VB (1940) The determination of characters at metamorphosis in *Rhodnius prolixus* (Hemiptera). J Exp Biol 17:201–222. https://doi.org/10.1242/jeb.17.2.201

Wigglesworth VB (1963) The juvenile hormone effect of farnesol and some related compounds: quantitative experiments. J Insect Physiol 9:105–119. https://doi.org/10.1016/0022-1910(63)90087-8

Williams CM (1956) The juvenile hormone of insects. Nature 178:212–213. https://doi.org/10.1038/178212b0

Williams CM (1963) The juvenile hormone. III. Its accumulation and storage in the abdomens of certain male moths. Biol Bull 124:355–367. https://doi.org/10.2307/1539485

Williams CM (1967) Third-generation pesticides. Sci Am 217:13–17. https://doi.org/10.1038/scientificamerican0767-13

Williams CM, Law JH (1965) The juvenile hormone—IV. Its extraction, assay, and purification. J Insect Physiol 11:569–580. https://doi.org/10.1016/0022-1910(65)90139-3

Yang H, Kim HS, Jeong EJ, Khiev P, Chin Y-W, Sung SH (2013) Plant-derived juvenile hormone III analogues and other sesquiterpenes from the stem bark of *Cananga latifolia*. Phytochemistry 94:277–283. https://doi.org/10.1016/j.phytochem.2013.06.010

Chapter 3
Biosynthesis of Juvenile Hormone: The Role of Juvenile Hormone Acid Methyltransferase (JHAMT)

Tetsuro Shinoda

Abstract Juvenile hormone acid methyltransferase (JHAMT) catalyzes terminal methylation, which defines the structure of juvenile hormone (JH) and controls key processes in insect development, metamorphosis, reproduction, and polyphenism. Since its identification in the silkworm, *Bombyx mori,* in 2003, *JHAMT* has become central to elucidating the molecular basis of JH biosynthesis. Structural and biochemical studies have established a conserved SAM-dependent catalytic architecture and strict substrate specificity shared by insects and crustaceans. *JHAMT* expression is regulated by neural and hormonal cues, nutrient- and stress-responsive pathways, and CA-specific transcriptional and posttranscriptional mechanisms. Functional analyses using ribonucleic acid (RNA) interference (RNAi) and genome editing have demonstrated that JHAMT is indispensable for maintaining the larval state, coordinating metamorphic transitions, enabling fertility, and regulating diapause and other environmentally triggered phenotypes. Comparative genomics further revealed the strong conservation of canonical *JHAMT* orthologs and the extensive diversification of *JHAMT-like* paralogs. These insights highlight JHAMT as a key target for RNAi-based pest control, structure-guided inhibitor design, and metabolic engineering of sesquiterpenoid pathways, bridging fundamental insect endocrinology and applied biotechnological innovations.

Keywords Epoxidase · Evolution · FAMeT · JH biosynthesis · JHAMT · Juvenile hormone · Metamorphosis · Methyl farnesoate · Reproduction · RNA interference

---

T. Shinoda (✉)
Faculty of Food and Agricultural Sciences, Fukushima University, Fukushima, Japan
e-mail: tetsuro.shinoda@agri.fukushima-u.ac.jp

T. Shinoda et al. (eds.), *Juvenile Hormone*, Entomology Monographs,
https://doi.org/10.1007/978-981-95-8605-9_3

3.1 Introduction

Juvenile hormone (JH) was first identified as a factor preventing metamorphosis through classic physiological studies by Wigglesworth (1934), who demonstrated that the corpora allata (CA) secrete a hormone essential for maintaining juvenile characteristics. Subsequent chemical investigations established that insects synthesize multiple sesquiterpenoid JHs that share a methyl ester moiety and, in most cases, an epoxide group (see Chaps. 1 and 2). JH is now recognized to regulate a broad array of processes, including metamorphosis, reproduction, diapause, caste differentiation, and various polyphenisms (Chaps. 7, 11, 12, 13, and 14).

The action of JH depends on an integrated system that encompasses its biosynthesis, transport, degradation (Chap. 4), and intracellular signaling (Chaps. 8 and 10). Although major progress has been made in identifying JH degradation pathways, carrier proteins, and components of the Met–Kr-h1 signaling cascade, the biosynthetic pathway remained poorly understood for many years owing to the minute size and low synthetic output of the CA. Therefore, identifying the enzymes responsible for JH production has been one of the central challenges in JH biology.

A major breakthrough occurred with the molecular identification of *juvenile hormone acid O-methyltransferase* (*JHAMT*) (Shinoda and Itoyama 2003), which catalyzes the terminal methylation step required to generate the characteristic methyl ester present in all JH molecules, including methyl farnesoate (MF). This discovery stimulated rapid advances in elucidating the terminal reactions of JH biosynthesis across insect taxa and promoted extensive biochemical, developmental, and evolutionary research. In this chapter, we summarize the current knowledge regarding the discovery, structure, regulation, and function of JHAMT and briefly introduce the epoxidation step catalyzed by cytochrome P450 enzymes, such as CYP15, thereby situating JHAMT within the broader framework of JH biosynthesis.

3.2 Chemical Structures and Diversity of Juvenile Hormones

The structure of JH was first elucidated in 1967, when Röller and colleagues identified JH I from the "golden oil" of the cecropia moth, *Hyalophora cecropia* (Röller et al. 1967). Subsequent studies revealed additional forms, including JH II and JH III, with the latter being the principal JH in most insects. Lepidoptera synthesize unique variants such as JH 0 and iso-JH 0, higher Diptera produce the bisepoxide JHB_3, and Hemiptera were recently shown to generate JH III skipped bisepoxide ($JHSB_3$) (see Chap. 2). Crustaceans and certain insects, including the fruit fly, , produce MF, a non-epoxidized precursor of JH III (see Chaps. 7 and 14). Collectively, these findings illustrate the extensive structural diversification of JHs among arthropods.

Despite this diversity, all JHs share two conserved chemical features: a terminal methyl ester and a stereospecific *10R* epoxide (MF lacks the latter). These functional groups correspond to the terminal steps of JH biosynthesis: the methyl ester introduced by JHAMT and the epoxide added by a cytochrome P450 epoxidase, such as CYP15A1 or its Dipteran equivalents. Thus, the structural variations observed across insect lineages reflect the evolutionary modifications of these final biosynthetic reactions.

3.3 JH Biosynthetic Pathway

3.3.1 Early (Mevalonate) Pathway

The JH biosynthetic pathway was largely elucidated by biochemical studies in the mid-1980s and is conventionally divided into early and late steps (Schooley and Baker 1985). The early steps, collectively known as the mevalonate pathway, correspond to the upstream portion of the biosynthetic cascades for cholesterol and steroid hormones in other organisms. This conserved pathway consists of eight enzyme-catalyzed reactions that convert acetyl-CoA into farnesyl pyrophosphate (FPP) (Fig. 3.1; see also Chaps. 5 and 7). In Lepidoptera, JHs with ethyl-branched side chains utilize propionyl-CoA in addition to acetyl-CoA, proceeding through homomevalonate; however, the overall framework remains the same as that leading to JH III.

The rapid expansion of insect genomic resources has greatly facilitated the identification of early pathway genes. Because the mevalonate pathway is broadly conserved, orthologs can be readily recovered using homology-based searches. The silkworm *Bombyx mori* provided the first comprehensive example: using sequence information from other organisms, all early step enzyme genes—from acetyl-CoA to FPP—were systematically identified, and their cDNAs were cloned (Kinjoh et al. 2007). Most genes were present as single-copy orthologs, except for farnesyl pyrophosphate synthase (FPPS), which had three paralogs in *B. mori*. Expression profiling further revealed that nearly all early pathway genes are strongly enriched in the CA (Kinjoh et al. 2007; Ueda et al. 2009), indicating that the mevalonate enzymes function as a tightly co-regulated gene set dedicated to JH biosynthesis in Lepidoptera.

3.3.2 Late Pathway: JH-Specific Reactions

The late steps of the pathway (Fig. 3.1), which convert FPP into active JH, are specific to JH biosynthesis. FPP is first dephosphorylated to farnesol by a phosphatase or a pyrophosphatase. Farnesol is then sequentially oxidized to farnesal and farnesoic acid (FA) by a dehydrogenase or oxidase. Although the genes encoding these enzymes have been cloned in several insect species, many aspects of their molecular identities, substrate specificities, and regulatory mechanisms remain unclear (see Chap. 5).

In the terminal phase, FA undergoes two key modifications, methylation and epoxidation, to generate biologically active JH. Notably, the order of these reactions

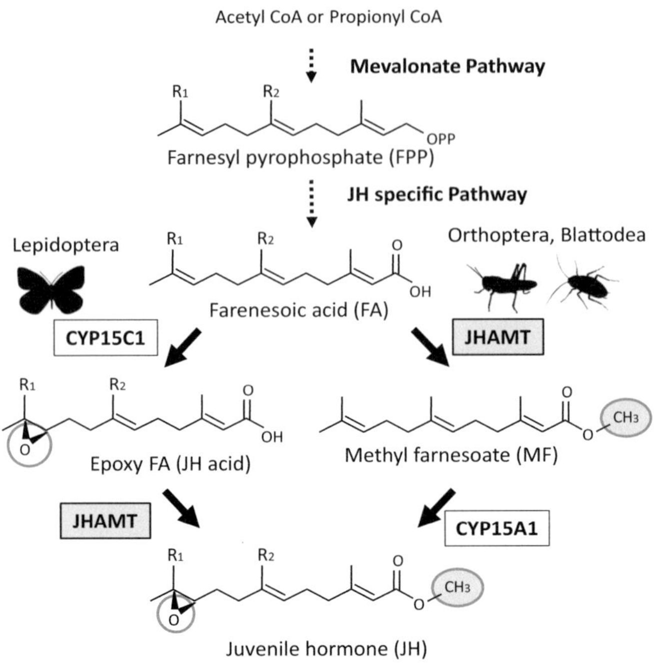

Fig. 3.1 Final steps of the juvenile hormone (JH) biosynthetic pathway. The characteristic epoxide and methyl ester groups of JH are introduced in the terminal steps of the pathway by CYP15 and juvenile hormone acid *O*-methyltransferase (JHAMT), respectively. See the main text for details. JH I: $R_1 = R_2 = C_2H_5$; JH II: $R_1 = C_2H_5$, $R_2 = CH_3$; JH III: $R_1 = R_2 = CH_3$. (Modified from Shinoda (2016), with permission)

varies among insect lineages and contributes to the structural diversity of JHs. In Lepidoptera, FA is first epoxidized to epoxy-FA (JH acid) and then methylated, whereas locusts and cockroaches methylate FA to produce MF and subsequently epoxidize MF to JH. This taxon-specific inversion of the reaction order represents a major evolutionary divergence. The genes and enzymes governing these reactions are described in detail in the following sections.

3.4 Juvenile Hormone Epoxidase

Although MF functions as a hormone in crustaceans and retains limited regulatory roles in insects, epoxidation is essential for generating the fully active form of JH in most insects. MF shows much lower activity than JH III in inducing *Kr-h1* expression in *B. mori* (Kayukawa et al. 2012) and exhibits only weak JH activity in the reproduction of the yellow fever mosquito, *Aedes aegypti* (Nouzova et al. 2021). These observations underscore the functional importance of the epoxidation step and its catalyst, JH epoxidase.

The first JH epoxidase gene, *CYP15A1*, was isolated from the CA of the Pacific beetle cockroach, *Diploptera punctata* (Helvig et al. 2004). The encoded protein stereospecifically epoxidizes MF, but not FA, to produce JH III with the natural *10R* configuration. In *B. mori*, the ortholog *CYP15C1* underlies the *dimolting* (*mod*) mutant, which exhibits precocious metamorphosis due to reduced JH signaling (Daimon et al. 2012). CYP15C1 epoxidizes FA but not MF, reflecting the reverse order of terminal reactions in Lepidoptera. In the red flour beetle, *Tribolium castaneum*, CYP15A1 epoxidizes both FA and MF (Minakuchi et al. 2015). These lineage-specific differences in substrate preference suggest that evolutionary shifts in epoxidase specificity contributed to the establishment of a taxon-specific order in the final biosynthetic steps.

CYP15 genes are absent in *D. melanogaster*. However, a recent study identified *CYP6G2* as the gene encoding a functional JH epoxidase in *D. melanogaster*, demonstrating that non-*CYP15* P450s can acquire epoxidase activity in higher Diptera (Jia et al. 2024; see Chap. 7). In crustaceans, *CYP15* genes were previously believed to be absent (Daimon and Shinoda 2013), but *CYP15* homologs have now been reported in several crustacean genomes (see Chap. 14). Although the enzymatic functions of these crustacean homologs remain unclear and epoxidized JHs have not yet been detected in crustaceans, these findings suggest that the diversity and evolution of JH epoxidases—and JH biosynthesis more broadly—may require future reevaluation.

3.5 Juvenile Hormone Acid Methyltransferase

JHAMT catalyzes the terminal methylation step, producing a methyl ester common to all insect JHs and crustacean MF. This reaction defines the chemical identity of JH and represents a central regulatory point in JH biosynthesis. Although biochemical studies had long inferred the presence of a S-adenosyl-L-methionine (SAM)-dependent enzyme responsible for this step, direct molecular identification remained elusive due to the minute size and low enzymatic output of CA. The cloning of *Bombyx mori JHAMT* in 2003 (Shinoda and Itoyama 2003) provided definitive genetic evidence for the terminal step of JH biosynthesis, enabling mechanistic, structural, and evolutionary analyses across arthropods. This section outlines the historical progression from early biochemical detection to molecular cloning and introduces the structural, catalytic, and evolutionary properties that establish JHAMT as the pivotal enzyme in JH biosynthesis.

3.5.1 Historical Background: Biochemical Identification of JHAMT Activity (1970s–1990s)

Early biochemical studies in the 1970s provided the first functional evidence for methylation in JH biosynthesis. Using radiochemical assays (RCAs; see Chap. 2), researchers detected an SAM-dependent methyltransferase capable of converting JH acid to JH in the CA of the tobacco hornworm, *Manduca sexta* (Reibstein and Law

1973), as well as in the male accessory glands of *H. cecropia*, the source of the "golden oil" (Weirich and Culver 1979). These pioneering studies firmly established that the terminal step of JH biosynthesis is catalyzed by a SAM-dependent methyltransferase, which is now recognized as JHAMT.

Biochemical work by Dr. G. Bhaskaran and colleagues provided the first developmental evidence that this methylation step represents a regulated point in the JH pathway. In *M. sexta*, they demonstrated that SAM-dependent methyltransferase activity corresponding to JHAMT fluctuates markedly during the final larval instar, peaking immediately after ecdysis and declining rapidly thereafter, in parallel with changes in JH synthetic capacity (Sparagana et al. 1984). Their subsequent conceptual analyses highlighted this enzyme as a key regulatory node in metamorphosis, anticipating the later molecular confirmation that JHAMT serves as the pivotal terminal regulator of JH biosynthesis (Bhaskaran et al. 1986).

3.5.2 Molecular Identification of JHAMT

Because the CA are extremely small and produce only trace amounts of JHAMT, classical biochemical purification was not feasible. To overcome this limitation, Shinoda and Itoyama (2003) used a developmental-expression–based cloning strategy in *B. mori*. In this species, JH synthetic activity is high throughout the fourth larval instar but declines sharply in the early fifth instar. Reasoning that the terminal enzyme in the pathway would exhibit a matching expression profile, we performed differential display on CA collected at 24 developmental time points from the fourth instar to pupation. One transcript showed strong expression during the entire fourth instar, a rapid decline at the onset of the fifth instar, and complete disappearance by the mid-fifth instar.

Cloning of the corresponding cDNA revealed a 278-amino-acid protein bearing conserved SAM-dependent methyltransferase motifs, despite showing only ~15% identity to *ubiquinone/menaquinone biosynthesis methyltransferase E* (*UbiE*) of *Escherichia coli*. Its highly restricted expression in the CA (Fig. 3.2a) was consistent with the expectations for the terminal JH-biosynthetic enzyme. Enzyme assays using recombinant proteins confirmed this assignment: the enzyme efficiently methylated the carboxyl group of both JH acids and FA but displayed only very weak activity toward straight-chain fatty acids. These results demonstrated that the cloned gene encodes a bona fide *juvenile hormone acid O-methyltransferase*, *JHAMT*.

3.5.3 Primary Structure of JHAMT

Following the cloning of *B. mori* JHAMT, homology-based approaches rapidly facilitated the identification and isolation of canonical *JHAMT* and *JHAMT-like* genes from a wide range of insects and crustaceans, leading to substantial progress

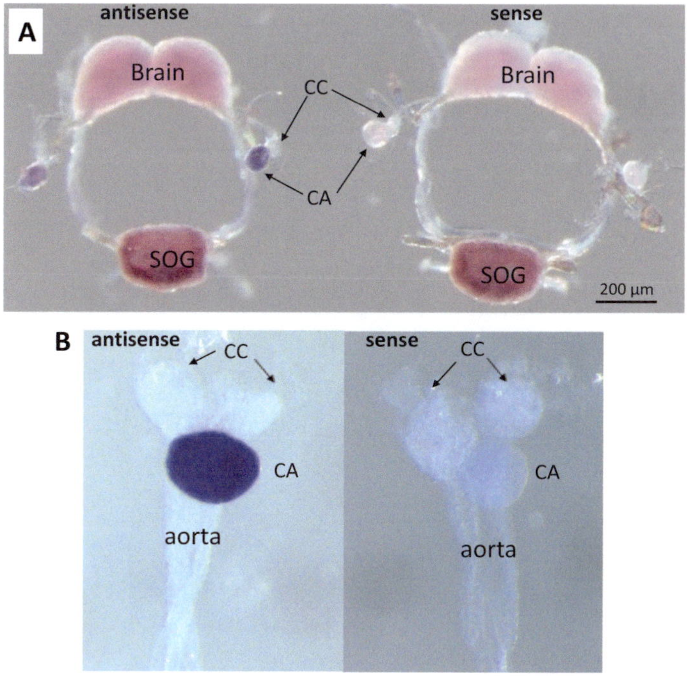

Fig. 3.2 Corpora allata (CA)-specific expression of the *juvenile hormone acid O-methyltransferase* (*JHAMT*) genes. In situ hybridization using antisense RNA probes for *JHAMT* detected a distinct CA-specific hybridization signal (purple) in the fourth-instar larvae of the silkworm *Bombyx mori* (**a**) and in non-diapause adult females of the brown-winged greenbug, *Plautia stali* (**b**), whereas the sense probes showed no detectable signal. The pink coloration observed in the brain and the subesophageal ganglion (SOG) of *B. mori* represents intrinsic tissue pigmentation rather than hybridization signals. The photograph in panel B was taken by Hiroto Ueda. Abbreviations: CC corpora cardiacum, CA corpora allatum, SOG subesophageal ganglion

in understanding their biochemical and physiological functions. Throughout this chapter, "canonical *JHAMT*" refers to the ortholog responsible for FA/JHA methylation in JH/MF biosynthesis, whereas "*JHAMT-like*" denotes paralogous SAM-dependent methyltransferases with uncertain functions.

Figure 3.3 shows the deduced amino-acid sequence alignment of JHAMT proteins from *B. mori*, several other insects, and the crustacean water flea *Daphnia pulex*. *B. mori* JHAMT comprises 278 amino acids with a molecular mass of approximately 34 kDa, comparable to that of other insect and crustacean JHAMTs. Conserved residues are distributed throughout the alignment, with particularly strong conservation observed in the SAM-binding motif (hh(D/E)hGXGXG; h = hydrophobic amino acid). Despite this conserved core, the overall sequence identity remains modest—for example, only ~45% between *D. melanogaster* and *A. aegypti* and ~27% between *B. mori* and *D. pulex*.

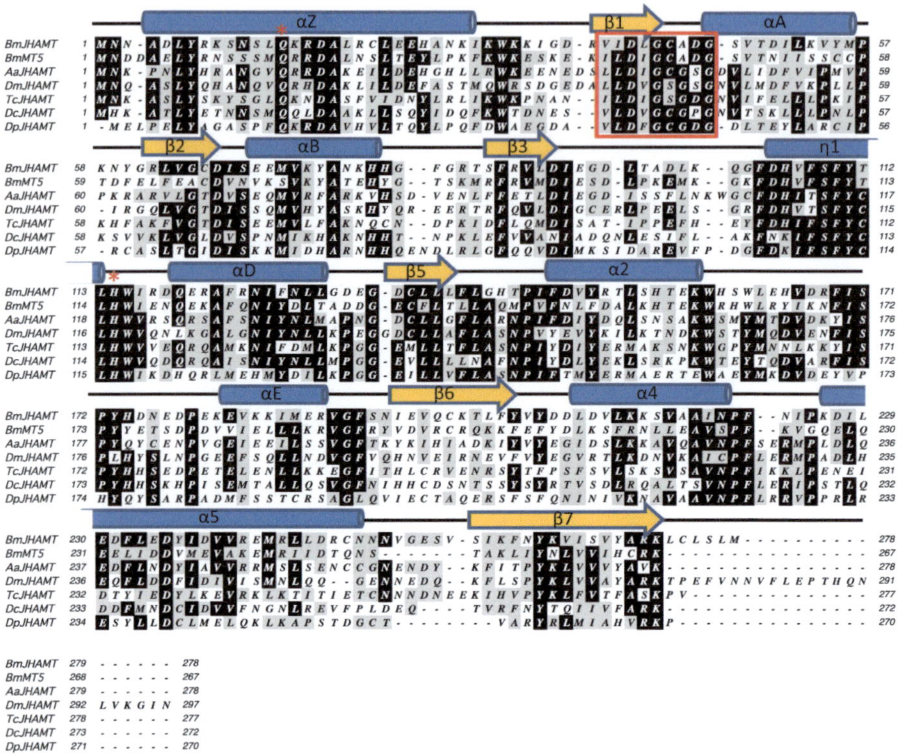

Fig. 3.3 Amino-acid sequence alignment of insect and crustacean juvenile hormone acid *O*-methyltransferase (JHAMT) proteins. Bm *Bombyx mori*, Aa *Aedes aegypti*, Dm *Drosophila melanogaster*, Tc *Tribolium castaneum*, Dc *Diaphorina citri*, Dp *Daphnia pulex*. All entries except *B. mori* MT6 correspond to recombinant proteins verified to possess catalytic activity. Blue bars: α-helices; yellow arrows: β-sheets; numbering follows Defelipe et al. (2011). Red boxes: SAM-binding motifs; *: Strictly conserved catalytic pair (Guo et al. 2021). (Modified from Shinoda (2016) with permission)

3.5.4 Biochemical Properties of Recombinant JHMATs

The biochemical properties of JHAMT, characterized using recombinant proteins, are summarized in Table 3.1. Early studies showed that JHAMT from the male accessory glands of *H. cecropia* preferentially methylates JH I and JH II acids over JH III acid and exhibits strict stereospecificity toward the natural (*10R,11S*) configuration (Weirich and Culver 1979). Recombinant *B. mori* JHAMT displays a similar substrate preference, with reactions using racemic JH III acid predominantly yielding the *10R* enantiomer (Shinoda, unpublished). Strong *10R* stereoselectivity has also been observed in recombinant enzymes from *T. castaneum* (Minakuchi et al. 2008), *A. aegypti* (Mayoral et al. 2009), and *D. melanogaster* (Niwa et al. 2008), indicating that this property is broadly conserved among holometabolous insects.

Table 3.1 Comparative enzymatic properties of recombinant juvenile hormone acid *O*-methyltransferase (JHAMT) proteins from various insect species

Species	Substrate						Stereospecificity to racemic JH III acid[c] *10R*: *10S*	References
	Farnesoic acid	JH III acid	JH II acid	JH I acid	Lauric acid	Palmitic acid		
Bombyx mori	4.8×10^{-1}[a]	7.9×10^{-1}	1.7	1.8	ND	ND	93: 7	Shinoda and Itoyama (2003)
Tribolium castaneum	5.9×10^{-1}	5.6×10^{-1}	–	–	1.6×10^{-2}	2.0×10^{-3}	87: 13	Minakuchi et al. (2008)
Drosophila melanogaster	10.1	7.7	–	–	1.0×10^{-2}	4.1×10^{-4}	80: 20	Niwa et al. (2008)
Aedes aegypti	2.6×10^{-1}	1.0	–	–	1.0×10^{-4}	8.9×10^{-4}	99.5: 0.5	Mayoral et al. (2009)
Aedes aegypti	1.3×10^{-2} Km 2.23[b]	6.9×10^{-2} Km 14.68	–	5.3×10^{-2} Km 10.27	–	–	–	Van Ekert et al. (2014a)
Samia cynthica ricini	–	3.7×10^{-3} Km 55.3	–	–	–	–	–	Sheng et al. (2008)
Diaphorina citri	4.5×10^{-2} Km 0.46	7.8×10^{-4} Km 2.53	–	–	–	–	–	Van Ekert et al. (2015)
Daphnia pulex	2.6×10^{-1} Km 1.13	ND	–	–	–	–	–	Toyota et al. (2015)

[a]Units of enzyme activity (mol product/mol JHAMT protein/min)
[b]Units of Km (μM)
[c]Ratio of (*10R*)- to (*10S*)-JH III produced when racemic JH III acid was used as the substrate
Modified from Shinoda 2016 with permission
ND not detected

Recombinant enzyme assays have also demonstrated that insect JHAMTs generally methylate both JH acids and FA, although the relative activities vary considerably among taxa. *D. melanogaster* and *T. castaneum* enzymes showed comparable activities toward JH III acid and FA, whereas *B. mori* JHAMT exhibited only ~25% of the FA-methylating activity observed for JH I acid. The *A. aegypti* enzyme similarly shows low activity toward FA, approximately one-fifth to one-fourth of that toward JH III acid (Mayoral et al. 2009; Van Ekert et al. 2014a). In contrast, the hemimetabolous psyllid *Diaphorina citri* displays ~60-fold higher activity toward FA than toward JH III acid, with a correspondingly higher affinity (Van Ekert et al. 2015). In crustaceans, which utilize non-epoxidized MF as their JH, recombinant *D. pulex* JHAMT efficiently methylates FA but shows no detectable activity toward JH III acid (Toyota et al. 2015).

Overall, biochemical studies across insects and crustaceans have revealed a conserved SAM-dependent methyltransferase architecture with strict *10R* stereoselectivity for JH acids, accompanied by lineage-specific differences in FA versus JH-acid utilization. These substrate preferences parallel the evolutionary diversification of JH structures and biosynthetic pathways, underscoring the central role of JHAMT in shaping the chemical and functional diversity of JH.

3.5.5 Structural Basis of Substrate Recognition and Catalysis in JHAMT

Understanding how JHAMT recognizes FA and JHA and achieves strict natural (*10R*) enantioselectivity requires knowledge of its three-dimensional structure, which is also essential for rational inhibitor design. However, recombinant JHAMTs have generally proven difficult to crystallize, and structural insights have long been limited to homology models (Defelipe et al. 2011; Van Ekert et al. 2014a; Van Ekert et al. 2015).

Recently, Guo et al. (2021) successfully crystallized *B. mori* JHAMT2, a paralog with weak methyltransferase activity, and determined its apo, cofactor-bound, and substrate-bound structures (Guo et al. 2021). JHAMT2 adopts a classical Rossmann-fold core, undergoes sequential conformational changes upon ligand binding, and contains a strictly conserved Gln/His catalytic pair essential for activity (Fig. 3.3), providing the first molecular framework for understanding JHAMT substrate recognition. Subsequently, Zhang et al. (2022) showed that the structure of the catalytically inactive paralog JHAMT3 closely resembles that of JHAMT2, displaying similar cofactor-induced conformational changes but with a more rigid substrate-binding pocket. Based on its strong expression in the silk gland and lack of enzymatic activity, JHAMT3 was proposed to function as a JH-acid-binding protein, offering additional structural insight into JHAMT family diversification.

These crystallographic studies of *B. mori* JHAMTs represent a major advance, providing reliable templates for structural prediction across insects and informing future efforts to elucidate substrate-recognition mechanisms and design selective JHAMT inhibitors.

3.5.6 Farnesoic Acid O-Methyltransferase (FAMeT) and Unknown Methyltransferase

JHAMT has been identified as the enzyme responsible for the *O*-methylation step in the biosynthesis of JH and MF in insects and crustaceans. In contrast, farnesoic acid *O*-methyltransferase (FAMeT), originally cloned from the shrimp *Metapenaeus ensis* (Silva Gunawardene et al. 2001), was once proposed as a candidate enzyme, but available evidence suggests that it is unlikely to be directly involved in JH/MF biosynthesis. FAMeT belongs to a completely distinct protein family and lacks the canonical SAM-binding motif required for SAM-dependent methyltransferase activity. Instead, FAMeT proteins contain two CF (CPAMD8/FAOMeT) or pfam12248 domains of unknown function (Holford et al. 2004; Smykal and Dolezel 2023). A report of weak FA-methylating activity in crude *E. coli* extracts expressing *M. ensis FAMeT* (Gunawardene et al. 2002) was never supported by analytical verification of MF production, and subsequent studies in other crustaceans, including the edible crab *Cancer pagurus* (Ruddell et al. 2003) and the American lobster, *Homarus americanus* (Holford et al. 2004), found no catalytic activity in recombinant FAMeTs. In contrast, bona fide JHAMT of *D. pulex* exhibits robust FA-methylating activity (Toyota et al. 2015).

FAMeT homologs also occur in insects, typically consisting of a pfam12248 domain paired with a DUF3421 domain (Smykal and Dolezel 2023). However, recombinant FAMeT from *D. melanogaster* shows no detectable enzymatic activity (Burtenshaw et al. 2008), and *FAMeT* loss-of-function mutants exhibit no defects in JH or MF synthesis (Wen et al. 2015). Consistently, RNA interference (RNAi) targeting *JHAMT*—but not *FAMeT*—reduces JH titers and delays ovarian development in the desert locust, *Schistocerca gregaria* (Marchal et al. 2011). Together, these results suggest that FAMeT is unlikely to function as an FA-methylating enzyme in insects or crustaceans and likely has a distinct, as-yet-unidentified physiological function.

Notably, both MF and JH III persist in *D. melanogaster* larvae even when *JHAMT* is genetically ablated, implying the existence of an as-yet-unidentified methyltransferase capable of catalyzing residual sesquiterpenoid methylation (Marchal et al. 2011; Wen et al. 2015). The identification of this alternative enzyme remains an important priority for future research.

3.6 Regulation of *JHAMT* Expression

The highly specific expression of *JHAMT* in the CA, the primary organ for JH biosynthesis, is conserved across insects, including *B. mori* (Shinoda and Itoyama 2003), *D. melanogaster* (Niwa et al. 2008), *T. castaneum* (Minakuchi et al. 2008), *A. aegypti* (Mayoral et al. 2009), and *B. germanica* (Dominguez and Maestro 2018). As JHAMT catalyzes the terminal and rate-limiting steps of JH biosynthesis, the transcriptional regulation of *JHAMT* constitutes a major control point governing JH titers. Accumulating evidence indicates that *JHAMT* expression is modulated by multiple regulatory layers, including intrinsic developmental programs, systemic neuropeptides and hormones, and intracellular signaling pathways, thereby integrating physiological and environmental cues within the CA. These coordinated mechanisms ensure the precise stage-, sex-, and tissue-specific production of JH, which is required for growth, metamorphosis, and reproduction. The following sections summarize the current knowledge of these regulatory systems and highlight the outstanding questions regarding CA-specific transcriptional control.

3.6.1 Developmental and Sexual Regulation

Molecular identification of *B. mori JHAMT* enabled precise developmental profiling, revealing a close correlation between *JHAMT* transcription and JH biosynthetic activity. Whereas transcripts of early mevalonate-pathway enzymes persist at low levels after JH synthesis ceases (Kinjoh et al. 2007; Shinoda and Itoyama 2003), *JHAMT* expression declines abruptly at the onset of the fifth instar and disappears by day 3, matching the developmental shutdown of JH production. This pattern parallels earlier biochemical studies in *M. sexta*, in which JHAMT activity peaked immediately after ecdysis and rapidly declined during the final larval instar (Sparagana et al. 1984).

In adults, *JHAMT* transcription reappears exclusively in females, consistent with sustained JH production in the female CA but not in males (Kinjoh et al. 2007). Together, these findings demonstrate that the developmental and sex-specific control of *JHAMT* expression is a primary determinant of JH biosynthetic capacity.

3.6.2 Regulation by Neuropeptides and Systemic Hormones

Allatoregulatory neuropeptides modulate both CA activity and *JHAMT* expression. Allatostatins (AST-A/B/C) generally inhibit JH biosynthesis, whereas allatotropin (AT) stimulates it (Audsley et al. 2008). Their effects on *JHAMT* vary by species and stage: AST knockdown increases *JHAMT* in the Colorado potato beetle, *Leptinotarsa decemlineata* (Meng et al. 2015); AST-C/CC upregulates *JHAMT* in larvae but downregulates it in adult females of the poplar tip moth, *Clostera anastomosis* (Dong et al. 2017); and AST depletion elevates *JHAMT* in all stages of the Chinese white pine beetle, *Dendroctonus armandi* (Sun et al. 2022). In adult *B. mori*, AT paradoxically suppresses JH biosynthesis by reducing the expression of multiple biosynthetic genes, including *JHAMT* (Kaneko and Hiruma 2015; see Chap. 6).

Systemic hormones provide an additional layer of hormonal control. In *B. mori*, pulsatile 20-hydroxyecdysone (20E) transiently induces *JHAMT* during the penultimate instar via brain-mediated signaling (Kaneko et al. 2011b). In *A. aegypti*, ecdysis-triggering hormone (ETH) (Areiza et al. 2014) and 20E (Areiza et al. 2015) strongly upregulate both *JHAMT* transcripts and enzymatic activity, coordinating JH biosynthesis with molting and reproductive activation.

3.6.3 Intracellular Signaling and Transcriptional Control

Within CA cells, multiple intracellular pathways transduce environmental and physiological inputs into the precise regulation of *JHAMT* expression. Transforming Growth Factor-beta (TGF-β)/Bone Morphogenetic Protein (BMP) ligands, such as Decapentaplegic (Dpp) and Glass bottom boat (Gbb), enhance *JHAMT* transcription through the Tkv–Mad pathway in *D. melanogaster* (Huang et al. 2011), while Myoglianin (Myo) represses *JHAMT* in the cricket *Gryllus bimaculatus* (Ishimaru et al. 2016; see Chap. 15). Nutritional signals act through the target of rapamycin (TOR) pathway: perturbation of TOR components alters *JHAMT* expression in the German cockroach, *Blattella germanica* (Dominguez and Maestro 2018), and the cricket *Modicogryllus siamensis* (Miki et al. 2020), reflecting a tight coupling between nutrient status and JH biosynthesis.

Multiple transcriptional regulators further refine the CA-specific expression. POU-domain proteins, such as *Ventral veins lacking* (*Vvl*) in *T. castaneum* (Cheng et al. 2014) and *POU-M2* in *B. mori* (Cai et al. 2022), function as positive regulators. *Dead ringer* (*dri*) acts as a CA-specific activator of JH pathway genes, including *JHAMT*, in *B. mori* (Kayukawa et al. 2024). Posttranscriptionally, microRNAs, including miR-276 and miR-182,013-5p in the migratory locust, *Locusta migratoria* (Song et al. 2024), and miR-2161 in the honeybee, *Apis mellifera* (Song et al. 2025), modulate JHAMT output.

3.6.4 Stage-Specific Regulation: Outstanding Questions

Although developmental studies clearly establish *JHAMT* transcription as the determinant of CA competence, the mechanisms underlying its abrupt and irreversible silencing during metamorphosis remain largely unknown. Allatostatins rapidly but reversibly regulate JH biosynthesis, suggesting that they cannot fully account for the permanent premetamorphic shutdown. A long-acting inhibitory factor, "allatinhibin," was proposed in *M. sexta* decades ago (Bhaskaran et al. 1990), but its molecular identity remains unknown. Its relationship with Myo represents an intriguing question and remains an important issue to be elucidated in future studies.

In addition, CA-specific transcriptional repressors required to extinguish *JHAMT* expression during metamorphosis have not been identified, nor have the *cis*-regulatory elements of the *JHAMT* promoter been fully mapped. Resolving these fundamental questions is essential for understanding the transcriptional architecture governing JH biosynthesis.

3.7 Physiological Roles of *JHAMT*

JH titers are closely associated with developmental, reproductive, and polyphenic processes. Because *JHAMT* encodes the terminal and rate-limiting enzyme of JH biosynthesis, its regulation is a key determinant of these physiological effects. The spatial and temporal modulation of *JHAMT* expression closely parallels JH-dependent events, and functional analyses have demonstrated that JHAMT activity is indispensable for major life-history transitions.

3.7.1 JHAMT in Development and Metamorphosis

RNAi-mediated silencing of *JHAMT* in *T. castaneum* induces precocious pupation and miniature adults, a phenotype fully rescued by exogenous JH, providing direct evidence that JHAMT-dependent JH biosynthesis is required to maintain larval status (Fig. 3.4a) (Minakuchi et al. 2008). Conversely, ectopic *JHAMT* expression disrupts adult morphogenesis in *D. melanogaster* (Niwa et al. 2008) and blocks pupation in *B. mori* (Fig. 3.4b; Shinoda unpublished), indicating that both activation and repression of *JHAMT* must be precisely coordinated.

Viral manipulation further underscores the importance of *JHAMT*: an entomopoxvirus encodes a functional *JHAMT* capable of synthesizing JH, presumably altering host JH titers and thereby disrupting normal metamorphic progression (Takatsuka et al. 2017).

Additional studies have shown that *JHAMT* knockdown impairs larval performance in *L. decemlineata* (Fu et al. 2016) and disrupts embryonic development and hatchability in *B. germanica* (Fernandez-Nicolas and Belles 2017). It also perturbs JH–ecdysone crosstalk in *Tribolium freemani*, which undergoes repeated supernumerary molting under crowded conditions, resulting in a lethal developmental arrest (Ruang-Rit and Park 2018). Genome editing in *B. mori* further revealed that JH is dispensable for embryogenesis but essential for proper hatching and preventing

Fig. 3.4 Phenotypes induced by artificial manipulation of juvenile hormone acid *O*-methyltransferase (JHAMT) expression. (**a**) Precocious metamorphosis induced by JHAMT RNA interference (JHAMTi) in *Tribolium castaneum*. Under normal conditions, *T. castaneum* pupates at the seventh or eighth instar. Injection of double-stranded RNA targeting JHAMT into day-0 third-instar larvae induces precocious pupation at the fifth or sixth instar, resulting in smaller adults. (**b**) Forced overexpression of JHAMT in transgenic *Bombyx mori* generated using the GAL4/UAS system. JHAMT-overexpressing larvae (JHAMT O/E) exhibit no apparent morphological differences from the controls—wild type (WT), UAS-JHAMT alone, or A3-GAL4 alone—during the early fifth instar. Control larvae normally pupate on day 10 of the fifth instar. In contrast, JHAMT-overexpressing larvae fail to reduce body weight, fail to spin cocoons, and ultimately undergo pupation arrest, resulting in lethality. The photograph in panel (**b**) was taken by Mika Murata

premature metamorphosis, supporting the classical "competence hypothesis" (Daimon et al. 2012; Smykal et al. 2014).

3.7.2 JHAMT and Reproductive Physiology

JHAMT is essential for reproduction in both sexes of insects. Knockdown of *JHAMT* markedly reduced female fertility in *L. decemlineata*, *A. aegypti*, *D. punctata*, *B. germanica*, and *Aethina tumida* (Dominguez and Maestro 2018; Fu et al. 2016; Gu et al. 2025; Huang et al. 2015; Van Ekert et al. 2014b). In males, *JHAMT* depletion impairs mating activity and accessory gland function in *T. castaneum* (Parthasarathy et al. 2009) and *D. melanogaster* (Wijesekera et al. 2016).

In crustaceans, short-day–induced male production in *Daphnia* correlates with elevated *JHAMT* expression (Toyota et al. 2015), suggesting a conserved MF/JH-based mechanism underlying reproductive polyphenism (see Chap. 14).

3.7.3 *JHAMT* in Diapause, Polyphenism, and Caste Differentiation

JHAMT frequently links environmental cues to JH-dependent polyphenic outcomes, including diapause, wing morphs, and caste determination (Hartfelder and Engels 1998; Nijhout 1994). The representative cases are as follows:

Diapause
Adult reproductive diapause in the leaf beetles *Colaphellus bowringi* and *Galeruca daurica* is correlated with strong *JHAMT* downregulation (Ma et al. 2021; Tian et al. 2021). In addition, maternal diapause in the jewel wasp, *Nasonia vitripennis*, requires reduced maternal *JHAMT* expression (Mukai et al. 2022; see Chap. 12). In contrast, nymphal diapause in the small brown planthopper, *Laodelphax striatellus*, involves *JHAMT* upregulation (Zhai et al. 2017), illustrating stage-specific differences in JH/JHAMT-mediated diapause regulation.

Polyphenism
In Hemiptera, *JHAMT* regulates wing-morph determination and contributes to the pronounced sexual dimorphism characteristic of scale insects (Vea et al. 2016; see Chap. 9).

Caste Differentiation
In eusocial insects, caste-specific JH titers correspond to differential *JHAMT* expression. Queen-destined *A. mellifera* larvae exhibit elevated *JHAMT* (Li et al. 2013), and termites employ JH/*JHAMT* signaling to regulate soldier and reproductive caste differentiation (Yaguchi et al. 2015; see Chap. 13).

3.7.4 Additional Physiological Roles and Responses to Stressors

Transcriptomic analyses have revealed broader physiological functions of JHAMT. In *D. melanogaster*, fat body–derived JHAMT contributes to iron homeostasis (Ye et al. 2024).

Exposure to insecticides or natural compounds can strongly alter *JHAMT* expression, affecting growth and reproduction. Sublethal doses of several insecticides upregulate *JHAMT* in major pest species, including the rice stem borer, *Chilo suppressalis* (Xu et al. 2017), the tobacco cutworm, *Spodoptera litura* (Liu et al. 2018), and the fall armyworm, *Spodoptera frugiperda* (Lu et al. 2023; Wang et al. 2025).

Sublethal stress may also induce compensatory *JHAMT* overexpression, as observed in the melon aphid, *Aphis gossypii*, feeding on genetically modified cotton (Zhang et al. 2024). Conversely, green tea proanthocyanidins suppress *JHAMT* and impair the development of the malaria mosquito *Anopheles gambiae* (Muema et al. 2017), and ribavirin reduces JH production in *S. litura* (Li et al. 2022).

Collectively, these findings position JHAMT as a central integrator of environmental and physiological cues, translating them into adaptive developmental, reproductive, and polyphenic responses in insects and other arthropods.

3.8 Evolutionary Perspectives of JH Biosynthesis and JHAMT

JH and MF are sesquiterpenoid hormones widely utilized by insects and crustaceans, both synthesized from the common precursor FA. The evolutionary acquisition of the methyltransferase JHAMT enabled the production of methylated sesquiterpenoids, representing a major innovation in arthropod endocrinology. This section summarizes the distribution of JH/MF across the animal kingdom and examines the evolutionary diversification of the *JHAMT* gene family.

3.8.1 Distribution and Diversity of JH/MF in the Animal Kingdom

FA, the immediate precursor of JH and MF, is synthesized from FPP through dephosphorylation followed by two oxidation steps (see Chap. 5). Because these enzymatic reactions are thought to be broadly conserved, FA is presumed to occur widely across metazoans and has been detected even in basal cnidarians (So et al. 2022a). FA itself exhibits hormonal activity in some organisms; for example, in the red crab *Charybdis feriatus*, FA induces *vitellogenin* expression (Mak et al. 2005).

MF, generated by JHAMT-mediated methylation of FA, is the principal reproductive hormone in crustaceans (Laufer et al. 1992). MF has also been reported in several insect groups, including *D. melanogaster* and cockroaches (Bendena et al. 2011; Jones et al. 2010), as well as in other arthropods, such as the two spotted spider mite, *Tetranychus urticae* (Grbic et al. 2011), and the centipede *Thereuonema tuberculata* (So et al. 2022b). Beyond Arthropoda, MF has been detected in polychaete annelids, and the presence of the JH receptor Met in these taxa (Schenk et al. 2016) suggests that sesquiterpenoid signaling originated prior to the divergence of arthropods.

In contrast, epoxidized JHs appear to be restricted to insects. Phylogenomic analyses have indicated that insects evolved from within a crustacean lineage and that ancestral hexapods colonized land approximately 480 million years ago (Misof et al. 2014). Phylogenomics robustly places insects within Pancrustacea, making MF the likely ancestral sesquiterpenoid hormone. Given this origin, early insects likely relied primarily on MF to regulate reproduction and embryogenesis, functions that are hypothesized to represent the ancestral roles of JH in insects (Truman et al. 2024; Wyatt 2013). However, as insects diversified, the emergence of epoxidized JHs represented a major evolutionary innovation. Structural and functional studies suggest that the transition from MF to epoxidized JHs coincides with modifications in the ligand–receptor system (Miyakawa et al. 2013). The acquisition of epoxidized JHs in insects may have facilitated the evolution of more complex life-history traits, including advanced metamorphosis, diapause regulation, and diverse forms of phenotypic polyphenism.

Together, these observations indicate that the methylation of FA to MF evolved early in metazoan evolution, whereas the emergence of epoxidized JHs represents a derived innovation restricted to insects. This sets the stage for understanding how the *JHAMT* gene family has diversified across lineages.

3.8.2 Evolutionary Expansion and Putative Functions of JHAMT/JHAMT-like Gene Families

JJHAMT is indispensable for MF/JH biosynthesis in both insects and crustaceans. *JHAMT* homologs have also been proposed to participate in MF biosynthesis in other arthropods, including *T. urticae* (Grbic et al. 2011), the pond wolf spider, *Pardosa pseudoannulata* (Yang et al. 2021), and centipedes (So et al. 2022b), although direct biochemical validation of these enzymes remains lacking.

Genomic surveys have revealed that *JHAMT-like* genes are widely distributed across the animal kingdom. To assess their taxonomic range, the National Center for Biotechnology Information (NCBI) protein database was queried for entries annotated as "juvenile hormone acid methyltransferase." These searches recovered *JHAMT* and *JHAMT-like* sequences from a broad array of hexapod groups, including Collembola (Fig. 3.5c). They were also detected across diverse arthropod lineages, including Crustacea and Chelicerata, such as Arachnida (mites and ticks, spiders, scorpions, and false scorpions), Pycnogonida (sea spiders), and Xiphosura (horseshoe crabs) (Fig. 3.5b). Beyond Arthropoda, *JHAMT-like* genes are present in Ecdysozoa, such as Nematomorpha (horsehair worms), and in several lineages of Lophotrochozoa, including Brachiopoda (lingula) and Nemertea (ribbon worms). They also occur within Deuterostomia—Echinodermata (starfish, crinoids, and sea urchins), Hemichordata (acorn worms), and Chordata (lancelets)—and have even been detected in more basal metazoan groups, such as Cnidaria (stony corals) and Porifera (sponges) (Fig. 3.5a). This remarkably broad phylogenetic distribution suggests that canonical *JHAMT*s likely originated from ancient SAM-dependent methyltransferases that were present early in metazoan evolution.

A

```
Animalia 2816(446)
├── Porifera 3(2)
├── Ctenophora ND
├── Placozoa ND
├── Cnidaria 20(6)
└── Bilateria
    ├── Deuterostomia
    │   ├── Echinodermata 47(8)
    │   ├── Hemichordata 35(1)
    │   └── Chordata 23(3)
    │
    └── Protostomia
        ├── Ecdysozoa
        │   ├── Arthropoda 2658(421)
        │   ├── Onychophora ND
        │   ├── Tardigrada ND
        │   ├── Nematomorpha 2(1)
        │   └── Nematoda ND
        └── Lophotrochozoa
            ├── Annelida ND
            ├── Mollusca ND
            ├── Brachiopoda 2(1)
            ├── Nemertea 26(1)
            └── Platyhelminthes ND
```

B

```
Arthropoda
├── Chelicerata
│   ├── Arachnida 859(40)
│   │   ├── Mites and Ticks 698(26)
│   │   ├── Spiders 130(12)
│   │   ├── Scorpions 30(2)
│   │   └── False scorpions 1(1)
│   ├── Pycnogonida 2(1)
│   └── Xiphosura 45(2)
│
├── Myriapoda ND
│
├── Crustacea 47(25)
│
└── Hexapoda 1705(352)
```

C

```
Hexapoda
├── Entognatha
│   ├── Collembola 10(2)
│   ├── Protura ND
│   └── Diplura ND
│
└── Insecta
    ├── Paleoptera
    │   ├── Odonata 2(2)
    │   └── Ephemeroptera 4(2)
    │
    └── Neoptera
        ├── Polyneoptera
        │   ├── Orthoptera 160(10)
        │   ├── Phasmatodea 1(1)
        │   ├── Blattodea 41(5)
        │   ├── Mantodea ND
        │   ├── Dermaptera ND
        │   ├── Plecoptera ND
        │   └── Embioptera ND
        │
        ├── Paraneoptera
        │   ├── Hemiptera 146(28)
        │   ├── Thysanoptera 4(2)
        │   └── Psocodea ND
        │
        └── Holometabola
            ├── Hymenoptera 453(97)
            ├── Coleoptera 282(31)
            ├── Lepidoptera 245(54)
            ├── Diptera 345(115)
            ├── Trichoptera 7(1)
            ├── Neuroptera 3(1)
            ├── Siphonaptera 2(1)
            ├── Mecoptera ND
            └── Strepsiptera ND
```

Fig. 3.5 Phylogenetic distribution of *juvenile hormone acid O-methyltransferase* (*JHAMT*)*-like* genes in metazoans (**a**), arthropods (**b**), and hexapods (**c**). The protein database in NCBI was searched using the keyword "juvenile hormone acid methyltransferase." (accessed 8 December 2025) The numbers following each taxonomic group represent the number of predicted proteins registered as *JHAMT* or *JHAMT-like* genes, and the numbers in parentheses indicate the number of source species. "ND" indicates that no registered genes were found

In contrast, several major metazoan lineages, including Nematoda, Mollusca, Annelida, and Vertebrata, lack detectable *JHAMT-like* genes. The presence of multiple *JHAMT-like* genes in lancelets implies that the ancestral deuterostome possessed such genes, which were subsequently lost in vertebrates. A similar pattern has been observed within Arthropoda: in myriapods, *JHAMT-like* genes are present in centipedes but absent in millipedes (So et al. 2022b), indicating lineage-specific loss.

Notably, the occurrence of MF does not consistently correspond to the presence of *JHAMT-like* genes. Annelids synthesize MF (Schenk et al. 2016) but possess no identifiable *JHAMT-like* sequences, suggesting that MF may be produced by alternative methyltransferases. Conversely, ticks contain numerous *JHAMT-like* genes (Zhu et al. 2016), even though neither JH nor MF has been chemically confirmed in these species. Therefore, the functional characterization of these *JHAMT-like* genes remains an important priority for future research.

3.8.3 Duplication and Functional Divergence of JHAMT-like Genes

Given the indispensable role of JH in insect development and reproduction, all insect species are presumed to encode at least one bona fide *JHAMT* ortholog. *D. melanogaster*, which possesses a single *JHAMT* locus (Niwa et al. 2008), appears to be rather exceptional. In contrast, many insects harbor multiple paralogs, for example, three in *T. castaneum* (Minakuchi et al. 2008), six in *B. mori* (Consortium 2008), and more than 30 in the desert locust *Schistocerca gregaria* (Smykal and Dolezel 2023).

Although most paralogs remain functionally uncharacterized, two broad evolutionary trajectories are plausible: (1) subfunctionalization, whereby duplicated genes partition ancestral functions and acquire tissue-, stage-, or sex-specific roles; and (2) neofunctionalization, whereby certain paralogs evolve novel catalytic capacities, including the synthesis of low-molecular-weight esters distinct from JH or MF.

Evidence for scenario (1) comes from the extra-CA enzymatic activity. Although *JHAMT* transcription in most insects is CA-enriched, JHAMT activity has been detected in the male accessory glands of *H. cecropia* (Weirich and Culver 1979), the longhorned beetle *Apriona germari* (Tian et al. 2010), and the wing imaginal discs of *M. sexta* (Sparagana et al. 1985). These cases suggest that certain paralogs may have acquired tissue-, stage-, or sex-specific roles in the biosynthesis of JH.

Scenario (2) is supported by the occurrence of bioactive metabolites that possess a carboxyl methyl ester functional group in diverse insects. Representative examples include the aggregation pheromone methyl *(E,E,Z)-2,4,6*-decatrienoate in the brown-winged green bug, *Plautia stali* (Sugie et al. 1996), and methyl salicylate, an anti-aphrodisiac pheromone in pierid butterflies (Andersson et al. 2003). In *P. stali*, seven *JHAMT-like* genes have been identified: one is CA-specific (Fig. 3.2b) and likely responsible for JH biosynthesis, whereas another is expressed exclusively in the abdomens of nondiapause males and may participate in aggregation pheromone production (Shinoda and Taira, unpublished). These findings support the possibility that certain *JHAMT-like* paralogs have been co-opted to synthesize bioactive molecules other than JH or MF.

Extensive *JHAMT-like* gene duplication has also occurred in noninsect arthropods (e.g., ticks) and basal chordates (e.g., lancelets). Functional divergence in these taxa may reflect changes in spatiotemporal expression, catalytic residue, or substrate specificity. Comprehensive biochemical, structural, and in vivo functional studies are essential for elucidating the evolutionary forces that shape the diversification of this enzyme family.

3.9 Applications for Insect Pest Control and Biotechnology

3.9.1 JHAMT Inhibitors

Although JH analogs are widely used as insect growth regulators (IGRs) for medical and agricultural pests (see Chap. 16), inhibitors of JH action or biosynthesis are often more suitable for controlling species whose damaging stages are larvae (see Chap. 17). Because JHAMT is indispensable for JH biosynthesis and is absent in vertebrates, it represents an attractive and highly selective target for next-generation IGRs (Minakuchi and Riddiford 2006; Cusson et al. 2012; Yin et al. 2020). Inhibiting JHAMT may disrupt JH production, potentially leading to precocious metamorphosis, reduced fecundity, and a substantial decline in population fitness.

Classical SAM-dependent methyltransferase inhibitors, including *S*-adenosylhomocysteine (SAH) and sinefungin, strongly inhibit JHAMT activity (Feyereisen and Farnsworth 1987). The SAH-hydrolase inhibitor 3-deazaneplanocin A (DZNep) similarly reduces JHAMT activity (Nouzova et al. 2019). However, these compounds act broadly on SAM-dependent enzymes and are rapidly metabolized in vivo, limiting their applicability as IGRs. More selective inhibition has been observed with imidazole derivatives, which suppress JH biosynthesis in the CA of *D. punctata*, at least in part, through JHAMT inhibition (Pratt et al. 1990). The anti-JH compound KF-13S downregulates the transcription of *JHAMT* and other JH-biosynthetic genes, inducing precocious metamorphosis (Kaneko et al. 2011a).

Structure-guided analyses of *T. castaneum* JHAMT recently enabled the discovery of the first JHAMT-specific inhibitor (JI3), providing proof of concept for the rational design of JHAMT-targeted IGRs (Yang et al. 2025). Moving forward, the combination of high-throughput screening with structure-based in silico approaches, facilitated by improved modeling of SAM-binding and substrate-recognition motifs, should accelerate the development of potent and selective JHAMT inhibitors.

3.9.2 RNA Interference and Transgenic Approaches

Gene-silencing strategies targeting *JHAMT* show considerable promise for controlling both medical and agricultural pests. Feeding RNA interference (RNAi) effectively suppresses egg maturation and larval development in *A. aegypti* (Van Ekert et al. 2014b). Comparable knockdown produces severe larval defects in major crop pests, including the tobacco budworm, *Helicoverpa armigera* (Asokan et al. 2013), the spongy moth, *Lymantria dispar* (Ji et al. 2025), and the oriental fruit fly, *Bactrocera dorsalis* (Zhou et al. 2022). Transgenic crops expressing double-stranded RNA (dsRNA) against *JHAMT* efficiently inhibit *L. decemlineata*, reducing larval growth and oviposition (Guo et al. 2018). The incorporation of dsRNA traits into Bt cotton further delays resistance development in *H. armigera* (Ni et al. 2017). Recent studies have expanded these approaches to metabolic engineering. Expression of only three genes—*JHAMT*, *CYP15B1* from *A. aegypti*, and an upregulated mevalonate-pathway enzyme (HMGR)—is sufficient to reconstitute JH III biosynthesis in the wild tobacco *Nicotiana benthamiana*, generating JH levels that exert insecticidal activity (Stathaki et al. 2025).

Together, these studies illustrate the growing potential of RNAi-based suppression and transgenic engineering of JH biosynthesis pathways as next-generation pest control technologies.

3.9.3 Biotechnological Applications

In addition to pest control, *JHAMT* has significant potential for metabolic engineering. Engineered *E. coli* (Sherkhanov et al. 2016), cyanobacteria (Yunus et al. 2020), and *Saccharomyces cerevisiae* (Wang et al. 2024) expressing *D. melanogaster JHAMT* produce fatty acid methyl esters (FAMEs) without exogenous methanol and with enhanced catalytic efficiency. These microbial platforms may provide sustainable routes for biodiesel production. These studies demonstrate that the distinctive regio- and stereoselective methylation catalyzed by JHAMT can be harnessed for industrial bioprocesses, expanding its relevance well beyond its established physiological roles in insects.

3.10 Conclusions and Perspectives

Research on JH biosynthesis has progressed from indirect biochemical detection to molecular dissection centered on JHAMT, the terminal enzyme that installs the methyl ester shared by all insect JHs and crustacean MF. The cloning of *B. mori JHAMT* provided the first definitive molecular evidence for this step and enabled comparative analyses, revealing a conserved SAM-dependent methyltransferase fold across insects and crustaceans. Genomic surveys have further shown that canonical *JHAMT* orthologs are broadly conserved, whereas *JHAMT-like* paralogs have repeatedly expanded in multiple arthropod lineages, although their functions remain largely unknown.

JHAMT is a central regulatory node that integrates allatoregulatory neuropeptides, ecdysteroids, nutrient- and stress-responsive signaling pathways, and diverse transcriptional and posttranscriptional mechanisms. However, several critical gaps remain. The molecular identity of the long-hypothesized allatinhibin-like factor, the full complement of CA-intrinsic transcription factors controlling *JHAMT* expression, and the nature of the alternative methyltransferase(s) suggested by *D. melanogaster* genetics remain unresolved. Likewise, the mechanisms that permanently extinguish *JHAMT* expression at metamorphosis—versus those that sustain it in reproductively active adults—have yet to be fully defined.

From an applied perspective, JHAMT is an exceptionally promising insect-specific target for next-generation pest management and synthetic biology applications. RNAi-based suppression, structure-guided inhibitor design, and engineered JH/MF biosynthetic pathways in heterologous hosts all demonstrate their translational potential. Future progress will require systematic biochemical and structural characterization of diverse orthologs and paralogs, clarification of species-specific catalytic properties, and a deeper understanding of the regulatory circuitry operating across developmental and environmental contexts.

Resolving these outstanding questions will sharpen the mechanistic framework of JH biology and enable the rational development of selective insect-control technologies, while also expanding the biotechnological applications of sesquiterpenoid biosynthesis.

References

Andersson J, Borg-Karlson A-K, Wiklund C (2003) Antiaphrodisiacs in pierid butterflies: a theme with variation! J Chem Ecol 29:1489–1499. https://doi.org/10.1023/A:1024277823101

Areiza M, Nouzova M, Rivera-Perez C, Noriega FG (2014) Ecdysis triggering hormone ensures proper timing of juvenile hormone biosynthesis in pharate adult mosquitoes. Insect Biochem Mol Biol 54:98–105. https://doi.org/10.1016/j.ibmb.2014.09.006

Areiza M, Nouzova M, Rivera-Perez C, Noriega FG (2015) 20-hydroxyecdysone stimulation of juvenile hormone biosynthesis by the mosquito corpora allata. Insect Biochem Mol Biol 64:100–105. https://doi.org/10.1016/j.ibmb.2015.08.001

Asokan R, Chandra GS, Manamohan M, Kumar NK (2013) Effect of diet delivered various concentrations of double-stranded RNA in silencing a midgut and a non-midgut gene of *Helicoverpa armigera*. Bull Entomol Res 103:555–563. https://doi.org/10.1017/S0007485313000138

Audsley N, Matthews HJ, Price NR, Weaver RJ (2008) Allatoregulatory peptides in Lepidoptera, structures, distribution and functions. J Insect Physiol 54:969–980. https://doi.org/10.1016/j.jinsphys.2008.01.012

Bendena WG, Zhang J, Burtenshaw SM, Tobe SS (2011) Evidence for differential biosynthesis of juvenile hormone (and related) sesquiterpenoids in *Drosophila melanogaster*. Gen Comp Endocrinol 172:56–61. https://doi.org/10.1016/j.ygcen.2011.02.014

Bhaskaran G, Dahm KH, Barrera P, Pacheco JL, Peck KE, Muszynska-Pytel M (1990) Allatinhibin, a neurohormonal inhibitor of juvenile hormone biosynthesis in *Manduca sexta*. Gen Comp Endocrinol 78:123–136. https://doi.org/10.1016/0016-6480(90)90053-O

Bhaskaran G, Sparagana SP, Barrera P, Dahm KH (1986) Change in corpus allatum function during metamorphosis of the tobacco hornworm *Manduca sexta*: regulation at the terminal step in juvenile hormone biosynthesis. Arch Insect Biochem Physiol 3:321–338. https://doi.org/10.1002/arch.940030402

Burtenshaw SM, Su PP, Zhang JR, Tobe SS, Dayton L, Bendena WG (2008) A putative farnesoic acid *O*-methyltransferase (FAMeT) orthologue in *Drosophila melanogaster* (CG10527): relationship to juvenile hormone biosynthesis? Peptides 29:242–251. https://doi.org/10.1016/j.peptides.2007.10.030

Cai R, Tao G, Zhao P, Xia Q, He H, Wang Y (2022) POU-M2 promotes juvenile hormone biosynthesis by directly activating the transcription of juvenile hormone synthetic enzyme genes in *Bombyx mori*. Open Biol 12:220031. https://doi.org/10.1098/rsob.220031

Cheng C, Ko A, Chaieb L, Koyama T, Sarwar P, Mirth CK et al (2014) The POU factor ventral veins lacking/drifter directs the timing of metamorphosis through ecdysteroid and juvenile hormone signaling. PLoS Genet 10:e1004425. https://doi.org/10.1371/journal.pgen.1004425

Consortium TISG (2008) The genome of a lepidopteran model insect, the silkworm *Bombyx mori*. Insect Biochem Mol Biol 38:1036–1045. https://doi.org/10.1016/j.ibmb.2008.11.004

Cusson M, Sen SE, Shinoda T (2012) Juvenile hormone biosynthetic enzymes as targets for insecticide discovery. In: Ishaaya I, Palli SR, Horowitz AR (eds) Advanced technologies for managing insect pests. Springer, Dordrecht, pp 31–55. https://doi.org/10.1007/978-94-007-4497-4_3

Daimon T, Kozaki T, Niwa R, Kobayashi I, Furuta K, Namiki T et al (2012) Precocious metamorphosis in the juvenile hormone-deficient mutant of the silkworm, *Bombyx mori*. PLoS Genet 8:e1002486. https://doi.org/10.1371/journal.pgen.1002486

Daimon T, Shinoda T (2013) Function, diversity, and application of insect juvenile hormone epoxidases (CYP15). Biotechnol Appl Biochem 60:82–91. https://doi.org/10.1002/bab.1058

Defelipe LA, Dolghih E, Roitberg AE, Nouzova M, Mayoral JG, Noriega FG et al (2011) Juvenile hormone synthesis: "esterify then epoxidize" or "epoxidize then esterify"? Insights from the structural characterization of juvenile hormone acid methyltransferase. Insect Biochem Mol Biol 41:228–235. https://doi.org/10.1016/j.ibmb.2010.12.008

Dominguez CV, Maestro JL (2018) Expression of juvenile hormone acid *O*-methyltransferase and juvenile hormone synthesis in *Blattella germanica*. Insect Sci 25:787–796. https://doi.org/10.1111/1744-7917.12467

Dong YQ, Wang ZY, Jing TZ (2017) Functional characterization of ASTC (allatostatin C) and ASTCC (allatostatin double C) in *Clostera anastomosis* (Lepidoptera: Notodontidae). Gene 598:1–8. https://doi.org/10.1016/j.gene.2016.10.038

Fernandez-Nicolas A, Belles X (2017) Juvenile hormone signaling in short germ-band hemimetabolan embryos. Development 144:4637–4644. https://doi.org/10.1242/dev.152827

Feyereisen R, Farnsworth DE (1987) Inhibition of insect juvenile hormone synthesis by phorbol 12-myristate 13-acetate. FEBS Lett 222:345–348. https://doi.org/10.1016/0014-5793(87)80399-X

Fu KY, Li Q, Zhou LT, Meng QW, Lu FG, Guo WC et al (2016) Knockdown of *juvenile hormone acid methyl transferase* severely affects the performance of *Leptinotarsa decemlineata* (Say) larvae and adults. Pest Manag Sci 72:1231–1241. https://doi.org/10.1002/ps.4103

Grbic M, Van Leeuwen T, Clark RM, Rombauts S, Rouze P, Grbic V et al (2011) The genome of *Tetranychus urticae* reveals herbivorous pest adaptations. Nature 479:487–492. https://doi.org/10.1038/nature10640

Gu Y, Yang X, Liu S, Chen X, Liu R, Gao J et al (2025) RNAi-mediated knockdown of juvenile hormone acid methyltransferase depresses reproductive performance in female *Aethina tumida*. Pestic Biochem Physiol 211:106420. https://doi.org/10.1016/j.pestbp.2025.106420

Gunawardene YI, Tobe SS, Bendena WG, Chow BK, Yagi KJ, Chan SM (2002) Function and cellular localization of farnesoic acid *O*-methyltransferase (FAMeT) in the shrimp, *Metapenaeus ensis*. Eur J Biochem 269:3587–3595. https://doi.org/10.1046/j.1432-1033.2002.03048.x

Guo P, Zhang Y, Zhang L, Xu H, Zhang H, Wang Z et al (2021) Structural basis for juvenile hormone biosynthesis by the juvenile hormone acid methyltransferase. J Biol Chem 297:101234. https://doi.org/10.1016/j.jbc.2021.101234

Guo W, Bai C, Wang Z, Wang P, Fan Q, Mi X et al (2018) Double-stranded RNAs high-efficiently protect transgenic potato from *Leptinotarsa decemlineata* by disrupting juvenile hormone biosynthesis. J Agric Food Chem 66:11990–11999. https://doi.org/10.1021/acs.jafc.8b03914

Hartfelder K, Engels W (1998) Social insect polymorphism: hormonal regulation of plasticity in development and reproduction in the honeybee. Curr Top Dev Biol 40:45–77. https://doi.org/10.1016/S0070-2153(08)60364-6

Helvig C, Koener JF, Unnithan GC, Feyereisen R (2004) CYP15A1, the cytochrome P450 that catalyzes epoxidation of methyl farnesoate to juvenile hormone III in cockroach corpora allata. Proc Natl Acad Sci USA 101:4024–4029. https://doi.org/10.1073/pnas.0306980101

Holford KC, Edwards KA, Bendena WG, Tobe SS, Wang Z, Borst DW (2004) Purification and characterization of a mandibular organ protein from the American lobster, *Homarus americanus*: a putative farnesoic acid *O*-methyltransferase. Insect Biochem Mol Biol 34:785–798. https://doi.org/10.1016/j.ibmb.2004.04.003

Huang J, Marchal E, Hult EF, Tobe SS (2015) Characterization of the juvenile hormone pathway in the viviparous cockroach, *Diploptera punctata*. PLoS One 10:e0117291. https://doi.org/10.1371/journal.pone.0117291

Huang J, Tian L, Peng C, Abdou M, Wen D, Wang Y et al (2011) DPP-mediated TGFβ signaling regulates juvenile hormone biosynthesis by activating the expression of juvenile hormone acid methyltransferase. Development 138:2283–2291. https://doi.org/10.1242/dev.057687

Ishimaru Y, Tomonari S, Matsuoka Y, Watanabe T, Miyawaki K, Bando T et al (2016) TGF-β signaling in insects regulates metamorphosis via juvenile hormone biosynthesis. Proc Natl Acad Sci USA 113:5634–5639. https://doi.org/10.1073/pnas.1600612113

Ji W, Xie Q, Chang X, Shi J (2025) Silencing of juvenile hormone-related genes through RNA interference leads to molt failure and high mortality in the spongy moth. Insect Sci. https://doi.org/10.1111/1744-7917.70117

Jia Q, Yang L, Wen J, Liu S, Wen D, Luo W et al (2024) Cyp6g2 is the major P450 epoxidase responsible for juvenile hormone biosynthesis in *Drosophila melanogaster*. BMC Biol 22:111. https://doi.org/10.1186/s12915-024-01910-4

Jones G, Jones D, Li X, Tang L, Ye L, Teal P et al (2010) Activities of natural methyl farnesoids on pupariation and metamorphosis of *Drosophila melanogaster*. J Insect Physiol 56:1456–1464. https://doi.org/10.1016/j.jinsphys.2010.06.001

Kaneko Y, Furuta K, Kuwano E, Hiruma K (2011a) An anti-juvenile hormone agent, ethyl 4-(2-benzylhexyloxy)benzoate, inhibits juvenile hormone synthesis through the suppression of the transcription of juvenile hormone biosynthetic enzymes in the corpora allata in *Bombyx mori*. Insect Biochem Mol Biol 41:788–794. https://doi.org/10.1016/j.ibmb.2011.05.009

Kaneko Y, Hiruma K (2015) Allatotropin inhibits juvenile hormone biosynthesis by the corpora allata of adult *Bombyx mori*. J Insect Physiol 80:15–21. https://doi.org/10.1016/j.jinsphys.2015.01.005

Kaneko Y, Kinjoh T, Kiuchi M, Hiruma K (2011b) Stage-specific regulation of juvenile hormone biosynthesis by ecdysteroid in *Bombyx mori*. Mol Cell Endocrinol 335:204–210. https://doi.org/10.1016/j.mce.2011.01.012

Kayukawa T, Minakuchi C, Namiki T, Togawa T, Yoshiyama M, Kamimura M et al (2012) Transcriptional regulation of juvenile hormone-mediated induction of Krüppel homolog 1, a repressor of insect metamorphosis. Proc Natl Acad Sci USA 109:11729–11734. https://doi.org/10.1073/pnas.1204951109

Kayukawa T, Nagamine K, Inui T, Yokoi K, Kobayashi I, Nakao H et al (2024) *Dead ringer* acts as a major regulator of juvenile hormone biosynthesis in insects. PNAS nexus 3:pgae435. https://doi.org/10.1093/pnasnexus/pgae435

Kinjoh T, Kaneko Y, Itoyama K, Mita K, Hiruma K, Shinoda T (2007) Control of juvenile hormone biosynthesis in *Bombyx mori*: cloning of the enzymes in the mevalonate pathway and assessment of their developmental expression in the corpora allata. Insect Biochem Mol Biol 37:808–818. https://doi.org/10.1016/j.ibmb.2007.03.008

Laufer H, Sagi A, Ahl J, Homola E (1992) Methyl farnesoate appears to be a crustacean reproductive hormone. Invert Reprod Dev 22:17–19. https://doi.org/10.1080/07924259.1992.9672252

Li D, He C, Wang M, Liu H, Liu R, Xu L (2022) Toxicity of ribavirin to *Spodoptera litura* by inhibiting the juvenile hormone. J Agric Food Chem 70:3117–3126. https://doi.org/10.1021/acs.jafc.1c06172

Li W, Huang ZY, Liu F, Li Z, Yan L, Zhang S et al (2013) Molecular cloning and characterization of juvenile hormone acid methyltransferase in the honey bee, *Apis mellifera*, and its differential expression during caste differentiation. PLoS One 8:e68544. https://doi.org/10.1371/journal.pone.0068544

Liu D, Jia ZQ, Peng YC, Sheng CW, Tang T, Xu L et al (2018) Toxicity and sublethal effects of fluralaner on *Spodoptera litura* Fabricius (Lepidoptera: Noctuidae). Pestic Biochem Physiol 152:8–16. https://doi.org/10.1016/j.pestbp.2018.08.004

Lu Y, Tian J, Ullah F, Desneux N, Guo J, Wang S et al (2023) Sublethal and transgenerational effects of lufenuron on biological characteristics and expression of reproductive related genes in the fall armyworm, *Spodoptera frugiperda*. Pestic Biochem Physiol 196:105593. https://doi.org/10.1016/j.pestbp.2023.105593

Ma HY, Li YY, Li L, Tan Y, Pang BP (2021) Regulation of juvenile hormone on summer diapause of *Geleruca daurica* and its pathway analysis. Insects 12:237. https://doi.org/10.3390/insects12030237

Mak AS, Choi CL, Tiu SH, Hui JH, He JG, Tobe SS et al (2005) Vitellogenesis in the red crab *Charybdis feriatus*: Hepatopancreas-specific expression and farnesoic acid stimulation of vitellogenin gene expression. Mol Reprod Dev 70:288–300. https://doi.org/10.1002/mrd.20213

Marchal E, Zhang J, Badisco L, Verlinden H, Hult EF, Van Wielendaele P et al (2011) Final steps in juvenile hormone biosynthesis in the desert locust, *Schistocerca gregaria*. Insect Biochem Mol Biol 41:219–227. https://doi.org/10.1016/j.ibmb.2010.12.007

Mayoral JG, Nouzova M, Yoshiyama M, Shinoda T, Hernandez-Martinez S, Dolghih E et al (2009) Molecular and functional characterization of a juvenile hormone acid methyltransferase expressed in the corpora allata of mosquitoes. Insect Biochem Mol Biol 39:31–37. https://doi.org/10.1016/j.ibmb.2008.09.010

Meng QW, Liu XP, Lu FG, Fu KY, Guo WC, Li GQ (2015) Involvement of a putative allatostatin in regulation of juvenile hormone titer and the larval development in *Leptinotarsa decemlineata* (Say). Gene 554:105–113. https://doi.org/10.1016/j.gene.2014.10.033

Miki T, Shinohara T, Chafino S, Noji S, Tomioka K (2020) Photoperiod and temperature separately regulate nymphal development through JH and insulin/TOR signaling pathways in an insect. Proc Natl Acad Sci USA 117:5525–5531. https://doi.org/10.1073/pnas.1922747117

Minakuchi C, Ishii F, Washidu Y, Ichikawa A, Tanaka T, Miura K et al (2015) Expressional and functional analysis of CYP15A1, a juvenile hormone epoxidase, in the red flour beetle *Tribolium castaneum*. J Insect Physiol 80:61–70. https://doi.org/10.1016/j.jinsphys.2015.04.008

Minakuchi C, Namiki T, Yoshiyama M, Shinoda T (2008) RNAi-mediated knockdown of *juvenile hormone acid O-methyltransferase* gene causes precocious metamorphosis in the red flour beetle *Tribolium castaneum*. FEBS J 275:2919–2931. https://doi.org/10.1111/j.1742-4658.2008.06428.x

Minakuchi C, Riddiford LM (2006) Insect juvenile hormone action as a potential target of pest management. J Pestic Sci 31:77–84. https://doi.org/10.1584/jpestics.31.77

Misof B, Liu S, Meusemann K, Peters RS, Donath A, Mayer C et al (2014) Phylogenomics resolves the timing and pattern of insect evolution. Science 346:763–767. https://doi.org/10.1126/science.1257570

Miyakawa H, Toyota K, Hirakawa I, Ogino Y, Miyagawa S, Oda S et al (2013) A mutation in the receptor Methoprene-tolerant alters juvenile hormone response in insects and crustaceans. Nat Commun 4:1856. https://doi.org/10.1038/ncomms2868

Muema JM, Nyanjom SG, Mutunga JM, Njeru SN, Bargul JL (2017) Green tea proanthocyanidins cause impairment of hormone-regulated larval development and reproductive fitness via repression of juvenile hormone acid methyltransferase, insulin-like peptide and cytochrome P450 genes in *Anopheles gambiae sensu stricto*. PLoS One 12:e0173564. https://doi.org/10.1371/journal.pone.0173564

Mukai A, Mano G, Des Marteaux L, Shinada T, Goto SG (2022) Juvenile hormone as a causal factor for maternal regulation of diapause in a wasp. Insect Biochem Mol Biol 144:103758. https://doi.org/10.1016/j.ibmb.2022.103758

Ni M, Ma W, Wang X, Gao M, Dai Y, Wei X et al (2017) Next-generation transgenic cotton: pyramiding RNAi and Bt counters insect resistance. Plant Biotechnol J 15:1204–1213. https://doi.org/10.1111/pbi.12709

Nijhout HF (1994) Polyphenisms. In: Nijhout HF (ed) Insect hormones. Princeton University Press, Princeton, NJ, pp 176–196

Niwa R, Niimi T, Honda N, Yoshiyama M, Itoyama K, Kataoka H et al (2008) Juvenile hormone acid *O*-methyltransferase in *Drosophila melanogaster*. Insect Biochem Mol Biol 38:714–720. https://doi.org/10.1016/j.ibmb.2008.04.003

Nouzova M, Edwards MJ, Michalkova V, Ramirez CE, Ruiz M, Areiza M et al (2021) Epoxidation of juvenile hormone was a key innovation improving insect reproductive fitness. Proc Natl Acad Sci USA 118. https://doi.org/10.1073/pnas.2109381118

Nouzova M, Michalkova V, Ramirez CE, Fernandez-Lima F, Noriega FG (2019) Inhibition of juvenile hormone synthesis in mosquitoes by the methylation inhibitor 3-deazaneplanocin A (DZNep). Insect Biochem Mol Biol 113:103183. https://doi.org/10.1016/j.ibmb.2019.103183

Parthasarathy R, Tan A, Sun Z, Chen Z, Rankin M, Palli SR (2009) Juvenile hormone regulation of male accessory gland activity in the red flour beetle, *Tribolium castaneum*. Mech Dev 126:563–579. https://doi.org/10.1016/j.mod.2009.03.005

Pratt G, Kuwano E, Farnsworth D, Feyereisen R (1990) Structure/activity studies on 1, 5-disubstituted imidazoles as inhibitors of juvenile hormone biosynthesis in isolated corpora allata of the cockroach *Diploptera punctata*. Pestic Biochem Physiol 38:223–230. https://doi.org/10.1016/0048-3575(90)90094-I

Reibstein D, Law JH (1973) Enzymatic synthesis of insect juvenile hormones. Biochem Biophys Res Commun 55:266–272. https://doi.org/10.1016/0006-291x(73)91083-8

Röller H, Dahm K, Sweely C, Trost B (1967) The structure of the juvenile hormone. Angew Chem Int Ed Engl 6:179–180. https://doi.org/10.1002/anie.196701792

Ruang-Rit K, Park Y (2018) Endocrine system in supernumerary molting of the flour beetle, *Tribolium freemani*, under crowded conditions. Insect Biochem Mol Biol 101:76–84. https://doi.org/10.1016/j.ibmb.2018.08.002

Ruddell CJ, Wainwright G, Geffen A, White MR, Webster SG, Rees HH (2003) Cloning, characterization, and developmental expression of a putative farnesoic acid *O*-methyl transferase in the female edible crab *Cancer pagurus*. Biol Bull 205:308–318. https://doi.org/10.2307/1543294

Schenk S, Krauditsch C, Frühauf P, Gerner C, Raible F (2016) Discovery of methylfarnesoate as the annelid brain hormone reveals an ancient role of sesquiterpenoids in reproduction. elife 5:e17126. https://doi.org/10.7554/eLife.17126

Schooley DA, Baker FC (1985) Juvenile hormone biosynthesis. In: Kerkut GA, Gilbert LI (eds) Comprehensive insect physiology, biochemistry and pharmacology. Pergammon Press, Oxford, pp 363–389

Sheng Z, Ma L, Cao MX, Jiang RJ, Li S (2008) Juvenile hormone acid methyl transferase is a key regulatory enzyme for juvenile hormone synthesis in the Eri silkworm, *Samia cynthica ricini*. Arch Insect Biochem Physiol 69:143–154. https://doi.org/10.1002/arch.20268

Sherkhanov S, Korman TP, Clarke SG, Bowie JU (2016) Production of FAME biodiesel in *E. coli* by direct methylation with an insect enzyme. Sci Rep 6:24239. https://doi.org/10.1038/srep24239

Shinoda T (2016) Juvenile hormone acid methyltransferase (JHAMT): function, diversification, and applications. Sanshi Konchu Biotec 85:87–99. (in Japanese). https://doi.org/10.11416/konchubiotec.85.3_087

Shinoda T, Itoyama K (2003) Juvenile hormone acid methyltransferase: a key regulatory enzyme for insect metamorphosis. Proc Natl Acad Sci USA 100:11986–11991. https://doi.org/10.1073/pnas.2134232100

Silva Gunawardene YI, Chow BK, He JG, Chan SM (2001) The shrimp FAMeT cDNA is encoded for a putative enzyme involved in the methylfarnesoate (MF) biosynthetic pathway and is temporally expressed in the eyestalk of different sexes. Insect Biochem Mol Biol 31:1115–1124. https://doi.org/10.1016/s0965-1748(01)00060-1

Smykal V, Daimon T, Kayukawa T, Takaki K, Shinoda T, Jindra M (2014) Importance of juvenile hormone signaling arises with competence of insect larvae to metamorphose. Dev Biol 390:221–230. https://doi.org/10.1016/j.ydbio.2014.03.006

Smykal V, Dolezel D (2023) Evolution of proteins involved in the final steps of juvenile hormone synthesis. J Insect Physiol 145:104487. https://doi.org/10.1016/j.jinsphys.2023.104487

So WL, Kai Z, Qu Z, Bendena WG, Hui JHL (2022a) Rethinking Sesquiterpenoids: a widespread hormone in animals. Int J Mol Sci 23. https://doi.org/10.3390/ijms23115998

So WL, Nong W, Xie Y, Baril T, Ma HY, Qu Z et al (2022b) Myriapod genomes reveal ancestral horizontal gene transfer and hormonal gene loss in millipedes. Nat Commun 13:3010. https://doi.org/10.1038/s41467-022-30690-0

Song J, Li W, Gao L, Yan Q, Zhang X, Liu M et al (2024) miR-276 and miR-182013-5p modulate insect metamorphosis and reproduction via dually regulating juvenile hormone acid methyltransferase. Commun Biol 7:1604. https://doi.org/10.1038/s42003-024-07285-0

Song YX, Ren YP, Ran YY, Fan N, Wu T, Zang H et al (2025) Ame-miR-2161 affects the survival and development of honeybee larvae through the juvenile hormone acid methyltransferase gene. Insect Mol Biol. https://doi.org/10.1111/imb.70009

Sparagana SP, Bhaskaran G, Barrera P (1985) Juvenile hormone acid methyltransferase activity in imaginal discs of *Manduca sexta* prepupae. Arch Insect Biochem Physiol 2:191–202. https://doi.org/10.1002/jez.1402300216

Sparagana SP, Bhaskaran G, Dahm KH, Riddle V (1984) Juvenile hormone production, juvenile hormone esterase, and juvenile hormone acid methyltransferase in corpora allata of *Manduca sexta*. J Exp Zool 230:309–313. https://doi.org/10.1002/jez.1402300216

Stathaki A, Alam RM, Köllner TG, O'Connor SE (2025) Engineering of insect juvenile hormone III biosynthesis in the plant *Nicotiana benthamiana*. Metab Eng 88:77–84. https://doi.org/10.1016/j.ymben.2024.12.005

Sugie H, Yoshida M, Kawasaki K, Noguchi H, Moriya S, Takagi K et al (1996) Identification of the aggregation pheromone of the brown-winged green bug, *Plautia stali* Scott (Heteroptera: Pentatomidae). Appl Entmol Zool 31:427–431. https://doi.org/10.1303/aez.31.427

Sun Y, Fu D, Liu B, Wang L, Chen H (2022) Functional characterization of allatostatin C (PISCF/AST) and juvenile hormone acid O-methyltransferase in *Dendroctonus armandi*. Int J Mol Sci 23. https://doi.org/10.3390/ijms23052749

Takatsuka J, Nakai M, Shinoda T (2017) A virus carries a gene encoding juvenile hormone acid methyltransferase, a key regulatory enzyme in insect metamorphosis. Sci Rep 7:13522. https://doi.org/10.1038/s41598-017-14059-8

Tian L, Ji B, Liu S, Jin F, Gao J, Li S (2010) Juvenile hormone III produced in male accessory glands of the longhorned beetle, *Apriona germari*, is transferred to female ovaries during copulation. Arch Insect Biochem Physiol 75:57–67. https://doi.org/10.1002/arch.20385

Tian Z, Guo S, Li JX, Zhu F, Liu W, Wang XP (2021) Juvenile hormone biosynthetic genes are critical for regulating reproductive diapause in the cabbage beetle. Insect Biochem Mol Biol 139:103654. https://doi.org/10.1016/j.ibmb.2021.103654

Toyota K, Miyakawa H, Hiruta C, Furuta K, Ogino Y, Shinoda T et al (2015) Methyl farnesoate synthesis is necessary for the environmental sex determination in the water flea *Daphnia pulex*. J Insect Physiol 80:22–30. https://doi.org/10.1016/j.jinsphys.2015.02.002

Truman JW, Riddiford LM, Konopova B, Nouzova M, Noriega FG, Herko M (2024) The embryonic role of juvenile hormone in the firebrat, *Thermobia domestica*, reveals its function before its involvement in metamorphosis. elife 12. https://doi.org/10.7554/eLife.92643

Ueda H, Shinoda T, Hiruma K (2009) Spatial expression of the mevalonate enzymes involved in juvenile hormone biosynthesis in the corpora allata in *Bombyx mori*. J Insect Physiol 55:798–804. https://doi.org/10.1016/j.jinsphys.2009.04.013

Van Ekert E, Heylen K, Rouge P, Powell CA, Shatters RG Jr, Smagghe G et al (2014a) *Aedes aegypti* juvenile hormone acid methyl transferase, the ultimate enzyme in the biosynthetic pathway of juvenile hormone III, exhibits substrate control. J Insect Physiol 64:62–73. https://doi.org/10.1016/j.jinsphys.2014.03.001

Van Ekert E, Powell CA, Shatters RG, Borovsky D (2014b) Control of larval and egg development in *Aedes aegypti* with RNA interference against juvenile hormone acid methyl transferase. J Insect Physiol 70:143–150. https://doi.org/10.1016/j.jinsphys.2014.08.001

Van Ekert E, Shatters RG, Rouge P, Powell CA, Smagghe G, Borovsky D (2015) Cloning and expressing a highly functional and substrate specific farnesoic acid o-methyltransferase from the Asian citrus psyllid (*Diaphorina citri* Kuwayama). FEBS Open Bio 5:264–275. https://doi.org/10.1016/j.fob.2015.03.012

Vea IM, Tanaka S, Shiotsuki T, Jouraku A, Tanaka T, Minakuchi C (2016) Differential juvenile hormone variations in scale insect extreme sexual dimorphism. PLoS One 11:e0149459. https://doi.org/10.1371/journal.pone.0149459

Wang A, Yang Y, Zhang Y, Xue C, Cheng Y, Zhang Y et al (2025) Insecticide-induced sublethal effect in the fall armyworm is mediated by miR-9993/miR-2a-3p - *FPPS/JHAMT* - JH molecular module. Pestic Biochem Physiol 210:106400. https://doi.org/10.1016/j.pestbp.2025.106400

Wang L, Liu B, Meng Q, Yang C, Hu Y, Wang C et al (2024) *Saccharomyces cerevisiae* cellular engineering for the production of FAME biodiesel. AMB Express 14:42. https://doi.org/10.1186/s13568-024-01702-7

Weirich GF, Culver MG (1979) S-Adenosylmethionine: juvenile hormone acid methyltransferase in male accessory reproductive glands of *Hyalophora cecropia* (L.). Arch Biochem Biophys 198:175–181. https://doi.org/10.1016/0003-9861(79)90408-9

Wen D, Rivera-Perez C, Abdou M, Jia Q, He Q, Liu X et al (2015) Methyl farnesoate plays a dual role in regulating *Drosophila* metamorphosis. PLoS Genet 11:e1005038. https://doi.org/10.1371/journal.pgen.1005038

Wigglesworth VB (1934) The physiology of ecdysis in *Rhodnius prolixus* (Hemiptera). II Factors controlling moulting and 'metamorphosis'. Q J Microsc Sci 77:191–222. https://doi.org/10.1242/jcs.s2-77.306.191

Wijesekera TP, Saurabh S, Dauwalder B (2016) Juvenile hormone is required in adult males for *Drosophila* courtship. PLoS One 11:e0151912. https://doi.org/10.1371/journal.pone.0151912

Wyatt G (2013) Juvenile hormone in insect reproduction-a paradox? Eur J Entomol 94:323–333

Xu B, Qian K, Zhang N, Miao L, Cai J, Lu M et al (2017) Sublethal effects of chlorantraniliprole on juvenile hormone levels and mRNA expression of *JHAMT* and *FPPS* genes in the rice stem borer, *Chilo suppressalis*. Pest Manag Sci 73:2111–2117. https://doi.org/10.1002/ps.4586

Yaguchi H, Masuoka Y, Inoue T, Maekawa K (2015) Expressions of juvenile hormone biosynthetic genes during presoldier differentiation in the incipient colony of *Zootermopsis nevadensis* (Isoptera: Archotermopsidae). Appl Ent Zool 50:497–508. https://doi.org/10.1007/s13355-015-0358-3

Yang Y, Li Q, Zheng H, Shi J, Hu Y, Xia Y et al (2025) Unveiling the molecular architecture of *Tc*JHAMT: structural basis for catalytic efficiency and targeted inhibitor screening in the red flour beetle, Tribolium castaneum. Pest Manag Sci. https://doi.org/10.1002/ps.70267

Yang ZM, Wu Y, Li FF, Zhou ZJ, Yu N, Liu ZW (2021) Genomic identification and functional analysis of JHAMTs in the pond wolf spider, *Pardosa pseudoannulata*. Int J Mol Sci 22. https://doi.org/10.3390/ijms222111721

Ye YY, Liu ZH, Wang HL (2024) Fat body-derived juvenile hormone acid methyltransferase functions to maintain iron homeostasis in *Drosophila melanogaster*. FASEB J 38:e23805. https://doi.org/10.1096/fj.202400119RR

Yin Y, Qiu YW, Huang J, Tobe SS, Chen SS, Kai ZP (2020) Enzymes in the juvenile hormone biosynthetic pathway can be potential targets for pest control. Pest Manag Sci 76:1071–1077. https://doi.org/10.1002/ps.5617

Yunus IS, Palma A, Trudeau DL, Tawfik DS, Jones PR (2020) Methanol-free biosynthesis of fatty acid methyl ester (FAME) in *Synechocystis* sp. PCC 6803. Metab Eng 57:217–227. https://doi.org/10.1016/j.ymben.2019.12.001

Zhai Y, Zhang Z, Gao H, Chen H, Sun M, Zhang W et al (2017) Hormone signaling regulates nymphal diapause in *Laodelphax striatellus* (Hemiptera: Delphacidae). Sci Rep 7:13370. https://doi.org/10.1038/s41598-017-13879-y

Zhang L, Li Y, Xu X, Feng M, Turak R, Liu X et al (2024) Functional analysis of *AgJHAMT* gene related to developmental period in *Aphis gossypii* Glover. Bull Entomol Res 114:707–716. https://doi.org/10.1017/S000748532400049X

Zhang L, Xu H, Zhang Y, Zhang H, Wang Z, Guo P et al (2022) Structural characterization and functional analysis of juvenile hormone acid methyltransferase JHAMT3 from the silkworm, *Bombyx mori*. Insect Biochem Mol Biol 151:103863. https://doi.org/10.1016/j.ibmb.2022.103863

Zhou QH, Zhang Q, Yang RL, Yuan GR, Wang JJ, Dou W (2022) RNAi-mediated knockdown of juvenile hormone acid *O*-methyltransferase disrupts larval development in the oriental fruit fly, *Bactrocera dorsalis* (Hendel). Pestic Biochem Physiol 188:105285. https://doi.org/10.1016/j.pestbp.2022.105285

Zhu J, Khalil SM, Mitchell RD, Bissinger BW, Egekwu N, Sonenshine DE et al (2016) Mevalonate-farnesal biosynthesis in ticks: comparative synganglion transcriptomics and a new perspective. PLoS One 11:e0141084. https://doi.org/10.1371/journal.pone.0141084

Chapter 4
Degradation and Transport of Juvenile Hormones

Takahiro Shiotsuki

Abstract This chapter focuses on the enzymes and proteins involved in the regulation of juvenile hormone (JH) activity through degradation and transport. Degradative enzymes play a critical role in rapidly "switching off" hormone activity in specific organs or throughout the body, in coordination with the cessation of biosynthesis. In contrast, JH-binding proteins (JHBPs) in the hemolymph are essential for facilitating the transport of JH from its site of biosynthesis, the corpus allatum, to target organs and tissues in the form of a JH–JHBP complex. This complex protects JH from nonspecific degradation by hydrolytic enzymes, such as general esterases, present in the hemolymph. Once JH reaches its target organs or tissues, it binds to the JH receptor to elicit biological responses in the target cells. In this chapter, the functions, characteristics, and regulatory mechanisms of these enzymes and proteins are discussed.

Keywords JH esterase · JH epoxide hydrolase · JH diol kinase · JH-binding protein · Hexamerin · Lipophorin

4.1 Degradation

The degradation of hydrophobic compounds, including endogenous metabolites and xenobiotics, typically occurs in two distinct phases. Phase I reactions introduce polar functional groups into molecules, often through oxidation, reduction, or hydrolysis. Phase II reactions involve the conjugation of these functional groups with hydrophilic moieties, such as glucose, phosphate, or glutathione, thereby increasing the solubility of the compound and facilitating its excretion (Fig. 4.1).

T. Shiotsuki (✉)
Shimane University, Matsue, Japan
e-mail: shiotsuk@life.shimane-u.ac.jp

T. Shinoda et al. (eds.), *Juvenile Hormone*, Entomology Monographs,
https://doi.org/10.1007/978-981-95-8605-9_4

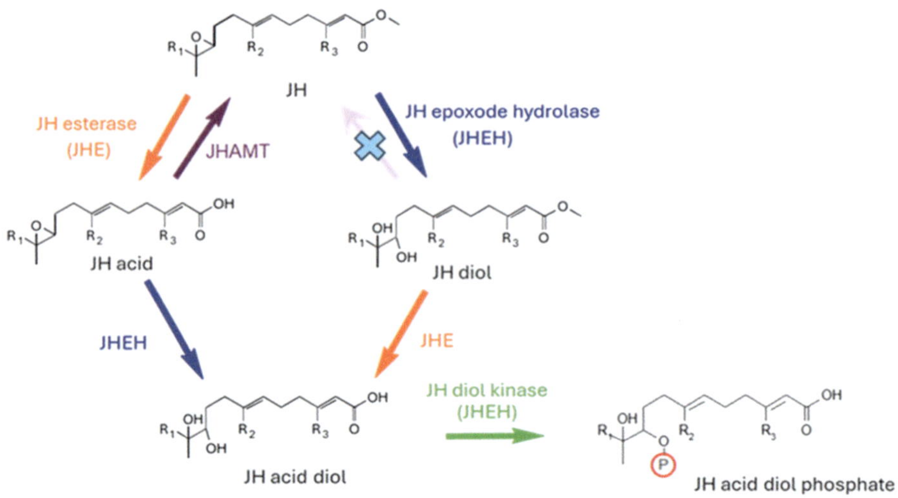

Fig. 4.1 Metabolic degradation of juvenile hormone (JH). Degradation by juvenile hormone esterase (JHE) is reversible because juvenile hormone acid (JHA) generated by JHE can be re--methylated back to JH by juvenile hormone acid *O*-methyltransferase (JHAMT). In contrast, degradation by juvenile hormone epoxide hydrolase (JHEH) and juvenile hormone diol kinase (JHDK) is irreversible and leads to metabolically inactive products

During phase I of juvenile hormone (JH) degradation, the oxygen-containing functional groups at both termini of the JH molecule, the carboxyl ester and epoxide, are targeted by two key enzymes: juvenile hormone esterase (JHE) and juvenile hormone epoxide hydrolase (JHEH). These enzymes play pivotal roles in initiating JH catabolism and have distinct biochemical properties. Following hydrolysis by JHEH, the resulting diol metabolites are further processed by juvenile hormone diol kinase (JHDK), which converts them into more water-soluble phosphate conjugates. This transformation facilitates subsequent excretion and is a critical step in the phase II metabolic pathway.

4.1.1 JH Esterase

Juvenile hormone esterase (JHE) is a substrate-specific carboxylesterase that hydrolyzes juvenile hormone (JH) into JH acids and is a member of the α/β-hydrolase superfamily (EC 3.1.1.1). This hydrolytic reaction is considered reversible, as JH acids can be converted back into bioactive JH by juvenile hormone acid *O*-methyltransferase (JHAMT) (Fig. 4.1). JHE exhibits a high degree of substrate specificity toward JH, enabling the precise regulation of JH levels without significantly affecting other ester-containing compounds within the insect system. Its high catalytic efficiency allows for the rapid modulation of JH concentrations, which is essential for ensuring the proper timing of developmental transitions (Kamita and Hammock 2010).

During insect development, sustained high JH levels maintain the larval state. A decline in JH titer, often mediated by increased JHE activity, facilitates the transition to pupation during the final larval instar and ultimately allows adult emergence. This reduction in JH is tightly regulated, as premature or delayed declines can result in developmental abnormalities and/or lethality. For example, Tan et al. (2005) demonstrated that overexpression of JHE in the silkworm *Bombyx mori* using the binary Gal4/UAS system induced precocious metamorphosis after the third instar, mimicking the effects of allatectomy or chemical ablation of JH. Conversely, targeted disruption of the *JHE* gene via CRISPR/Cas9 extended the duration of the larval stage (Zhang et al. 2017).

In adult insects, JH plays an essential role in the regulation of reproductive functions. In females, JH promotes vitellogenesis, which is the synthesis of yolk proteins that are critical for oocyte development. In males, it regulates the development and secretory activity of accessory glands, thereby influencing fertility and mating behavior. JHE contributes to the fine-tuning of these reproductive processes by precisely modulating JH titers to ensure the appropriate timing of these events.

JH titers are also involved in diapause regulation. Knockdown of *JHE* and *JHEH* in lady beetles *Coccinella septempunctata* (Li et al. 2022) and *Harmonia axyridis* (Gao et al. 2022) led to elevated expression of vitellogenin and suppression of diapause-associated phenotypes. In contrast, in the orange wheat blossom midge, *Sitodiplosis mosellana*, larval diapause is maintained under conditions of high JH titers, with low expression levels of *JHE* and *JHEH* observed during this period (Wan et al. 2020).

Beyond development and reproduction, JH influences a range of behaviors, including foraging, aggression, and social interactions, particularly in eusocial insects, such as bees, ants, and termites. Through the regulation of JH levels, JHE indirectly modulates these behaviors, thereby contributing to colony cohesion, survival, and environmental adaptation. Recent studies have further demonstrated the role of JHE in thermal adaptation. In the diamondback moth, *Plutella xylostella*, CRISPR/Cas9-mediated knockout of the *JHE* gene revealed that JHE regulates lipid and sugar metabolism by modulating JH titers, highlighting its broader physiological significance (Lei et al. 2025).

The enzyme contains five conserved sequence motifs: GQSAG at the catalytic site (Ser221 in the tobacco budworm, *Heliothis virescens*), GxxHxxD (including His467), a glutamate residue (Glu352), an RF motif (Arg69), and a DM motif (Asp193). Mutations in these residues typically abolish or severely impair enzymatic activity (Ward et al. 1992; Tsubota et al. 2010a).

The three-dimensional structure of the tobacco hornworm, *Manduca sexta* JHE, has been resolved in complex with the JHE-specific transition-state inhibitor 3-octylthio-1,1,1-trifluoro-2-propanone (OTFP), confirming the critical interactions at the active site (Wogulis et al. 2006; Fig. 4.2). In addition to the catalytic triad, Thr259 and Thr314 contribute to JH binding via π-stacking and hydrogen bonding with the α,β-unsaturated ester and epoxide moieties of the substrate, respectively.

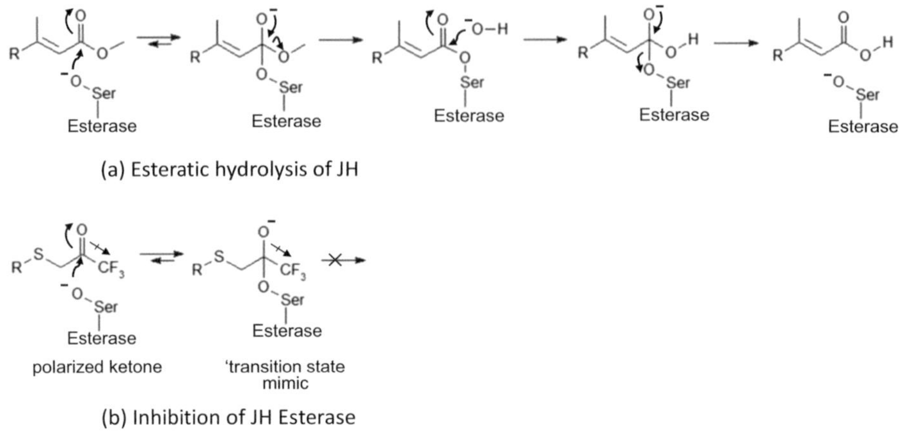

Fig. 4.2 Interaction of juvenile hormone esterase (JHE) with (**a**) its substrate, juvenile hormone (JH), and (**b**) the inhibitor, OTFP. The hydrolytic mechanism of JH by JHE, and inhibition by OTFP are illustrated (Shiotsuki 1995)

Exposure to OTFP delays molting and pupation, increases pupal weight, and suppresses both JHE enzymatic activity and *JHE* gene transcription (Duan et al. 2016).

The gene structure of *JHE* typically comprises multiple exons and introns, although the number and organization vary among insect species. For example, *B. mori* possesses six exons (Tsubota and Shiotsuki 2010), whereas the red flour beetle, *Tribolium castaneum,* and the fruit fly, *Drosophila melanogaster,* have six and eight exons, respectively (Tsubota et al. 2010b; Campbell et al. 2001). Spatially, JHE is predominantly expressed in the fat body, and the protein is secreted into the hemolymph following the cleavage of its N-terminal signal peptide.

The promoter region of the *JHE* gene was analyzed and found containing various cis-regulatory elements, including binding sites for transcription factors that modulate the timing and level of gene expression (Jones et al. 2000; Kethidi et al. 2004). The presence of hormone-responsive elements provides a mechanistic basis for the precise regulation of JHE expression in response to developmental cues.

In contrast, viral infection has been shown to downregulate JHE activity—without affecting JHEH—in the smaller tea tortrix, *Adoxophyes honmai*, thereby disrupting normal larval development (Saito et al. 2015).

4.1.2 JH Epoxide Hydrolase

JH epoxide hydrolase (JHEH) irreversibly inactivates JH by catalyzing the hydrolytic opening of its terminal epoxide ring through the addition of water. In contrast to JHE-mediated hydrolysis, which is reversible, the reaction catalyzed by JHEH leads to permanent deactivation of JH. JHEH is a membrane-associated protein that influences both its access to substrates and its interactions with other cellular components (Roe et al. 1996). In *M. sexta*, photoaffinity labeling identified JHEH as a ~50 kDa protein (Touhara and Prestwich 1994) and cDNA was cloned by Wojtasek and Prestwich (1996).

JHEH expression is tightly regulated at both the transcriptional and posttranscriptional levels in response to developmental and environmental stimuli. Typically, its expression increases during the mid-instar stages and decreases during molting (Seino et al. 2010). Moreover, tissue-specific expression patterns have been reported in several species, including *B. mori*, the spongy moth, *Lymantria dispar* (Wen et al. 2018), *T. castaneum*, and the Colorado potato beetle, *Leptinotarsa decemlineata* (Tusun et al. 2017), reflecting the localized nature of JH degradation. Unlike secreted JHE, which acts systemically via the hemolymph, JHEH primarily functions within specific tissues.

Functional studies have underscored the essential role of JHEH in insect development. In the small green plant bug, *Apolygus lucorum*, RNA interference (RNAi) targeting JHEH resulted in molting defects (Tusun et al. 2017). Similarly, RNAi-mediated knockdown of *LdJHEH1* and/or *LdJHEH2* in *L. decemlineata* delayed larval development and impaired adult emergence (Lu et al. 2015). In the boll weevil, *Anthonomus grandis*, the combined knockdown of *JHEH* and *JHDK* proved more effective than *JHEH* knockdown alone (Vasquez et al. 2023). Comparable developmental disruptions have been observed in several lepidopteran species, including *P. xylostella* (Chaitanya et al. 2017), *L. dispar* (Wen et al. 2018), and the common cutworm, *Spodoptera litura* (Sarkar and Kalia 2023).

Furthermore, the terminal transcription factor *Forkhead box O* (*FOXO*) represses both *JHE* and *JHEH* expression, thereby sustaining high JH titers. In *B. mori*, CRISPR/Cas9-mediated knockout of *FOXO* leads to precocious pupation, a phenotype that can be rescued by treatment with 20-hydroxyecdysone or JH analogs (Zeng et al. 2017).

JHEH plays a critical role in diapause regulation. In the cabbage beetle, *Colaphellus bowringi*, RNAi-mediated knockdown of *JHEH* and *JHDK* resulted in increased *vitellogenin* expression and a reduction in diapause-associated traits (Guo et al. 2019). Similar phenotypic effects have been reported in lady beetles, such as *C. septempunctata* (Li et al. 2022) and *H. axyridis* (Gao et al. 2022). In the true bug *Aspongopus chinensis*, RNAi targeting *JHEH* in diapause-stage adults accelerated reproductive development and promoted mating behavior (Zhou et al. 2022).

In the rice brown planthopper, *Nilaparvata lugens*, silencing of the *JHEH* gene enhanced the formation of short-winged morphs, thereby influencing wing polymorphism (Zhao et al. 2017). Furthermore, *JHEH*-deficient strains generated in *D. melanogaster* exhibited growth retardation and reduced tolerance to cold stress, phenotypes attributed to impaired glucose metabolism caused by the downregulation of key genes involved in glycolysis and the tricarboxylic acid (TCA) cycle (Rogalski et al. 2025).

Structurally, JHEH belongs to the α/β-hydrolase fold superfamily, characterized by an alternating arrangement of α-helices and β-strands. The crystal structure of *B. mori* JHEH reveals a homodimer composed of two distinct domains: an epoxide hydrolase domain (amino acids 53–166) and an α/β-hydrolase domain (residues 180–442) [PDB ID: 4QLA] (Zhou et al. 2014). The catalytic pocket is located at the interface between these domains and accommodates the JH substrate.

The catalytic triad, comprising Asp226, His430, and Glu403, mediates nucleophilic attack on the epoxide ring of JH, resulting in the formation of a diol. Glu403 and His430 facilitate proton transfer and stabilize the transition state during catalysis (Harris et al. 1999). Site-directed mutagenesis confirmed the essential roles of these residues in enzymatic activity.

The substrate-binding pocket includes an oxyanion hole formed by the conserved HGWP motif, along with Tyr298 and Tyr373, which are critical for stabilizing the JH molecule through hydrogen bonding and hydrophobic interactions (Tsubota et al. 2010c) (Fig. 4.3). A similar binding architecture has been modeled for JHEH from *A. lucorum* (Tusun et al. 2017). Notably, related en'zymes, such as *T. castaneum* TcJHEHr-3, possess a modified catalytic triad (Asp-Asp-His) but still exhibit robust activity toward JH III (Tsubota et al. 2010c).

The hydrolyzed JH diol product is significantly more hydrophilic than the parent hormone, facilitating its excretion and preventing residual hormonal activity. JHEH exhibits species-dependent substrate specificities. For example, in the cabbage looper, *Trichoplusia ni* (Kallapur et al. 1996), *D. melanogaster* (Borovsky et al. 2022a), and the yellow fever mosquito, *Aedes aegypti* (Borovsky et al. 2022b), JHEH preferentially hydrolyzes JH rather than JH acid, reflecting evolutionary adaptations in substrate selectivity.

Given its critical role in insect development and reproduction, JHEH is a promising target for pest management strategies. Inhibitors of JHEH activity can artificially elevate endogenous JH levels, resulting in developmental arrest and disruption of reproductive processes. Candidate inhibitors include *N*-[(*Z*)-9-octadecenyl]- and *N*-hexadecyl-*N*′-propylurea derivatives (Severson et al. 2002) and phurealipids (Nollmann et al. 2015), which offer species-specific alternatives to conventional chemical pesticides.

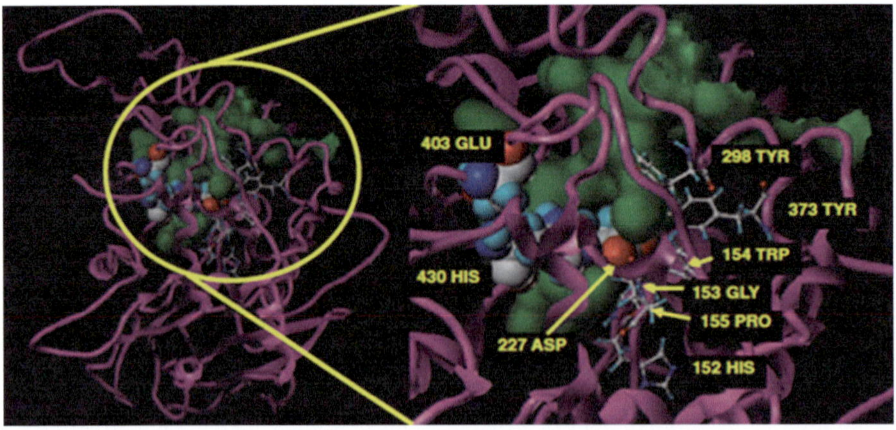

Fig. 4.3 Predicted active site of juvenile hormone epoxide hydrolase from *Bombyx mori* (BmJHEH). The catalytic triad (Asp227-His430-Gln403) and the oxyanion hole-forming residues HGWP (152–155), Tyr298, and Tyr373 are labeled in the expanded right panel (Seino et al. 2010)

4.1.3 JH Diol Kinase

JHDK catalyzes the phosphorylation of JH 10,11-diols, metabolites generated via hydrolysis by JHEH, by transferring a phosphate group to the C10 position. This reaction produces JH diol phosphate, a more polar and hydrophilic compound that facilitates efficient excretion and prevents the reactivation of hormones. JHDK was first purified from *M. sexta* (Maxwell et al. 2002a), with subsequent identification in *B. mori* (Li et al. 2005) and other insect species. In *M. sexta*, JHDK functions as a homodimer composed of two 20 kDa subunits, each consisting of 179 amino acids (Maxwell et al. 2002b). Homologous enzymes from *B. mori* and *S. litura* possess slightly longer sequences of 183 aa (Li et al. 2005; Zeng et al. 2015).

The expression of JHDK is regulated in a developmental and tissue-specific manner. In *B. mori*, the highest expression levels are observed in the midgut, with lower expression detected in the Malpighian tubules and hemocytes (Hua-jun et al. 2011), even though the enzyme was initially isolated from the Malpighian tubules. This expression profile suggests that the midgut is the primary site of JH catabolism. Interestingly, JHDK transcription does not appear to be directly regulated by JH, implying that other hormonal or environmental factors modulate its expression.

Structurally, JHDK is characterized by a core comprising eight α-helices connected by loops that form four helix–loop–helix motifs. This arrangement is evident in the crystal structures of *B. mori* JHDK (PDB ID: 7PJD; Guo et al. 2021), the JHDK-like isoform JHDK-L2 (PDB ID: 6KTH; Xu et al. 2021), and *S. litura* JHDK (Zeng et al. 2015). This folding pattern is critical for kinase activity. Notably, JHDK contains calcium-binding motifs, indicating that its enzymatic activity is calcium-dependent. Additionally, ATP-Mg^{2+} binding sites are present, consistent with ATP serving as a phosphate donor during phosphorylation.

Functional studies have confirmed the essential role of JHDK in JH clearance and in developmental timing. In the moth *Heortia vitessoides*, RNAi targeting *JHDK* resulted in significantly elevated JH titers, causing developmental delays, particularly in the larval-to-pupal transition (Lyu et al. 2019). These findings highlight the importance of JHDK in maintaining appropriate JH levels necessary for timely metamorphosis and normal physiological development.

4.2 Transport

JHs synthesized in the corpora allata are secreted into the hemolymph and transported to the peripheral target organs and tissues. During this transit, JHs are bound by specific carrier proteins known as juvenile hormone-binding proteins (JHBPs), which stabilize the hormones by shielding their physicochemical properties, particularly their relatively hydrophobic nature and low concentrations in the hemolymph. Several classes of JHBPs have been identified, including low-molecular-weight takeout-type proteins found in the hemolymph of Lepidoptera and mosquitoes, as well as high-molecular-weight carriers, such as hexamerins and lipophorins.

4.2.1 Takeout Type of Lepidopteran Hemolymph JH-Binding Proteins

In lepidopteran larvae, JH is stabilized in the hemolymph by binding to a monomeric carrier protein known as hJHBP. hJHBP belongs to the "takeout" family within the tubular lipid-binding (TULIP) superfamily, which also includes mammalian lipid-binding proteins and bactericidal/permeability-increasing proteins (Kolodziejczyk et al. 2008; Alva and Lupas 2016; Wong and Levine 2017). Although insects possess approximately 20 takeout homologs (23 in *D. melanogaster*; Saurabh et al. 2018), hJHBP is unique in its specific binding to JH, thereby protecting it from nonspecific esterases. Paradoxically, hJHBP also facilitates JH hydrolysis when complexed with juvenile hormone esterase (Touhara et al. 1995).

hJHBP was first purified and cloned from *M. sexta* (Touhara et al. 1993) and characterized by Park and Goodman (1993). It was subsequently identified in *H. virescens* (Wojtasek and Prestwich 1995), *B. mori* (Vermunt et al. 2001), and the greater wax moth *Galleria mellonella* (Rodriguez Parkitna et al. 2002). The sequence identity among lepidopteran hJHBPs ranges from 40% to 50%, whereas they share less than 30% identity with other takeout family proteins (Shiotsuki et al. 2023). In *M. sexta* larvae, hJHBP mRNA expression is strongly upregulated upon JH application (Orth et al. 1999) and is predominantly localized in the fat body, peaking at mid-instar and declining during the molting stages (Shiotsuki et al. 2023).

The JH-binding affinities of lepidopteran hJHBPs have been characterized using dextran-coated charcoal (DCC) assays, which yield dissociation constants (Kd) in the low nanomolar range (Touhara et al. 1993), and isothermal titration calorimetry (Dupas et al. 2020). These proteins preferentially bind JH I and JH II over JH III, which is consistent with the hormonal repertoire typical of this order (Vermunt et al. 2001).

The crystal structures of hJHBP from *G. mellonella* (PDB ID: 2RCK) and *B. mori* (PDB ID: 2RQF) reveal a tubular β-barrel architecture wrapped around a long central α-helix (Kolodziejczyk et al. 2008; Suzuki et al. 2011) (Fig. 4.4). Two conserved disulfide bonds, Cys9-Cys16 and Cys151-Cys194, stabilize the structure, with the N-terminal bond specifically reinforcing the JH-binding pocket (Suzuki et al. 2011). In *B. mori* hJHBP, the β-barrel is constricted to form two hydrophobic cavities: one binding JH and the other accommodating 2-methyl-2,4-pentanediol (MPD), a crystallization artifact (PDB ID: 3AZ1; Fujimoto et al. 2013). Although the physiological ligand for the second cavity remains unknown, MPD binding induces significant expansion of the pocket, suggesting a potential regulatory function.

In contrast, "classic" takeout proteins, such as takeout 1 protein in the light brown apple moth, *Epiphyas postvittana*, adopt elongated tubular structures that bind lipids, such as ubiquinone-8 in *Escherichia coli* expression systems (PDB ID: 3E8T; Hamiaux et al. 2009) or myristic acid in Sf9 expression systems (PDB ID: 4G0S; Hamiaux et al. 2013). In hJHBP, aromatic residues line the JH-binding site;

Fig. 4.4 Crystal structure of juvenile hormone II (JH II)-bound hemolymph juvenile hormone-binding protein from *Bombyx mori* (hJHBP). JH II (magenta) binds at the N-terminal region, where a short α-helix (blue) covers the JH and stabilizes the hormone

notably, the mutation of key residues, especially Tyr128, which forms a hydrogen bond with the JH epoxide, abolishes the JH-binding affinity (Suzuki et al. 2011; Shiotsuki et al. 2023). An acidic pH at the target cell membrane may induce conformational changes that facilitate JH release from hJHBP (Dupas et al. 2020). The JH fluorescent biosensor based on Förster resonance energy transfer (FRET) is composed of the *Bombyx* hJHBP, which is fused with two fluorescent protein pairs' core structure and N-terminal short α-helix (Kikuta et al. 2025). Given its crucial role in JH stabilization, hJHBP has emerged as a promising target for insecticidal interventions. Destruxin A, a cyclic depsipeptide, inhibits *B. mori* growth by binding to the hJHBP (Hu et al. 2023). Moreover, RNAi knockdown of hJHBP effectively controlled the South American tomato pinworm, *Tuta absoluta*, through leaf feeding (Ramkumar et al. 2021) and reduced the fitness of the Bt-resistant cotton bollworm, *Helicoverpa armigera*, on transgenic cotton expressing hJHBP-targeting dsRNA (Ni et al. 2017).

4.2.2 Hexamerins and Lipophorins

High-molecular-weight hemolymph proteins that can bind JH include hexamerins and lipophorins (Goodman and Granger 2005). These proteins are primarily synthesized in the fat body during the larval stages and accumulate in the hemolymph, serving as storage proteins.

Hexamerins belong to the hemocyanin superfamily but lack the ability to bind copper ions or transport oxygen (Martins et al. 2010). The molecular cloning of a hexamerin gene was first reported in *G. mellonella*, which encodes a 706-residue preprotein that includes an 18-amino-acid signal peptide (Memmel et al. 1994). Hexamerins typically assemble into hexameric structures composed of subunits with molecular weights of 70–80 kDa and are widely distributed across arthropods (Hathaway et al. 2009). JH-binding activity has been confirmed in several orthopteran species, in which hexamerins function as JH transporters. The reported dissociation constants (Kd) for JH III range from 3.7 to 0.6 nM in the migratory locust, *Locusta migratoria* (Koopmanschap and de Kort 1988), 19 nM in the desert locust, *Schistocerca gregaria*, and 28 nM in the two-spotted cricket, *Gryllus bimaculatus* (Tawfik et al. 2006). Although the crystal structure of hexamerin from *T. castaneum* has been elucidated (Hřebík et al. 2022), its capacity to bind to JH remains unconfirmed.

Functionally, hexamerins are involved in the regulation of JH activity during developmental transitions and under varying nutritional conditions (Burmester 1999; Hathaway et al. 2009; Martins et al. 2010). In social insects, hexamerins have been implicated in caste differentiation, as demonstrated in species such as the eastern subterranean termite, *Reticulitermes flavipes* (Zhou et al. 2007); the European honey bee, *Apis mellifera* (Martins et al. 2010); the red imported fire ant, *Solenopsis invicta* (Hawkings et al. 2016); the buff-tailed bumblebee, *Bombus terrestris* (Tian et al. 2021); and the spurred ceratina, *Ceratina calcarata* (Huisken and Rehan 2023). In *R. flavipes*, Hex-1 and Hex-2 are highly expressed in workers and soldiers, and the JH-binding activity of Hex-1 has been demonstrated using an anti-JH antibody (Zhou et al. 2006).

Another high-molecular-weight hemolymph protein capable of binding to JH is lipophorin. Lipophorin is a lipid-transport protein found across multiple insect orders, including Coleoptera, Isoptera, Diptera, Hymenoptera, and Blattodea, and was identified through photoaffinity labeling (Sevala et al. 1997). Lipophorin consists of two subunits, apolipophorin-I and apolipophorin-II, both derived from a single gene (Gondim and Majerowicz 2024). In the blowfly *Lucilia cuprina*, these subunits have molecular weights of approximately 228 and 70 kDa, respectively. Lipophorin binds JH III in an enantioselective manner, with reported Kd values of 3 nM in the Pacific beetle cockroach, *Diploptera punctata* (King and Tobe 1993), 21 nM in the American cockroach, *Periplaneta americana* (de Kort and Koopmanschap 1989), and 33 nM in *L. cuprina* (Trowell et al. 1994). As a multifunctional lipid carrier, the precise role of lipophorin in regulating JH action remains to be elucidated.

4.2.3 Mosquito JH-Binding Proteins

A member of the hemolymph odorant-binding protein (OBP) family in mosquitoes, termed mosquito juvenile hormone-binding protein (mJHBP), has been identified as a high-affinity binder of JH II and JH III, exhibiting a dissociation constant (Kd) of 1.0×10^{-7} M, as determined by isothermal titration calorimetry (Kim et al. 2017). This binding is highly selective for the biologically active *10R* enantiomer of JH, preserving its native stereochemistry. No binding activity was detected with structurally related compounds, such as 20-hydroxyecdysone, methyl farnesoate, or methoprene.

The expression of mJHBP is restricted to the pupal and adult stages in both male and female mosquitoes and is absent in larvae. This stage-specific expression pattern suggests a potential role in adult physiological processes, including reproduction and nutrient transport associated with blood feeding. Structural analysis by X-ray crystallography of *A. aegypti* mJHBP (PDB ID: 5V13) revealed that JH binds within a hydrophobic pocket formed by residues in the N-terminal domain, which is subsequently capped by a C-terminal α-helix, effectively sealing the hormone within the binding cavity (Kim et al. 2017).

In addition to its role in hormone transport, mJHBP is implicated in immune function. CRISPR/Cas9-mediated knockout of the *mJHBP* gene increased susceptibility to bacterial infection, resulting in higher rates of septicemia and mortality following bacterial challenge compared to wild-type mosquitoes (Kim et al. 2020), indicating its involvement in the innate immune response of insects.

Given its dual role in hormone regulation and immunity, mJHBP has emerged as a promising target for mosquitocide development, particularly for the vector control of disease-transmitting species, such as those spreading the dengue virus. Structure-based drug design approaches utilizing binding energy calculations with the mJHBP structure (PDB ID: 5V13) have identified potential inhibitory compounds. These include silafluofen, a silicon-containing pyrethroid (Ononamadu et al. 2021); α-phellandrene, a terpene derived from the pepper tree *Schinus molle* (Herrera-Calderon et al. 2022); and synthetic molecules such as quinazoline derivatives (Costa et al. 2023) and indolizine derivatives (Singh et al. 2025).

4.3 Future Perspectives

JH activity is precisely controlled by its synthesis, signaling, degradation, and transport mechanisms. Understanding these processes is crucial for advancing insect endocrinology and developing novel pest control strategies. Enzymes such as JHE, JHEH, and JHDK tightly regulate JH levels in a tissue- and stage-specific manner. Additionally, JHBP in the hemolymph stabilizes and delivers the hormone efficiently to target tissues, despite its hydrophobic nature and low concentration.

Recent advances in structural biology and gene-editing techniques, such as CRISPR/Cas9, have revealed detailed functional insights into these enzymes and transport proteins, enabling the potential development of species-specific inhibitors. These compounds can disrupt JH signaling by inhibiting degradation enzymes or blocking hormone transport, leading to developmental arrest or reproductive failure in target pests.

Unlike broad-spectrum neurotoxic insecticides, JH-targeted approaches are highly selective, offering environmentally sustainable pest control with minimal impact on humans and beneficial insects, including pollinators and natural predators of pests. These methods are particularly promising in the face of increasing resistance to conventional insecticides. Overall, research on JH degradation and transport opens new avenues for precision pest management in the agricultural, public health, and environmental domains.

References

Alva A, Lupas AN (2016) The TULIP superfamily of eukaryotic lipid-binding proteins as a mediator of lipid sensing and transport. Biochim Biophys Acta 1861:913–923. https://doi.org/10.1016/j.bbalip.2016.01.016

Borovsky D, Breyssens H, Buytaert E, Peeters T, Laroye C, Stoffels K et al (2022a) Cloning and characterization of *Drosophila melanogaster* juvenile hormone epoxide hydrolases (JHEH) and their promoters. Biomolecules 12:991. https://doi.org/10.3390/biom12070991

Borovsky D, Ekert EV, Buytaert E, Peeters T, Rougé P (2022b) Cloning and characterization of *Aedes aegypti* juvenile hormone epoxide hydrolases (JHEHs). Arch Insect Biochem Physiol 112:e21977. https://doi.org/10.1002/arch.21977

Burmester T (1999) Evolution and function of the insect hexamerins. Eur J Entomol 96:213–225

Campbell PM, Harcourt RL, Crone EJ, Claudianos C, Hammock BD, Russell RJ et al (2001) Identification of a juvenile hormone esterase gene by matching its peptide mass fingerprint with a sequence from the *Drosophila* genome project. Insect Biochem Mol Biol 31:513–520. https://doi.org/10.1016/s0965-1748(01)00035-2

Chaitanya BN, Asokan R, Sita T, Rebijith KB, Kumar PR, Kumar NKK (2017) Silencing of JHEH and EcR genes of *Plutella xylostella* (Lepidoptera: Plutellidae) through double stranded RNA oral delivery. J Asia Pac Entomol 20:637–643. https://doi.org/10.1016/j.aspen.2017.03.020

Costa RA, Costa ASS, Rocha JAP, Lima MRC, Rocha ECM, Nascimento FCA et al (2023) Exploring natural alkaloids from Brazilian biodiversity as potential inhibitors of the *Aedes aegypti* juvenile hormone enzyme: a computational approach for vector mosquito control. Molecules 28:6871. https://doi.org/10.3390/molecules28196871

De Kort S, Koopmanschap B (1989) Binding of juvenile hormone III to lipophorin from the american cockroach *Periplaneta americana*. Arch Insect Biochem Physiol 11:159–172. https://doi.rog/10.1002/arch.v11:3, https://doi.org/10.1002/arch.940110304

Duan D, Zheng R, Lin S, Chen Y, Tian H, Zhao J et al (2016) Modulation of juvenile hormone esterase gene expression against development of *Plutella xylostella* (Lepidoptera: Plutellidae). J Econ Entomol 109:865–872. https://doi.org/10.1093/jee/tow003

Dupas S, Neiers F, Granon E, Rougeux E, Dupont S, Beney L et al (2020) Collisional mechanism of ligand release by *Bombyx mori* JHBP, a member of the TULIP/takeout family of lipid transporters. Insect Biochem Molec Biol 117:103293. https://doi.org/10.1016/j.ibmb.2019.103293

Fujimoto Z, Suzuki R, Shiotsuki T, Tsuchiya W, Tase A, Momma M et al (2013) Crystal structure of silkworm *Bombyx mori* JHBP in complex with 2-methyl-2,4-pentanediol: plasticity of JH

binding pocket and ligand-induced conformational change of the second cavity in JHBP. PLoS One 8:e56261. https://doi.org/10.1371/journal.pone.0056261
Gao Q, Li B, Tian Z, De Loof A, Wang JL, Wang XP et al (2022) Key role of juvenile hormone in controlling reproductive diapause in females of the Asian lady beetle *Harmonia axyridis*. Pest Manag Sci 78:193–204. https://doi.org/10.1002/ps.6619
Gondim KC, Majerowicz D (2024) Lipophorin: the lipid shuttle. In: Advances in experimental medicine and biology. Springer, Cham, pp 1–18. https://doi.org/10.1007/5584_2024_806
Goodman WG, Granger NA (2005) The juvenile hormones. In: Gilbert LI (ed) Comprehensive molecular insect science, Endocrinology, vol 3. Elsevier, Amsterdam, pp 319–408. https://doi.org/10.1016/B0-44-451924-6/00039-9
Guo S, Sun D, Tian Z, Liu W, Zhu F, Wang XP (2019) The limited regulatory roles of juvenile hormone degradation pathways in reproductive diapause preparation of the cabbage beetle, *Colaphellus bowringi*. J Insect Physiol 119:103967. https://doi.org/10.1016/j.jinsphys.2019.103967
Guo J, Keegan RM, Rigden DJ, Erskine PT, Wood SP, Li S et al (2021) The X-ray structure of juvenile hormone diol kinase from the silkworm *Bombyx mori*. Acta Crystallogr F Struct Biol Commun 77:465–472. https://doi.org/10.1107/S2053230X21012012
Hamiaux C, Stanley D, Greenwood DR, Baker EN, Newcomb RD (2009) Crystal structure of *Epiphyas postvittana* takeout 1 with bound ubiquinone supports a role as ligand carriers for takeout proteins in insects. J Biol Chem 284:3496–3503. https://doi.org/10.1074/jbc.m807467200
Hamiaux C, Basten L, Greenwood DR, Baker EN, Newcomb RD (2013) Ligand promiscuity within the internal cavity of *Epiphyas postvittana* takeout 1 protein. J Struct Biol 182:259–263. https://doi.org/10.1016/j.jsb.2013.03.013
Harris SV, Thompson DM, Linderman RJ, Tomalski MD, Roe RM (1999) Cloning and expression of a novel juvenile hormone-metabolizing epoxide hydrolase during larval–pupal metamorphosis of the cabbage looper, *Trichoplusia ni*. Insect Mol Biol 8:85–96. https://doi.org/10.1046/j.1365-2583.1999.810085.x
Hathaway M, Hatle J, Li S, Ding X, Barry T, Hong F et al (2009) Characterization of hexamerin proteins and their mRNAs in the adult lubber grasshopper: the effects of nutrition and juvenile hormone on their levels. Comp Biochem Physiol A 154:323–332. https://doi.org/10.1016/j.cbpa.2009.06.018
Hawkings C, Calkins TL, Pietrantonio PV, Tamborindeguy CT (2016) Caste-based differential transcriptional expression of hexamerins in response to a juvenile hormone analog in the red imported fire ant (*Solenopsis invicta*). PLoS One 20:e0216800. https://doi.org/10.1371/journal.pone.0216800
Herrera-Calderon O, Chavez H, Enciso-Roca EC, Común-Ventura PW, Hañari-Quispe RD, Figueroa-Salvador L et al (2022) GC-MS profile, antioxidant activity, and *in silico* study of the essential oil from *Schinus molle* L. leaves in the presence of mosquito juvenile hormone-binding protein (mJHBP) from *Aedes aegypti*. Biomed Res Int 2023:105654. https://doi.org/10.1155/2022/5601531
Hrebík D, Gondová M, Valentová L, Füzik T, Přidal A, Nováček J et al (2022) Polyelectrolyte coating of cryo-EM grids improves lateral distribution and prevents aggregation of macromolecules. Acta Crystallogr D 78:1337–1346. https://doi.org/10.1107/S2059798322009299
Hu H, Yin X, Pang S, Jiang Y, Weng Q, Hu Q et al (2023) Mechanism of destruxin a inhibits juvenile hormone binding protein transporting juvenile hormone to affect insect growth. Pestic Biochem Physiol 19:105654. https://doi.org/10.1016/j.pestbp.2023.105654
Hua-Jun Y, Fang Z, Awquib S, Malik FA, Roy B, Xing-Hua L et al (2011) Expression pattern of enzymes related to juvenile hormone metabolism in the silkworm, *Bombyx mori* L. Mol Biol Rep 38:4337–4342. https://doi.org/10.1007/s11033-010-0559-3
Huisken JL, Rehan SM (2023) Brain gene expression of foraging behavior and social environment in *Ceratina calcarata*. Genome Biol Evol 15:evad117. https://doi.org/10.1093/gbe/evad117

Jones G, Chu YX, Schelling D, Jones D (2000) Regulation of the juvenile hormone esterase gene by a composite core promoter. Biochem J 346:233–240. https://doi.org/10.1042/bj3460233

Kallapur VL, Majumder C, Roe RM (1996) *In vivo* and *in vitro*-tissue specific metabolism of juvenile hormone during the last stadium of the cabbage looper, *Trichoplusia ni*. J Insect Physiol 42:181–190. https://doi.org/10.1016/0022-1910(95)00057-7

Kamita SG, Hammock BD (2010) Juvenile hormone esterase: biochemistry and structure. J Pestic Sci 35:265–274. https://doi.org/10.1584/jpestics.R10-09

Kethidi DR, Perera SC, Zheng S, Feng QL, Krell P, Retnakaran A et al (2004) Identification and characterization of a juvenile hormone (JH) response region in the JH esterase gene from the spruce budworm *Choristoneura fumiferana*. J Biol Chem 279:19634–19642. https://doi.org/10.1074/jbc.M311647200

Kikuta S, Shiotsuki T, Tsuchiya W, Suzuki R, Kikawada T et al (2025) Genetically encoded FRET-based nanosensor for insect juvenile hormone. ACS Omega 10:53262–53268. https://doi.org/10.1021/acsomega.5c07961

Kim IH, Pham V, Jablonka W, Goodman WG, Ribeiro JMC, Andersen JF (2017) A mosquito hemolymph odorant-binding protein family member specifically binds juvenile hormone. J Biol Chem 292:15329–15339. https://doi.org/10.1074/jbc.M117.802009

Kim IH, Castillo JC, Aryan A, Martin-Martin I, Nouzova M, Noriega FG et al (2020) A mosquito juvenile hormone binding protein (mJHBP) regulates the activation of innate immune defenses and hemocyte development. PLoS Pathog 16:e1008288. https://doi.org/10.1371/journal.ppat.1008288

King LE, Tobe SS (1993) Changes in the titre of a juvenile hormone ill binding lipophorin in the hemolymph of *Diploptera punctata* during development and reproduction: functional significance. J Insect Physiol 39:241–251. https://doi.org/10.1016/0022-1910(93)90095-9

Kolodziejczyk R, Bujacz G, Jakób M, Ozyhar A, Jaskolski M, Kochman M (2008) Insect juvenile hormone binding protein shows ancestral fold present in human lipid-binding proteins. J Mol Biol 377:870–881. https://doi.org/10.1016/j.jmb.2008.01.026

Koopmanschap AB, de Kort CAD (1988) Isolation and characterization of a high molecular weight JH-III transport protein in the hemolymph of *Locusta migratoria*. Arch Insect Biochem Physiol 7:105–118. https://doi.org/10.1002/arch.940070204

Lei G, Zhou H, Yu G, Miao X, Zhang Y, Ma Z et al (2025) Metabolic network remodeling through PxJHE modulates temperature adaptation in a cosmopolitan insect. Insect Biochem Mol Biol 182:104348. https://doi.org/10.1016/j.ibmb.2025.104348

Li S, Zhang QR, Xu WH, Schooley DA (2005) Juvenile hormone diol kinase, a calcium-binding protein with kinase activity, from the silkworm, *Bombyx mori*. Insect Biochem Mol Biol 35:235–1248. https://doi.org/10.1016/j.ibmb.2005.06.005

Li Y, Chen J, Liu M, He W, Reynolds JA, Wang Y (2022) Enhanced degradation of juvenile hormone promotes reproductive diapause in the predatory ladybeetle *Coccinella septempunctata*. Front Physiol 13:877153. https://doi.org/10.3389/fphys.2022.877153

Lü FG, Fu KY, Guo WC, Li GQ (2015) Characterization of two juvenile hormone epoxide hydrolases by RNA interference in the Colorado potato beetle. Gene 570:264–271. https://doi.org/10.1016/j.gene.2015.06.032

Lyu Z, Li Z, Cheng J, Wang C, Chen J, Lin T (2019) Suppression of gene juvenile hormone diol kinase delays pupation in *Heortia vitessoides* Moore. Insects 10:278. https://doi.org/10.3390/insects10090278

Martins JR, Nunes FM, Cristino AS, Simões ZL, Bitondi MM (2010) The four hexamerin genes in the honeybee: structure, molecular evolution and function deduced from expression patterns in queens, workers and drones. BMC Mol Biol 11:23. https://doi.org/10.1186/1471-2199-11-23

Maxwell RA, Welch WH, Schooley DA (2002a) JH diol kinase: part I-purification, characterization and substrate specificity of juvenile hormone selective diol kinase from *Manduca sexta*. J Biol Chem 277:21874–21881. https://doi.org/10.1074/jbc.M201510200

Maxwell RA, Welch WH, Horodyski FM, Schegg KM, Schooley DA (2002b) JH diol kinase: part II- sequencing, cloning, and molecular modeling of juvenile hormone-selective diol kinase from *Manduca sexta*. J Biol Chem 277:21882–21890. https://doi.org/10.1074/jbc.m201509200

Memmel NA, Trewitt PM, Grzelak K, Rajaratnam VS, Kumaran KA (1994) Nucleotide sequence structure and developmental regulation of LHP82 a juvenile hormone-suppressible hexamerin gene from the waxmoth *Galleria mellonella*. Insect Biochem Molec Biol 24:133–144. https://doi.org/10.1016/0965-1748(94)90079-5

Ni M, Ma W, Wang X, Gao M, Dai Y, Wei X et al (2017) Next-generation transgenic cotton: pyramiding RNAi and Bt counters insect resistance. Plant Biotechnol J 15:1204–1213. https://doi.org/10.1111/pbi.12709

Nollmann FI, Heinrich AK, Brachmann AO, Morisseau C, Mukherjee K, Ángel M et al (2015) A *Photorhabdus* natural product inhibits insect juvenile hormone epoxide hydrolase. Chembiochem 16:766–771. https://doi.org/10.1002/cbic.201402650

Ononamadu CJ, Abdalla M, Ihegboro GO, Li J, Owolarafe TA, John TD et al (2021) *In silico* identification and study of potential anti-mosquito juvenile hormone binding protein (MJHBP) compounds as candidates for dengue virus – vector insecticides. Biochem Biophys Rep 28:101178. https://doi.org/10.1016/j.bbrep.2021.101178

Orth AP, Lan Q, Goodman WG (1999) Ligand regulation of juvenile hormone binding protein mRNA in mutant *Manduca sexta*. Mol Cell Endocrinol 149:61–69. https://doi.org/10.1016/S0303-7207(98)00258-5

Park YC, Goodman WG (1993) Analysis and modification of thiols in the hemolymph juvenile hormone binding protein of *Manduca sexta*. Arch Biochem Biophys 302:12–18. https://doi.org/10.1006/abbi.1993.1174

Ramkumar G, Asokan R, Prasannakumar NR, Kariyanna B, Karthi S, Alwahibi MS et al (2021) RNAi Interference suppression of v-ATPase B and juvenile hormone binding protein genes through topically applied dsRNA on tomato leaves: developing biopesticides to control the South American pinworm, *Tuta absoluta* (Lepidoptera: Gelechiidae). Front Physiol 12:742871. https://doi.org/10.3389/fphys.2021.742871

Rodriguez Parkitna JM, Ozyhar A, Wiśniewski JR, Kochman M (2002) Cloning and sequence analysis of *Galleria mellonella* juvenile hormone binding protein A search for ancestors and relatives. Biol Chem 383:1343–1355. https://doi.org/10.1515/BC.2002.153

Roe RM, Kallapur V, Linderman RJ, Viviani F, Harris SV, Walker EA et al (1996) Mechanism of action and cloning of epoxide hydrolase from the cabbage looper, *Trichoplusia ni*. Arch Insect Biochem Physiol 32:527–535. https://doi.org/10.1002/(SICI)1520-6327(1996)32:3/4<527::AID-ARCH24>3.0.CO;2-D

Rogalski F, Tsuda M, Takeo S, Asano T, Sakai T, Aigaki T (2025) Epoxide hydrolases JHEH1 and JHEH2 deficiency impairs glucose metabolism in *Drosophila*. Biochem Biophys Res Commun 748:151313. https://doi.org/10.1016/j.bbrc.2025.151313

Saito Y, Kamita SG, Hammock BD, Kunimi Y, Inoue MN, Nakai M (2015) Juvenile hormone (JH) esterase activity but not JH epoxide hydrolase activity is downregulated in larval *Adoxophyes honmai* following nucleopolyhedroviruses infection. J Insect Physiol 80:71–80. https://doi.org/10.1016/j.jinsphys.2015.02.005

Sarkar S, Kalia VK (2023) Silencing of juvenile hormone epoxide hydrolase gene in *Spodoptera litura* (Lepidoptera: Noctuidae) by oral delivery of double-stranded RNA. Biologia 78:429–442. https://doi.org/10.1007/s11756-022-01150-3

Saurabh S, Vanaphan N, Wen W, Dauwalder B (2018) High functional conservation of *takeout* family members in a courtship model system. PLoS One 13:e204615. https://doi.org/10.1371/journal.pone.0204615

Seino A, Ogura T, Tsubota T, Shimomura M, Nakakura T, Tan A et al (2010) Characterization of juvenile hormone epoxide hydrolase and related genes in the larval development of the silkworm *Bombyx mori*. Biosci Biotechnol Biochem 74:1421–1429. https://doi.org/10.1271/bbb.100104

Sevala V, Shu S, Ramaswamy SB, Schal C (1997) Lipophorin: a hemolymph juvenile hormone binding protein in the German cockroach, *Blattella germanica*. Insect Biochem Mol Biol 27:663–670. https://doi.org/10.1016/S0965-1748(97)00042-8

Severson TF, Goodrow MH, Morisseau C, Dowdy DL, Hammock BD (2002) Urea and amide-based inhibitors of the juvenile hormone epoxide hydrolase of the tobacco hornworm (*Manduca sexta*: Sphingidae). Insect Biochem Mol Biol 32:1741–1756. https://doi.org/10.1016/S0965-1748(02)00115-7

Shiotsuki T (1995) New approaches for studies on carboxylesterases. J Pestic Chem 20:181–185. (in Japanese). https://doi.org/10.1584/jpestics.20.181

Shiotsuki T, Suzuki R, Tsuchiya W, Yamazaki T, Shimomura M, Tsubota T et al (2023) Characteristics of the takeout protein ce-0330 in the silkworm, *Bombyx mori* (Lepidoptera: Bombycidae). Appl Entomol Zool 58:237–244. https://doi.org/10.1007/s13355-023-00827-8

Singh HR, Tiwari P, Deb PK, Rakshit G, Maity P, Mohanlall V et al (2025) Larvicidal activity, molecular docking, and molecular dynamics studies of 7-(trifluoromethyl)-indolizine derivatives against *Anopheles arabiensis*. Mol Divers 28:2323–2340. https://doi.org/10.1007/s11030-024-10994-7

Suzuki R, Fujimoto Z, Shiotsuki T, Tsuchiya W, Momma M, Tase A et al (2011) Structural mechanism of JH delivery in hemolymph by JHBP of silkworm, Bombyx mori. Sci Rep 1:133. https://doi.org/10.1038/srep00133

Tan A, Tanaka H, Tamura T, Shiotsuki T (2005) Precocious metamorphosis in transgenic silkworms overexpressing juvenile hormone esterase. Proc Natl Acad Sci USA 102:11751–11756. https://doi.org/10.1073/pnas.0500954102

Tawfik AI, Kellner R, Hoffmann KH, Lorenz MW (2006) Purification, characterisation and titre of the haemolymph juvenile hormone binding proteins from *Schistocerca gregaria* and *Gryllus bimaculatus*. J Insect Physiol 52:255–268. https://doi.org/10.1016/j.jinsphys.2005.11.005

Tian Y, Qu Y, Dong K, He S, Jie W, Huang J (2021) Characterization and developmental expression patterns of four hexamerin genes in the bumble bee, *Bombus terrestris* (Hymenoptera: Apidae). J Insect Sci 21:13. https://doi.org/10.1093/jisesa/ieab078

Touhara K, Prestwich GD (1994) Juvenile hormone epoxide hydrolase. Photoaffinity labeling of juvenile hormone epoxide hydrolase and JH-binding proteins during ovarian and egg development in *Manduca sexta*. Insect Biochem Mol Biol 24:633–640. https://doi.org/10.1016/0965-1748(94)90100-7

Touhara K, Lerro KA, Bonning BC, Hammock BD, Prestwich GD (1993) Ligand binding by a recombinant insect juvenile hormone binding protein. Biochemistry 32:2068–2075. https://doi.org/10.1021/bi00059a026

Touhara K, Bonning BC, Hammock BD, Prestwich GD (1995) Action of juvenile hormone (JH) esterase on the JH-JH binding protein complex. An *in vitro* model of JH metabolism in a caterpillar. Insect Biochem Mol Biol 25:727–734. https://doi.org/10.1016/0965-1748(95)00012-K

Trowell SC, Hines ER, Herlt AJ, Rickards RW (1994) Characterization of a juvenile hormone binding lipophorin from the blowfly *Lucilia cuprina*. Comp Biochem Physiol B 109:339–357. https://doi.org/10.1016/0305-0491(94)90018-3

Tsubota T, Shiotsuki T (2010) Genomic analysis of carboxyl/cholinesterase genes in the silkworm *Bombyx mori*. BMC Genomics 11:377. https://doi.org/10.1186/1471-2164-11-377

Tsubota T, Shimomura M, Ogura T, Seino A, Nakakura T, Mita K et al (2010a) Molecular characterization and functional analysis of novel carboxyl/cholinesterases with GQSAG motif in silkworm *Bombyx mori*. Insect Biochem Molec Biol 40:100–112. https://doi.org/10.1016/j.ibmb.2009.12.015

Tsubota T, Minakuchi C, Nakakura T, Shinoda T, Shiotsuki T (2010b) Molecular characterization of a gene encoding juvenile hormone esterase in the red flour beetle, *Tribolium castaneum*. Insect Mol Biol 19:527–535. https://doi.org/10.1111/j.1365-2583.2010.01019.x

Tsubota T, Nakakura T, Shiotsuki T (2010c) Molecular characterization and enzymatic analysis of juvenile hormone epoxide hydrolase genes in the red flour beetle, *Tribolium castaneum*. Insect Mol Biol 19:389–408. https://doi.org/10.1111/j.1365-2583.2010.01019.x

Tusun A, Li M, Liang X, Yang T, Yang B, Wang G (2017) Juvenile hormone epoxide hydrolase: a promising target for Hemipteran pest management. Sci Rep 7:789. https://doi.org/10.1038/s41598-017-00907-0

Vasquez DDN, Pinheiro DH, Teixeira LA, Moreira-Pinto CE, Macedo LLP, Salles-Filho ALO et al (2023) Simultaneous silencing of juvenile hormone metabolism genes through RNAi interrupts metamorphosis in the cotton boll weevil. Front Mol Biosci 10:1073721. https://doi.org/10.3389/fmolb.2023.1073721

Vermunt AM, Kamimura M, Hirai M, Kiuchi M, Shiotsuki T (2001) The juvenile hormone binding protein of silkworm haemolymph: gene and functional analysis. Insect Mol Biol 10:147–154. https://doi.org/10.1046/j.1365-2583.2001.00249.x

Wan J, Liang T, Cheng W, Zhu K (2020) Cloning of juvenile hormone esterase and juvenile hormone epoxide hydrolase genes and their expression dynamics during diapause and metamorphosis in *Sitodiplosis mosellana* (Diptera: Cecidomyiidae). Acta Entomol Sinica 63:401–412. (in Chinese with English abstract). https://doi.org/10.16380/j.kcxb.2020.04.003

Ward VK, Bonning BC, Huang T, Shiotsuki T, Griffeth VN, Hammock BD (1992) Analysis of the catalytic mechanism of juvenile hormone esterase by site-directed mutagenesis. Int J Biochem 24:1933–1941. https://doi.org/10.1016/0020-711x(92)90289-d

Wen R, Wang B, Wang B, Ma L (2018) Characterization and expression profiles of juvenile hormone epoxide hydrolase from *Lymantria dispar* (Lepidoptera: Lymantridae) and RNA interference by ingestion. J Insect Sci 18:13. https://doi.org/10.1093/jisesa/iey002

Wogulis M, Wheelock CE, Kamita SG, Hinton AC, Whetstone PA, Hammock BD et al (2006) Structural studies of a potent insect maturation inhibitor bound to the juvenile hormone esterase of *Manduca sexta*. Biochemistry 45:4045–4057. https://doi.org/10.1021/bi0521644

Wojtasek H, Prestwich GD (1995) Key disulfide bonds in an insect hormone binding protein: cDNA cloning of a juvenile hormone binding protein of *Heliothis virescens* and ligand binding by native and mutant forms. Biochemistry 34:5234–5241. https://doi.org/10.1021/bi00015a037

Wojtasek H, Prestwich GD (1996) An insect juvenile hormone-specific epoxide hydrolase is related to vertebrate microsomal epoxide hydrolases. Biochem Biophys Res Commun 220:323–329. https://doi.org/10.1006/bbrc.1996.0404

Wong JH, Levine TP (2017) Tubular lipid binding proteins (TULIPs) growing everywhere. Biochim Biophys Acta, Mol Cell Res 1864:1439–1449. https://doi.org/10.1016/j.bbamcr.2017.05.019

Xu H, Zhang Y, Zhang L, Wang Z, Guo P, Zhao P (2021) Structural characterization and functional analysis of juvenile hormone diol kinase from the silkworm, *Bombyx mori*. Int J Biol Macromol 167:570–577. https://doi.org/10.1016/j.ijbiomac.2020.11.138

Zeng BJ, Lv Y, Zhang LL, Huang LH, Feng QL (2015) Cloning and structural characterization of juvenile hormone diol kinase in *Spodoptera litura*. Insect Sci 23:819–828. https://doi.org/10.1111/1744-7917.12553

Zeng B, Huang Y, Xu J, Shiotsuki T, Bai H, Palli SR et al (2017) The FOXO transcription factor controls insect growth and development by regulating juvenile hormone degradation in the silkworm, *Bombyx mori*. J Biol Chem 292:1659–11669. https://doi.org/10.1074/jbc.M117.777797

Zhang Z, Liu X, Shiotsuki T, Wang Z, Xu X, Huang Y et al (2017) Depletion of juvenile hormone esterase extends larval growth in *Bombyx mori*. Insect Biochem Molec Biol 81:72–79. https://doi.org/10.1016/j.ibmb.2017.01.001

Zhao J, Zhou Y, Li X, Cai W, Hua H (2017) Silencing of juvenile hormone epoxide hydrolase gene (*Nljheh*) enhances short wing formation in a macropterous strain of the brown planthopper, *Nilaparvata lugens*. J Insect Physiol 102:18–26. https://doi.org/10.1016/j.jinsphys.2017.08.012

Zhou X, Tarver MR, Bennett GW, Oi FM, Scharf ME (2006) Two hexamerin genes from the termite *Reticulitermes flavipes*: sequence, expression, and proposed functions in caste regulation. Gene 376:47–58. https://doi.org/10.1016/j.gene.2006.02.002

Zhou X, Tarver MR, Scharf ME (2007) Hexamerin-based regulation of juvenile hormone-dependent gene expression underlies phenotypic plasticity in a social insect. Development 134:601–610. https://doi.org/10.1242/dev.02755

Zhou K, Jia N, Hu C, Jiang YL, Yang JP, Chen Y et al (2014) Crystal structure of juvenile hormone epoxide hydrolase from the silkworm *Bombyx mori*. Proteins 82:3224–3229. https://doi.org/10.1002/prot.24676

Zhou WZ, Wu YF, Yin ZY, Guo JJ, Li HY (2022) Juvenile hormone is an important factor in regulating *Aspongopus chinensis* Dallas diapause. Front Physiol 13:873580. https://doi.org/10.3389/fphys.2022.873580

Chapter 5
Genes Involved in Unknown Juvenile Hormone Biosynthetic Pathway

Toru Togawa

Abstract Juvenile hormones (JHs) are synthesized in a pair of tiny endocrine organs, the corpora allata (CA), through 13 enzymatic steps. This JH biosynthetic pathway is conventionally divided into two phases, the early mevalonate pathway and the late JH-branch pathway. The early pathway produces farnesyl pyrophosphate (FPP) from acetyl-CoA, and the late pathway catalyzes the conversion from FPP to JHs. There are several variations in the JH biosynthetic pathway depending on insect groups. In lepidopteran insects, the order of the final two steps of the late pathway is reversed, and they utilize propionyl-CoA as well as acetyl-CoA as starting molecules to produce ethyl-branched JHs. Although many insects use single epoxidized JHs, some of the Hemiptera and higher Diptera also use double epoxidized JHs with an additional epoxidation possibly catalyzed by unknown epoxidases. Enzymes catalyzing 13 steps of JH biosynthesis have been identified. Although most of them are found in all insects studied, orthologs of the enzymes catalyzing the first three steps of the late pathway are not always found, or genes of their most similar homologs are hardly expressed in the CA in some species. In these species, other pathways from FPP to farnesoic acid are proposed.

Keywords Glucose-methanol-choline oxidoreductase · Juvenile hormone biosynthesis · Mevalonate pathway · Protein tyrosine phosphatase, mitochondrial 1

T. Togawa (✉)
Department of Biosciences, College of Humanities and Sciences, Nihon University, Tokyo, Japan
e-mail: togawa.toru@nihon-u.ac.jp

T. Shinoda et al. (eds.), *Juvenile Hormone*, Entomology Monographs,
https://doi.org/10.1007/978-981-95-8605-9_5

5.1 A Variety of Juvenile Hormone Biosynthetic Pathways and Unsolved Issues

Juvenile hormones (JHs) are acyclic sesquiterpenoids synthesized mainly in a pair of tiny endocrine organs, the corpora allata (CA), which are connected to the brain through the corpora cardiaca (CC) (Goodman and Granger 2005; Nijhout 1994; Tobe and Stay 1985). The JH biosynthetic pathway is conventionally divided into the early mevalonate and the late JH-branch pathways (Belles et al. 2005; Noriega 2014; Fig. 5.1). The early pathway produces farnesyl pyrophosphate (FPP) from acetyl-CoA via mevalonate, and the late pathway catalyzes the conversion of FPP to JHs. The early pathway is shared with the first half of the biosynthetic pathway of cholesterol and other isoprenoids and is conserved among most eukaryotes (Dellas et al. 2013; Lombard and Moreira 2011); however, insects and other arthropods do not synthesize cholesterol de novo because of the lack of squalene synthase (Belles et al. 2005). In contrast to the early pathway, the late pathway is considered to be specific to insects, although the partial pathway up to methyl farnesoate (MF) is likely conserved in other arthropods (Miyakawa et al. 2018). MF shows several physiological functions in Crustacea and is thought to be a crustacean JH (Laufer et al. 1987; Olmstead and Leblanc 2002; see Chap. 14).

The insect JH biosynthetic pathway is composed of 13 enzymatic steps (Fig. 5.1). Because the mevalonate pathway is conserved among eukaryotes as mentioned above, the insect genes for the early pathway have been identified as orthologs of other organisms (Belles et al. 2005; Kinjoh et al. 2007; Villalobos-Sambucaro et al. 2020). The enzyme genes responsible for the late pathway have also been identified. Firstly, FPP is dephosphorylated to produce farnesol by FPP pyrophosphatase (FPPP). It is a member of the NagD family of the Class IIA C2 cap-containing haloalkanoic acid dehalogenase (HAD) superfamily (Cao et al. 2009; Nyati et al. 2013). Farnesol is then oxidized to farnesal by farnesol dehydrogenase (FolD), which is a member of the cP2 subfamily of $NADP^+$-dependent classical short-chain dehydrogenase/reductase (SDR) (Huang et al. 2015; Kumar et al. 2022; Mayoral et al. 2009b). Farnesal is further oxidized to farnesoic acid (FA) by farnesal dehydrogenase (FalD), which is a member of the aldehyde dehydrogenase 3 (ALDH3) family (Rivera-Perez et al. 2013). FA is then methylated to form methyl farnesoate (MF) by juvenile hormone acid *O*-methyltransferase (JHAMT) (Mayoral et al. 2009a; Minakuchi et al. 2008; Niwa et al. 2008; Shinoda and Itoyama 2003). As the final step of JH biosynthesis, MF is epoxidized to JH III by MF epoxidase (CYP15A1) (Helvig et al. 2004; Minakuchi et al. 2015; Fig. 5.1).

There are several variations in the JH biosynthetic pathway. The order of the last two steps is reversed in Lepidoptera, where FA is first epoxidized to JH acid by FA epoxidase (CYP15C1) (Daimon et al. 2012), and JH acid is then methylated to form JHs by JHAMT (Shinoda and Itoyama 2003; Fig. 5.1). This reversal is presumed to be ascribed to the substrate specificity and affinity of the epoxidase, because all JHAMTs are likely able to methylate FA and JH acid (Defelipe et al. 2011). In contrast, epoxidases have a narrow substrate specificity and epoxidize MF much more

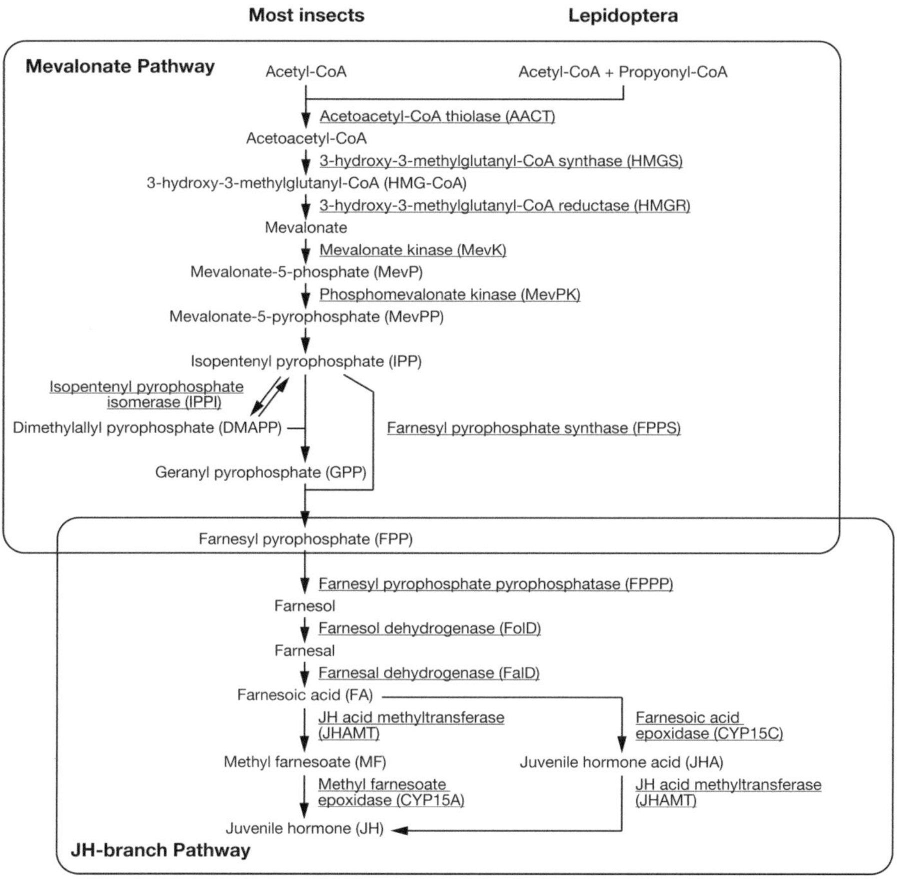

Fig. 5.1 Juvenile hormone (JH) biosynthetic pathway. The early mevalonate pathway produces farnesyl pyrophosphate (FPP) from acetyl-CoA, and the late JH-branch pathway subsequently converts FPP into JHs. Enzymes catalyzing each step are underlined. In the final step of the mevalonate pathway, farnesyl pyrophosphate synthase (FPPS) catalyzes sequential condensation reactions of isopentenyl pyrophosphate (IPP) units: IPP is first isomerized to dimethylallyl pyrophosphate (DMAPP) by IPP isomerase (IPPI), followed by condensation of IPP with DMAPP to form geranyl pyrophosphate (GPP), and a second IPP molecule is then added to GPP to generate FPP. Alternative pathways reported in Lepidoptera are shown on the right

effectively than JH acid in many insects. In Lepidoptera, however, epoxidase may have a higher affinity for FA than JHAMT does. Therefore, epoxidation precedes methylation (Defelipe et al. 2011).

Lepidopteran insects use the ethyl-branched JHs (Goodman and Granger 2005; see Chap. 2). To produce these molecules, they utilize propionyl-CoA as well as acetyl-CoA as starting molecules (Schooley and Baker 1985; Fig. 5.1). Because several steps in the mevalonate pathway do not show strict substrate specificity between isoprenoid and ethyl-branched isoprenoid (Baker and Schooley 1981), the utilization of propionyl-CoA in Lepidoptera is thought to depend on the availability of branched-chain amino acids such as isoleucine and valine, precursors of propionyl-CoA (Brindle et al. 1988). It was also reported that lepidopteran isopentenyl pyrophosphate isomerases (IPPI) utilize ethyl-branched substrates effectively (Sen et al. 2012). The homologous JHs with two epoxides, JHB_3 and $JHSB_3$, have been found in higher Diptera (Richard et al. 1989; Yin et al. 1995) and Hemiptera (Kotaki et al. 2009; Villalobos-Sambucaro et al. 2020), respectively (see Chap. 2). Although the common epoxidation of JHs at the C10-C11 position is mediated by MF/FA epoxidase (CYP15A/C), these extra epoxidations might be catalyzed by other cytochromes P450 (CYPs). In addition, CYP15 is lacking in Cyclorrhapha, including *Drosophila* (Dermauw et al. 2020); therefore, it is especially possible that other CYPs are involved in JH biosynthesis in these insects. This issue is discussed in detail in Chap. 7.

As JHs play essential roles in insect physiology, including development, reproduction, and polyphenism, it is generally assumed that JHs arose only once during insect evolution. If so, their biosynthetic pathways are expected to be conserved among insects. Indeed, JHAMT and CYP15A/C of the late pathway, as well as enzymes of the early mevalonate pathway, have been found in most studied species (Defelipe et al. 2011; Dermauw et al. 2020). However, there are some exceptions to this rule. CYP15 is absent in Cyclorrhapha, and multiple paralogs of CYP15A/C and JHAMT are sometimes observed in some other species (Dermauw et al. 2020; Minakuchi et al. 2008; see Chap. 3). The situation of the enzymes catalyzing the first three steps of the late pathway, i.e., FPPP, FolD, and FalD, is more complicated. Their clear orthologs are not always found in each species (Mayoral et al. 2009b; Nyati et al. 2013; Rivera-Perez et al. 2013). Even candidates of their orthologs are not found in some species, such as the Pacific beetle cockroach, *Diploptera punctata*, and the bed bug, *Cimex hemipterus* (Huang et al. 2015; Law et al. 2025). There might be three possibilities to explain this situation: (1) the orthologs simply have not yet been found in those species; (2) these genes were replicated independently in each species after speciation, resulting in multiple paralogs; and (3) this pathway is not conserved among insects, and other enzymes catalyze these steps in other insect species. In this chapter, I would like to discuss further the third possibility.

5.2 Other Candidates of Enzymes Responsible for Conversion of Farnesyl Pyrophosphate into Farnesol and Farnesoic Acid

We previously performed DNA microarray analysis to assess the expression of known JH biosynthetic genes and to identify new genes involved in JH biosynthesis in the CA of the silkworm, *Bombyx mori* (Homma et al. 2020a). The genes encoding enzymes of the mevalonate pathway, FA epoxidase, and JHAMT were highly expressed in the CC-CA complexes of day 2 fourth-instar larvae, where JH is actively synthesized. In contrast, homologous genes for FPPP, FolD, and FalD showed very low expression levels in the CC-CA. Instead, genes encoding for two other proteins, protein tyrosine phosphatase, mitochondrial 1 (PTPMT1) and a member of the glucose-methanol-choline oxidoreductase family (GMC oxidoreductase), showed abundant expression in CA (Homma et al. 2020a). In addition, the expression of these genes in CA was much higher than that in other tissues and fluctuated corresponding to the JH biosynthetic activity of *B. mori* larval CA, similar to the *JHAMT* gene (Homma et al. 2020a).

5.2.1 Protein Tyrosine Phosphatase, Mitochondrial 1 (PTPMT1)

PTPMT1 is a mitochondrial phosphatase belonging to the protein tyrosine phosphatase (PTP) superfamily and is conserved among eukaryotes and some bacteria (Pagliarini et al. 2005; Pagliarini et al. 2004). Although many PTPs dephosphorylate the phosphorylated tyrosine/serine/threonine residues of specific proteins, members of a subfamily of PTPs catalyze the dephosphorylation of phospholipids (Tonks and Neel 2001). A well-known example is phosphatase and tensin homolog (PTEN), which dephosphorylates phosphatidylinositol (3,4,5)-trisphosphate (PIP_3) and plays an important role in cellular signal transduction (Maehama et al. 2001; Tonks and Neel 2001). PTPMT1 was initially named PTEN-like phosphatase (PLIP) because of its PTEN-like active site (Pagliarini et al. 2004) and was renamed PTPMT1 later (Pagliarini et al. 2005).

PTPMT1 dephosphorylates phospholipids similar to PTEN. In mice, it has been reported that PTPMT1 dephosphorylates phosphatidylglycerol phosphate into phosphatidylglycerol, which is a precursor of cardiolipin, an important component of the inner mitochondrial membrane (Xiao et al. 2011; Zhang et al. 2011). Knockout of the *PTPMT1* gene in mice leads to embryonic lethality because of the lack of normal mitochondrial activity (Zhang et al. 2011). Phosphoproteins can also be substrates of PTPMT1. In zebrafish, PTPMT1 has been reported to dephosphorylate phosphorylated succinate dehydrogenase, thereby regulating the tricarboxylic acid (TCA) cycle in mitochondria; knockout of *PTPMT1* resulted in larval lethality (Nath et al. 2015).

Because *PTPMT1* was expressed abundantly in CA with a correlating pattern to JH biosynthetic activity, it might be involved in JH biosynthesis in CA of *B. mori* (Homma et al. 2020a). To assess this possibility, we knocked out the *PTPMT1* gene in *B. mori* by using TALEN (Homma et al. 2020b). The homozygotes of a null mutant allele showed embryonic lethality. The lower temperature condition partially rescued the embryonic lethality, but hatched escapers did not grow after the first molt, and all of them died before becoming fourth-instar larvae (Homma et al. 2020b). The decreased expression levels of genes encoded by the mitochondrial genome indicated dysfunction of mitochondria in the homozygous mutants. These phenotypes are similar to those seen in vertebrate mutants (Nath et al. 2015; Zhang et al. 2011). Because PTPMT1 showed a clear single ortholog in each insect species (Homma et al. 2020b), we also tried knockdown of *PTPMT1* in the red flour beetle, *Tribolium castaneum,* by injection of double-stranded RNA into middle instar larvae (Homma et al. 2022). The knockdown larvae showed a delayed growth rate with the reduction of mitochondrial gene expression. Papakyrikos et al. (2020) studied PTPMT1 function in the fruit fly *Drosophila melanogaster* and reported that *PTPMT1*-deficient flies showed mitochondrial aggregation in tracheal epithelial cells, leading to air-filling failure and tracheal melanization at the first molt (Papakyrikos et al. 2020). These data indicate that PTPMT1 plays an essential role in insect mitochondria, similar to its role in vertebrates.

Defects in JH biosynthesis and signaling generally cause precocious metamorphosis (Daimon et al. 2012; Daimon et al. 2015; Minakuchi et al. 2009; Minakuchi et al. 2008). *PTPMT1*-deficient insects, however, have not shown any signs of precocious metamorphosis. This means possibly that *PTPMT1*-deficient insects might die before showing signs of precocious metamorphosis, or *PTPMT1* might contribute more strongly to mitochondrial function in non-CA organs than to JH biosynthesis. Nevertheless, considering the abundant expression of the gene and its correlation with JH biosynthetic activity in the *B. mori* CA and its ability to dephosphorylate phospholipids, PTPMT1 may function as an FPPP dephosphorylating FPP into farnesol in *B. mori* CA (Fig. 5.2). Our preliminary experiments showed decreased expression of *Krüppel-homolog 1* (*Kr-h1*), a major JH signal mediator, in *PTPMT1*-knockout *B. mori* larvae, supporting the idea that PTPMT1 is involved in JH biosynthesis in *B. mori*.

5.2.2 Glucose-Methanol-Choline (GMC) Oxidoreductase

GMC oxidoreductases are a family of flavin adenine dinucleotide (FAD) dependent oxidoreductase utilizing a variety of substrates (Cavener 1992). These enzymes oxidize hydroxyl groups to carbonyl groups, e.g., alcohol to aldehyde. In an aqueous solution, water can be incorporated into the carbonyl groups of aldehydes spontaneously, producing geminal diols. Some of GMC oxidoreductases have been reported to further oxidize hydrated aldehydes, whose carbonyl groups are converted into geminal diols, into their corresponding carboxylic acids (Romero and Gadda 2014; Fig. 5.2b). For example, choline oxidase from an actinomyces bacterium, *Arthrobacter globiformis*, catalyzes the oxidation of choline to betaine aldehyde and then the oxidation of betaine aldehyde to glycine betaine (Gadda 2003; Romero and Gadda 2014). Therefore, the member of GMC oxidoreductase found in *B. mori* might catalyze the two sequential oxidations in the late pathway of JH biosynthesis, i.e., farnesol to farnesal and farnesal to farnesoic acid (Fig. 5.2). The possible involvement of a GMC oxidoreductase in farnesol oxidation is consistent with the previous observation using CA homogenates of another lepidopteran insect, the tobacco hornworm, *Manduca sexta*, where this reaction was likely mediated by FAD-dependent alcohol oxidase rather than NAD^+-dependent alcohol

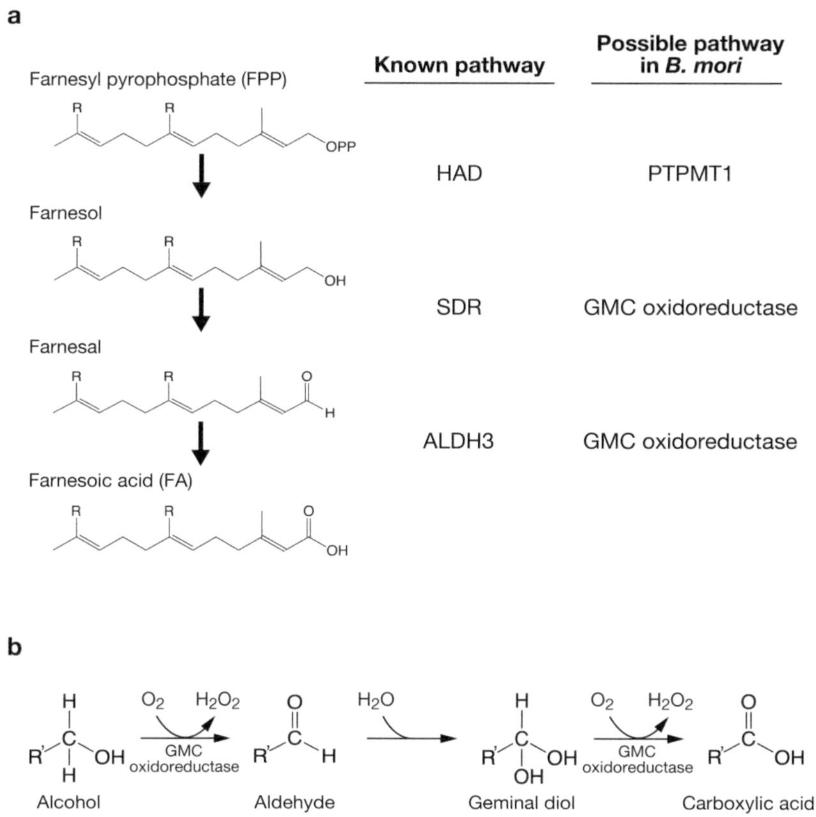

Fig. 5.2 Known and proposed pathway of the conversion from farnesyl pyrophosphate (FPP) to farnesoic acid (FA). (**a**) Enzymes catalyzing each step of the previously reported pathway and an alternative putative pathway in *Bombyx mori* are indicated. (**b**) GMC oxidoreductases oxidize alcohols to aldehydes via spontaneously formed geminal diol intermediates. This reaction scheme illustrates a possible oxidation mechanism for the conversion of farnesol to farnesal and subsequently to FA catalyzed by a GMC oxidoreductase. *R* represents a methyl or ethyl group. *R'* represents various substituents or the sesquiterpenoid backbone in the context of juvenile hormone biosynthesis. *Abbreviations:* HAD haloalkanoic acid dehalogenase, SDR short-chain dehydrogenase/reductase, ALDH3 aldehyde dehydrogenase 3, PTPMT1 protein tyrosine phosphatase, mitochondrial 1, GMC oxidoreductase glucose-methanol-choline oxidoreductase

dehydrogenase (Sperry and Sen 2001). The GMC oxidoreductase gene, identified as a putative JH biosynthetic gene in *B. mori*, has numerous paralogs in the *B. mori* genome as well as in other lepidopteran genomes, but no clear orthologs have been identified in other insect groups (Homma et al. 2020a). Therefore, if this GMC oxidoreductase is indeed involved in JH biosynthesis, this role is likely restricted to Lepidoptera.

5.3 Involvement of Protein Prenylation

Protein prenylation is a ubiquitous covalent posttranslational modification of proteins found in all eukaryotic cells. It comprises the linkage of either a farnesyl or geranylgeranyl isoprenoid to conserved cysteine residues at or near the C-terminus of proteins (Jung and Bachmann 2023; Palsuledesai and Distefano 2015; Zhang and Casey 1996). Prenylated proteins include various proteins, such as the small GTPases of the Ras superfamily, the γ-subunit of heterotrimeric G proteins, and nuclear lamins, which are involved in many cellular processes (Schafer and Rine 1992). It has been proposed that farnesal is provided as a by-product of prenylated protein metabolism (Law et al. 2025; So et al. 2022; Fig. 5.3).

Proteins are prenylated by four prenyltransferases: farnesyltransferase (FTase) and geranylgeranyl transferase types I, II, and III (GGTase-I, II, and III) (Kuchay et al. 2019; Zhang and Casey 1996). FTase and GGTase-I catalyze the transfer of a single farnesyl or geranylgeranyl group from the corresponding pyrophosphate

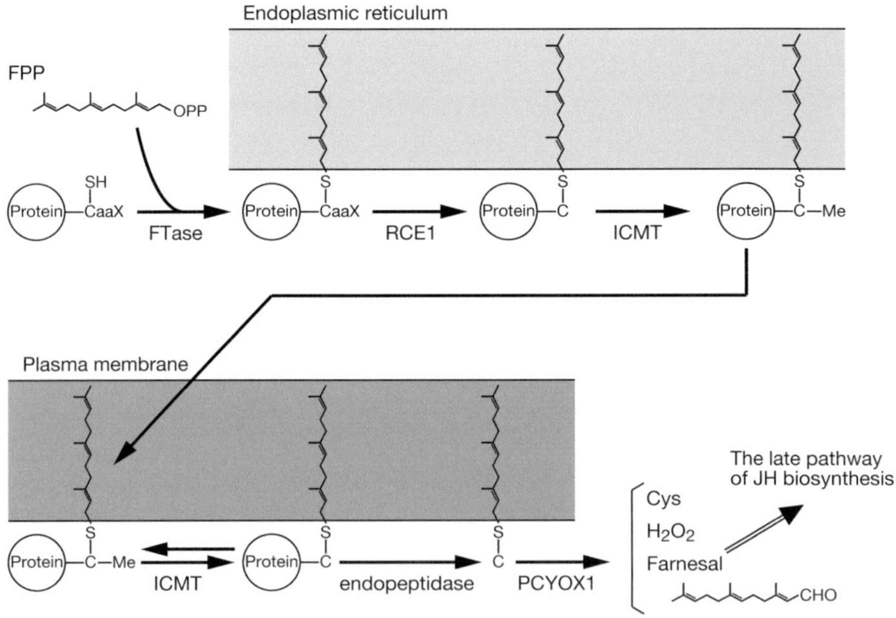

Fig. 5.3 Proposed pathway for the production of farnesal from farnesyl pyrophosphate (FPP) as a by-product of protein prenylation. FPP is utilized for the prenylation of various proteins bearing a *CaaX* motif at the C-terminus, where *C* denotes cysteine and "*a*" generally represents an aliphatic amino acid. During the degradation of farnesylated proteins, the polypeptide moieties are proteolyzed, and the remaining farnesylcysteines are subsequently oxidized by PCYOX1 to yield cysteine, hydrogen peroxide (H_2O_2), and farnesal, which can then serve as a precursor for juvenile hormone biosynthesis. See the main text for details. *Abbreviations:* FTase farnesyltransferase, RCE1 RAS-converting CaaX endopeptidase 1, ICMT isoprenylcystein carboxylmethyltransferase, PCYOX1 prenylcysteine oxidase 1

(FPP or geranylgeranyl pyrophosphate, respectively) to the cysteine residue at the C-terminal consensus sequence of target proteins. The consensus sequence is known as the CaaX motif, in which "C" denotes cysteine, "a" generally represents an aliphatic amino acid, and "X" is the amino acid residue that typically determines whether the protein is targeted by FTase or GGTase-1 (Zhang and Casey 1996). FTase and GGTase-I are composed of two α and β subunits. They share an identical α subunit and use distinct β subunits, which represent the catalytic domain (Jung and Bachmann 2023; Zhang and Casey 1996). Farnesylated (and geranylgeranylated) proteins are transported into the membrane of the endoplasmic reticulum (ER) (Schmidt et al. 1998). The C-terminal aaX peptide is then cleaved by the RAS-converting CaaX endopeptidase 1 (RCE1)/STE24 (Boyartchuk et al. 1997; Jung and Bachmann 2023; Otto et al. 1999), and the carboxyl group of the farnesylated cysteine residue remaining at the C-terminal is methylated by the isoprenylcystein carboxylmethyltransferase (ICMT)/STE14 (Hrycyna et al. 1991; Jung and Bachmann 2023; Stephenson and Clarke 1992). The processed proteins are transported into the plasma membrane via the Golgi apparatus and play their specific roles (Palsuledesai and Distefano 2015; Resh 2006; Seabra 1998). In the process of degradation, polypeptide portions of the farnesylated proteins are proteolyzed by various proteinases, and the farnesylcysteines remain on the membrane (Zhang and Casey 1996). The farnesylcysteines are then cleaved by prenylcysteine lyase, now referred to as prenylcysteine oxidase 1 (PCYOX1), producing cysteine, hydrogen peroxide, and farnesal (Digits et al. 2002; Herrera-Marcos et al. 2018). The farnesal is retrieved from the plasma membrane by an unknown mechanism and can be recruited into the late pathway of JH biosynthesis (Fig. 5.3).

Because protein prenylation is common among eukaryotes, the enzymes involved in this pathway are expected to be conserved among insects. Therefore, this could be an alternative pathway from the endpoint product of the early pathway, FPP, to the middle intermediate of the late pathway, farnesal. However, this pathway is closely associated with protein prenylation, which is essential in many cellular processes; consequently, JH biosynthesis via this pathway depends on a variety of other physiological events. Therefore, the actual involvement of this pathway in JH biosynthesis should be examined with caution.

5.4 Conclusion

The JH biosynthetic pathway is composed of 13 enzymatic steps divided into the early mevalonate pathway and the late JH-branch pathway. The genes encoding the enzymes that catalyze these 13 steps have been identified. Because the mevalonate pathway is conserved among eukaryotes, the genes responsible for this pathway are also conserved in insects. In contrast, the conservation of the late pathway in insects appears to be more uncertain. The five genes responsible for the late pathway often occur as paralogs in some species and are occasionally absent in others. In some cases, homologous genes show very low expression in the CA. Instead, the involvement of other genes in the late pathway has been proposed. Therefore, the late JH-branch pathway may be more diverse than previously expected. This might be related to the existence of multiple JH molecules among different phylogenetic groups. If some steps in the late pathway are catalyzed by group-specific enzymes, these enzymes could be ideal targets for group-specific insecticides. Nevertheless, it is important not to regard homologs as JH biosynthetic genes solely on the basis of sequence homology but to evaluate their functional relevance, such as expression in the CA and the effects of their functional disruption by gene knockout or knockdown in each species.

References

Baker FC, Schooley DA (1981) Biosynthesis of 3-hydroxy-3-methylglutaryl-CoA, 3-hydroxy-3-ethylglutaryl-CoA, mevalonate and homomevalonate by insect corpus allatum and mammalian hepatic tissues. Biochim Biophys Acta 664:356–372. https://doi.org/10.1016/0005-2760(81)90058-8

Belles X, Martin D, Piulachs MD (2005) The mevalonate pathway and the synthesis of juvenile hormone in insects. Annu Rev Entomol 50:181–199. https://doi.org/10.1146/annurev.ento.50.071803.130356

Boyartchuk VL, Ashby MN, Rine J (1997) Modulation of Ras and a-factor function by carboxyl-terminal proteolysis. Science 275:1796–1800. https://doi.org/10.1126/science.275.5307.1796

Brindle PA, Schooley DA, Tsai LW, Baker FC (1988) Comparative metabolism of branched-chain amino acids to precursors of juvenile hormone biogenesis in corpora allata of lepidopterous versus nonlepidopterous insects. J Biol Chem 263:10653–10657. https://doi.org/10.1016/S0021-9258(18)38021-9

Cao L, Zhang P, Grant DF (2009) An insect farnesyl phosphatase homologous to the N-terminal domain of soluble epoxide hydrolase. Biochem Biophys Res Commun 380:188–192. https://doi.org/10.1016/j.bbrc.2009.01.079

Cavener DR (1992) GMC oxidoreductases. A newly defined family of homologous proteins with diverse catalytic activities. J Mol Biol 223:811–814. https://doi.org/10.1016/0022-2836(92)90992-s

Daimon T, Kozaki T, Niwa R, Kobayashi I, Furuta K, Namiki T et al (2012) Precocious metamorphosis in the juvenile hormone-deficient mutant of the silkworm, *Bombyx mori*. PLoS Genet 8:e1002486. https://doi.org/10.1371/journal.pgen.1002486

Daimon T, Uchibori M, Nakao H, Sezutsu H, Shinoda T (2015) Knockout silkworms reveal a dispensable role for juvenile hormones in holometabolous life cycle. Proc Natl Acad Sci USA 112:E4226–E4235. https://doi.org/10.1073/pnas.1506645112

Defelipe LA, Dolghih E, Roitberg AE, Nouzova M, Mayoral JG, Noriega FG et al (2011) Juvenile hormone synthesis: "esterify then epoxidize" or "epoxidize then esterify"? Insights from the structural characterization of juvenile hormone acid methyltransferase. Insect Biochem Mol Biol 41:228–235. https://doi.org/10.1016/j.ibmb.2010.12.008

Dellas N, Thomas ST, Manning G, Noel JP (2013) Discovery of a metabolic alternative to the classical mevalonate pathway. eLife 2:e00672. https://doi.org/10.7554/eLife.00672

Dermauw W, Van Leeuwen T, Feyereisen R (2020) Diversity and evolution of the P450 family in arthropods. Insect Biochem Mol Biol 127:103490. https://doi.org/10.1016/j.ibmb.2020.103490

Digits JA, Pyun HJ, Coates RM, Casey PJ (2002) Stereospecificity and kinetic mechanism of human prenylcysteine lyase, an unusual thioether oxidase. J Biol Chem 277:41086–41093. https://doi.org/10.1074/jbc.M208069200

Gadda G (2003) Kinetic mechanism of choline oxidase from *Arthrobacter globiformis*. Biochim Biophys Acta 1646:112–118. https://doi.org/10.1016/s1570-9639(03)00003-7
Goodman WG, Granger NA (2005) The juvenile hormones. In: Gilbert LI, Iatrou K, Gill SS (eds) Comprehensive molecular insect science, vol 3. Elsevier, Oxford, pp 319–408
Helvig C, Koener JF, Unnithan GC, Feyereisen R (2004) CYP15A1, the cytochrome P450 that catalyzes epoxidation of methyl farnesoate to juvenile hormone III in cockroach corpora allata. Proc Natl Acad Sci USA 101:4024–4029. https://doi.org/10.1073/pnas.0306980101
Herrera-Marcos LV, Lou-Bonafonte JM, Martinez-Gracia MV, Arnal C, Navarro MA, Osada J (2018) Prenylcysteine oxidase 1, a pro-oxidant enzyme of low density lipoproteins. Front Biosci (Landmark Ed) 23:1020–1037. https://doi.org/10.2741/4631
Homma Y, Inui T, Kayukawa T, Toga K, Shinoda T, Togawa T (2022) The mitochondrial phosphatase PTPMT1 is required for the proper growth rate in the red flour beetle, *Tribolium castaneum*. Zool Sci 39:236–241. https://doi.org/10.2108/zs210092
Homma Y, Mita K, Nakamura Y, Namiki T, Noda H, Shinoda T et al (2020a) Identification of novel genes expressed highly and selectively in the corpora allata of the silkworm, *Bombyx mori* (Lepidoptera: Bombycidae). Appl Entomol Zool 55:45–54. https://doi.org/10.1007/s13355-019-00652-y
Homma Y, Toga K, Daimon T, Shinoda T, Togawa T (2020b) A mitochondrial phosphatase PTPMT1 is essential for the early development of silkworm, *Bombyx mori*. Biochem Biophys Res Commun 530:713–718. https://doi.org/10.1016/j.bbrc.2020.07.124
Hrycyna CA, Sapperstein SK, Clarke S, Michaelis S (1991) The *Saccharomyces cerevisiae STE14* gene encodes a methyltransferase that mediates C-terminal methylation of a-factor and RAS proteins. EMBO J 10:1699–1709. https://doi.org/10.1002/j.1460-2075.1991.tb07694.x
Huang J, Marchal E, Hult EF, Tobe SS (2015) Characterization of the juvenile hormone pathway in the viviparous cockroach, *Diploptera punctata*. PLoS One 10:e0117291. https://doi.org/10.1371/journal.pone.0117291
Jung D, Bachmann HS (2023) Regulation of protein prenylation. Biomed Pharmacother 164:114915. https://doi.org/10.1016/j.biopha.2023.114915
Kinjoh T, Kaneko Y, Itoyama K, Mita K, Hiruma K, Shinoda T (2007) Control of juvenile hormone biosynthesis in *Bombyx mori*: cloning of the enzymes in the mevalonate pathway and assessment of their developmental expression in the corpora allata. Insect Biochem Mol Biol 37:808–818. https://doi.org/10.1016/j.ibmb.2007.03.008
Kotaki T, Shinada T, Kaihara K, Ohfune Y, Numata H (2009) Structure determination of a new juvenile hormone from a heteropteran insect. Org Lett 11:5234–5237. https://doi.org/10.1021/ol902161x
Kuchay S, Wang H, Marzio A, Jain K, Homer H, Fehrenbacher N et al (2019) GGTase3 is a newly identified geranylgeranyltransferase targeting a ubiquitin ligase. Nat Struct Mol Biol 26:628–636. https://doi.org/10.1038/s41594-019-0249-3
Kumar R, Das J, Mahto JK, Sharma M, Vivek S, Kumar P et al (2022) Crystal structure and molecular characterization of NADP(+)-farnesol dehydrogenase from cotton bollworm, *Helicoverpaarmigera*. Insect Biochem Mol Biol 147:103812. https://doi.org/10.1016/j.ibmb.2022.103812
Laufer H, Borst D, Baker FC, Reuter CC, Tsai LW, Schooley DA et al (1987) Identification of a juvenile hormone-like compound in a crustacean. Science 235:202–205. https://doi.org/10.1126/science.235.4785.202
Law STS, Nong W, Li C, Chong TK, Yip HY, Swale T et al (2025) Genome of tropical bed bug *Cimex hemipterus* (Cimicidae, Hemiptera) reveals tetraspanin expanded in bed bug ancestor. Insect Sci 32:42–54. https://doi.org/10.1111/1744-7917.13388
Lombard J, Moreira D (2011) Origins and early evolution of the mevalonate pathway of isoprenoid biosynthesis in the three domains of life. Mol Biol Evol 28:87–99. https://doi.org/10.1093/molbev/msq177
Maehama T, Taylor GS, Dixon JE (2001) PTEN and myotubularin: novel phosphoinositide phosphatases. Annu Rev Biochem 70:247–279. https://doi.org/10.1146/annurev.biochem.70.1.247

Mayoral JG, Nouzova M, Navare A, Noriega FG (2009b) NADP+-dependent farnesol dehydrogenase, a corpora allata enzyme involved in juvenile hormone synthesis. Proc Natl Acad Sci USA 106:21091–21096. https://doi.org/10.1073/pnas.0909938106

Mayoral JG, Nouzova M, Yoshiyama M, Shinoda T, Hernandez-Martinez S, Dolghih E et al (2009a) Molecular and functional characterization of a juvenile hormone acid methyltransferase expressed in the corpora allata of mosquitoes. Insect Biochem Mol Biol 39:31–37. https://doi.org/10.1016/j.ibmb.2008.09.010

Minakuchi C, Ishii F, Washidu Y, Ichikawa A, Tanaka T, Miura K et al (2015) Expressional and functional analysis of CYP15A1, a juvenile hormone epoxidase, in the red flour beetle *Tribolium castaneum*. J Insect Physiol 80:61–70. https://doi.org/10.1016/j.jinsphys.2015.04.008

Minakuchi C, Namiki T, Shinoda T (2009) *Kruppel homolog 1*, an early juvenile hormone-response gene downstream of Methoprene-tolerant, mediates its anti-metamorphic action in the red flour beetle *Tribolium castaneum*. Dev Biol 325:341–350. https://doi.org/10.1016/j.ydbio.2008.10.016

Minakuchi C, Namiki T, Yoshiyama M, Shinoda T (2008) RNAi-mediated knockdown of *juvenile hormone acid O-methyltransferase* gene causes precocious metamorphosis in the red flour beetle *Tribolium castaneum*. FEBS J 275:2919–2931. https://doi.org/10.1111/j.1742-4658.2008.06428.x

Miyakawa H, Sato T, Song Y, Tollefsen KE, Iguchi T (2018) Ecdysteroid and juvenile hormone biosynthesis, receptors and their signaling in the freshwater microcrustacean *Daphnia*. J Steroid Biochem Mol Biol 184:62–68. https://doi.org/10.1016/j.jsbmb.2017.12.006

Nath AK, Ryu JH, Jin YN, Roberts LD, Dejam A, Gerszten RE et al (2015) PTPMT1 inhibition lowers glucose through succinate dehydrogenase phosphorylation. Cell Rep 10:694–701. https://doi.org/10.1016/j.celrep.2015.01.010

Nijhout HF (1994) Insect hormones. Princeton University Press, Princeton

Niwa R, Niimi T, Honda N, Yoshiyama M, Itoyama K, Kataoka H et al (2008) Juvenile hormone acid *O*-methyltransferase in *Drosophila melanogaster*. Insect Biochem Mol Biol 38:714–720. https://doi.org/10.1016/j.ibmb.2008.04.003

Noriega FG (2014) Juvenile hormone biosynthesis in insects: what is new, what do we know, and what questions remain? Int Sch Res Notices 2014:967361. https://doi.org/10.1155/2014/967361

Nyati P, Nouzova M, Rivera-Perez C, Clifton ME, Mayoral JG, Noriega FG (2013) Farnesyl phosphatase, a *Corpora allata* enzyme involved in juvenile hormone biosynthesis in *Aedes aegypti*. PLoS One 8:e71967. https://doi.org/10.1371/journal.pone.0071967

Olmstead AW, Leblanc GA (2002) Juvenoid hormone methyl farnesoate is a sex determinant in the crustacean *Daphnia magna*. J Exp Zool 293:736–739. https://doi.org/10.1002/jez.10162

Otto JC, Kim E, Young SG, Casey PJ (1999) Cloning and characterization of a mammalian prenyl protein-specific protease. J Biol Chem 274:8379–8382. https://doi.org/10.1074/jbc.274.13.8379

Pagliarini DJ, Wiley SE, Kimple ME, Dixon JR, Kelly P, Worby CA et al (2005) Involvement of a mitochondrial phosphatase in the regulation of ATP production and insulin secretion in pancreatic beta cells. Mol Cell 19:197–207. https://doi.org/10.1016/j.molcel.2005.06.008

Pagliarini DJ, Worby CA, Dixon JE (2004) A PTEN-like phosphatase with a novel substrate specificity. J Biol Chem 279:38590–38596. https://doi.org/10.1074/jbc.M404959200

Palsuledesai CC, Distefano MD (2015) Protein prenylation: enzymes, therapeutics, and biotechnology applications. ACS Chem Biol 10:51–62. https://doi.org/10.1021/cb500791f

Papakyrikos AM, Kim MJ, Wang X (2020) *Drosophila* PTPMT1 has a function in tracheal air filling. iScience 23:101285. https://doi.org/10.1016/j.isci.2020.101285

Resh MD (2006) Trafficking and signaling by fatty-acylated and prenylated proteins. Nat Chem Biol 2:584–590. https://doi.org/10.1038/nchembio834

Richard DS, Applebaum SW, Sliter TJ, Baker FC, Schooley DA, Reuter CC et al (1989) Juvenile hormone bisepoxide biosynthesis *in vitro* by the ring gland of *Drosophila melanogaster*: a putative juvenile hormone in the higher Diptera. Proc Natl Acad Sci USA 86:1421–1425. https://doi.org/10.1073/pnas.86.4.1421

Rivera-Perez C, Nouzova M, Clifton ME, Garcia EM, LeBlanc E, Noriega FG (2013) Aldehyde dehydrogenase 3 converts farnesal into farnesoic acid in the *corpora allata* of mosquitoes. Insect Biochem Mol Biol 43:675–682. https://doi.org/10.1016/j.ibmb.2013.04.002

Romero E, Gadda G (2014) Alcohol oxidation by flavoenzymes. Biomol Concepts 5:299–318. https://doi.org/10.1515/bmc-2014-0016

Schafer WR, Rine J (1992) Protein prenylation: genes, enzymes, targets, and functions. Annu Rev Genet 26:209–237. https://doi.org/10.1146/annurev.ge.26.120192.001233

Schmidt WK, Tam A, Fujimura-Kamada K, Michaelis S (1998) Endoplasmic reticulum membrane localization of Rce1p and Ste24p, yeast proteases involved in carboxyl-terminal CAAX protein processing and amino-terminal a-factor cleavage. Proc Natl Acad Sci USA 95:11175–11180. https://doi.org/10.1073/pnas.95.19.11175

Schooley DA, Baker FC (1985) Juvenile hormone biosynthesis. In: Kerkut GA, Gilbert LI (eds) Comprehensive insect physiology, biochemistry and pharmacology, vol 7. Pergamon Press, Oxford, pp 363–389

Seabra MC (1998) Membrane association and targeting of prenylated Ras-like GTPases. Cell Signal 10:167–172. https://doi.org/10.1016/s0898-6568(97)00120-4

Sen SE, Tomasello A, Grasso M, Denton R, Macor J, Beliveau C et al (2012) Cloning, expression and characterization of lepidopteran isopentenyl diphosphate isomerase. Insect Biochem Mol Biol 42:739–750. https://doi.org/10.1016/j.ibmb.2012.07.001

Shinoda T, Itoyama K (2003) Juvenile hormone acid methyltransferase: a key regulatory enzyme for insect metamorphosis. Proc Natl Acad Sci USA 100:11986–11991. https://doi.org/10.1073/pnas.2134232100

So WL, Kai Z, Qu Z, Bendena WG, Hui JHL (2022) Rethinking sesquiterpenoids: a widespread hormone in animals. Int J Mol Sci 23:5998. https://doi.org/10.3390/ijms23115998

Sperry AE, Sen SE (2001) Farnesol oxidation in insects: evidence that the biosynthesis of insect juvenile hormone is mediated by a specific alcohol oxidase. Insect Biochem Mol Biol 31:171–178. https://doi.org/10.1016/S0965-1748(00)00115-6

Stephenson RC, Clarke S (1992) Characterization of a rat liver protein carboxyl methyltransferase involved in the maturation of proteins with the -CXXX C-terminal sequence motif. J Biol Chem 267:13314–13319

Tobe SS, Stay B (1985) Structure and regulation of the corpus allatum. Adv Insect Physiol 18:305–432. https://doi.org/10.1016/S0065-2806(08)60043-0

Tonks NK, Neel BG (2001) Combinatorial control of the specificity of protein tyrosine phosphatases. Curr Opin Cell Biol 13:182–195. https://doi.org/10.1016/s0955-0674(00)00196-4

Villalobos-Sambucaro MJ, Nouzova M, Ramirez CE, Eugenia Alzugaray M, Fernandez-Lima F, Ronderos JR et al (2020) The juvenile hormone described in *Rhodnius prolixus* by Wigglesworth is juvenile hormone III skipped bisepoxide. Sci Rep 10:3091. https://doi.org/10.1038/s41598-020-59495-1

Xiao J, Engel JL, Zhang J, Chen MJ, Manning G, Dixon JE (2011) Structural and functional analysis of PTPMT1, a phosphatase required for cardiolipin synthesis. Proc Natl Acad Sci USA 108:11860–11865. https://doi.org/10.1073/pnas.1109290108

Yin C-M, Zou B-X, Jiang M, Li M-F, Qin W, Potter TL et al (1995) Identification of juvenile hormone III bisepoxide (JHB3), juvenile hormone III and methyl farnesoate secreted by the corpus allatum of *Phormia regina* (Meigen), *in vitro* and function of JHB_3 either applied alone or as a part of a juvenoid blend. J Insect Physiol 41:473–479. https://doi.org/10.1016/0022-1910(94)00134-3

Zhang FL, Casey PJ (1996) Protein prenylation: molecular mechanisms and functional consequences. Annu Rev Biochem 65:241–269. https://doi.org/10.1146/annurev.bi.65.070196.001325

Zhang J, Guan Z, Murphy AN, Wiley SE, Perkins GA, Worby CA et al (2011) Mitochondrial phosphatase PTPMT1 is essential for cardiolipin biosynthesis. Cell Metab 13:690–700. https://doi.org/10.1016/j.cmet.2011.04.007

Chapter 6
Regulation of Juvenile Hormone Synthesis in the Corpora Allata: Lessons from *Bombyx mori*

Yu Kaneko and Kiyoshi Hiruma

Abstract Insect development is primarily governed by two hormones, i.e., ecdysteroids and juvenile hormones (JHs). Larval molt is initiated through the combined action of ecdysteroid and high levels of JH, and the decline in JH levels is necessary to program pupal metamorphosis by the action of ecdysteroid. Thus, JH serves as a critical modulator of ecdysteroid action, orchestrating the timing of molting and metamorphosis in insects. Precise fluctuations in hemolymph JH levels are maintained through the tightly regulated synthesis of JH by the corpora allata (CA) in response to developmental cues. In this chapter, we elucidated the regulatory mechanisms of JH synthesis during the final two larval stadiums based on studies in the silkworm, *Bombyx mori*, a well-established model species. CA activity is regulated in a stage-specific manner by multiple factors, including ecdysteroids, peptide hormones, and dopamine. In particular, the inactivation of CA, which triggers pupal metamorphosis, is achieved through the coordinated action of three different mechanisms, each operating via a different mode of action.

Keywords Juvenile hormone synthesis · Corpora allata · Allatostatin · Allatotropin · Short neuropeptide F · Ecdysis triggering hormone · Ecdysteroid · Dopamine · *Bombyx mori*

6.1 Introduction

Juvenile hormone (JH) is a pleiotropic hormone that is not only necessary for insect molting and metamorphosis but also for many other events, such as suppression of imaginal disc growth (Truman et al. 2006), pigmentation (Truman et al. 1973; Hiruma et al. 1984, 1985; Futahashi and Fujiwara 2008), phase polyphenism

Y. Kaneko (✉) · K. Hiruma
Faculty of Agriculture and Life Sciences, Hirosaki University, Hirosaki, Japan

Graduate School of Agricultural Sciences, Iwate University, Morioka, Japan
e-mail: yukaneko@hirosaki-u.ac.jp

T. Shinoda et al. (eds.), *Juvenile Hormone*, Entomology Monographs,
https://doi.org/10.1007/978-981-95-8605-9_6

(Nijhout 1999; Hartfelder and Emlen 2012), ovarian development (Leyria 2024), behavior (Nijhout 1994; Bilen et al. 2013; Wijesekera et al. 2016), and caste differentiation (Oguchi and Miura 2024). The precise timing of JH elevation and decline in the hemolymph during insect development is crucial for ensuring the proper execution of these events.

Insects must periodically shed their rigid exoskeletons to grow. After hatching, they undergo a series of larval molts without any major structural changes. Upon reaching a certain size, they metamorphose into pupae and subsequently adults in holometabolous insects or directly into adults in hemimetabolous insects. Molting and metamorphosis are primarily regulated by two hormones: JH, secreted from the corpora allata (CA), and ecdysteroid (in this chapter, "ecdysteroid" is used as a generic name), secreted from the prothoracic glands (PG). The balance between JH and ecdysteroids in the hemolymph determines whether insects undergo larval, pupal, or adult molt (Riddiford 1985; Hiruma 2003).

When the JH level is high, ecdysteroid secretion repeatedly induces larval molt, the "status quo" action of JH (Riddiford 1996). However, once the JH level declines, ecdysteroids trigger metamorphosis. In hemimetabolous insects, a drop in JH titers to undetectable levels in the final nymphal instar allows a subsequent ecdysteroid surge to initiate adult metamorphosis. However, hormonal regulation in holometabolous insects is more complex than that in hemimetabolous species. Holometabolous insects undergo an additional pupal stage between larval and adult metamorphosis, which involves two steps: pupal commitment and pupal differentiation (Hiruma 2003). In the tobacco hornworm, *Manduca sexta*, the pupal commitment of the epidermis is induced by a small increase in ecdysteroid levels in the absence of JH (Riddiford 1976, 1985). Following this commitment, a large surge of ecdysteroids induces pupal differentiation of the epidermis. A similar but slightly different hormonal regulation occurs in the silkworm, *Bombyx mori*, where epidermal commitment proceeds gradually over 3 days, induced by a modest increase in ecdysteroid under conditions of very low JH titers (Muramatsu et al. 2008).

Hormone titers are controlled by the balance between synthesis, release, degradation, and excretion. In the case of JH, synthesis is generally regarded as the primary step in determining hemolymph levels, rather than the other steps (Tobe and Stay 1985; Gilbert et al. 2000; Goodman and Cusson 2012; Hernández-Martínez et al. 2015). During the final two larval instars of *B. mori*, fluctuations in hemolymph JH titers closely mirror JH synthetic activity (Kinjoh et al. 2007). CA activity is regulated by both neural and humoral factors transmitted from the brain via the associated nerves (Tobe and Stay 1985; Stay 2000; Hiruma and Kaneko 2013; Kaneko and Hiruma 2011, 2014, 2016; Orchard and Lange 2024). 20-Hydroxyecdysone (20E), an active form of ecdysteroid, is also known to be a regulatory factor of JH production (Kaneko et al. 2011). Therefore, elucidating the regulation of JH synthesis during development is essential for understanding the endocrine control of molting and metamorphosis.

Early studies on CA regulation have primarily focused on identifying potential regulatory factors. Various tropic and static factors, including peptide hormones, neurotransmitters, and 20E, have been identified in both hemimetabolous and holometabolous species (Whisenton et al. 1985, 1987; Gu and Chow 1996; Goodman and Granger 2005; Stay and Tobe 2007; Weaver and Audsley 2009; Kaneko et al. 2011; Kaneko and Hiruma 2014; Areiza et al. 2015; Hult et al. 2015; Bendena et al. 2020). However, these studies did not focus on the precise regulation of JH synthetic activity, that is, how these factors are involved in determining JH fluctuations in the hemolymph during insect development. Consequently, the mechanisms underlying stage-specific JH production by the CA remain poorly understood. To clarify how JH synthesis is regulated during larval development, we performed a comprehensive analysis using *B. mori* as a model. We demonstrated that multiple factors orchestrate CA activity in a stage-specific manner. In this chapter, we describe how these factors coordinately control JH synthesis in a stage-specific manner, thereby elucidating the hormonal regulation of larval and pupal molt. The fluctuation patterns of the transcriptional profiles of most JH synthetic enzyme genes closely match those of JH titer (Kinjoh et al. 2007; Nouzova et al. 2011; Huang et al. 2015; Yin et al. 2020). Therefore, to elucidate the modes of action of these regulatory factors, we examined their effects on the transcriptional regulation of JH synthetic enzyme genes.

6.2 Changes in JH Titer and Its Role During the Penultimate (Fourth) and Last (Fifth) Instar Stadiums in *B. mori*

In this chapter, the term "stadium" is used to denote the period between two successive molts, whereas "stage" refers to morphological developmental phases such as larval, pupal, and adult stages. During the penultimate (fourth) larval instar stadium, JH titers fluctuate at relatively high levels (Fig. 6.1). An ecdysteroid peak was observed midway through the instar. Allatectomy performed before this peak induces pupal metamorphosis rather than larval molt (Sakurai 1983). This indicates that JH is required for larval molting during this period.

After the last larval molt, JH titers increased and then declined to trace levels by the end of day 2 of the fifth instar, remaining very low until the onset of wandering behavior (Fig. 6.1). The decline in JH serves as a major signal to initiate pupal metamorphosis. In early fifth instar larvae, the prothoracic glands (PG) are kept inactive by JH; however, the subsequent decline in JH titers reactivates the PG, leading to the production of a small amount of ecdysteroid between days 3 and 5 of the final instar (Sakurai 1984; Okuda et al. 1985; Takaki and Sakurai 2003). This modest level of ecdysteroid, acting in the presence of only trace amounts of JH, induces pupal commitment of the epidermis (Muramatsu et al. 2008). After the wandering stage, the JH titer rises again, followed by a large ecdysteroid peak that triggers pupal development. The role of this high JH level is unclear in *B. mori*, but comparable JH elevations prevent the formation of adultoids in the cecropia moth, *Hyalophora cecropia* (Williams 1961); *M. sexta* (Kiguchi and Riddiford 1978); and the cabbage armyworm, *Mamestra brassicae* (Hiruma 1980).

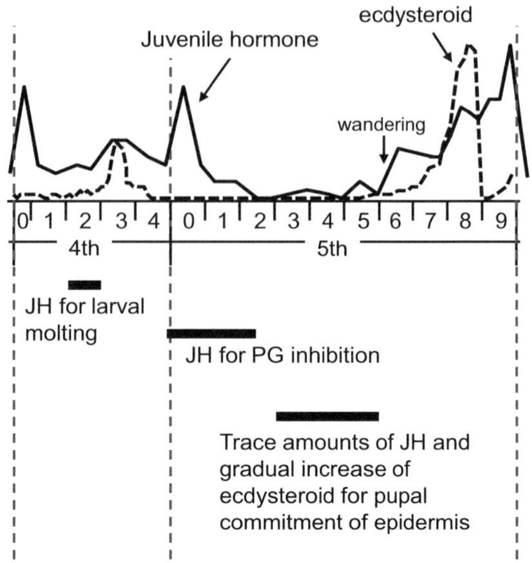

Fig. 6.1 Juvenile hormone (JH) titer, ecdysteroid titer, and various events regulated by JH during the fourth and fifth instar stadiums of *Bombyx mori*. The approximate duration of the JH sensitive period for each event is indicated by the horizontal bars (Based on data from Sakurai 1983; Óhashi et al. 1983; Sakurai 1984; Takaki and Sakurai 2003; Mizoguchi et al. 2015). The JH titer (solid line) and ecdysteroid titer (dashed line) are from Niimi and Sakurai (1997), Koyama et al. (2004), Sakurai et al. (1998), and Kaneko et al. (2006)

6.3 Control of JH Synthesis During the Fourth and Early Fifth Instar Stadiums in *B. mori*: Roles of Brain, CC, and CA

The developmental fluctuation in JH titers is primarily governed by changes in JH synthetic activity within the CA. To elucidate the mechanisms regulating the CA, we focused on the brain and corpora cardiaca (CC). The brain is connected to the CA, and the CC is anatomically positioned between them (Stay 2000; Goodman and Cusson 2012). We compared JH synthetic activity among the brain-CC-CA complex, the CC-CA complex, and isolated CA in vitro (Kaneko and Hiruma 2014; Fig. 6.2). The JH synthetic activity of the brain-CC-CA complex was high during the fourth instar stadium but declined soon after the final larval molt, and on day 2 of the fifth instar, the brain-CC-CA complex exhibited very low JH synthetic activity. These fluctuations in the activity of the brain-CC-CA complex closely correspond with changes in hemolymph JH titer. When examined in isolation, the CA exhibited higher synthetic activity than the brain-CC-CA complex, although the temporal patterns of fluctuation across the fourth and fifth instar stadiums were similar. Thus, the brain exerts an inhibitory influence on JH production by the CA throughout the fourth and fifth instar stadiums via its neural connections. The isolated CA showed high JH synthetic activity, but when connected to the CC, its activity decreased on day 3 of the fourth instar. This indicates that CC exerts a stage-dependent inhibitory effect on JH synthesis.

During the fourth and early fifth instar stadiums of *B. mori*, JH synthesis in the CA is regulated by both the brain and the CC. CA activity during the fourth and fifth

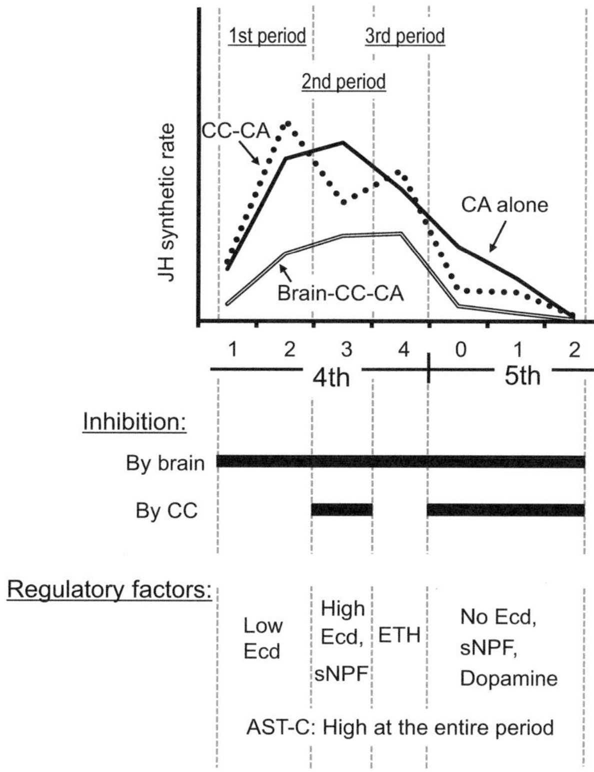

Fig. 6.2 Fluctuations in juvenile hormone (JH) synthetic activity during the fourth and early fifth instar stadiums and their regulatory factors in *Bombyx mori*. The top panel shows JH synthetic rates in the brain-corpora cardiaca (CC)-corpora allata (CA) complex (double line), CC-CA complex (dotted line), and CA alone (solid line). The middle panel shows the duration of the inhibitory effect mediated by the brain or CC. The lower panel shows the regulators in each stadium. Ecd indicates ecdysteroid. (Modified from Hiruma and Kaneko (2013) and Kaneko and Hiruma (2014). See text for detailed explanations)

instar stadiums might be suppressed by allatostatin-C (AST-C), a neuropeptide produced in the brain. Short neuropeptide F (sNPF), produced in the CC, is another candidate inhibitory factor acting on the CA. In addition to these inhibitory factors, 20E, ecdysis triggering hormone (ETH), and dopamine have been identified as regulatory factors that modulate JH synthesis in the CA (Hiruma and Kaneko 2013; Kaneko and Hiruma 2011, 2014, 2016; Kaneko et al. 2011).

In the following sections, we will discuss how these factors regulate JH synthesis during development, with particular emphasis on their effects on the transcriptional regulation of JH-synthetic enzyme genes.

6.4 Allatostatin-C: A Brain-Derived Inhibitory Factor Acting Throughout the Fourth and Fifth Instar Stadiums in *B. mori*

ASTs are neuropeptides produced in the brain and well-known inhibitors of JH synthesis in a wide range of insect species (Stay and Tobe 2007; Weaver and Audsley 2009; Verlinden et al. 2015). They are transported directly to the CA via axons or secreted from the CC into the hemolymph to inhibit CA activity. These peptides are classified into three major groups—AST-A, AST-B, and AST-C—based on their amino acid sequences and receptor specificities. AST-A was first identified in the Pacific beetle cockroach, *Diploptera punctata*, and its inhibitory activity on JH synthesis was subsequently demonstrated in cockroaches, crickets, and termites (Bendena et al. 2020). AST-B is the second group of ASTs, which inhibits CA activity in the two-spotted cricket, *Gryllus bimaculatus* (Lorenz et al. 1995). In *B. mori*, AST-B is known as a prothoracicostatic peptide that inhibits PG activity (Yamanaka et al. 2005). In lepidopteran and dipteran species, the functional AST is AST-C, which belongs to the PISCF family (Weaver and Audsley 2009; Bendena et al. 2020). The JH synthetic activities of the *B. mori* CA cultured with 1 μM AST-C were inhibited by less than 50% during the fourth and fifth instar stadiums (Kaneko and Hiruma 2014).

In the case of *B. mori* CA, mRNA expression of *HMG-CoA reductase* (*HMGR*), *Mevalonate kinase* (*MevK*), and *Isopentenyl diphosphate isomerase* (*IPPI*), which encode enzymes involved in JH synthesis, was suppressed by exposure to AST-C for 6 h in vitro. This indicates that AST-C regulates CA activity through the transcriptional regulation of JH-synthetic enzyme genes (Kaneko and Hiruma 2014). In addition to transcriptional regulation, metabolic regulation has been reported in the yellow fever mosquito, *Aedes aegypti* (Nouzova et al. 2015). The first step of JH synthesis is the transportation of citrate from the mitochondria to the cytosol. Citrate is then metabolized to acetyl-CoA, which is one of the sources of JH synthesis by CA. Nouzova et al. (2015) demonstrated that AST-C inhibited CA activity within 2 h of culture by blocking the transport of citrate from the mitochondria to the cytosol. A similar inhibition of CA activity was also reported with AST-A in the CA of *D. punctata* (Sutherland and Feyereisen 1996). Based on the results of the *B. mori* and *A. aegypti* studies, it can be speculated that AST-C inhibits CA activity in at least two different ways.

6.5 Stage-Specific Regulatory Factors in the Penultimate Instar: Roles of 20E, sNPF, and ETH

During the fourth instar stadium, the rate of JH synthesis is relatively high due to the action of a combination of multiple regulatory factors. In this section, to elucidate the mechanisms of high JH synthetic rate, we divided the experiment into three periods according to the regulatory mechanisms: days 1 and 2 as the first period, day 3 as the second period, and day 4 as the third period (Fig. 6.2).

6.5.1 First Period (Day 1 and Day 2): A Small Amount of Ecdysteroid Stimulates JH Synthesis

Ecdysteroids are known to regulate JH synthesis. For example, in *M. sexta*, JH synthetic activity in the brain-CC-CA complex of day 2 penultimate instar larvae is stimulated by 20E (Whisenton et al. 1987). In this first period of *B. mori*, between days 1 and 2 of the fourth instar stadium, JH synthesis increased rapidly. Simultaneously, the level of ecdysteroid increased, suggesting that this small increase in ecdysteroid enhanced CA activity during this period.

Feeding a diet containing 20E increased ecdysteroid levels in the hemolymph of *B. mori* (Kamimura et al. 2003; Gu and Chow 1996). To confirm whether ecdysteroid induces JH synthesis, the concentration of hemolymph ecdysteroids in fourth-instar larvae was manipulated by feeding an artificial diet containing 20E (Kaneko et al. 2011). By feeding a diet containing 400 ppm 20E for 6 h, the rate of JH synthesis rapidly increased. The transcript levels of *Farnesyl diphosphate synthase 1* (*FPPS1*) and *JH acid O-methyltransferase* (*JHAMT*) dramatically increased when 20E was added. Thus, we concluded that in the fourth instar stadium on days 1 and 2, the increase in JH synthesis is caused by the stimulation of the transcription of *FPPS1* and *JHAMT* in response to a small increase in ecdysteroid titer.

6.5.2 Second Period (Day 3): Large Amounts of Ecdysteroid and sNPF Inhibit JH Synthesis

In day 3 fourth instar larvae, JH synthesis by the CC-CA decreased slightly, but not by the Br-CC-CA complexes and CA alone (Fig. 6.2). During this period, the ecdysteroid titer increased dramatically, reaching approximately 700 ng/mL (Koyama et al. 2004; Kaneko et al. 2011). Thus, it is suggested that high concentrations of ecdysteroids inhibit JH synthesis by the CC-CA. To confirm this hypothesis, we forced the ecdysteroid peak to appear 1.5 days earlier than that in normal larvae by feeding a 20E-containing diet (Kaneko et al. 2011). Consequently, the decrease in JH synthetic levels by CC-CA shifted earlier. This suppression in both normal and 20E-fed larvae was caused by >250 ng/mL 20E in the hemolymph. Therefore, ecdysteroids play two roles in the fourth instar stadium: a small amount of ecdysteroid acts as a stimulator of JH synthesis, whereas a large amount of ecdysteroid acts as a suppressor.

Although low levels of ecdysteroids upregulated *FPPS1* and *JHAMT* expression, high levels of ecdysteroids downregulated *IPPI*, *FPPS1*, and *JHAMT* expression, which were apparently responsible for the decrease in JH synthesis.

In this period, on day 3 of the fourth instar stadium, the JH synthetic rate in the CA alone was approximately twofold higher than that in the CC-CA complex in vitro. This indicates that CC contains inhibitory factors for JH synthesis. We found that this inhibitory factor was the sNPF peptide (Kaneko and Hiruma 2014). The sNPF peptide translated in the CC is transported via the axon to the CA, where sNPF receptors are expressed. Three forms of sNPF have been identified in *B. mori*, two of which, sNPF-1 and -2, showed inhibitory effects on the CA, whereas sNPF-3 did not (Yamanaka et al. 2008; Kaneko and Hiruma 2014). The inhibitory effect of sNPF was achieved by inhibiting the transcript levels of *HMGR* and *MevK* (Kaneko and Hiruma 2014), which differ from the ecdysteroid targets described above. Therefore, the CA at day 3 of the fourth instar stadium is inhibited by the cooperation of sNPF and the high concentration of ecdysteroid in different fashions.

6.5.3 Third Period (Day 4): ETH Stimulates JH Synthesis

The level of ecdysteroids decreases rapidly before the last larval ecdysis, which refers to the shedding of old cuticles. This is followed by an increase in the JH titer, which determines the feeding period of the last instar larvae. In *B. mori*, the JH synthesis by CC-CA began to decline 12 h before ecdysis, followed by a rapid increase at 5.5 h before ecdysis, and then declined again (Hiruma and Kaneko 2013; Kaneko and Hiruma 2016; Fig. 6.3A). This transient increase apparently results in an increase in JH concentration in the hemolymph immediately after ecdysis (Fig. 6.1).

ETH is a peptide hormone involved in the sequence of ecdysis behavior and is produced in a specific endocrine gland, the epitracheal cells (Inka cells), located near the tracheal systems in each spiracle (Žitňan and Adams 2012). The release of ETH into the hemolymph is stimulated by corazonin, a peptide hormone secreted from the brain in response to a decrease in ecdysteroid titer prior to ecdysis (Žitňan and Adams 2005; Daubnerová et al. 2021). There are two isoforms of the ETH receptor identified in *B. mori* that are expressed in the CA (Yamanaka et al. 2008), and these receptors reach their peak expression on day 4 of the fourth larval stadium (Hiruma and Kaneko 2013; Kaneko and Hiruma 2016; Fig. 6.3B).

Culturing CA from various time points with 0.1 μM ETH stimulated JH synthesis during 6–9 h before ecdysis, just before the peak of CA activity (Hiruma and Kaneko 2013; Kaneko and Hiruma 2016; Fig. 6.3C). Apparently, the narrow window for the ETH-sensitive period of CA induces a very sharp peak of JH synthesis just before ecdysis. The stimulatory effect of ETH was observed when the CA was cultured with as low as 0.01 nM ETH (Fig. 6.3D). Thus, it is suggested that the increase in JH synthesis in this period is due to the secretion of ETH.

A similar transient increase in JH titer at ecdysis has been observed in other lepidopteran insects (Fain and Riddiford 1975; Hsiao and Hsiao 1977), and this increase is presumably due to the action of ETH.

ETH also stimulates JH synthesis in adult CA in the fruit fly, *Drosophila melanogaster*, *A. aegypti*, and the Oriental fruit fly, *Bactrocera dorsalis* (Meiselman et al. 2017; Areiza et al. 2014; Shi et al. 2019). In these cases, ETH caused calcium mobilization in CA cells, suggesting that calcium acts as a downstream signal of ETH. In *A. aegypti* CA, the activation of JH synthesis is also due to the activation of JHAMT enzymatic activity (Areiza et al. 2014).

In the fourth instar stadium, the JH synthetic activity of the CA is regulated by two stimulatory factors (20E and ETH) and two inhibitory factors (20E and sNPF)

Fig. 6.3 (continued) with various concentrations of ETH for 6 h, and then the amount of JH was determined by RCA. The amount of JH produced by the CA cultured with or without ETH is indicated by the black or white bars, respectively. The CA activity was significantly stimulated by as low as 0.01 nM ETH. The P values are 0.01 nM; $P = 0.0379$, 10 nM; $P = 0.0353$, 100 nM; $P = 0.0073$, and 1 μM; $P = 0.0015$. Each point represents Mean ± S.E. ($N = 3–8$). (Adapted from Kaneko and Hiruma (2016) with permission from the Japanese Society of Sericultural Science)

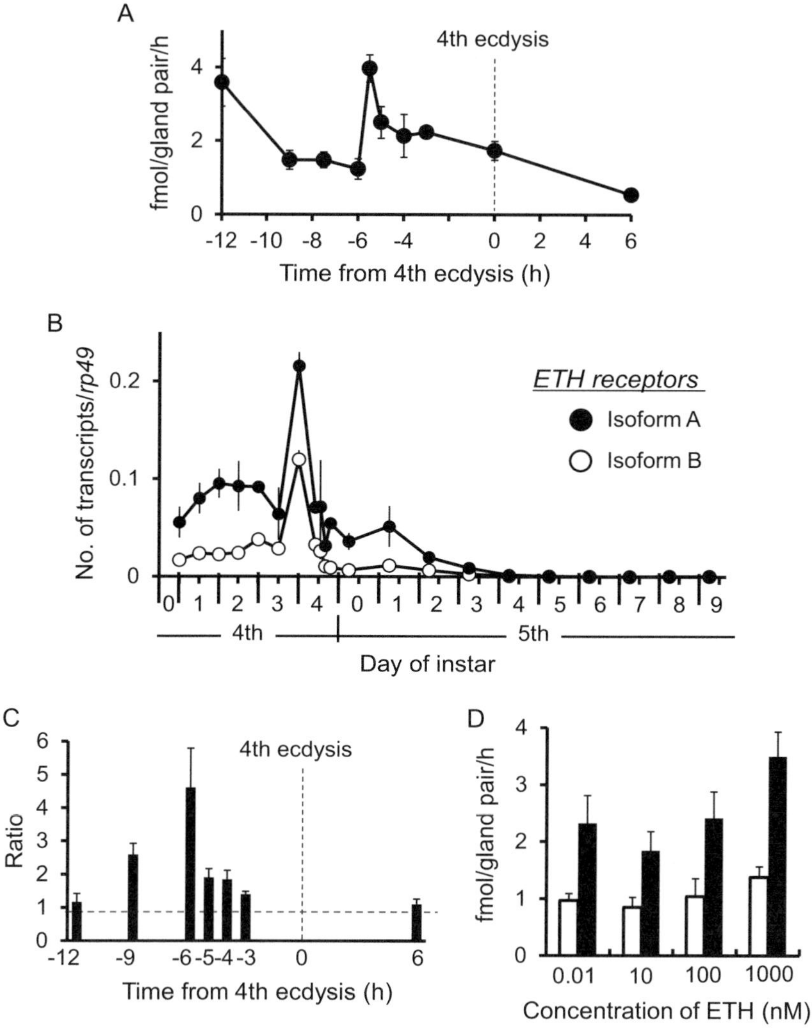

Fig. 6.3 Ecdysis triggering hormone (ETH) stimulates the juvenile hormone (JH) synthesis in *Bombyx mori* just prior to the fourth larval ecdysis. (**A**) The fluctuation of JH synthetic activity in the corpora allata (CA) during the fourth larval ecdysis. The CA from 12 h before to 6 h after ecdysis were cultured for 6 h, and then the JH synthetic activities were determined by radiochemical assay (RCA). The synthetic activity peaked 5.5 h prior to the ecdysis, followed by a rapid decline. Each point represents Mean ± S.E. (N = 3–4). (**B**) The mRNA expression of the ETH receptors, isoform A (*BNGR-A6-A*) and isoform B (*BNGR-A6-B*), in the CA during the fourth and fifth instar stadiums. The expression levels were analyzed by quantitative PCR. Each point represents Mean ± S.D. (N = 3). (**C**) The stimulation of JH synthesis by ETH during the fourth larval ecdysis. The CA from various time points during the ecdysis were cultured for 6 h with or without 0.1 μM ETH, and then the CA activities were determined by radio chemical assays (RCAs). The ratio represents the quantity of the amount of JH synthesized by the CA cultured with ETH divided by the amount without ETH. Note that the stimulation of the JH synthesis by ETH was well correlated with the increase in JH synthesis. Each point represents Mean ± S.E. (N = 3–4). (**D**) Dose-dependent stimulation of JH synthesis in the CA of day 4 fourth instar larvae. The CA was cultured

in a stage-specific manner. The combinatorial action of these factors is crucial during the fourth instar stadium to produce the appropriate amount of JH. However, the regulatory mechanisms of JH synthesis are completely different in the last larval instar stadium, as shown in the next section.

6.6 Factors Shutting Down JH Synthesis for Pupal Metamorphosis in the Fifth Instar: Roles of 20E, Dopamine, sNPF, and AT

At the beginning of the last larval stadium, the JH titer is high due to ETH stimulation in the CA, which occurs shortly before ecdysis (Fig. 6.4), followed by a decline in the JH titer. This decline activates the PG and triggers pupal commitment in epidermal cells. Preventing JH depletion by applying JH or its analogs induces additional larval molts or dauer larvae that never pupate (Sakurai 1984; Muramatsu et al. 2008; Reynolds et al. 2009). Therefore, shutting off JH synthesis during the final larval instar stadium—approximately 60 h after the last larval ecdysis in *B. mori*—is essential for pupal metamorphosis. The CA terminates JH production through at least three different mechanisms: (1) decline of ecdysteroid titer, (2) dopamine signaling, and (3) inhibition by sNPF (Fig. 6.4).

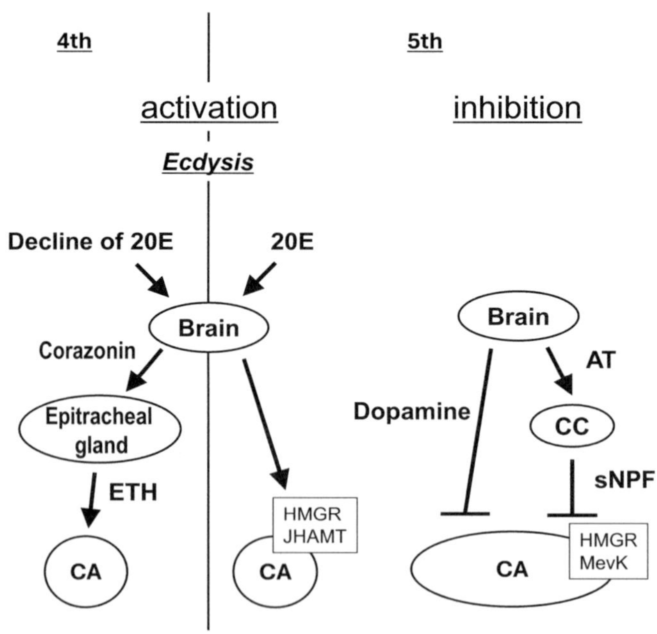

Fig. 6.4 Schematic representation of the mode of action of ecdysis triggering hormon (ETH), 20-hydroxyecdysone (20E), dopamine, allatotropin (AT) and short neuropeptide F (sNPF) on juvenile hormone (JH) synthetic regulation in the corpora allata (CA) of *Bombyx mori* during the fourth ecdysis and early fifth instar stadium. (See text for detailed explanations)

6.6.1 Decline of Ecdysteroid Titer

After the last larval molt, PG activity ceases completely, and ecdysteroid levels in the hemolymph drop within 12 h (Sakurai et al. 1998). This decline likely inactivates JH synthesis, as exogenous 20E during this period induces an extra larval molt in *B. mori* (Kamimura et al. 2003; Kaneko et al. 2011). Artificially maintaining high 20E (20–400 ppm) in the diet immediately after molting enhances CA activity (Kaneko et al. 2011). Feeding 20 ppm of 20E kept the hemolymph ecdysteroid titer at ~30 ng/mL for several days—the level of newly molted fifth instar larvae—and maintained CA activity (~3 fmol/pair/h), equivalent to that in the fourth instar. Thus, the post-ecdysial decline in ecdysteroid titer contributes to JH shutdown.

In *M. sexta*, 20E activates the CA connected to the brain-CC at similar concentrations (~20 ng/mL). The brain-CC-CA complex of *A. aegypti* also responds to 20E (Areiza et al. 2015). Hence, 20E-dependent CA activation may be common in both Lepidoptera and Diptera.

In *B. mori* fed with 20E, the expression of *HMGR* and *JHAMT* in the CA remains high, indicating transcriptional upregulation of JH synthetic enzyme genes (Kaneko et al. 2011). Knockdown of the ecdysone receptor reduces *HMGR* and *JHAMT* expression in *D. melanogaster* (Liu et al. 2018), suggesting a direct action of ecdysteroid signaling on these enzymes.

Conversely, 20E had no effect on the isolated CC-CA complex, although brain-CC-CA preparations were stimulated (Kaneko et al. 2011). The CC-CA of day 0 fifth instar larvae was activated by incubation with brain extracts but not by those from the frontal ganglion (Kaneko and Hiruma, unpublished), implying that an unidentified brain factor(s) mediates the 20E signal to the CA.

6.6.2 Dopamine

Projections of neurons from the brain innervate the CA (Tobe and Stay 1985; Stay 2000), and neurotransmitters such as dopamine, octopamine, and L-glutamine modulate CA function in many insects (Goodman and Granger 2005; Marchal et al. 2013; Zhang et al. 2022).

In *B. mori*, dopamine is localized in neurosecretory cells that project into the CA (Hiruma and Kaneko 2013). The activity of CA in day 0 fifth instar larvae was inhibited by above 5 μM dopamine, whereas the CA of day 2 fourth instar larvae was unaffected. In *M. sexta*, dopamine stimulates CA activity early in the final instar, but later becomes inhibitory (Granger et al. 1996). Two dopamine receptor isoforms are expressed in *B. mori*, peaking at the final larval molt and declining thereafter (Hiruma and Kaneko 2013). Their expression pattern matches the stage-dependent inhibition of JH synthesis, indicating that the regulation of receptor expression mediates dopamine sensitivity. Dopamine may be released from the brain in response to declining ecdysteroids to inhibit CA activity. However, it does not affect the transcription of JH synthetic enzyme genes (Kaneko and Hiruma, unpublished), suggesting that a neuronal signal is involved in the action of dopamine.

6.6.3 sNPF and AST

In early fifth instar larvae, JH synthetic activity in the CC-CA complex was lower than that in the isolated CA, suggesting inhibitory control by sNPF produced in the CC. Incubation of CA with sNPF suppresses activity, and this inhibition increases with larval age (Kaneko and Hiruma 2014), likely contributing to the complete shutdown required for pupal commitment.

Two sNPF receptors (*BNGR-A10* and *A11*) have been identified in *B. mori* (Yamanaka et al. 2008), but their stage-dependent expression patterns are different. *BNGR-A10* is expressed in the fourth instar CA, whereas both receptors appear in the early fifth instar CA, peaking on day 0 and gradually declining until day 3 (Kaneko and Hiruma 2014). The expression of receptors does not fully explain stage-dependent inhibition, implying the presence of additional regulatory layers.

In *B. mori*, allatotropin (AT), which is synthesized in the brain, stimulates *sNPF* expression in the CC. *sNPF* mRNA and AT receptor mRNA are co-expressed in CC cells (Yamanaka et al. 2008), and their expression patterns during the fourth and fifth instar stadiums are similar. AT enhances sNPF transcription in the CC of day 1 fifth instar larvae (Kaneko and Hiruma 2014). High AT receptor expression during the early fifth instar stadium correlated with reduced JH synthesis, indicating that AT indirectly inhibits CA activity via sNPF. In adults, AT also inhibits CA activity in *B. mori* through an attached esophagus (Kaneko and Hiruma 2015). Interestingly, although the AT sequences are identical between *B. mori* and *M. sexta* (Park et al. 2002), AT stimulates adult CA in *M. sexta* and *A. aegypti* (Kataoka et al. 1989; Li et al. 2003), suggesting species-specific functional divergence.

sNPF suppresses *MevK* expression independently of 20E or dopamine (Kaneko and Hiruma 2014), indicating that the AT-sNPF pathway inhibits JH synthesis through a distinct mechanism (Fig. 6.4).

In summary, during the early fifth instar, 20E, dopamine, and sNPF act via different pathways to cooperatively suppress JH synthesis in the CA, ensuring its complete shutdown and enabling metamorphosis into the pupal stage.

6.7 Myoglianin as a Candidate Suppressor of JH Synthesis Induced by the Critical Weight

The timing of JH synthesis cessation is well correlated with body mass. During the final larval instar stadium, when the body weight reaches a threshold size, the CA ceases to synthesize JH (Nijhout and Williams 1974; Truman et al. 2006). The threshold size is determined by the critical weight for pupal metamorphosis. JH is a critical factor in regulating the onset of metamorphosis in the critical weight system. Although the molecular mechanism underlying critical weight has long been unclear, a candidate molecule of this process has recently been identified: myoglianin (Myo) (He et al. 2020).

Myo belongs to the transforming growth factor-ß (TGF-ß) superfamily, which consists of evolutionarily conserved cytokines. In *G. bimaculatus*, RNAi of *Myo* activated the CA and suppressed adult metamorphosis, indicating that Myo is a suppressor of CA activity (Ishimaru et al. 2016). A similar role of Myo in CA regulation has been demonstrated in the German cockroach, *Blattella germanica* (Kamsoi and Belles 2019). In the larvae of the red flour beetle, *Tribolium catstaneum*, RNAi targeting either *myo* or its receptor gene, *baboon*, inhibited pupal metamorphosis; the larvae continued to grow until they eventually died without pupation (Chafino et al. 2023). Thus, the inhibitory function of Myo on CA activity is thought to be conserved in hemimetabolous and holometabolous insects.

In *G. bimaculatus*, *myo* and *baboon* are also expressed in the CA, suggesting that Myo signaling is transmitted in an autocrine manner. This observation is consistent with the general principle that members of the TGF-ß superfamily often act via autocrine mechanisms. However, in *D. melanogaster* and *M. sexta*, Myo appears to act as an endocrine factor (Upadhyay et al. 2020; He et al. 2020). In *M. sexta*, *myo* expression in the muscles of penultimate instar larvae increases significantly once they reach the threshold size, which is the size ready to enter the last larval ecdysis (He et al. 2020). Because skeletal muscle mass closely correlates with total body mass, muscles could be key sensors of body size if Myo indeed represents a "body size mediating molecule" for the regulation of CA activity. Taken together, these findings suggest that skeletal muscle may act as an additional endocrine source of Myo to regulate CA. The function of Myo in *B. mori* has not yet been analyzed, but confirming its function in more insect species, including *B. mori*, will establish a new molecular mechanism of metamorphosis.

6.8 Conclusions and Perspectives

Our series of comprehensive studies on the synthetic regulation of JH in *B. mori* has opened the door to revealing the endocrine mechanisms of regulation of insect molting and metamorphosis. In other species, 20E and peptide hormones control JH synthetic activity by regulating the transcription of JH synthetic enzyme genes (Hult et al. 2015; Shi et al. 2019; Kamruzzaman et al. 2020; Sun et al. 2022; Leyria et al. 2023; Yu et al. 2023). In addition, several transcription factors have been identified as regulators of the transcription of enzyme genes involved in JH synthesis in CA cells, such as *ventral veins lacking*, *POU-M2*, *pointed*, and *Dead ringer* (Cheng et al. 2014; Sarwar et al. 2020; Cai et al. 2022; Li et al. 2022; Kayukawa et al. 2024).

Nutritional status is also important in the development of insects and is involved in the regulation of JH synthesis by CA. In *M. sexta*, starvation in the early fifth larval stage induces the activation of CA and suppression of JH esterase activity (Cymborowski et al. 1982; Lee and Horodyski 2006). Starvation also affects the transcription of JH-synthetic enzyme genes in the fourth instar larvae of *B. mori* (Taguchi et al. 2017), indicating that nutritional signals affect the transcription of JH-synthetic enzyme genes. Additionally, when the enzyme PI3 kinase, which is involved in insulin signaling, was inhibited by incubation with the PI3 kinase inhibitor wortmannin, the synthesis of JH increased up to 2.64-fold in the CA of day 0 fifth instar larvae of *B. mori* (Kaneko and Hiruma, unpublished). These results indicate that CA is stimulated by a decrease in insulin signaling under starvation and that insulin signaling may inhibit the transcriptional activity of JH synthetic enzyme genes. The insulin and target of rapamycin (TOR) pathways are also involved in regulating JH synthesis in several insect species (Tu et al. 2005; Mutti et al. 2011; Pérez-Hedo et al. 2013; Abrisqueta et al. 2014; Fu et al. 2016; Yin et al. 2018). However, the opposite effect of starvation on JH synthesis has been reported in other species. For instance, in *A. aegypti* and *B. germanica*, JH synthetic activity decreases with starvation in adults (Perez-Hedo et al. 2014; Maestro et al. 2009). In both species, treatment with LY294002 or rapamycin, a PI3 kinase and TOR pathway inhibitor, respectively, induces a decrease in the mRNA levels of JH synthetic enzyme genes (Pérez-Hedo et al. 2013; Zhu et al. 2020), indicating that the nutritional signal stimulates JH synthesis.

As described in this chapter, peptide hormones and 20E are transported from other tissues to the CA to regulate JH synthesis. In addition to endocrine systems, paracrine or autocrine factors have also been identified as CA regulatory factors. A molecule is considered to act on CA cells in a paracrine/autocrine manner if it is expressed in the same cells and they respond to it. For instance, decapentaplegic, belonging to the TGF-ß superfamily, is produced in CA cells and stimulates the expression of *JHAMT*, which is thought to act in an autocrine manner (Huang et al. 2011; Ishimaru et al. 2016). Spitz and Vein in the epidermal growth factor (EGF) family also act similarly; they and their receptors are expressed in the CA cells of *B. germanica* (Li et al. 2022). Since EGF and TGF are part of a very large family, there may be other unidentified factors involved in JH synthesis.

In the past, it was difficult to analyze transcription in CA due to its small size and the small amount of total RNA. However, recent high-throughput sequencing has made it possible to analyze a small amount of RNA extracted from CA to identify other factors regulating CA, especially paracrine/autocrine factors. Analyses using omics technologies, including high-throughput sequencing, may provide more detailed information on the coordination of JH synthesis.

References

Abrisqueta M, Süren-Castillo S, Maestro JL (2014) Insulin receptor-mediated nutritional signalling regulates juvenile hormone biosynthesis and vitellogenin production in the German cockroach. Insect Biochem Mol Biol 49:14–23. https://doi.org/10.1016/j.ibmb.2014.03.005

Areiza M, Nouzova M, Rivera-Perez C, Noriega FG (2014) Ecdysis triggering hormone ensures proper timing of juvenile hormone biosynthesis in pharate adult mosquitoes. Insect Biochem Mol Biol 54:98–105. https://doi.org/10.1016/j.ibmb.2014.09.006

Areiza M, Nouzova M, Rivera-Perez C, Noriega FG (2015) 20-Hydroxyecdysone stimulation of juvenile hormone biosynthesis by the mosquito corpora allata. Insect Biochem Mol Biol 64:100–105. https://doi.org/10.1016/j.ibmb.2015.08.001

Bendena WG, Hui JHL, Chin-Sang I, Tobe SS (2020) Neuropeptide and microRNA regulators of juvenile hormone production. Gen Comp Endocrinol 295:113507. https://doi.org/10.1016/j.ygcen.2020.113507

Bilen J, Atallah J, Azanchi R, Levine JD, Riddiford LM (2013) Regulation of onset of female mating and sex pheromone production by juvenile hormone in *Drosophila melanogaster*. Proc Natl Acad Sci USA 110:18321–18326. https://doi.org/10.1073/pnas.1318119110

Cai R, Tao G, Zhao P, Xia Q, He H, Wang Y (2022) POU-M2 promotes juvenile hormone biosynthesis by directly activating the transcription of juvenile hormone synthetic enzyme genes in *Bombyx mori*. Open Biol 12:220031. https://doi.org/10.1098/rsob.220031

Chafino S, alvia R, Cruz J, Martín D, Franch-Marro X (2023) TGFß/activin-dependent activation of Torso controls the timing of the metamorphic transition in the red flour beetle *Tribolium castaneum*. PLoS Genet 19:e1010897. https://doi.org/10.1371/journal.pgen.1010897

Cheng C, Ko A, Chaieb L, Koyama T, Sarwar P, Mirth CK et al (2014) The POU factor ventral veins lacking/Drifter directs the timing of metamorphosis through ecdysteroid and juvenile hormone signaling. PLoS Genet 10:e1004425. https://doi.org/10.1371/journal.pgen.1004425

Cymborowski B, Bogus M, Beckage NE, Williams CM, Riddiford LM (1982) Juvenile hormone titres and metabolism during starvation-induced supernumerary larval moulting of the tobacco hornworm, *Manduca sexta* (L.). J Insect Physiol 28:129–135. https://doi.org/10.1016/0022-1910(82)90120-2

Daubnerová I, Roller L, Satake H, Zhang C, Kim YJ, Žitňan D (2021) Identification and function of ETH receptor networks in the silkworm *Bombyx mori*. Sci Rep 11:11693. https://doi.org/10.1038/s41598-021-91022-8

Fain MJ, Riddiford LM (1975) Juvenile hormone titers in the hemolymph during late larval development of the tobacco hornworm, *Manduca sexta* (L.). Biol Bull 149:506–521. https://doi.org/10.2307/1540383

Fu KY, Zhu TT, Guo WC, Ahmat T, Li GQ (2016) Knockdown of a putative insulin-like peptide gene *LdILP2* in *Leptinotarsa decemlineata* by RNA interference impairs pupation and adult emergence. Gene 581:170–177. https://doi.org/10.1016/j.gene.2016.01.037

Futahashi R, Fujiwara H (2008) Juvenile hormone regulates butterfly larval pattern switches. Science 319:1061. https://doi.org/10.1126/science.1149786

Gilbert LI, Granger NA, Roe RM (2000) The juvenile hormones: historical facts and speculations on future research directions. Insect Biochem Mol Biol 30:617–644. https://doi.org/10.1016/S0965-1748(00)00034-5

Goodman WG, Cusson M (2012) The juvenile hormones. In: Gilbert LI (ed) Insect endocrinology. Elsevier, Amsterdam, pp 310–365. https://doi.org/10.1016/B978-0-12-384749-2.10008-1

Goodman WG, Granger NA (2005) The juvenile hormones. In: Gilbert LI, Iatrou K, Gill SS (eds) Comprehensive molecular insect science, Endocrinology, vol 3. Elsevier, Oxford, pp 319–408. https://doi.org/10.1016/B0-44-451924-6/00039-9

Granger NA, Sturgis SL, Ebersohl R, Geng CX, Sparks TC (1996) Dopaminergic control of corpora allata activity in the larval tobacco hornworm *Manduca sexta*. Arch Insect Biochem Physiol 32:449–466. https://doi.org/10.1002/(SICI)1520-6327(1996)32:3/4<449::AID-ARCH17>3.0.CO;2-C

Gu SH, Chow YS (1996) Regulation of juvenile hormone biosynthesis by ecdysteroid levels during the early stages of the last two larval instars of *Bombyx mori*. J Insect Physiol 42:625–632. https://doi.org/10.1016/0022-1910(96)00025-X

Hartfelder K, Emlen DJ (2012) Endocrine control of insect polyphenism. In: Gilbert LI (ed) Insect endocrinology. Elsevier, Amsterdam, pp 464–522. https://doi.org/10.1016/B978-0-12-384749-2.10011-1

He LL, Shin SH, Wang Z, Yuan I, Weschler R, Chiou A et al (2020) Mechanism of threshold size assessment: metamorphosis is triggered by the TGF-beta/Activin ligand Myoglianin. Insect Biochem Mol Biol 126:103452. https://doi.org/10.1016/j.ibmb.2020.103452

Hernández-Martínez S, Rivera-Perez C, Nouzova M, Noriega FG (2015) Coordinated changes in JH biosynthesis and JH hemolymph titers in *Aedes aegypti* mosquitoes. J Insect Physiol 72:22–27. https://doi.org/10.1016/j.jinsphys.2014.11.003

Hiruma K (1980) Possible roles of juvenile hormone in the prepupal stage of *Mamestra brassicae*. Gen Comp Endocrinol 41:392–399. https://doi.org/10.1016/0016-6480(80)90084-2

Hiruma K (2003) Juvenile hormone action in insect development. In: Henry H, Norman AW (eds) Encyclopedia of hormones, vol 2. Academic Press, San Diego, pp 528–535. https://doi.org/10.1016/B0-12-341103-3/00184-4

Hiruma K, Kaneko Y (2013) Hormonal regulation of insect metamorphosis with special reference to juvenile hormone biosynthesis. Curr Top Dev Biol 103:73–100. https://doi.org/10.1016/B978-0-12-385979-2.00003-4

Hiruma K, Matsumoto S, Isogai A, Suzuki A (1984) Control of ommochrome synthesis by both juvenile hormone and melanization hormone in the cabbage armyworm, *Mamestra brassicae*. J Comp Physiol B 154:13–21. https://doi.org/10.1007/BF00683211

Hiruma K, Riddiford LM, Hopkins TL, Morgan TD (1985) Roles of dopa decarboxylase and phenoloxidase in the melanization of the tobacco hornworm and their control by 20-hydroxyecdysone. J Comp Physiol 155B:659–669. https://doi.org/10.1007/BF00694579

Hsiao TH, Hsiao C (1977) Simultaneous determination of molting and juvenile hormone titers of the greater wax moth. J Insect Physiol 23:89–93. https://doi.org/10.1016/0022-1910(77)90113-5

Huang J, Marchal E, Hult EF, Tobe SS (2015) Characterization of the juvenile hormone pathway in the viviparous cockroach, *Diploptera punctata*. PLoS One 10:e0117291. https://doi.org/10.1371/journal.pone.0117291

Huang J, Tian L, Peng C, Abdou M, Wen D, Wang Y et al (2011) DPP-mediated TGFβ signaling regulates juvenile hormone biosynthesis by activating the expression of juvenile hormone acid methyltransferase. Development 138:2283–2291. https://doi.org/10.1242/dev.057687

Hult EF, Huang J, Marchal E, Lam J, Tobe SS (2015) RXR/USP and EcR are critical for the regulation of reproduction and the control of JH biosynthesis in *Diploptera punctata*. J Insect Physiol 80:48–60. https://doi.org/10.1016/j.jinsphys.2015.04.006

Ishimaru Y, Tomonari S, Matsuoka Y, Watanabe T, Miyawaki K, Bando T et al (2016) TGF-β signaling in insects regulates metamorphosis via juvenile hormone biosynthesis. Proc Natl Acad Sci USA 113:5634–5639. https://doi.org/10.1073/pnas.1600612113

Kamimura M, Shimura S, Kiuchi M (2003) Simple manipulation of silkworm molting by an artificial diet cotaining plant-derived 20-hydroxyecdysone. J Insect Biothech Sericol 72:197–201. https://doi.org/10.11416/jibs.72.197

Kamruzzaman ASM, Mikani A, Mohamed AA, Elgendy AM, Takeda M (2020) Crosstalk among indoleamines, neuropeptides and JH/20E in regulation of reproduction in the American cockroach, *Periplaneta americana*. Insects 11:155. https://doi.org/10.3390/insects11030155

Kamsoi Q, Belles X (2019) Myoglianin triggers the premetamorphosis stage in hemimetabolan insects. FASEB J 33:3659–3669. https://doi.org/10.1096/fj.201801511R

Kaneko Y, Hiruma K (2011) The mechanisms of stage specific regulation of JH biosynthsis. Hikakunaibunpitsugaku 37:204–211. https://doi.org/10.5983/nl2008jsce.37.204. (in Japanese)

Kaneko Y, Hiruma K (2014) Short neuropeptide F (sNPF) is a stage-specific suppressor for juvenile hormone biosynthesis by corpora allata, and a critical factor for the initiation of insect metamorphosis. Dev Biol 393:312–319. https://doi.org/10.1016/j.ydbio.2014.07.014

Kaneko Y, Hiruma K (2015) Allatotropin inhibits juvenile hormone biosynthesis by the corpora allata of adult *Bombyx mori*. J Insect Physiol 80:15–21. https://doi.org/10.1016/j.jinsphys.2015.01.005

Kaneko Y, Hiruma K (2016) The mechanisms of regulation of JH biosynthesis. Sanshi-Konchu Biotec 85:117–125. https://doi.org/10.11416/konchubiotec.85.3_117. (in Japanese)

Kaneko Y, Kinjoh T, Kiuchi M, Hiruma K (2011) Stage-specific regulation of juvenile hormone biosynthesis by ecdysteroid in *Bombyx mori*. Mol Cell Endocrinol 335:204–210. https://doi.org/10.1016/j.mce.2011.01.012

Kaneko Y, Takaki K, Iwami M, Sakurai S (2006) Developmental profile of annexin IX and its possible role in programmed cell death of the *Bombyx mori* anterior silk gland. Zool Sci 23:533–542. https://doi.org/10.2108/zsj.23.533

Kataoka H, Toschi A, Li JP, Carney RL, Schooley DA, Kramer SJ (1989) Identification of an allatotropin from adult *Manduca sexta*. Science 243:1481–1483. https://doi.org/10.1126/science.243.4897.1481

Kayukawa T, Nagamine K, Inui T, Yokoi K, Kobayashi I, Nakao H et al (2024) *Dead ringer* acts as a major regulator of juvenile hormone biosynthesis in insects. PNAS Nexus 3:pgae435. https://doi.org/10.1093/pnasnexus/pgae435

Kiguchi K, Riddiford LM (1978) A role of juvenile hormone in pupal development of the tobacco hornworm, *Manduca sexta*. J Insect Physiol 24:673–680. https://doi.org/10.1016/0022-1910(78)90063-X

Kinjoh T, Kaneko Y, Itoyama K, Mita K, Hiruma K, Shinoda T (2007) Control of juvenile hormone biosynthesis in *Bombyx mori*: cloning of the enzymes in the mevalonate pathway and assessment of their developmental expression in the corpora allata. Insect Biochem Mol Biol 37:808–818. https://doi.org/10.1016/j.ibmb.2007.03.008

Koyama T, Obara Y, Iwami M, Sakurai S (2004) Commencement of pupal commitment in late penultimate instar and its hormonal control in wing imaginal discs of the silkworm, *Bombyx mori*. J Insect Physiol 50:123–133. https://doi.org/10.1016/j.jinsphys.2003.09.009

Lee KY, Horodyski FM (2006) Effects of starvation and mating on corpora allata activity and allatotropin (Manse-AT) gene expression in *Manduca sexta*. Peptides 27:567–574. https://doi.org/10.1016/j.peptides.2005.08.024

Leyria J (2024) Endocrine factors modulating vitellogenesis and oogenesis in insects: an update. Mol Cell Endocrinol 587:112211. https://doi.org/10.1016/j.mce.2024.112211

Leyria J, Benrabaa S, Nouzova M, Noriega FG, Tose LV, Fernandez-Lima F et al (2023) Crosstalk between nutrition, insulin, juvenile hormone, and ecdysteroid signaling in the classical insect model, *Rhodnius prolixus*. Int J Mol Sci 24(7). https://doi.org/10.3390/ijms24010007

Li Y, Unnithan GC, Veenstra JA, Feyereisen R, Noriega FG (2003) Stimulation of JH biosynthesis by the corpora allata of adult female *Aedes aegypti in vitro*: effect of farnesoic acid and *Aedes* allatotropin. J Exp Biol 206:1825–1832. https://doi.org/10.1242/jeb.00371

Li Z, Zhou C, Chen Y, Ma W, Cheng Y, Chen J et al (2022) Egfr signaling promotes juvenile hormone biosynthesis in the German cockroach. BMC Biol 20:278. https://doi.org/10.1186/s12915-022-01484-z

Liu S, Li K, Gao Y, Liu X, Chen W, Ge W et al (2018) Antagonistic actions of juvenile hormone and 20-hydroxyecdysone within the ring gland determine developmental transitions in *Drosophila*. Proc Natl Acad Sci USA 115:139–144. https://doi.org/10.1073/pnas.1716897115

Lorenz MW, Kellner R, Hoffmann KH (1995) A family of neuropeptides that inhibit juvenile hormone biosynthesis in the cricket, *Gryllus bimaculatus*. J Biol Chem 270:21103–21108. https://doi.org/10.1074/jbc.270.36.21103

Maestro JL, Cobo J, Bellés X (2009) Target of rapamycin (TOR) mediates the transduction of nutritional signals into juvenile hormone production. J Biol Chem 284:5506–5513. https://doi.org/10.1074/jbc.M807042200

Marchal E, Hult EF, Huang J, Stay B, Tobe SS (2013) *Diploptera punctata* as a model for studying the endocrinology of arthropod reproduction and development. Gen Comp Endocrinol 188:85–93. https://doi.org/10.1016/j.ygcen.2013.04.018

Meiselman M, Lee SS, Tran RT, Dai H, Ding Y, Rivera-Perez C et al (2017) Endocrine network essential for reproductive success in *Drosophila melanogaster*. Proc Natl Acad Sci USA 114:E3849–E3858. https://doi.org/10.1073/pnas.1620760114

Mizoguchi A, Kamimura M, Kiuchi M, Kataoka H (2015) Positive feedback regulation of prothoracicotropic hormone secretion by ecdysteroid—a mechanism that determines the timing of metamorphosis. Insect Biochem Mol Biol 58:39–45. https://doi.org/10.1016/j.ibmb.2015.01.001

Muramatsu D, Kinjoh T, Shinoda T, Hiruma K (2008) The role of 20-hydroxyecdysone and juvenile hormone in pupal commitment of the epidermis of the silkworm, *Bombyx mori*. Mech Dev 125:411–420. https://doi.org/10.1016/j.mod.2008.02.001

Mutti NS, Dolezal AG, Wolschin F, Mutti JS, Gill KS, Amdam GV (2011) IRS and TOR nutrient-signaling pathways act via juvenile hormone to influence honey bee caste fate. J Exp Biol 214:3977–3984. https://doi.org/10.1242/jeb.061499

Niimi S, Sakurai S (1997) Development changes in juvenile hormone and juvenile hormone acid titers in the hemolymph and *in vitro* juvenile hormone synthesis by corpora allata of the silkworm, *Bombyx mori*. J Insect Physiol 43:875–884. https://doi.org/10.1016/S0022-1910(97)00021-8

Nijhout HF (1994) Insect hormones. Princeton University Press, Princeton, NJ

Nijhout HF (1999) Control mechanisms of polyphenic development in insects: in polyphenic development, environmental factors alter some aspects of development in an orderly and predictable way. Bioscience 49:181–192. https://doi.org/10.2307/1313508

Nijhout HF, Williams CM (1974) Control of moulting and metamorphosis in the tobacco hornworm, *Manduca sexta* (L.): cessation of juvenile hormone secretion as a trigger for pupation. J Exp Biol 61:493–501. https://doi.org/10.1242/jeb.61.2.493

Nouzova M, Edwards MJ, Mayoral JG, Noriega FG (2011) A coordinated expression of biosynthetic enzymes controls the flux of juvenile hormone precursors in the corpora allata of mosquitoes. Insect Biochem Mol Biol 41:660–669. https://doi.org/10.1016/j.ibmb.2011.04.008

Nouzova M, Rivera-Perez C, Noriega FG (2015) Allatostatin-C reversibly blocks the transport of citrate out of the mitochondria and inhibits juvenile hormone synthesis in mosquitoes. Insect Biochem Mol Biol 57:20–26. https://doi.org/10.1016/j.ibmb.2014.12.003

Oguchi K, Miura T (2024) Body part-specific development in termite caste differentiation: crosstalk between hormonal actions and developmental toolkit genes. Curr Opin Insect Sci 63:101183. https://doi.org/10.1016/j.cois.2024.101183

Óhashi M, Tsusué M, Kiguchi K (1983) Juvenile hormone control of larval colouration in the silkworm, *Bombyx mori*: characterization and determination of epidermal brown colour induced by the hormone. Insect Biochem 13:123–127. https://doi.org/10.1016/0020-1790(83)90074-4

Okuda M, Sakurai S, Ohtaki T (1985) Activity of the prothoracic gland and its sensitivity to prothoracicotropic hormone in the penultimate and last-larval instar of *Bombyx mori*. J Insect Physiol 31:455–461. https://doi.org/10.1016/0022-1910(85)90092-7

Orchard I, Lange AB (2024) The neuroendocrine and endocrine systems in insect—historical perspective and overview. Mol Cell Endocrinol 580:112108. https://doi.org/10.1016/j.mce.2023.112108

Park C, Hwang JS, Kang SW, Lee BH (2002) Molecular characterization of a cDNA from the silk moth *Bombyx mori* encoding *Manduca sexta* allatotropin peptide. Zool Sci 19:287–292. https://doi.org/10.2108/zsj.19.287

Pérez-Hedo M, Rivera-Perez C, Noriega FG (2013) The insulin/TOR signal transduction pathway is involved in the nutritional regulation of juvenile hormone synthesis in *Aedes aegypti*. Insect Biochem Mol Biol 43:495–500. https://doi.org/10.1016/j.ibmb.2013.03.008

Perez-Hedo M, Rivera-Perez C, Noriega FG (2014) Starvation increases insulin sensitivity and reduces juvenile hormone synthesis in mosquitoes. PLoS One 9:e86183. https://doi.org/10.1371/journal.pone.0086183

Reynolds SE, Brown AM, Seth RK, Riddiford LM, Hiruma K (2009) Induction of supernumerary larval moulting in the tobacco hornworm *Manduca sexta*: interaction of bisacylhydrazine ecdysteroid agonists with endogenous juvenile hormone. Physiol Entomol 34:30–38. https://doi.org/10.1111/j.1365-3032.2008.00648.x

Riddiford LM (1976) Hormonal control of insect epidermal cell commitment *in vitro*. Nature 259:115–117. https://doi.org/10.1038/259115a0

Riddiford LM (1985) Hormone action at the cellular level. In: Kerkut GA, Gilbert LI (eds) Comprehensive insect physiology, biochemistry and pharmacology, Endocrinology II, vol 8. Pergamon Press, Oxford, pp 37–84. https://doi.org/10.1016/B978-0-08-030809-8.50008-3

Riddiford LM (1996) Juvenile hormone: the status of its "status quo" action. Arch Insect Biochem Physiol 32:271–286. https://doi.org/10.1002/(SICI)1520-6327(1996)32:3/4<271::AID-ARCH2>3.0.CO;2-W

Sakurai S (1983) Temporal organization of endocrine events underlying larval-larval ecdysis in the silkworm, *Bombyx mori*. J Insect Physiol 29:919–932. https://doi.org/10.1016/0022-1910(83)90055-0

Sakurai S (1984) Temporal organization of endocrine events underlying larval-pupal metamorphosis in the silkworm, *Bombyx mori*. J Insect Physiol 30:657–664. https://doi.org/10.1016/0022-1910(84)90051-9

Sakurai S, Kaya M, Satake S (1998) Hemolymph ecdysteroid titer and ecdysteroid-dependent developmental events in the last-larval stadium of the silkworm, *Bombyx mori*: role of low ecdysteroid titer in larval-pupal metamorphosis and a reappraisal of the head critical period. J Insect Physiol 44:867–881. https://doi.org/10.1016/S0022-1910(98)00075-4

Sarwar PF, McDonald IR, Wang VR, Suzuki Y (2020) The POU factor Ventral veins lacking regulates ecdysone and juvenile hormone biosynthesis during development and reproduction of the milkweed bug, *Oncopeltus fasciatus*. Dev Biol 459:181–193. https://doi.org/10.1016/j.ydbio.2019.12.001

Shi Y, Liu TY, Jiang HB, Liu XQ, Dou W, Park Y et al (2019) The ecdysis triggering hormone system, via ETH/ETHR-B, is essential for successful reproduction of a major pest insect, *Bactrocera dorsalis* (Hendel). Front Physiol 10:151. https://doi.org/10.3389/fphys.2019.00151

Stay B (2000) A review of the role of neurosecretion in the control of juvenile hormone synthesis: a tribute to Berta Scharrer. Insect Biochem Mol Biol 30:653–662. https://doi.org/10.1016/S0965-1748(00)00036-9

Stay B, Tobe SS (2007) The role of allatostatins in juvenile hormone synthesis in insects and crustaceans. Annu Rev Entomol 52:277–299. https://doi.org/10.1146/annurev.ento.51.110104.151050

Sun Y, Fu D, Liu B, Wang L, Chen H (2022) Functional characterization of allatostatin C (PISCF/AST) and juvenile hormone acid O-methyltransferase in *Dendroctonus armandi*. Int J Mol Sci 23:2749. https://doi.org/10.3390/ijms23052749

Sutherland TD, Feyereisen R (1996) Target of cockroach allatostatin in the pathway of juvenile hormone biosynthesis. Mol Cell Endocrinol 120:115–123. https://doi.org/10.1016/0303-7207(96)03825-7

Taguchi S, Iwami M, Kiya T (2017) Identification and characterization of novel genes expressed preferentially in the corpora allata or corpora cardiaca during the juvenile hormone synthetic period in the silkworm, *Bombyx mori*. Zool Sci 34:398–405. https://doi.org/10.2108/zs170069

Takaki K, Sakurai S (2003) Regulation of prothoracic gland ecdysteroidogenic activity leading to pupal metamorphosis. Insect Biochem Mol Biol 33:1189–1199. https://doi.org/10.1016/j.ibmb.2003.06.006

Tobe SS, Stay B (1985) Structure and regulation of the corpus allatum. Adv Insect Physiol 18:305–432. https://doi.org/10.1016/S0065-2806(08)60043-0

Truman JW, Riddiford LM, Safranek L (1973) Hormonal control of cuticle coloration in the tobacco hornworm, *Manduca sexta*: basis of an ultrasensitive bioassay for juvenile hormone. J Insect Physiol 19:195–203. https://doi.org/10.1016/0022-1910(73)90232-1

Truman JW, Hiruma K, Allee JP, Macwhinnie SG, Champlin DT, Riddiford LM (2006) Juvenile hormone is required to couple imaginal disc formation with nutrition in insects. Science 312:1385–1388. https://doi.org/10.1126/science.1123652

Tu MP, Yin CM, Tatar M (2005) Mutations in insulin signaling pathway alter juvenile hormone synthesis in *Drosophila melanogaster*. Gen Comp Endocrinol 142:347–356. https://doi.org/10.1016/j.ygcen.2005.02.009

Upadhyay A, Peterson AJ, Kim MJ, O'Connor MB (2020) Muscle-derived Myoglianin regulates *Drosophila* imaginal disc growth. eLife 9:e51710. https://doi.org/10.7554/eLife.51710

Verlinden H, Gijbels M, Lismont E, Lenaerts C, Vanden Broeck J, Marchal E (2015) The pleiotropic allatoregulatory neuropeptides and their receptors: a mini-review. J Insect Physiol 80:2–14. https://doi.org/10.1016/j.jinsphys.2015.04.004

Weaver RJ, Audsley N (2009) Neuropeptide regulators of juvenile hormone synthesis: structures, functions, distribution, and unanswered questions. Ann N Y Acad Sci 1163:316–329. https://doi.org/10.1111/j.1749-6632.2009.04459.x

Whisenton LR, Bowen MF, Granger NA, Gilbert LI, Bollenbacher WE (1985) Brain-mediated 20-hydroxyecdysone regulation of juvenile hormone synthesis by the corpora allata of the tobacco hornworm, *Manduca sexta*. Gen Comp Endcrinol 58:311–318. https://doi.org/10.1016/0016-6480(85)90347-8

Whisenton LR, Watson RD, Granger NA, Bollenbacher WE (1987) Regulation of juvenile hormone biosynthesis by 20-hydroxyecdysone during the fourth larval instar of the tobacco hornworm, *Manduca sexta*. Gen Comp Endocrinol 66:62–70. https://doi.org/10.1016/0016-6480(87)90350-9

Wijesekera TP, Saurabh S, Dauwalder B (2016) Juvenile hormone is required in adult males for *Drosophila* courtship. PLoS One 11:e0151912. https://doi.org/10.1371/journal.pone.0151912

Williams CM (1961) The juvenile hormone II. Its role in the endocrine control of moulting, pupation and adult development in the Cecropia silkworm. Biol Bull 121:572–585. https://doi.org/10.2307/1539456

Yamanaka N, Hua YJ, Mizoguchi A, Watanabe K, Niwa R, Tanaka Y et al (2005) Identification of a novel prothoracicostatic hormone and its receptor in the silkworm *Bombyx mori*. J Biol Chem 280:14684–14690. https://doi.org/10.1074/jbc.M500308200

Yamanaka N, Yamamoto S, Žitňan D, Watanabe K, Kawada T, Satake H et al (2008) Neuropeptide receptor transcriptome reveals unidentified neuroendocrine pathways. PLoS One 3:e3048. https://doi.org/10.1371/journal.pone.0003048

Yin ZJ, Dong XL, Kang K, Chen H, Dai XY, Wu GA et al (2018) FoxO transcription factor regulate hormone mediated signaling on nymphal diapause. Front Physiol 9:1654. https://doi.org/10.3389/fphys.2018.01654

Yin Y, Qiu YW, Huang J, Tobe SS, Chen SS, Kai ZP (2020) Enzymes in the juvenile hormone biosynthetic pathway can be potential targets for pest control. Pest Manag Sci 76:1071–1077. https://doi.org/10.1002/ps.5617

Yu Z, Shi J, Jiang X, Song Y, Du J, Zhao Z (2023) Neuropeptide F regulates feeding via the juvenile hormone pathway in *Ostrinia furnacalis* larvae. Pest Manag Sci 79:1193–1203. https://doi.org/10.1002/ps.7289

Zhang X, Li S, Liu S (2022) Juvenile hormone studies in *Drosophila melanogaster*. Front Physiol 12:785320. https://doi.org/10.3389/fphys.2021.785320

Zhu S, Liu F, Zeng H, Li N, Ren C, Su Y et al (2020) Insulin/IGF signaling and TORC1 promote vitellogenesis via inducing juvenile hormone biosynthesis in the American cockroach. Development 147:dev188805. https://doi.org/10.1242/dev.188805

Žitňan D, Adams ME (2005) Neuroendocrine regulation of insect ecdysis. In: Gilbert LI, Iatrou K, Gill SS (eds) Comprehensive molecular insect science, Endocrinology, vol 3. Elsevier, Oxford, pp 1–60. https://doi.org/10.1016/B0-44-451924-6/00032-6

Žitňan D, Adams ME (2012) Neuroendocrine regulation of ecdysis. In: Gilbert LI (ed) Insect endocrinology. Elsevier, Amsterdam, pp 253–309. https://doi.org/10.1016/B978-0-12-384749-2.10007-X

Chapter 7
Juvenile Hormone Biosynthesis and Its Regulation in *Drosophila melanogaster*

Daiki Fujinaga and Naoki Yamanaka

Abstract While early studies on juvenile hormone (JH) functions during metamorphosis were mainly conducted using classical models of insect physiology such as the kissing bug *Rhodnius prolixus* and tobacco hornworm *Manduca sexta*, the fruit fly *Drosophila melanogaster*, has also made unique contributions to JH research, as exemplified by the isolation of the first *Methoprene-tolerant* (*Met*) mutant and the identification of *Krüppel homolog 1* (*Kr-h1*) as a major mediator of JH action. After a key enzyme in JH biosynthesis was cloned and Met was characterized as the long-sought JH receptor in other insect species, the powerful molecular genetic tools in *D. melanogaster* have greatly advanced our understanding of JH signaling and its regulation. In contrast, some unique aspects of JH signaling in higher dipteran insects, such as the lack of precocious metamorphosis in JH-deficient conditions and the loss of a conserved epoxidase enzyme, pose a significant challenge for JH research in *D. melanogaster*. Consequently, many JH biosynthetic enzymes remain to be characterized in flies, which has also limited our understanding of how JH biosynthesis is regulated at the molecular level. In this chapter, we aim to summarize our current understanding of JH signaling in *D. melanogaster* and propose future research directions.

Keywords Juvenile hormone · Sesquiterpenoids · Mevalonate pathway · *Drosophila melanogaster*

D. Fujinaga · N. Yamanaka (✉)
Department of Entomology, Institute for Integrative Genome Biology, University of California, Riverside, Riverside, CA, USA
e-mail: naoki.yamanaka@ucr.edu

T. Shinoda et al. (eds.), *Juvenile Hormone*, Entomology Monographs,
https://doi.org/10.1007/978-981-95-8605-9_7

7.1 JH Production in *Drosophila*

7.1.1 Corpus Allatum

Similar to other insects, JH is synthesized in a pair of endocrine organs called the corpus allatum (CA, plural: corpora allata) in dipteran species (Jones and Jones 2007). In higher Diptera (Cyclorrhapha), such as the fruit fly *Drosophila melanogaster*, larvae have the CA as a single structure (Scharrer and Hadorn 1938). Adult *D. melanogaster* flies have a fused CA, whereas some other flies, such as the blue bottle fly *Calliphora vomitoria*, have a paired CA structure (Burtt 1937, 1938). Approximately 20 CA cells are located at the center of the ring gland (RG), a composite endocrine organ consisting of the CA, prothoracic gland, and corpora cardiaca in larval Drosophilidae (Scharrer and Hadorn 1938; Fig. 7.1a). CA cells are differentiated during mid-embryogenesis (Tobe and Stay 1985). No additional mitosis occurs in the CA until the pupal stage, when CA cells begin to form the adult CA (Day 1943; Vogt 1943; King et al. 1966a). The RG undergoes migration and drastic morphological changes during metamorphosis (Dai and Gilbert 1991). The larval RG lies above the brain hemispheres until pupation, after which it migrates toward the ventral ganglion and eventually lands on the esophagus (Fig. 7.1b). Meanwhile, the prothoracic gland degenerates and separates from the CA and corpora cardiaca. After adult eclosion, the volume of the CA increases almost twice within 12 h as it functionally matures (King et al. 1966b).

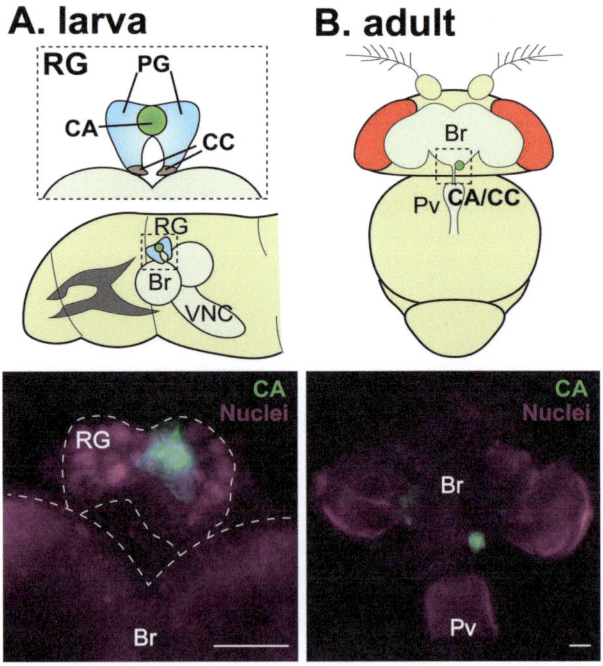

Fig. 7.1 Corpus allatum in *Drosophila melanogaster*. (**a**) The larval corpus allatum (CA) composes the ring gland (RG) along with the prothoracic gland (PG) and corpora cardiaca (CC). The RG is attached to the brain (Br). VNC, ventral nerve cord. (**b**) The adult CA-CC complex is attached to the esophagus between the brain and proventriculus (Pv). The CA was visualized by *jhamt-LexA*-driven expression of membrane-localized GFP. Nuclei were stained with DAPI. Scale bars: 50 μm

In *D. melanogaster*, powerful genetic tools, such as the Gal4-UAS system, have been used to investigate the mechanisms of JH synthesis. *Jhamt-Gal4* and *Aug21-Gal4* drivers are widely used to visualize the CA, as well as to manipulate gene expression levels in the CA by inducing UAS-mediated expression of transgenes or double-stranded RNA (Siegmund and Korge 2001; Wijesekera et al. 2016; Huang et al. 2011). However, a few studies have reported that they are not CA-specific Gal4 drivers, as they are also highly expressed in the salivary glands during larval development (Riddiford et al. 2010). Therefore, another binary expression system is used to manipulate gene expression in the CA (Ohhara et al. 2018; Fujinaga et al. 2026; Fig. 7.1). In addition to the targeted manipulation of gene expression, CA-specific drivers have also been used to ablate the CA by overexpressing apoptosis-inducing genes to investigate developmental functions of the CA (Liu et al. 2009; Riddiford et al. 2010). CA-ablated *Drosophila* larvae do not exhibit precocious metamorphosis, which is commonly observed in CA-ablated lepidopteran and coleopteran insects (Bounhiol 1938; Radtke 1942; Riddiford 2020). Instead, CA-ablated fly larvae develop almost normally until pupariation with a smaller body size, after which most of them die before pupation (Mirth et al. 2014). Since JH receptor mutants show similar lethal phenotypes (Abdou et al. 2011), JH signaling is necessary for the prepupal-pupal transition in *D. melanogaster*. Importantly, a few studies have reported that JH-deficient flies exhibit precocious differentiation of some tissues, such as the fat body and optic lobe, in *D. melanogaster* (Liu et al. 2009; Riddiford et al. 2010). Therefore, the "status quo" function of JH to maintain the larval form seems to be conserved in Diptera (Riddiford 2020). Recent studies have shown that JH is also produced in the gut cells of adults and regulates gut homeostasis in an autocrine manner in *D. melanogaster* (Rahman et al. 2017; Shianiou et al. 2023).

7.1.2 JHs in Drosophila

Three sesquiterpenoids, methyl farnesoate (MF), JH III, and JH III bisepoxide (JHB_3), are produced in the CA of higher dipterans. JH III has been detected in several Drosophilidae, such as *Drosophila hydei* and *D. melanogaster* (Rembold and Llackner 1985), whereas other types of JHs, JH I and JH II, are undetectable in dipteran insects (Rembold and Llackner 1985; Sliter et al. 1987). JHB_3 was initially discovered in *D. melanogaster* and *Musca domestica* as a dominant product of the cultured larval ring gland (Richard et al. 1989a). Some reports have confirmed higher titers of JHB_3 than JH III during larval development (Jones et al. 2013; Wen et al. 2015). In contrast to the larval period, JH III appears to be the dominant JH in adults. JHB_3 (and MF) was undetectable in the hemolymph of adult females in some reports (Reiff et al. 2015; Sugime et al. 2017; Kurogi et al. 2023), whereas other studies detected JHB_3 in adults (Zhang et al. 2021b; Chen et al. 2024). It should be noted that many studies have demonstrated that JHB_3 is predominantly produced in the cultured adult CA (Altaratz et al. 1991; Bendena et al. 2011). Therefore, it is possible that JHB_3 production is inhibited in adult flies in vivo.

The activity of JHs has been analyzed using multiple bioassays (Jones and Jones 2007). Topical application of JHs and their analogs (e.g., methoprene) on white prepupae caused multiple developmental defects, including eclosion failure, reduction in the number of adult bristles, and abnormal rotation of the male genitalia (Ashburner 1970; Richard et al. 1989a). Richard and colleagues demonstrated that the application of 0.5 μg JHB_3 per animal caused developmental abnormalities, which was approximately a ten-fold higher dose of JH III to achieve similar effects (Richard et al. 1989a). This indicates that the hormonal activity of JHB_3 is much lower than that of JH III, although one report suggests higher efficiency of JHB_3 in breaking reproductive dormancy in adult females (Saunders et al. 1990).

A precursor of JH III, MF, is also released into the larval hemolymph (Jones et al. 2013; Wen et al. 2015). Although the application of MF on white prepupae hardly causes any developmental disorders, the application of a high dose of MF on larvae disrupts normal larval development, which is not caused by the application of JH III or JHB_3 (Harshman et al. 2010; Jones et al. 2010a). MF directly activates the JH response element through JH receptors, although its activity is significantly lower than that of JH III (Jindra et al. 2015). Taken together, MF potentially functions as JH, although its physiological function in vivo remains unclear. As MF also interacts with the nuclear receptor Ultraspiracle (USP), the retinoid-X receptor ortholog in insects (Jones et al. 2006, 2013), it is possible that the larval lethality induced by MF application is due to the disruption of normal USP function.

7.1.3 Timing of JH Biosynthesis

Because of their small body size, previous studies mostly analyzed whole-body JH titers in Drosophilidae rather than their titers in the hemolymph (Jones and Jones 2007). For example, Sliter et al. (1987) analyzed whole-body JH III titers from millions of embryos and thousands of larvae, pupae, and adults of *D. melanogaster* using gas chromatography-mass spectrometry (Sliter et al. 1987). The JH III titer is extremely low during embryogenesis, after which it is increased to approximately 20–30 fg/larva through the second and third instars. The titer then forms a peak in wandering larvae (~300 fg/larva), followed by a sudden drop to low levels after pupation (~30 fg/pupa). JH III becomes detectable again in pharate adults (~60 fg/pupa) and reaches its highest level in early adults (~1000 fg/adult). These results are consistent with other studies on *D. hydei* and *D. melanogaster* (Bührlen et al. 1984; Bownes and Rembold 1987). Although detailed developmental changes in JHB_3 titers have not been closely investigated so far, the change in JHB_3 biosynthetic activity of the isolated CA is consistent with the JH III titer during the third instar (Richard et al. 1989b). JHB_3 biosynthetic activity is drastically upregulated in the CA of wandering larvae, followed by a decrease after puparium formation. In contrast, Jones et al. (2013) detected high levels of JHB_3 in the hemolymph of early third instar larvae, which decreased by puparium formation (Jones et al. 2013). This study also detected a large amount of MF in the hemolymph during larval development. In adult flies, the hemolymph JH titer is increased in mated females to induce vitellogenesis (see Chap. 11).

7.2 Enzymes of JH Biosynthesis

7.2.1 The JH Biosynthetic Pathway

JH is synthesized downstream of the mevalonate pathway, which is a widely conserved metabolic pathway in eukaryotes, archaea, and some bacteria (Miziorko 2011; see Chap. 5). Farnesyl pyrophosphate (FPP), synthesized by the mevalonate pathway, is used in multiple cellular processes, such as cholesterol biosynthesis and protein isoprenylation in vertebrates (Göbel et al. 2020). In contrast, invertebrates, including insects, cannot synthesize cholesterol de novo because of the lack of synthetic enzymes required for producing squalene, a precursor of cholesterol biosynthesis (Bellés et al. 2005). In contrast, arthropods convert FPP into sesquiterpenoids such as JH III. These synthetic steps are known as the JH branch.

7.2.2 Early Steps (Mevalonate Pathway)

Enzymes that mediate the mevalonate pathway are well conserved among organisms (Bellés et al. 2005; Miziorko 2011; Dellas et al. 2013). Genes encoding enzymes involved in the mevalonate pathway in *D. melanogaster* are shown in Fig. 7.2. In the first step of the mevalonate pathway, acetoacetyl-coenzyme A (CoA) is synthesized by condensing two acetyl-CoA molecules. Acetyl-CoA acetyltransferase/thiolase 1 (Acat1) and Acat2 catalyze this reaction (Silva et al. 2022). One acetyl-CoA molecule and one acetoacetyl-CoA molecule are then condensed into β-Hydroxy β-methylglutaryl-CoA (HMG-CoA) by HMG-CoA synthase (Hmgs). In the next step, CoA is removed by the NADPH-dependent enzyme HMG-CoA reductase (Hmgcr). Hmgcr is a rate-limiting enzyme in the mevalonate pathway (Göbel et al. 2020). Some in vivo studies have shown that Hmgcr functions upstream of JH synthesis (Jones et al. 2010b; Wen et al. 2015; Santos and Lehmann 2004; Barton et al. 2024), and CA-specific Hmgcr knockdown suppresses JH synthesis (Wen et al. 2015). Mevalonate is then phosphorylated twice into mevalonate-pyrophosphate by two kinases, namely, Mevalonate kinase (Mvk) and Phosphomevalonate kinase (Pmvk). It is further converted into isopentenyl-pyrophosphate by Mevalonate diphosphate decarboxylase (Mvd). In subsequent steps, isopentenyl-pyrophosphate and its isomerized isomer, dimethylallyl-pyrophosphate, are condensed into geranyl-pyrophosphate by Farnesyl

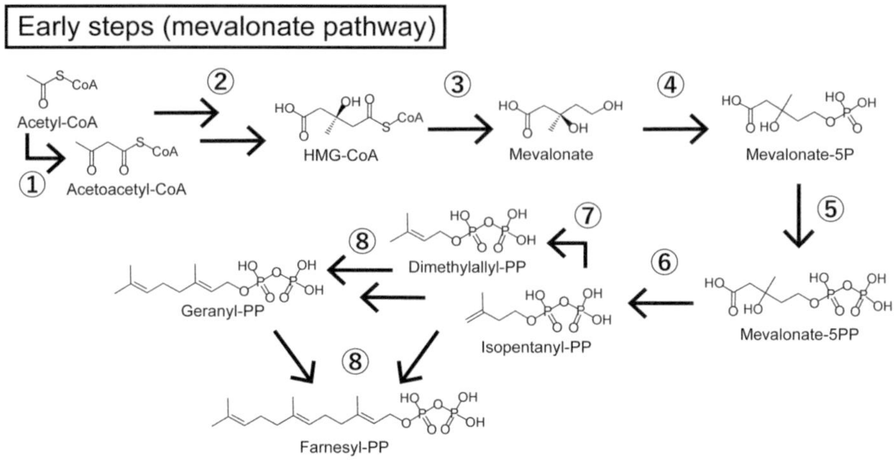

Step	Reaction	Enzymes in *H. sapiens*	Enzymes in *D. melanogaster*
①	Condensation	Acat1, Acat2	Acat1/CG10932, Acat2/CG9149
②	Condensation	Hmgcs1, Hmgcs2	Hmgs/CG4311
③	Reduction	Hmgcr	Hmgcr/CG10367
④	Phosphorylation	Mvk	Mvk/CG33671
⑤	Phosphorylation	Pmvk	Pmvk/CG10268
⑥	Decarboxylation	Mvd	Mvd/CG8239
⑦	Isomerization	Idi1, Idi2	Idi/CG5919
⑧	Condensation	Fdps	Fpps/CG12389

Fig. 7.2 Early steps in the juvenile hormone (JH) biosynthetic pathway (mevalonate pathway). Schematic diagram of the mevalonate pathway. The enzymes catalyzing each step are conserved between humans (*Homo sapiens*) and the fruit fly *Drosophila melanogaster*. Acat, Acetyl-CoA acetyltransferase; Hmgcs (Hmgs), 3-Hydroxy-3-methylglutaryl-CoA synthase; Hmgcr, 3-Hydroxy-3-methylglutaryl-CoA reductase; MvK, Mevalonate kinase; Pmvk, Phosphomevalonate kinase; Mvd, Mevalonate diphosphate decarboxylase; Idi, Isopentenyl-diphosphate delta isomerase; Fdps, Farnesyl diphosphate synthase; Fpps, Farnesyl pyrophosphate synthase

pyrophosphate synthase (Fpps). This enzyme also catalyzes the condensation of geranyl-pyrophosphate and isopentenyl-pyrophosphate into FPP.

7.2.3 Late Steps (JH Branch)

In the JH branch, JH III and JHB_3 are synthesized by insect-specific enzymes (Fig. 7.3). Enzymes in the JH branch have not been fully identified in *D. melanogaster*, although they have been identified in other insects, such as the yellow fever mosquito *Aedes aegypti* and the silkmoth *Bombyx mori*. In the first step, farnesyl pyrophosphate is dephosphorylated by farnesyl pyrophosphate diphosphatase (FPPase). A phosphatase domain-containing enzyme, CG15739, has been demonstrated as an FPPase in *D. melanogaster* using an in vitro conversion assay with a

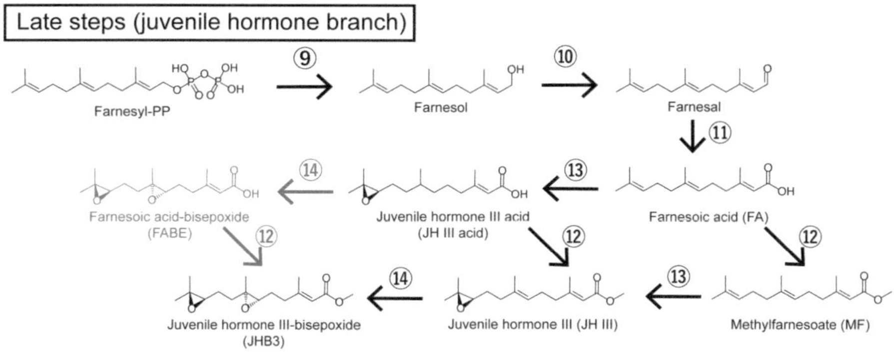

Step	Reaction	Enzymes in other insects	Enzymes in *D. melanogaster*
⑨	Dephosphorylation	FPPase (e.g. *A. aegypti*)	CG15739, other enzymes (?)
⑩	Alchol oxidation	SDR-1 (e.g. *A. aegypti*)	unknown
⑪	Aldehyde oxidation	ALDH3 (e.g. *A. aegypti*)	unknown
⑫	Transmethylation	JHAMT (e.g. *B. mori*), JHAMT-like	JHAMT/CG17330, other enzymes
⑬	10,11-epoxidation	CYP15 (e.g. *D. punctata*)	CYP6G2, CYP6A13
⑭	6,7-epoxidation	CYP6A1 (*M. domestica*)	CYP6G2

Fig. 7.3 Late steps in the juvenile hormone (JH) biosynthetic pathway (juvenile hormone branch). Schematic diagram of the JH branch. Most enzymes catalyzing each step have not been identified in the fruit fly *Drosophila melanogaster*, whereas some enzymes responsible for each step have been identified in other insects such as the yellow fever mosquito *Aedes aegypti*, the silkmoth *Bombyx mori*, the Pacific beetle cockroach *Diploptera punctata*, and the housefly *Musca domestica*. Although FABE has been proposed to be a JHB_3 precursor, it has not been detected in *D. melanogaster*. FPPase, Farnesyl pyrophosphatase; SDR-1, Short-chain dehydrogenase-1; ALDH3, Aldehyde dehydrogenase 3; JHAMT, Juvenile hormone acid *O*-methyltransferase; CYP, Cytochrome P450

recombinant protein (Cao et al. 2009). Knockdown of an orthologous enzyme in *A. aegypti* caused a reduction in JH biosynthesis (Nyati et al. 2013). In *D. melanogaster*, CG15739 is expressed in the larval ring gland (Cao et al. 2009), although its expression level is low in wandering larvae, in which high JH production has been reported (Sliter et al. 1987; Cao et al. 2009). Farnesyl alcohol (farnesol) is oxidized to farnesyl aldehyde (farnesal), and Short-chain dehydrogenase/reductase-1 (SDR-1) has been demonstrated as the farnesal synthase in *A. aegypti* (Mayoral et al. 2009). Phylogenetic analysis suggests that there are six enzymes highly similar to AaSDR-1 in *D. melanogaster* (Mayoral et al. 2009), although their functions have not yet been investigated. In the next step, farnesal is further oxidized to farnesoic acid (FA). A fatty aldehyde dehydrogenase named Aldehyde dehydrogenase 3 (ALDH3) in *A. aegypti* has been identified to catalyze this step (Rivera-Perez et al. 2013). There is a single orthologous enzyme (Aldh-III) in *D. melanogaster*, although no functional analysis has yet been reported.

7.2.4 Final Steps in the JH Branch

FA is converted to JH III through two reactions: methylation of the hydroxyl group and 10, 11-epoxidation (Defelipe et al. 2011). The order of these two steps differs among insect species (Tsang et al. 2020; Smykal and Dolezel, 2023). For example, the epoxidation of FA is first mediated by Cytochrome P450 15 (CYP15), and then the product, JH III acid, is methylated to JH III by juvenile hormone acid *O*-methyltransferase (JHAMT) in Lepidoptera (Daimon and Shinoda 2013; see Chap. 3). In contrast, in many other insects, such as cockroaches, FA is first methylated to MF, which is then epoxidized to JH (Defelipe et al. 2011). Functional assays have shown that CYP15C1 in *B. mori* has high FA epoxidation activity but low MF epoxidation activity (Daimon et al. 2012), whereas CYP15A1 in the cockroach *Diploptera punctata* does not epoxidize FA (Helvig et al. 2004). JHAMT can convert both FA and JH III acid into MF and JH III, respectively, in many insects (Shinoda and Itoyama 2003; Niwa et al. 2008; Defelipe et al. 2011). Thus, the order of the final JH synthetic reaction steps seems to be largely determined by the ligand affinity and enzymatic activity of epoxidases.

CYP15 enzymes are not conserved in higher Diptera (Daimon and Shinoda 2013; Smykal and Dolezel 2023). CYP6A1 in *M. domestica* has been demonstrated to be a terpenoid epoxidase (Andersen et al. 1997), which can catalyze both 6, 7- and 10, 11-epoxidation of MF to JHB3. In contrast, it does not epoxidize FA. Therefore, it is likely that JHs are synthesized in the order of FA-MF-JH III-JHB3 in *M. domestica* unless other terpenoid epoxidases exist.

Despite recent advances, the terminal steps of JH III and JHB_3 biosynthesis in *D. melanogaster* have not yet been fully elucidated (Dermauw et al. 2020; Tsang et al. 2020). Some studies have suggested that multiple JH synthetic pathways exist in *D. melanogaster* (Moshitzky and Applebaum 1995; Wen et al. 2015). Richard and colleagues described that JH III was converted to JHB_3 with high efficiency by the homogenate of larval ring glands (Richard et al. 1989a). High levels of MF are detected in the RG and hemolymph during larval development (Bendena et al. 2011; Jones et al. 2013). Furthermore, topical application of MF increases the JHB_3 titer in larvae (Wen et al. 2015). Collectively, these studies suggest that JHs are synthesized in the order of FA-MF-JH III-JHB_3 during *Drosophila* larval development. In contrast, one study demonstrated that chemically synthesized FA bisepoxide (FABE) was converted into JHB_3, whereas JH III was hardly epoxidized to JHB_3 by the adult CA cultured in vitro (Moshitzky and Applebaum 1995). Therefore, it is conceivable that JHB_3 is synthesized via the FA-FABE-JHB_3 axis in the CA of adults. However, this pathway is likely inactivated in vivo, as JHB_3 is rarely detected in the adult hemolymph (Reiff et al. 2015; Sugime et al. 2017; Kurogi et al. 2023), and FABE has never been detected in the CA.

Due to the absence of canonical CYP15 enzymes in *D. melanogaster*, CYP6G2 has been considered a strong candidate for the epoxidase because of its high expression in the CA (Chung et al. 2009). Recently, CYP6G2 was identified as an epoxidase in *D. melanogaster* (Jia et al. 2024). CYP6G2 can synthesize both JH III and

JHB_3, similar to *M. domestica* CYP6A1. CYP6G2 can use both FA and MF as substrates; however, CYP6G2 effectively converts FA, but not MF, into JH III and JHB_3. These results suggest that epoxidation of FA occurs first by CYP6G2, and then JH III acid or FABE is methylated into JH III or JHB_3, respectively, in *D. melanogaster* (Jia et al. 2024). As *Cyp6g2* mutants do not exhibit the severe developmental defects observed in JH receptor knockout flies (Abdou et al. 2011; Jia et al. 2024), the existence of additional epoxidase(s) has been postulated. In line with this, we identified another epoxidase, CYP6A13 (Fujinaga et al. 2026). *Cyp6a13* is also expressed in the CA, albeit at lower levels than *Cyp6g2*. Functional studies have indicated that *Cyp6a13* contributes to JH III production, but not to JHB_3 synthesis. These findings suggest that CYP6G2 serves as the primary epoxidase, whereas CYP6A13 plays a more limited role, likely because of its lower enzymatic activity and reduced expression (Fujinaga et al. 2026).

As described above, *Drosophila* JHAMT can convert both FA and JH III acid to MF and JH III, respectively (Niwa et al. 2008). JHAMT is considered a key enzyme in JH production because the fluctuation in *jhamt* expression levels coincides with that of the JH titer in the hemolymph of many insect species (Shinoda and Itoyama 2003; Niwa et al. 2008). *Drosophila jhamt* mutant larvae have reduced amounts of MF, JH III, and JHB_3, which is consistent with its critical function in JH production (Wen et al. 2015). However, these larvae rarely show the prepupal lethal phenotype, which is consistently observed in CA-ablated larvae (Liu et al. 2009; Riddiford et al. 2010). Collectively, these studies suggest the existence of other methyltransferase(s). Some studies have examined FA Methyltransferase (FAMeT)/CG10527, which is an orthologous enzyme of FAMeT in crustaceans (Silva Gunawardene et al. 2001; Burtenshaw et al. 2008; Wen et al. 2015). However, these studies failed to demonstrate its function as a methyltransferase. Thus, the identification of additional methyltransferase(s) is warranted to further elucidate JH synthetic pathways in *D. melanogaster*.

7.3 Degradation of JHs

Circulating JH III and JHB_3 are degraded in the hemolymph and peripheral tissues by hydrolysis of the methyl ester and epoxide groups (Wilson and Gilbert 1978; see Chap. 4). JH degradation reactions have been investigated in vitro using hemolymph and dissected tissues from larvae and pupae (Klages and Emmerich 1979; Casas et al. 1991; Campbell et al. 1992). In these studies, high hydrolytic activity on the methyl group was detected in both the hemolymph and supernatant of tissue homogenates, whereas high epoxide hydrolase activity was observed in both the mitochondrial and microsomal fractions of tissue homogenates. These results indicate that the former reaction takes place in the hemolymph and cytosol, whereas the latter reaction mainly occurs within organelles.

The enzyme for the former reaction, JH esterase (JHE)/CG8425, has been identified in *D. melanogaster* (Campbell et al. 1998, 2001; Crone et al. 2007). The expression of *jhe* is drastically upregulated after pupariation when the JH titer is low (Kethidi et al. 2005), which coincides with high JHE activity after pupariation in vivo (Campbell et al. 1992). Furthermore, *jhe* knockdown caused pharate adult lethality, which has been observed in JH-treated flies and flies ectopically expressing *jhamt* (Niwa et al. 2008; Jones et al. 2010a). Therefore, inactivation of JH by JHE is necessary for normal pupal-adult development.

JH epoxide hydrolase (JHEH) was first identified in Lepidoptera (Touhara and Prestwich 1993; Wojtasek and Prestwich 1996). Three orthologous *jheh* genes are tandemly arranged in the *Drosophila* genome (Taniai et al. 2003). However, their functions remain unclear, as recombinant JHEH2 does not metabolize JH III, although it metabolizes other epoxide-containing molecules (Taniai et al. 2003). Indeed, JHEH1/2 deficiency leads to impaired glucose metabolism rather than JH-related phenotypes (Rogalski et al. 2025). In contrast, another study indicated that JHEH3 has a binding pocket that can accommodate JH III in a computational model (Borovsky et al. 2022).

JH-diol and JH acid-diol produced by JH hydrolysis are further polarized by phosphorylation (Halarnkar et al. 1993). JH-diol kinase (JHDK) has been identified in many insects (Maxwell et al. 2002a, b; Li et al. 2005). In *D. melanogaster*, Sarcoplasmic calcium-binding protein 2 (Scp2) is considered a potential JH-diol kinase because of its high similarity to JHDK, although no functional analysis has been performed (Maxwell et al. 2002b).

7.4 Regulation of JH Biosynthesis in *Drosophila*

7.4.1 Peptide Signals that Regulate JH Production

Multiple humoral factors and neuronal pathways that regulate JH production have been identified in *D. melanogaster* (Bendena et al. 2020; Tsang et al. 2020; Zhang et al. 2022b). Many recent studies focus on reproductive regulation by JH. Detailed mechanisms of reproductive regulation by JH are described in Chap. 11.

Richard and colleagues demonstrated that the isolated brain-RG complex produces a much smaller amount of JHB_3 than the isolated RG (Richard et al. 1989b). They further found that the brain extract also suppressed JHB_3 production in a dose-dependent manner (Richard et al. 1990), suggesting the presence of humoral factors produced in the brain that inhibit JH production, in addition to neuronal regulation of JH biosynthesis.

Allatostatin C (AstC) is a somatostatin-like peptide in invertebrates that was first identified in Lepidoptera (Kramer et al. 1991). Knockdown of either AstC or its receptor-encoding genes (AstC-R1/R2) upregulates JH III production in the isolated adult CA, suggesting its inhibitory effect on JH production (Wang et al. 2012). Furthermore, the high activity of a group of AstC-producing neurons (AstC-mTh neurons) downregulates JH production in the adult CA of virgin females (Zhang et al. 2021a). AstC-mTh neurons do not innervate the CA, but they seem to function as neurosecretory cells. Therefore, it is likely that AstC functions as a hormone that activates AstC-R1/R2 in the CA to inhibit JH production.

Several positive regulators of JH biosynthesis have also been identified. In adult flies, Sex peptide (SP) is transferred from males to females upon mating. SP signaling inhibits AstC-mTh neurons so that copulation promotes JH production in mated females (Zhang et al. 2022a). Other studies have also suggested that SP in the hemolymph directly activates JH synthesis in the CA of adult females (Moshizky et al. 1996; Pilpel et al. 2008).

Ecdysis-triggering hormone (ETH) was first identified as a neuropeptide that induces ecdysis behavior in larvae (Zitnan et al. 1996). This hormone not only triggers ecdysis behavior but also promotes fecundity by inducing JH production in the CA (Meiselman et al. 2017). ETH expression is stimulated by the insect steroid hormone ecdysone (Meiselman et al. 2017). Because ecdysone synthesis is promoted by SP signaling (Ameku and Niwa 2016), ETH is released after mating to activate the CA (Zhang et al. 2022a). ETH also stimulates JH production in newly eclosed males (Lee et al. 2017), which prevents mating behavior (Zhang et al. 2021b). ETH-driven JH production is also required for the retention of courtship memory in early adult males (Lee et al. 2017).

Insulin-like peptides (ILPs) are another group of CA-stimulating factors. Adult flies carrying hypomorphic mutations of the ILP receptor-encoding gene (*insulin-like receptor* or *InR*) have low JH-producing activities (Tatar et al. 2001; Tu et al. 2005). A recent study also showed that CA-specific overexpression of a dominant negative form of InR reduces systemic JH signaling (Chen et al. 2024). As down-regulation of ILP signaling makes CA smaller (Tu et al. 2005; Mirth et al. 2005; Zhang et al. 2021c), ILP signaling seems to promote JH production indirectly by inducing the growth of CA cells. However, it is also possible that ILPs stimulate JH production more directly; a study in cockroaches showed that the expression of some JH biosynthetic enzymes is downregulated by the knockdown of ILP signaling components (Zhu et al. 2020).

7.4.2 Regulation of JH Production by the Steroid Hormone

Ecdysone functions not only as a positive regulator of JH production by inducing ETH expression but also as a negative regulator of JH production during larval development. Knockdown of *ecdysone receptor* in the CA increases the expression of *jhamt* and *hmgcr*, suggesting negative regulation of JH production by ecdysone (Liu et al. 2018). Interestingly, ecdysone and JH suppress each other's production. JH signaling downregulates ecdysone production in the prothoracic gland (Richard and Gilbert 1991; Liu et al. 2018; Zhang et al. 2018), and CA-ablated larvae have higher levels of ecdysone in the hemolymph (Mirth et al. 2014). As ecdysone is the master regulator of developmental transitions in insects (Yamanaka 2021), the antagonistic action between ecdysone and JH is likely to be the central mechanism that regulates postembryonic development in insects.

7.4.3 Neuronal Regulation of JH Biosynthesis

The activity of the CA is controlled by neuronal inputs as well as humoral factors. CA-LP neurons are CA-projecting brain neurons (Siegmund and Korge 2001). These neurons have abnormal synapses in *Fasciclin 2* mutants, which show incomplete rotation of the male genitalia (Ádám et al. 2003). As this morphological defect is known to be caused by topical JH application during the pupal stage, it is considered that CA-LP neurons inhibit CA activation during pupal-adult development. Recently, Diuretic hormone 31 (DH31) was shown to be produced in CA-LP neurons (Kurogi et al. 2023). These neurons are activated under dormancy-inducing conditions and suppress the CA activity by stimulating cAMP production (see Chap. 11). The *Dh31* mutation does not cause abnormal genitalia rotation in males (Kurogi et al. 2023), suggesting that CA-LP neurons produce additional neuropeptides or neurotransmitters to regulate JH production.

Although other CA-projecting neurons have also been reported, their functions remain unclear (Mizuno et al. 2021). The N-methyl-D-aspartate-subtype glutamate receptor (NMDAR) is highly expressed in the larval CA (Chiang et al. 2002). A genetic study showed that NMDAR is responsible for the expression of the transforming growth factor-β ligand Decapentaplegic, which stimulates the expression of *jhamt* in an autocrine manner in the CA (Huang et al. 2011). Therefore, glutamatergic neurons are expected to induce JH production in the CA during larval development.

7.4.4 Signaling Components that Regulate Expression of JH Biosynthetic Enzymes

One class of noncoding RNAs, microRNAs (miRNAs), inhibits target gene expression by forming double-stranded RNAs or hairpin molecules. They promote mRNA degradation or diminish the translation of target genes (Bellés et al. 2012; Bendena et al. 2020). Recent studies have identified four miRNAs involved in JH production (Qu et al. 2017; Zhang et al. 2021c; Chan et al. 2023). One miRNA, *bantam*, has been identified as a *jhamt*-targeting miRNA. *bantam* negatively regulates *jhamt* expression and JH production (Qu et al. 2017). This study also identified *miR-252* and *miR-304* as *jhamt*-targeting miRNAs, although no obvious phenotype was observed in flies when these miRNAs were overexpressed in the CA (Qu et al. 2017). Another miRNA, *miR-8*, is highly expressed in the CA, and its silencing or overexpression in the CA results in the reduction or upregulation of *jhamt* expression, respectively (Zhang et al. 2021c). Therefore, *miR-8* does not seem to control *jhamt* expression directly but rather targets a negative regulator of *jhamt*. Since the CA of *miR-8*-silenced flies is smaller, it is expected that *miR-8* targets cellular growth-related genes.

In contrast to the recent advances in our understanding of miRNAs controlling JH production, transcription factors that regulate the expression of JH biosynthetic enzymes have not been well investigated in *D. melanogaster* (Riddiford 2020; Zhang et al. 2022b). A Pit-Oct-Unc family transcription factor, ventral veins lacking (Vvl), has been reported as a positive regulator of JH synthesis in some insects (Cheng et al. 2014). Although no experimental results have been reported in *D. melanogaster*, studies in the red flour beetle, *Tribolium castaneum*, indicated that knockdown of *Vvl* drastically reduces the expression of *jhamt*, accompanied by a reduction in JH biosynthesis (Cheng et al. 2014). However, there is no direct evidence suggesting that Vvl directly regulates *jhamt* expression. Vvl regulates ecdysteroid biosynthesis in the prothoracic gland of *D. melanogaster* (Danielsen et al. 2014). Therefore, it is possible that Vvl regulates JH production through other factors such as ecdysone. Another transcription factor, retained/Dead ringer (Dri), was recently reported as a major positive regulator of the transcription of JH synthetic enzymes in several species, such as *B. mori* (Kayukawa et al. 2024). This gene is also highly expressed in the larval CA of *D. melanogaster* (Kayukawa et al. 2024).

7.5 Conclusion

Powerful *Drosophila* genetic tools have significantly contributed to our understanding of JH biosynthesis and signaling pathways. However, many JH biosynthetic enzymes have not yet been identified in *D. melanogaster*. As a result, signaling components, such as transcription factors that directly regulate the expression of JH biosynthetic enzymes, also remain largely unidentified. The identification and functional characterization of JH biosynthetic enzymes will be a critical next step to further advance our understanding of JH signaling and its regulation in *D. melanogaster*.

References

Abdou MA, He Q, Wen D, Zyaan O, Wang J, Xu J et al (2011) *Drosophila* Met and Gce are partially redundant in transducing juvenile hormone action. Insect Biochem Mol Biol 41:938–945. https://doi.org/10.1016/j.ibmb.2011.09.003

Ádám G, Perrimon N, Noselli S (2003) The retinoic-like juvenile hormone controls the looping of left-right asymmetric organs in *Drosophila*. Development 130:2397–2406. https://doi.org/10.1242/dev.00460

Altaratz M, Applebaum SW, Richard DS, Gilbert LI, Segal D (1991) Regulation of juvenile hormone synthesis in wild-type and apterous mutant *Drosophila*. Mol Cell Endocrinol 81:205–216. https://doi.org/10.1016/0303-7207(91)90219-i

Ameku T, Niwa R (2016) Mating-induced increase in germline stem cells via the neuroendocrine system in female *Drosophila*. PLoS Genet 12:e1006123. https://doi.org/10.1371/journal.pgen.1006123

Andersen JF, Walding JK, Evans PH, Bowers WS, Feyereisen R (1997) Substrate specificity for the epoxidation of terpenoids and active site topology of house fly cytochrome P450 6A1. Chem Res Toxicol 10:156–164. https://doi.org/10.1021/tx9601162

Ashburner M (1970) Effects of juvenile hormone on adult differentiation of *Drosophila melanogaster*. Nature 227:187–189. https://doi.org/10.1038/227187a0

Barton LJ, Sanny J, Packard Dawson E, Nouzova M, Noriega FG et al (2024) Juvenile hormones direct primordial germ cell migration to the embryonic gonad. Curr Biol 34:505–518. https://doi.org/10.1016/j.cub.2023.12.033

Bellés X, Cristino AS, Tanaka ED, Rubio M, Piulachs MD (2012) Insect microRNAs: from molecular mechanisms to biological roles. In: Gilbert LI (ed) Insect molecular biology and biochemistry. Elsevier, Amsterdam, pp 30–56. https://doi.org/10.1016/C2009-0-62118-8

Bellés X, Martín D, Piulachs MD (2005) The mevalonate pathway and the synthesis of juvenile hormone in insects. J Insect Physiol 50:181–199. https://doi.org/10.1146/annurev.ento.50.071803.130356

Bendena WG, Zhang J, Burtenshaw SM, Tobe SS (2011) Evidence for differential biosynthesis of juvenile hormone (and related) sesquiterpenoids in *Drosophila melanogaster*. Gen Comp Endocrinol 172:56–61. https://doi.org/10.1016/j.ygcen.2011.02.014

Bendena WG, Hui JHL, Chin-Sang I, Tobe SS (2020) Neuropeptide and microRNA regulators of juvenile hormone production. Gen Comp Endocrinol 295:113507. https://doi.org/10.1016/j.ygcen.2020.113507

Borovsky D, Breyssens H, Buytaert E, Peeters T, Laroye C, Stoffels K et al (2022) Cloning and characterization of *Drosophila melanogaster* juvenile hormone epoxide hydrolases (JHEH) and their promoters. Biomolecules 12:991. https://doi.org/10.3390/biom12070991

Bounhiol J (1938) Recherches experimentales sur le determinisme de la metamorphose chez les Lepidopteres. Bull Biol Fr Belg 24

Bownes M, Rembold H (1987) The titre of juvenile hormone during the pupal and adult stages of the life cycle of *Drosophila melanogaster*. Eur J Biochem 164:709–712. https://doi.org/10.1111/j.1432-1033.1987.tb11184.x

Bührlen U, Emmerich H, Rembold H (1984) Titer of juvenile hormone III in *Drosophila hydei* during metamorphosis determined by GC-MS-MIS. Z Naturforsch C 39:1150–1154. https://doi.org/10.1515/znc-1984-11-1227

Burtenshaw SM, Su PP, Zhang JR, Tobe SS, Dayton L, Bendena WG (2008) A putative farnesoic acid O-methyltransferase (FAMeT) orthologue in *Drosophila melanogaster* (CG10527): relationship to juvenile hormone biosynthesis? Peptides 29:242–251. https://doi.org/10.1016/j.peptides.2007.10.030

Burtt ET (1937) On the corpora allata of dipterous insects. Proc Royal Soc B 124:13–23. https://doi.org/10.1098/rspb.1937.0069

Burtt ET (1938) On the corpora allata of dipterous insects. II. Proc Royal Soc B 126(843):210–223. https://doi.org/10.1098/rspb.1938.0052

Campbell PM, Harcourt RL, Crone EJ, Claudianos C, Hammock BD, Russell RJ et al (2001) Identification of a juvenile hormone esterase gene by matching its peptide mass fingerprint with a sequence from the *Drosophila* genome project. Insect Biochem Mol Biol 31:513–520. https://doi.org/10.1016/s0965-1748(01)00035-2

Campbell PM, Oakeshott JG, Healy MJ (1998) Purification and kinetic characterisation of juvenile hormone esterase from *Drosophila melanogaster*. Insect Biochem Mol Biol 28:501–515. https://doi.org/10.1016/s0965-1748(98)00037-x

Campbell PM, Healy MJ, Oakeshott JG (1992) Characterisation of juvenile hormone esterase in *Drosophila melanogaster*. Insect Biochem Mol Biol 22:665–677. https://doi.org/10.1016/0965-1748(92)90045-G

Cao L, Zhang P, Grant DF (2009) An insect farnesyl phosphatase homologous to the N-terminal domain of soluble epoxide hydrolase. Biochem Biophys Res Commun 380:188–192. https://doi.org/10.1016/j.bbrc.2009.01.079

Casas J, Harshman LG, Messeguer A, Kuwano E, Hammock BD (1991) *In vitro* metabolism of juvenile hormone III and juvenile hormone III bisepoxide by *Drosophila melanogaster* and mammalian cytosolic epoxide hydrolase. Arch Biochem Biophys 286:153–158. https://doi.org/10.1016/0003-9861(91)90021-a

Chan KK, Chan TF, Bendena W, Hui JHL (2023) Noncoding RNA regulation of hormonal and metabolic systems in the fruit fly *Drosophila*. Metabolites 13:152. https://doi.org/10.3390/metabo13020152

Chen J, Nouzová M, Noriega FG, Tatar M (2024) Gut-to-brain regulation of *Drosophila* aging through neuropeptide F, insulin, and juvenile hormone. Proc Natl Acad Sci USA 121:e2411987121. https://doi.org/10.1073/pnas.2411987121

Cheng C, Ko A, Chaieb L, Koyama T, Sarwar P, Mirth CK et al (2014) The POU factor ventral veins lacking/drifter directs the timing of metamorphosis through ecdysteroid and juvenile hormone signaling. PLoS Genet 10:e1004425. https://doi.org/10.1371/journal.pgen.1004425

Chiang AS, Lin WY, Liu HP, Pszczolkowski MA, Fu TF, Chiu SL et al (2002) Insect NMDA receptors mediate juvenile hormone biosynthesis. Proc Natl Acad Sci USA 99:37–42. https://doi.org/10.1073/pnas.012318899

Chung H, Sztal T, Pasricha S, Sridhar M, Batterham P, Daborn PJ (2009) Characterization of Drosophila melanogaster cytochrome P450 genes. Proc Natl Acad Sci USA 106:5731–5736. https://doi.org/10.1073/pnas.0812141106

Crone EJ, Sutherland TD, Campbell PM, Coppin CW, Russell RJ, Oakeshott JG (2007) Only one esterase of *Drosophila melanogaster* is likely to degrade juvenile hormone *in vivo*. Insect Biochem Mol Biol 37:540–549. https://doi.org/10.1016/j.ibmb.2007.02.010

Dai JD, Gilbert LI (1991) Metamorphosis of the corpus allatum and degeneration of the prothoracic glands during the larval-pupal-adult transformation of *Drosophila melanogaster*: a cytophysiological analysis of the ring gland. Dev Biol 144:309–326. https://doi.org/10.1016/0012-1606(91)90424-2

Daimon T, Shinoda T (2013) Function, diversity, and application of insect juvenile hormone epoxidases (CYP15). Biotechnol Appl Biochem 60:82–91. https://doi.org/10.1002/bab.1058

Daimon T, Kozaki T, Niwa R, Kobayashi I, Furuta K, Namiki T et al (2012) Precocious metamorphosis in the juvenile hormone-deficient mutant of the silkworm, *Bombyx mori*. PLoS Genet 8:e1002486. https://doi.org/10.1371/journal.pgen.1002486

Danielsen ET, Moeller ME, Dorry E, Komura-Kawa T, Fujimoto Y, Troelsen JT et al (2014) Transcriptional control of steroid biosynthesis genes in the *Drosophila* prothoracic gland by ventral veins lacking and knirps. PLoS Genet 10:e1004343. https://doi.org/10.1371/journal.pgen.1004343

Day MF (1943) The homologies of the ring gland of diptera *Brachycera*. Ann Entomol Soc Am 36:1–10

Defelipe LA, Dolghih E, Roitberg AE, Nouzova M, Mayoral JG, Noriega FG et al (2011) Juvenile hormone synthesis: "esterify then epoxidize" or "epoxidize then esterify"? Insights from the structural characterization of juvenile hormone acid methyltransferase. Insect Biochem Mol Biol 41:228–235. https://doi.org/10.1016/j.ibmb.2010.12.008

Dellas N, Thomas ST, Manning G, Noel JP (2013) Discovery of a metabolic alternative to the classical mevalonate pathway. eLife 2:e00672. https://doi.org/10.7554/eLife.00672

Dermauw W, Van Leeuwen T, Feyereisen R (2020) Diversity and evolution of the P450 family in arthropods. Insect Biochem Mol Biol 127:103490. https://doi.org/10.1016/j.ibmb.2020.103490

Fujinaga D, Ohhara Y, Okamoto N, Chu H, Mauck KE, Yamanaka N (2026) Two cytochrome P450 epoxidases mediate juvenile hormone biosynthesis in *Drosophila melanogaster*. Insect Biochem Mol Biol 187:104475. https://doi.org/10.1016/j.ibmb.2025.104475

Göbel A, Rauner M, Hofbauer LC, Rachner TD (2020) Cholesterol and beyond - the role of the mevalonate pathway in cancer biology. Biochim Biophys Acta Rev Cancer 1873:188351. https://doi.org/10.1016/j.bbcan.2020.188351

Halarnkar PP, Jackson GP, Straub KM, Schooley DA (1993) Juvenile hormone catabolism in *Manduca sexta*: homologue selectivity of catabolism and identification of a diol-phosphate conjugate as a major end product. Experientia 49:988–994. https://doi.org/10.1007/BF02125646

Harshman LG, Song KD, Casas J, Schuurmans A, Kuwano E, Kachman SD et al (2010) Bioassays of compounds with potential juvenoid activity on *Drosophila melanogaster*: juvenile hormone III, bisepoxide juvenile hormone III and methyl farnesoates. J Insect Physiol 56:1465–1470. https://doi.org/10.1016/j.jinsphys.2010.06.003

Helvig C, Koener JF, Unnithan GC, Feyereisen R (2004) CYP15A1, the cytochrome P450 that catalyzes epoxidation of methyl farnesoate to juvenile hormone III in cockroach corpora allata. Proc Natl Acad Sci USA 101:4024–4029. https://doi.org/10.1073/pnas.0306980101

Huang J, Tian L, Peng C, Abdou M, Wen D, Wang Y et al (2011) DPP-mediated TGFβ signaling regulates juvenile hormone biosynthesis by activating the expression of juvenile hormone acid methyltransferase. Development 138:2283–2291. https://doi.org/10.1242/dev.057687

Jia Q, Yang L, Wen J, Liu S, Wen D, Luo W et al (2024) Cyp6g2 is the major P450 epoxidase responsible for juvenile hormone biosynthesis in *Drosophila melanogaster*. BMC Biol 22:111. https://doi.org/10.1186/s12915-024-01910-4

Jindra M, Uhlirova M, Charles JP, Smykal V, Hill RJ (2015) Genetic evidence for function of the bHLH-PAS protein Gce/met as a juvenile hormone receptor. PLoS Genet 11:e1005394. https://doi.org/10.1371/journal.pgen.1005394

Jones D, Jones G (2007) Farnesoid secretions of dipteran ring glands: what we do know and what we can know. Insect Biochem Mol Biol 37:771–798. https://doi.org/10.1016/j.ibmb.2007.05.014

Jones D, Jones G, Teal P, Hammac C, Messmer L, Osborne K et al (2010b) Suppressed production of methyl farnesoid hormones yields developmental defects and lethality in *Drosophila* larvae. Gen Comp Endocrinol 165:244–254. https://doi.org/10.1016/j.ygcen.2009.07.006

Jones G, Jones D, Teal P, Sapa A, Wozniak M (2006) The retinoid-X receptor ortholog, ultraspiracle, binds with nanomolar affinity to an endogenous morphogenetic ligand. FEBS J 273:4983–4996. https://doi.org/10.1111/j.1742-4658.2006.05498.x

Jones G, Jones D, Li X, Tang L, Ye L, Teal P et al (2010a) Activities of natural methyl farnesoids on pupariation and metamorphosis of *Drosophila melanogaster*. J Insect Physiol 56:1456–1464. https://doi.org/10.1016/j.jinsphys.2010.06.001

Jones G, Teal P, Henrich VC, Krzywonos A, Sapa A, Wozniak M et al (2013) Ligand binding pocket function of *Drosophila* USP is necessary for metamorphosis. Gen Comp Endocrinol 182:73–82. https://doi.org/10.1016/j.ygcen.2012.11.009

Kayukawa T, Nagamine K, Inui T, Yokoi K, Kobayashi I, Nakao H et al (2024) *Dead ringer* acts as a major regulator of juvenile hormone biosynthesis in insects. PNAS Nexus 3:435. https://doi.org/10.1093/pnasnexus/pgae435

Kethidi DR, Xi Z, Palli SR (2005) Developmental and hormonal regulation of juvenile hormone esterase gene in *Drosophila melanogaster*. J Insect Physiol 51:393–400. https://doi.org/10.1016/j.jinsphys.2004.12.007

King RC, Aggarwal SK, Bodenstein D (1966a) The comparative submicroscopic morphology of the ring gland of *Drosophila melanogaster* during the second and third larval instars. Z Zellforsch Mik Ana 73:272–285. https://doi.org/10.1007/BF00334868

King RC, Surinder AK, Dietrich B (1966b) The comparative submicroscopic cytology of the corpus allatum-corpus cardiacum complex of wild type and *fes* adult female *Drosophila melanogaster*. J Exp Zool 161:151–175

Klages G, Emmerich H (1979) Juvenile hormone metabolism and juvenile hormone esterase titer in hemolymph and peripheral tissues of *Drosophila hydei*. J Comp Physiol 132:319–325. https://doi.org/10.1007/BF00799045

Kramer SJ, Toschi A, Miller CA, Kataoka H, Quistad GB, Li JP et al (1991) Identification of an allatostatin from the tobacco hornworm *Manduca sexta*. Proc Natl Acad Sci USA 88:9458–9462. https://doi.org/10.1073/pnas.88.21.9458

Kurogi Y, Imura E, Mizuno Y, Hoshino R, Nouzova M, Matsuyama S et al (2023) Female reproductive dormancy in *Drosophila* is regulated by DH31-producing neurons projecting into the corpus allatum. Development 150:201186. https://doi.org/10.1242/dev.201186

Lee SS, Ding Y, Karapetians N, Rivera-Perez C, Noriega FG, Adams ME (2017) Hormonal signaling cascade during an early-adult critical period required for courtship memory retention in *Drosophila*. Curr Biol 27:2798–2809.e3. https://doi.org/10.1016/j.cub.2017.08.017

Li S, Zhang QR, Xu WH, Schooley DA (2005) Juvenile hormone diol kinase, a calcium-binding protein with kinase activity, from the silkworm, *Bombyx mori*. Insect Biochem Mol Biol 35:1235–1248. https://doi.org/10.1016/j.ibmb.2005.06.005

Liu S, Li K, Gao Y, Liu X, Chen W, Ge W et al (2018) Antagonistic actions of juvenile hormone and 20-hydroxyecdysone within the ring gland determine developmental transitions in *Drosophila*. Proc Natl Acad Sci USA 115:139–144. https://doi.org/10.1073/pnas.1716897115

Liu Y, Sheng Z, Liu H, Wen D, He Q, Wang S et al (2009) Juvenile hormone counteracts the bHLH-PAS transcription factors MET and GCE to prevent caspase-dependent programmed cell death in *Drosophila*. Development 136:2015–2025. https://doi.org/10.1242/dev.033712

Mayoral JG, Nouzova M, Navare A, Noriega FG (2009) NADP+-dependent farnesol dehydrogenase, a corpora allata enzyme involved in juvenile hormone synthesis. Proc Natl Acad Sci USA 106:21091–21096. https://doi.org/10.1073/pnas.0909938106

Maxwell RA, Welch WH, Schoole DA (2002a) Juvenile hormone diol kinase. I. Purification, characterization, and substrate specificity of juvenile hormone-selective diol kinase from *Manduca sexta*. J Biol Chem 277:21874–21881. https://doi.org/10.1074/jbc.M201510200

Maxwell RA, Welch WH, Horodyski FM, Schegg KM, Schooley DA (2002b) Juvenile hormone diol kinase. II. Sequencing, cloning, and molecular modeling of juvenile hormone-selective diol kinase from *Manduca sexta*. J Biol Chem 277:21882–21890. https://doi.org/10.1074/jbc.M201509200

Meiselman M, Lee SS, Tran RT, Dai H, Ding Y, Rivera-Perez C et al (2017) Endocrine network essential for reproductive success in *Drosophila melanogaster*. Proc Natl Acad Sci USA 114:E3849–E3858. https://doi.org/10.1073/pnas.1620760114

Mirth CK, Truman JW, Riddiford LM (2005) The role of the prothoracic gland in determining critical weight for metamorphosis in *Drosophila melanogaster*. Curr Biol 15:1796–1807. https://doi.org/10.1016/j.cub.2005.09.017

Mirth CK, Tang HY, Makohon-Moore SC, Salhadar S, Gokhale RH, Warner RD et al (2014) Juvenile hormone regulates body size and perturbs insulin signaling in *Drosophila*. Proc Natl Acad Sci USA 111:7018–7023. https://doi.org/10.1073/pnas.1313058111

Miziorko HM (2011) Enzymes of the mevalonate pathway of isoprenoid biosynthesis. Arch Biochem Biophys 505:131–143. https://doi.org/10.1016/j.abb.2010.09.028

Mizuno Y, Imura E, Kurogi Y, Shimada-Niwa Y, Kondo S, Tanimoto H et al (2021) A population of neurons that produce hugin and express the *diuretic hormone 44* receptor gene projects to the corpora allata in *Drosophila melanogaster*. Develop Growth Differ 63:249–261. https://doi.org/10.1111/dgd.12733

Moshitzky P, Applebaum SW (1995) Pathway and regulation of JHIII-bisepoxide biosynthesis in adult *Drosophila melanogaster* corpus allatum. Arch Insect Biochem Physiol 30:225–237. https://doi.org/10.1002/arch.940300211

Moshitzky P, Fleischmann I, Chaimov N, Saudan P, Klauser S, Kubli E et al (1996) Sex-peptide activates juvenile hormone biosynthesis in the *Drosophila melanogaster* corpus allatum. Arch Insect Biochem Physiol 32:363–374. https://doi.org/10.1002/(SICI)1520-6327(1996)32:3/4<363::AID-ARCH9>3.0.CO;2-T

Niwa R, Niimi T, Honda N, Yoshiyama M, Itoyama K, Kataoka H et al (2008) Juvenile hormone acid *O*-methyltransferase in *Drosophila melanogaster*. Insect Biochem Mol Biol 38:714–720. https://doi.org/10.1016/j.ibmb.2008.04.003

Nyati P, Nouzova M, Rivera-Perez C, Clifton ME, Mayoral JG, Noriega FG (2013) Farnesyl phosphatase, a corpora allata enzyme involved in juvenile hormone biosynthesis in *Aedes aegypti*. PLoS One 8:e71967. https://doi.org/10.1371/journal.pone.0071967

Ohhara Y, Kobayashi S, Yamakawa-Kobayashi K, Yamanaka N (2018) Adult-specific insulin-producing neurons in *Drosophila melanogaster*. J Comp Neurol 526:1351–1367. https://doi.org/10.1002/cne.24410

Pilpel N, Nezer I, Applebaum SW, Heifetz Y (2008) Mating-increases trypsin in female *Drosophila* hemolymph. Insect Biochem Mol Biol 38:320–330. https://doi.org/10.1016/j.ibmb.2007.11.010

Qu Z, Bendena WG, Nong W, Siggens KW, Noriega FG, Kai ZP et al (2017) MicroRNAs regulate the sesquiterpenoid hormonal pathway in *Drosophila* and other arthropods. Proc Royal Soc 284:20171827. https://doi.org/10.1098/rspb.2017.1827

Radtke A (1942) Hemmung der verpuppung beim mehlkäfer *Tenebroadio molitor* L. Naturwissenschafen 30:451–452

Rahman MM, Franch-Marro X, Maestro JL, Martin D, Casali A (2017) Local juvenile hormone activity regulates gut homeostasis and tumor growth in adult *Drosophila*. Sci Rep 7:11677. https://doi.org/10.1038/s41598-017-11199-9

Reiff T, Jacobson J, Cognigni P, Antonello Z, Ballesta E, Tan KJ et al (2015) Endocrine remodelling of the adult intestine sustains reproduction in *Drosophila*. eLife 4:e06930. https://doi.org/10.7554/eLife.06930

Rembold H, Llackner B (1985) Convenient method for the determination of picomole amounts of juvenile hormone. J Chromatogr A 323:355–361. https://doi.org/10.1016/S0021-9673(01)90397-8

Richard DS, Applebaum SW, Sliter TJ, Baker FC, Schooley DA, Reuter CC et al (1989a) Juvenile hormone bisepoxide biosynthesis *in vitro* by the ring gland of *Drosophila melanogaster*: a putative juvenile hormone in the higher Diptera. Proc Natl Acad Sci USA 86:1421–1425. https://doi.org/10.1073/pnas.86.4.1421

Richard DS, Applebaum SW, Gilbert LI (1989b) Developmental regulation of juvenile hormone biosynthesis by the ring gland of *Drosophila melanogaster*. J Comp Physiol B 159:383–387. https://doi.org/10.1007/BF00692410

Richard DS, Applebaum SW, Gilbert LI (1990) Allatostatic regulation of juvenile hormone production *in vitro* by the ring gland of *Drosophila melanogaster*. Mol Cell Endocrino 68:153–161. https://doi.org/10.1016/0303-7207(90)90188-e

Richard DS, Gilbert LI (1991) Reversible juvenile hormone inhibition of ecdysteroid and juvenile hormone synthesis by the ring gland of *Drosophila melanogaster*. Experientia 47:1063–1066. https://doi.org/10.1007/BF01923343

Riddiford LM, Truman JW, Mirth CK, Shen YC (2010) A role for juvenile hormone in the prepupal development of *Drosophila melanogaster*. Development 137:1117–1126. https://doi.org/10.1242/dev.037218

Riddiford LM (2020) *Rhodnius*, golden oil, and *Met*: a history of juvenile hormone research. Front Cell Dev Biol 8:679. https://doi.org/10.3389/fcell.2020.00679

Rivera-Perez C, Nouzova M, Clifton ME, Garcia EM, LeBlanc E, Noriega FG (2013) Aldehyde dehydrogenase 3 converts farnesal into farnesoic acid in the corpora allata of mosquitoes. Insect Biochem Mol Biol 43:675–682. https://doi.org/10.1016/j.ibmb.2013.04.002

Rogalski F, Tsuda M, Takeo S, Asano T, Sakai T, Aigaki T (2025) Epoxide hydrolases JHEH1 and JHEH2 deficiency impairs glucose metabolism in *Drosophila*. Biochem Biophys Res Commun 748:151313. https://doi.org/10.1016/j.bbrc.2025.151313

Santos AC, Lehmann R (2004) Isoprenoids control germ cell migration downstream of HMGCoA reductase. Dev Cell 6:283–293. https://doi.org/10.1016/s1534-5807(04)00023-1

Saunders DS, Richard DS, Applebaum SW, Ma M, Gilbert LI (1990) Photoperiodic diapause in *Drosophila melanogaster* involves a block to the juvenile hormone regulation of ovarian maturation. Gen Comp Endocrinol 79:174–184. https://doi.org/10.1016/0016-6480(90)90102-r

Scharrer B, Hadorn E (1938) The structure of the ring-gland (corpus allatum) in normal and lethal larvae of *Drosophila Melanogaster*. Proc Natl Acad Sci USA 24:236–242. https://doi.org/10.1073/pnas.24.6.236

Shianiou G, Teloni S, Apidianakis Y (2023) Intestinal immune deficiency and juvenile hormone signaling mediate a metabolic trade-off in adult *Drosophila* females. Metabolites 13:340. https://doi.org/10.3390/metabo13030340

Shinoda T, Itoyama K (2003) Juvenile hormone acid methyltransferase: a key regulatory enzyme for insect metamorphosis. Proc Natl Acad Sci USA 100:11986–11991. https://doi.org/10.1073/pnas.2134232100

Siegmund T, Korge G (2001) Innervation of the ring gland of *Drosophila melanogaster*. J Comp Neurol 431:481–491. https://doi.org/10.1002/1096-9861(20010319)431:4<481::aid-cne1084>3.0.co;2-7

Silva B, Mantha OL, Schor J, Pascual A, Plaçais PY, Pavlowsky A et al (2022) Glia fuel neurons with locally synthesized ketone bodies to sustain memory under starvation. Nat Metab 4:213–224. https://doi.org/10.1038/s42255-022-00528-6

Silva Gunawardene YI, Chow BK, He JG, Chan SM (2001) The shrimp FAMeT cDNA is encoded for a putative enzyme involved in the methylfarnesoate (MF) biosynthetic pathway and is temporally expressed in the eyestalk of different sexes. Insect Biochem Mol Biol 31:1115–1124. https://doi.org/10.1016/s0965-1748(01)00060-1

Sliter TJ, Sedlak BJ, Baker FC, Schooley DA (1987) Juvenile hormone in *Drosophila melanogaster*: identification and titer determination during development. Insect Biochem 17:161–165. https://doi.org/10.1016/0020-1790(87)90156-9

Smykal V, Dolezel D (2023) Evolution of proteins involved in the final steps of juvenile hormone synthesis. J Insect Physiol 145:104487. https://doi.org/10.1016/j.jinsphys.2023.104487

Sugime Y, Watanabe D, Yasuno Y, Shinada T, Miura T, Tanaka NK (2017) Upregulation of juvenile hormone titers in female *Drosophila melanogaster* through mating experiences and host food occupied by eggs and larvae. Zool Sci 34:52–57. https://doi.org/10.2108/zs160150

Taniai K, Inceoglu AB, Yukuhiro K, Hammock BD (2003) Characterization and cDNA cloning of a clofibrate-inducible microsomal epoxide hydrolase in *Drosophila melanogaster*. Eur J Biochem 270:4696–4705. https://doi.org/10.1046/j.1432-1033.2003.03868.x

Tatar M, Kopelman A, Epstein D, Tu MP, Yin CM, Garofalo RS (2001) A mutant *Drosophila* insulin receptor homolog that extends life-span and impairs neuroendocrine function. Science 292:107–110. https://doi.org/10.1126/science.1057987

Tobe S, Stay B (1985) Structure and regulation of the corpus allatum. Adv Insect Physiol 18:305–432. https://doi.org/10.1016/S0065-2806(08)60043-0

Touhara K, Prestwich GD (1993) Juvenile hormone epoxide hydrolase. Photoaffinity labeling, purification, and characterization from tobacco hornworm eggs. J Biol Chem 268:19604–19609. https://doi.org/10.1016/S0021-9258(19)36559-7

Tsang SSK, Law STS, Li C, Qu Z, Bendena WG, Tobe SS et al (2020) Diversity of insect sesquiterpenoid regulation. Front Genet 11:1027. https://doi.org/10.3389/fgene.2020.01027

Tu MP, Yin CM, Tatar M (2005) Mutations in insulin signaling pathway alter juvenile hormone synthesis in *Drosophila melanogaster*. Gen Comp Endocrinol 142:347–356. https://doi.org/10.1016/j.ygcen.2005.02.009

Vogt M (1943) Zur Kenntnis des larvalen und pupalen Corpus allatum von Calliphora. Biol Zbl 63:56–71

Wang C, Zhang J, Tobe SS, Bendena WG (2012) Defining the contribution of select neuropeptides and their receptors in regulating sesquiterpenoid biosynthesis by *Drosophila melanogaster* ring gland/corpus allatum through RNAi analysis. Gen Comp Endocrinol 176:347–353. https://doi.org/10.1016/j.ygcen.2011.12.039

Wen D, Rivera-Perez C, Abdou M, Jia Q, He Q, Liu X et al (2015) Methyl farnesoate plays a dual role in regulating *drosophila* metamorphosis. PLoS Genet 11:e1005038. https://doi.org/10.1371/journal.pgen.1005038

Wijesekera TP, Saurabh S, Dauwalder B (2016) Juvenile hormone is required in adult males for *Drosophila* courtship. PLoS One 11:e0151912. https://doi.org/10.1371/journal.pone.0151912

Wilson TG, Gilbert LI (1978) Metabolism of juvenile hormone I in *Drosophila melanogaster*. Comp Biochem Physiol Mol 60:85–89. https://doi.org/10.1016/0300-9629(78)90041-5

Wojtasek H, Prestwich GD (1996) An insect juvenile hormone-specific epoxide hydrolase is related to vertebrate microsomal epoxide hydrolases. Biochem Biophys Res Commun 220:323–329. https://doi.org/10.1006/bbrc.1996.0404

Yamanaka N (2021) Chapter one - Ecdysteroid signalling in insects—from biosynthesis to gene expression regulation. Adv Insect Physiol 60:1–36. https://doi.org/10.1016/bs.aiip.2021.03.002

Zhang C, Daubnerova I, Jang YH, Kondo S, Žitňan D, Kim YJ (2021a) The neuropeptide allatostatin C from clock-associated DN1p neurons generates the circadian rhythm for oogenesis. Proc Natl Acad Sci USA 118:e2016878118. https://doi.org/10.1073/pnas.2016878118

Zhang C, Kim AJ, Rivera-Perez C, Noriega FG, Kim YJ (2022a) The insect somatostatin pathway gates vitellogenesis progression during reproductive maturation and the post-mating response. Nat Commun 13:969. https://doi.org/10.1038/s41467-022-28592-2

Zhang J, Wen D, Li EY, Palli SR, Li S, Wang J et al (2021c) MicroRNA *miR-8* promotes cell growth of corpus allatum and juvenile hormone biosynthesis independent of insulin/IGF signaling in *Drosophila melanogaster*. Insect Biochem Mol Biol 136:103611. https://doi.org/10.1016/j.ibmb.2021.103611

Zhang SX, Glantz EH, Miner LE, Rogulja D, Crickmore MA (2021b) Hormonal control of motivational circuitry orchestrates the transition to sexuality in *Drosophila*. Sci Adv 7:eabg6926. https://doi.org/10.1126/sciadv.abg6926

Zhang T, Song W, Li Z, Qian W, Wei L, Yang Y et al (2018) Krüppel homolog 1 represses insect ecdysone biosynthesis by directly inhibiting the transcription of steroidogenic enzymes. Proc Natl Acad Sci USA 115:3960–3965. https://doi.org/10.1073/pnas.1800435115

Zhang X, Li S, Liu S (2022b) Juvenile hormone studies in *Drosophila melanogaster*. Front Physiol 12:785320. https://doi.org/10.3389/fphys.2021.785320

Zhu S, Liu F, Zeng H, Li N, Ren C, Su Y et al (2020) Insulin/IGF signaling and TORC1 promote vitellogenesis via inducing juvenile hormone biosynthesis in the American cockroach. Development 147:dev188805. https://doi.org/10.1242/dev.188805

Zitnan D, Kingan TG, Hermesman JL, Adams ME (1996) Identification of ecdysis-triggering hormone from an epitracheal endocrine system. Science 271:88–91. https://doi.org/10.1126/science.271.5245.88

Part III
Function

Chapter 8
Molecular Mechanism Underlying Repression of Insect Metamorphosis by Juvenile Hormone

Takumi Kayukawa and Tetsuro Shinoda

Abstract This chapter, centered on the authors' research using the silkworm *Bombyx mori,* elucidates recent advancements in our understanding of the fundamental mechanisms controlling insect metamorphosis through the regulation of key genes by juvenile hormone (JH), such as *Krüppel homolog 1* (*Kr-h1*), *Broad-Complex* (*BR-C*), *Ecdysone-induced protein 93F* (*E93*), and *chronologically inappropriate morphogenesis* (*Chinmo*). The chapter delves into the JH-mediated transcriptional control of these metamorphic genes. Synthesizing our latest findings, this chapter provides a comprehensive overview of the molecular intricacies that govern insect metamorphosis.

Keywords *BR-C* · *Chinmo* · *E93* · Juvenile hormone · *Kr-h1* · Metamorphosis · Molecular mechanism

8.1 Roles of JH in Metamorphosis

Insect metamorphosis is coordinated by ecdysone and juvenile hormone (JH), which are secreted by the prothoracic gland (PG) and corpora allata (CA), respectively (Fig. 8.1). In holometabolous insects, 20-hydroxyecdysone (20E), the active metabolite of ecdysone, induces larval-larval molt in the presence of JH, whereas in the absence of JH, it induces larval-pupal and pupal-adult molts (Riddiford 1994). In many insects, the deprivation of JH by the removal of the CA (allatectomy) from the larvae induces precocious metamorphosis (Wigglesworth 1936; Fukuda 1944;

T. Kayukawa (✉)
National Agriculture and Food Research Organization, Tsukuba, Japan
e-mail: kayukawa.takumi815@naro.go.jp

T. Shinoda
Faculty of Food and Agricultural Sciences, Fukushima University, Fukushima, Japan

T. Shinoda et al. (eds.), *Juvenile Hormone*, Entomology Monographs,
https://doi.org/10.1007/978-981-95-8605-9_8

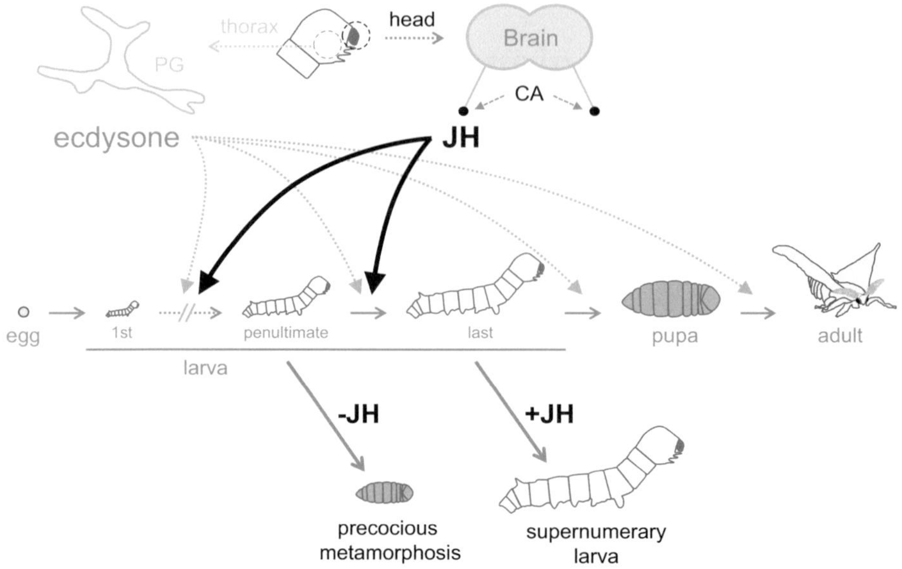

Fig. 8.1 Hormonal regulation of development in holometabolous insects. PG prothoracic gland, CA corpora allata

Kiguchi and Riddiford 1978). Additionally, the application of anti-JH compounds, which act on JH biosynthetic and signaling pathways (Bowers et al. 1976; Riddiford et al. 1983; Unnithan et al. 1995; Kayukawa et al. 2020; Furuta et al. 2023), and the knockdown or knockout of JH-related genes also cause precocious metamorphosis (Konopova and Jindra 2007; Minakuchi et al. 2008a, 2009; Daimon et al. 2015; Kayukawa et al. 2024). Conversely, implantation of the active CA and topical application of JH analog (JHA) in the last larval instar inhibit normal metamorphosis, leading to extension of the larval stage and supernumerary larval molt (Wigglesworth 1936; Sláma and Williams 1965; Bowers et al. 1966; Sehnal and Sláma 1966; Hatakoshi et al. 1986). In the pupal stage, JH treatment prevents complete pupal-adult transition and results in a supernumerary pupa (second pupa), pupal-adult intermediates, and/or adults preserving some pupal features (Wigglesworth 1958; Riddiford and Ajami 1973; Postlethwait 1974). Thus, the primary role of JH in metamorphosis is to prevent larvae from precociously turning into pupae and adults, which is termed the "*status quo*" action (Riddiford 1994).

8.2 Ecdysone-Related Key Genes in Insect Metamorphosis

8.2.1 Ecdysone Receptor (EcR)/Ultraspiracle (USP)

The molecular actions and mechanisms of ecdysone in insect metamorphosis have been elucidated in detail in the twentieth century (Ashburner et al. 1974). Ecdysone secreted from the PG is converted into its active form (20E) in peripheral tissues (Gilbert and King 1973; Grieneisen 1994). 20E carried to target cells is received by the ligand-binding domain (LBD) of the ecdysone receptor (EcR), and the 20E-liganded EcR forms a heterodimer with ultraspiracle (USP). EcR and USP belong to the nuclear receptor superfamily and have a DNA-binding domain (DBD) composed of two C4-type zinc fingers (Lee et al. 2000b; Fig. 8.2). The liganded EcR/USP complex directly binds to an ecdysone response element (EcRE) upstream of ecdysone-inducible genes, activating their transcription. The EcRE consensus sequence is RGKTCANTGAMCY, which is a palindromic structure consisting of two 6 bp half-sites (Riddihough and Pelham 1987; Cherbas et al. 1991; Antoniewski et al. 1994, 1995; Lehmann and Korge 1995; Lehmann et al. 1997). As a major recent development, the ecdysone importer (EcI) was identified in the fruit fly *Drosophila melanogaster* in 2018 (Okamoto et al. 2018). EcI is localized on the cell membrane and functions to transport 20E from the hemolymph into target cells (Okamoto et al. 2018; Yamanaka 2021).

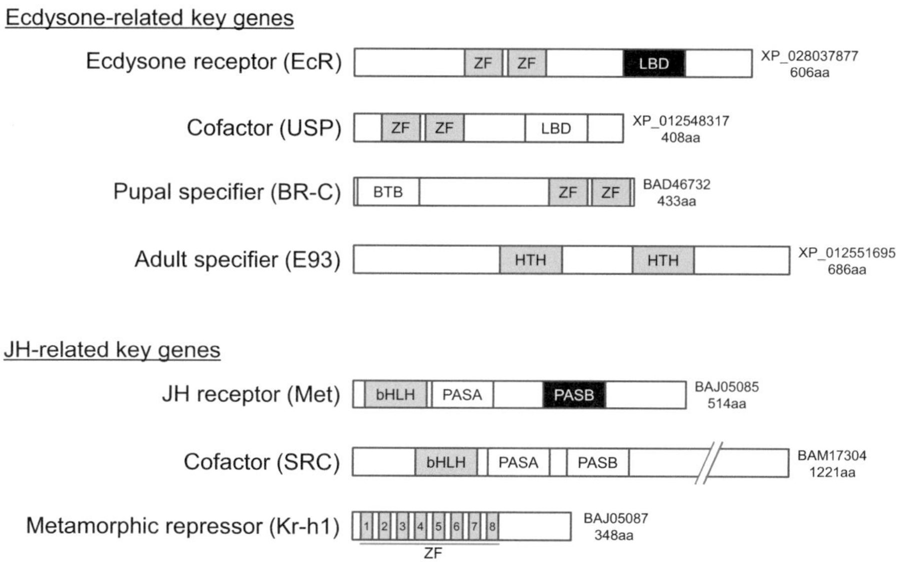

Fig. 8.2 Schematic representation of key genes involved in insect metamorphosis. Each gene has several isoforms, but one of them in *Bombyx mori* is shown here on a protein basis. Boxes represent domains, and black and gray boxes indicate ligand and DNA-binding domains, respectively. The accession numbers of the DDBJ/EMBL/GenBank database and the numbers of amino acid residues are shown to the right of the protein boxes

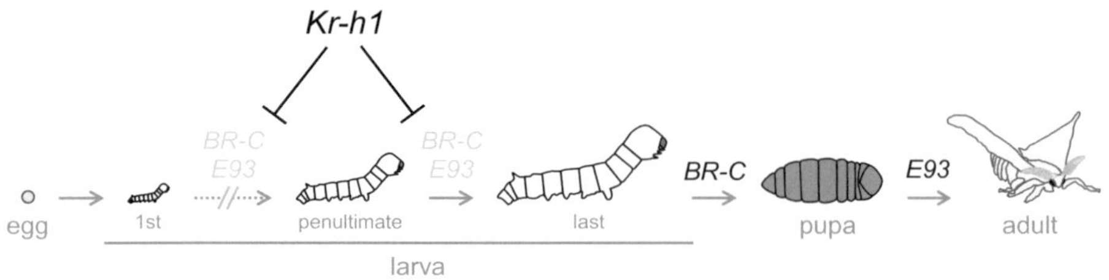

Fig. 8.3 Regulation of development by major metamorphic genes in holometabolous insects. *Kr-h1 Krüppel homolog 1*, *BR-C Broad-Complex*, *E93 Ecdysone-induced protein 93F*

8.2.2 Broad-Complex (BR-C)

The *Broad-Complex* (*BR-C*; also known as *Broad*, *BRC* and *Br*) is induced by 20E-liganded EcR/USP in the prepupal stage of the last larval instar and functions as a "pupal specifier" in the larval-pupal transition of holometabolous insects (Kiss et al. 1976, 1988; Zhou et al. 1998, 2004; Zhou and Riddiford 2002; Muramatsu et al. 2008; Suzuki et al. 2008; Konopova and Jindra 2008; Fig. 8.3). An amorphic *BR-C* mutant of *D. melanogaster* fails to achieve larval-pupal metamorphosis (Kiss et al. 1976, 1988), and overexpression of *BR-C* in penultimate instar larvae concomitantly suppresses the expression of larval cuticle genes and activates pupal cuticle genes (Zhou and Riddiford 2002). The BR-C protein is a transcription factor that comprises a Bric-a-brac/Tramtrack/Broad complex (BTB) domain and a paired C2H2 zinc-finger domain (DiBello et al. 1991; Bayer et al. 1996; Spokony and Restifo 2007; Piulachs et al. 2010; Nagamine et al. 2014; Fig. 8.2). Several isoforms of *BR-C*, which determine tissue-specific metamorphic outcomes during the larval-pupal transition, are synthesized by alternative splicing in the zinc-finger domain (Emery et al. 1994; Suzuki et al. 2008; Konopova and Jindra 2008).

8.2.3 Ecdysone-Induced Protein 93F (E93)

Ecdysone-induced protein 93F (*E93;* also known as *Eip93F*), originally identified as an ecdysone-induced gene from *D. melanogaster*, is a member of the helix-turn-helix (HTH) transcription factor family and has two HTH-DNA-binding domains (Baehrecke and Thummel 1995; Chafino et al. 2018; Fig. 8.2). *E93* regulates the expression of genes involved in programmed cell death, including apoptosis and autophagy, to facilitate the remodeling of larval to adult tissues (Lee et al. 2000a; Lee and Baehrecke 2001; Liu et al. 2014, 2015b). A subsequent study using the hemimetabolous German cockroach, *Blattella germanica*, identified that *E93* is an "adult specifier" because cockroaches treated with *E93* RNA interference (RNAi) continuously repeat nymphal molts and finally reach the tenth instar, whereas normal cockroaches only progress through six nymphal instars before adult metamorphosis (Ureña et al. 2014; Fig. 8.3). Similarly, *E93* RNAi prevents the pupal-adult transition and results in a supernumerary pupa in the red flour beetle, *Tribolium castaneum*, and *D. melanogaster*, suggesting that the function of *E93* as an adult specifier gene is conserved in holometabolous insects (Ureña et al. 2014).

8.3 JH-Related Key Genes in Insect Metamorphosis

8.3.1 *Methoprene tolerant (Met) and steroid receptor coactivator (SRC)*

The methoprene-tolerant (Met) protein belongs to a family of basic helix-loop-helix Per-ARNT-Sim (bHLH-PAS) transcription factors and has a DNA-binding domain (bHLH) and two PAS domains (PASA and PASB) (Ashok et al. 1998; Baumann et al. 2010; Fig. 8.2). Met was first identified in a *D. melanogaster* mutant that was resistant to toxic doses of JH and JHA (methoprene) (Wilson and Fabian 1986; Ashok et al. 1998) and functions as a core component of JH receptors (Jindra et al. 2013, 2015, 2021b). Using a JH binding assay (Touhara et al. 1993), Miura et al. (2005) were the first to demonstrate that the Met protein of *D. melanogaster*, which is synthesized in vitro using rabbit reticulocyte lysate, binds to JH with high affinity. Subsequent studies further confirmed that Met in *D. melanogaster* and *T. castaneum* binds to JH (Charles et al. 2011). Moreover, RNAi experiments have proven that Met mediates the anti-metamorphic action of JH in both holometabolous and hemimetabolous insects (Konopova and Jindra 2007; Parthasarathy et al. 2008; Konopova et al. 2011). Taken together, these results demonstrate that Met is a bona fide JH receptor.

bHLH-PAS family proteins often function as homodimers or heterodimers (Furness et al. 2007; Partch and Gardner 2010; Edwards and Gorelick 2022; Tumova et al. 2024). Met forms a complex with steroid receptor coactivator (SRC also known as FISC or taiman) and directly activates the transcription of the early trypsin gene during the reproduction of the yellow fever mosquito, *Aedes aegypti* (Li et al. 2011; Fig. 8.2). Thereafter, Met of *T. castaneum* was shown to sense the JH signal through direct, specific binding and interact with SRC, thus establishing itself as a JH receptor (Charles et al. 2011). Furthermore, other intriguing reports on Met include its phosphorylation (Liu et al. 2015a) and interaction with HSP83 (He et al. 2014; Jindra et al. 2024; see Chap. 10).

8.3.2 *Krüppel homolog 1 (Kr-h1)*

Krüppel homolog 1 (*Kr-h1*), a C2H2 zinc-finger type transcription factor, has eight C2H2 zinc-finger domains and was identified as a modulator of the prepupal ecdysone response in *D. melanogaster* (Pecasse et al. 2000; Fig. 8.2). Subsequently, microarray analysis revealed that *Kr-h1* is also a JH-early inducible gene in *D. melanogaster* pupae (Minakuchi et al. 2008b). *Kr-h1* in *T. castaneum* larvae is also rapidly induced by JH, and its knockdown causes precocious metamorphosis (Minakuchi et al. 2009), as seen in the knockdown of *Met* (Konopova and Jindra 2007; Parthasarathy et al. 2008). Moreover, RNAi silencing analysis showed that *Kr-h1* acts downstream of *Met* (Minakuchi et al. 2009). Taken together, the available information indicates that *Kr-h1* mediates JH action and plays a primary role as a "metamorphic repressor" in insects (Fig. 8.3). Later studies showed that Kr-h1 represses metamorphosis in both holometabolous and hemimetabolous insects (Konopova et al. 2011; Lozano and Belles 2011). The following sections introduce the molecular mechanisms underlying the induction of *Kr-h1* by JH to the repression of metamorphosis, based mainly on analyses using the silkworm *Bombyx mori*.

8.4 Induction Mechanism of *Kr-h1* by JH

8.4.1 Expression Profile of Kr-h1

In *B. mori*, *Kr-h1* is constitutively expressed until the penultimate instar larvae with some fluctuations, but its expression declines to a trace level early in the final larval instar and remains at this level until the prepupal stage (spinning stage in *B. mori*) (Kayukawa et al. 2012). The expression pattern of *Kr-h1* showed a good correlation with the changes in the JH titer in the hemolymph (Kayukawa et al. 2012; Niimi and Sakurai 1997). In allatectomized larvae, the *Kr-h1* transcript decreases prematurely to a trace level and is maintained at a high level in allatectomized larvae treated with JHA (methoprene) (Kayukawa et al. 2012). These results clearly show that the expression of *Kr-h1* is positively regulated by JH in *B. mori* larvae, as reported in other insect species (Minakuchi et al. 2008a, b, 2009; Konopova et al. 2011; Lozano and Belles 2011). Tissue culture experiments of the *B. mori* epidermis showed that 20E significantly synergizes the induction by JHA, whereas 20E alone induces *Kr-h1* very slightly. Therefore, *Kr-h1* is primarily induced by JH, and 20E acts secondarily to induce *Kr-h1* (Kayukawa et al. 2014). The synergistic induction of *K-h1* is likely required to firmly repress the expression of 20E-inducible metamorphic genes during the larval stage in the presence of JH and 20E.

In a *Bombyx* cell line, the *Kr-h1* transcript is barely detectable before JHA treatment; however, this level increases significantly within 0.5 h of treatment and approaches the peak level by 2 h (Kayukawa et al. 2012). Dose-dependent increases in *Kr-h1* transcript levels were observed in cells treated with JH I, JH II, JH III, JHA, or methyl farnesoate (MF); the median effective concentrations (EC50) were 1.6×10^{-10}, 1.2×10^{-10}, 2.6×10^{-10}, 6.0×10^{-8}, and 1.1×10^{-7} M, respectively, and farnesoic acid (FA) was ineffective (Kayukawa et al. 2012). This analysis clearly demonstrated that *Kr-h1* is responsive to subnanomolar levels of natural JHs in *B. mori* (i.e., JH I and JH II). Treatment with the protein synthesis inhibitor cycloheximide (CHX) had no significant effect on the level of the *Kr-h1* transcript, indicating that the JH-dependent induction of *Kr-h1* is not mediated by de novo synthesized proteins (Kayukawa et al. 2012). This high response of *Kr-h1* to JH was also observed in a *T. castaneum* cell line (Kayukawa et al. 2013).

8.4.2 Identification of JH Response Element (JHRE) in Kr-h1

To account for the rapid and extensive response of *Kr-h1* to JH, we searched for a JH response element (JHRE) in the *Kr-h1* promoter region using a combination of reporter assays, which indirectly measure the activity or expression level of a specific gene using a reporter gene that is easy to detect and quantify, and mutagenesis in the *Bombyx* cell line. These analyses identified a novel JHRE (JHRE of *Kr-h1*, *k*JHRE) located approximately 2 kb upstream of the transcription start site of *Kr-h1* (Kayukawa et al. 2012). The core region of *k*JHRE (*k*JHREc, GCCTC**CACGTG**) contains a canonical E-box sequence (CANNTG), to which bHLH transcription factors preferentially bind (Massari and Murre 2000; Sailsbery et al. 2012; Fig. 8.4). Therefore, *k*JHREc is a plausible sequence for the binding of Met, which possesses a bHLH domain.

In *Kr-h1* of *T. castaneum*, *k*JHREc is perfectly conserved and located 150 bp upstream of the transcription start site (Kayukawa et al. 2013; Fig. 8.4). When a mutation is introduced in *k*JHREc, reporter activity in the presence of JHA decreases significantly compared to the wild-type reporter, suggesting that this *k*JHREc contributes to the JH-dependent induction of *T. castaneum Kr-h1* (Kayukawa et al.

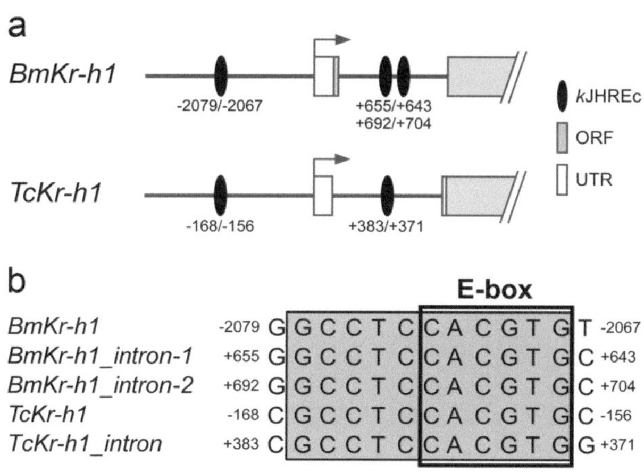

Fig. 8.4 Core sequence of juvenile hormone (JH) response element in *Kr-h1* (*k*JHREc). (**a**) Genomic structures of *Krüppel homolog 1* (*Kr-h1*) and *k*JHREc positions. *Bm*, *Bombyx mori*; *Tc*, *Tribolium castaneum*. White boxes, gray boxes, and arrows represent the UTRs, ORFs, and transcription start sites, respectively. Black ellipses indicate *k*JHREc, and the numbers represent distances from the transcription start sites. (**b**) Alignment of *k*JHREc sequences. Light gray shading indicates highly conserved nucleotides

2013). Intriguingly, the *k*JHREc sequences are also found in the first intron of *Kr-h1* in both *B. mori* and *T. castaneum* (Kayukawa et al. 2013; Fig. 8.4). A minor isoform with a transcription start site downstream of the *k*JHREc has been identified in *B. mori* (Kayukawa et al. 2012). In *T. castaneum*, such an isoform has been registered in the latest NCBI, suggesting that the *k*JHREc in the intron might be involved in the induction of these isoforms. A recent study using the fall armyworm, *Spodoptera frugiperda*, demonstrated that the isoform-specific *k*JHREc regulates the expression of *Kr-h1* in an isoform-specific manner (Chen and Palli 2022).

8.4.3 Transcriptional Activation of Kr-h1 by Met/SRC

The JH reception by Met and the interaction between Met and SRC have been introduced above. Comprehensive interactions of JH, Met, SRC, and *k*JHRE in the JH signaling pathway were examined using a two-hybrid system, a method that detects protein-protein interactions, in a mammalian cell line. A JHA-dependent increase in UAS reporter activity was observed in cells co-expressing Gal4DBD-Met and VP16AD-SRC, indicating that Met and SRC form a protein-protein complex in a JH-dependent manner (Kayukawa et al. 2012; Kayukawa and Shinoda 2015). Moreover, the reporter carrying the *k*JHRE region is activated by the application of JH when native Met and SRC are co-expressed, and mutations in the *k*JHRE result in the loss of JH-dependent induction (Kayukawa et al. 2012; Kayukawa and Shinoda 2015). Therefore, these results propose the following hypothesis for the mechanism of JH-mediated induction of *Kr-h1*: Met accepts JH as a ligand, JH-liganded Met interacts with SRC, and the JH/Met/SRC complex activates *Kr-h1* by interacting with *k*JHRE (Fig. 8.5). Subsequent analyses using an electrophoresis mobility shift assay (EMSA) demonstrated the direct binding of the JH/Met/SRC complex to *k*JHREc, strongly supporting this hypothesis (Li et al. 2014; Jindra et al. 2021a).

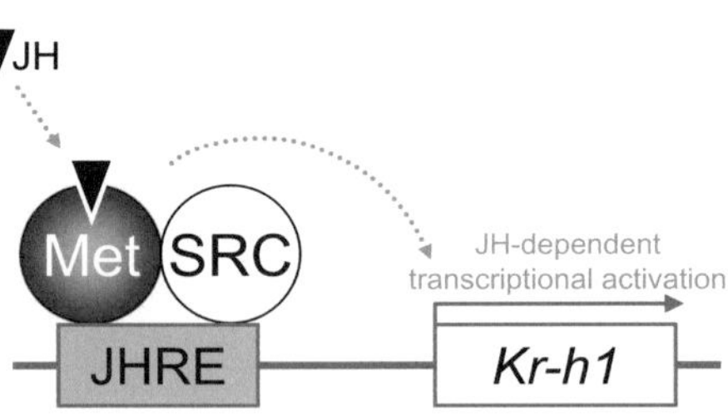

Fig. 8.5 A model for juvenile hormone (JH)-mediated transcriptional induction of *Krüppel homolog 1* (*Kr-h1*). Met Methoprene tolerant, SRC Steroid receptor coactivator, JHRE JH response element

8.5 Repression Mechanism of Larval-Pupal Metamorphosis by JH

8.5.1 BR-C Repression by JH

In the tobacco hornworm, *Manduca sexta* and *B. mori*, allatectomy in penultimate larvae decreases the JH titer in the hemolymph and induces *BR-C* expression, resulting in precocious pupation. The application of exogenous JH to allatectomized larvae inhibits *BR-C* expression and pupation in these larvae (Zhou et al. 1998; Reza et al. 2004). JH-mediated repression of the activation of *BR-C* by 20E has also been reported in the cultured epidermis, but the inhibitory effect of JH declines after the transition to the pupal commitment phase (Muramatsu et al. 2008; Zhou and Riddiford 2001). A study using the *Kr-h1* mutant of *D. melanogaster* showed that *Kr-h1* is required for the repression of *BR-C* expression in *D. melanogaster* larvae (Huang et al. 2011; Fig. 8.3).

In the *Bombyx* cell line, the *BR-C* transcript is activated by 20E, and this activation is repressed by JH (Nishita and Takiya 2009; Nishita 2014). The hormonal responses of a reporter carrying the approximately 5 kb region of the *BR-C* promoter (*BR-C*_P) exhibited the same pattern as those of the *BR-C* transcript. EcRE, which interacts with EcR/USP and is located approximately 5 kb upstream of the *BR-C* transcription start site, was identified from *BR-C*_P (Nishita and Takiya 2009; Nishita 2014). Therefore, we elucidated the following molecular mechanism of JH-mediated repression of larval-pupal metamorphosis using these responses in a *Bombyx* cell line.

8.5.2 Identification of Kr-h1 Binding Site (KBS) in BR-C

To examine the involvement of Kr-h1 in JH-mediated repression against the activation of *BR-C* by 20E, we performed reporter assays using the *Bombyx* cell line. Ectopic expression of Kr-h1 represses the 20E-dependent activation of *BR-C*_P (Kayukawa et al. 2016). Additionally, RNAi silencing of *Kr-h1* alleviates the repression of reporter activity by JHA, confirming that Kr-h1 represses the activation of *BR-C*_P (Kayukawa et al. 2016).

As mentioned above, Kr-h1 proteins commonly have eight C2H2 zinc-finger domains, which putatively bind to specific DNA sequences (Minakuchi et al. 2009; Kayukawa et al. 2012). Therefore, we hypothesized that Kr-h1 directly binds to *BR-C*_P and represses the activation of *BR-C*_P by 20E. We then searched for a putative Kr-h1 binding site (KBS) using a combination of reporter assays and mutagenesis. These experiments narrowed down a 60-bp sequence (ACGTCTATCCGT-**GACCTACGCTAACGCTAAATAGAGTTCCGA**CCTTGACTCTAACG-GTCG), located approximately 4.5 kb upstream of the *BR-C* transcription start site, and demonstrated that the bold 30-bp sequence is crucial for Kr-h1 binding

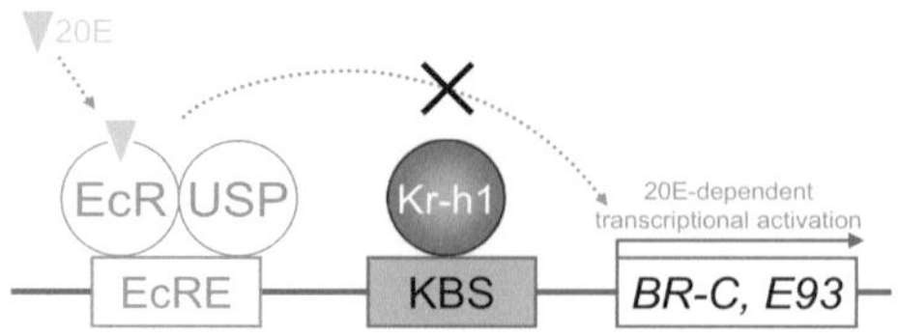

Fig. 8.6 A model for the repression mechanism of *Broad-complex* (*BR-C*) and *Ecdysone-induced protein 93F* (*E93*) by Krüppel homolog 1 (Kr-h1). EcR Ecdysone receptor, USP Ultraspiracle, EcRE ecdysone response element, KBS Kr-h1 binding site, 20E 20-hydroxyecdysone

(Kayukawa et al. 2016; Fig. 8.6). Moreover, EMSA demonstrated that Kr-h1 directly and specifically binds to the KBS sequence (Kayukawa et al. 2016).

8.5.3 Molecular Mechanism of BR-C Repression by Kr-h1

The reporter carrying *BR-C*_P, which contains EcRE and KBS, is activated by 20E, and this activation is repressed by JHA. Ectopic expression of Kr-h1 repressed the activation of the reporter in the absence of JHA. When KBS is deleted from the reporter, the repression of reporter activation is not repressed by JHA, even if Kr-h1 is ectopically expressed in the cells (Kayukawa et al. 2016). Taken together, we propose the following mechanism of hormonal regulation in *BR-C* as a pupal specifier: in the presence of JH, *Kr-h1* expression is induced by JH via Met/SRC; subsequently, Kr-h1 binds to the KBS sequence of the *BR-C* promoter, thereby repressing the 20E-dependent expression of *BR-C* (Fig. 8.6).

The response of *BR-C* in the cell line to 20E and/or JH is reminiscent of that observed in the larval epidermis and silk gland before pupal commitment (Zhou et al. 1998; Reza et al. 2004; Muramatsu et al. 2008). The expression of *Kr-h1* until the penultimate larval stage of *B. mori* is maintained at high levels by the continuous presence of JH (Kayukawa et al. 2012, 2014). The decline in JH titer at the beginning of the final instar larvae causes the temporal disappearance of *Kr-h1* (Kayukawa et al. 2012, 2014). This might facilitate the subsequent induction of *BR-C* by 20E after commitment to larval-pupal metamorphosis. Thus, this mechanism is likely to be applicable to immature larvae until pupal commitment.

8.6 Repression Mechanism of Larval-Adult Metamorphosis

8.6.1 E93 Repression by JH

In the expression of the adult specifier *E93*, 20E-dependent induction and JH-dependent repression are observed in *B. germanica*, similar to *BR-C* (Ureña et al. 2014). Moreover, JH-inducible Kr-h1 represses *E93* transcript levels in *B.*

germanica, and this repression has also been observed in *D. melanogaster* and *T. castaneum* (Belles and Santos 2014; Ureña et al. 2016; Fig. 8.3).

In *B. mori*, only trace expression of *E93* is detected during the last larval stage, but it gradually increases at low levels in the pupal stage. *E93* transcript levels peak during the early pupal stage (Kayukawa et al. 2017). The expression pattern of *E93* is correlated with changes in the ecdysone titer of the hemolymph when JH is absent (Calvez et al. 1976; Satake et al. 1998; Kaneko et al. 2006; Kamimura et al. 2012), suggesting that *E93* in *B. mori* is induced by 20E and suppressed by JH. Following this result, the effects of 20E and JHA on the cultured epidermis of pupae at day 0 were investigated in *B. mori*. *E93* expression in pupae on day 0 is induced by 20E, and JHA represses this induction. JH-dependent induction of *Kr-h1* has also been observed in the epidermis of fifth-instar larvae on day 5 and adults on day 0 (Kayukawa et al. 2014). Consistently, JHA induced *Kr-h1* expression in the pupal epidermis at day 0, and 20E superinduced it. In the *Bombyx* cell line, the *E93* transcript is also weakly detectable at baseline but is significantly upregulated by 20E, and this effect is repressed by JHA (Kayukawa et al. 2017).

8.6.2 Identification of KBS in E93

As described above, 20E-dependent *E93* transcription is repressed by JH in *B. mori*. Thus, we hypothesized that the hormonal response of *E93* is also regulated by the same molecular mechanism as *BR-C*. A KBS-like sequence (TGACCTTTCTCAAC) with similarity to the KBS of *BR-C* is found approximately 2 kb upstream of the transcription start site of *E93*. In addition, a putative EcRE (GGGTCGATGACCG) homologous to a consensus EcRE sequence (RGKTCANTGAMCY) is located near the KBS-like sequence (Kayukawa et al. 2017). To characterize the functions of the putative KBS and EcRE at the *E93* locus, reporter assays were performed in the *Bombyx* cell line using constructs carrying the *E93* promoter region (*E93*_P, approximately 3 kb). The wild-type *E93*_P reporter is strongly activated by 20E, and this induction is completely repressed by JHA (Kayukawa et al. 2017). Mutations in the putative KBS region abolish this JHA-dependent repression (Kayukawa et al. 2017). Deletion of the putative EcRE from the wild-type reporter abolished 20E-induced expression, indicating that the sequence represents an authentic EcRE (Kayukawa et al. 2017). Collectively, these results reveal that *E93* transcription is activated by 20E via the EcRE and that the 20E-induction is repressed by JHA via the putative KBS.

Further reporter assays in combination with RNAi were performed to examine whether Kr-h1 interacts with the putative KBS in *E93*. *Kr-h1* RNAi silencing abolishes JHA-dependent repression of *E93* at the transcript level (Kayukawa et al. 2017). Moreover, *Kr-h1* RNAi alleviates JHA-dependent reporter suppression, indicating that Kr-h1 is involved in the repression of 20E-induced *E93* transcription (Kayukawa et al. 2017). These results indicate that the KBS of *E93* is an authentic KBS that functionally interacts with Kr-h1.

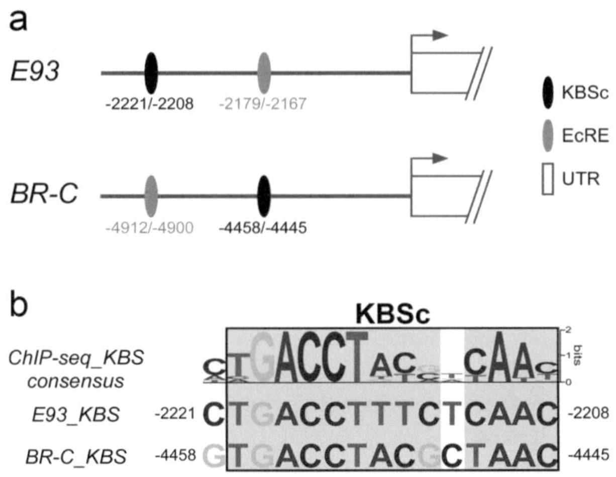

Fig. 8.7 Consensus sequence of Krüppel homolog 1 (Kr-h1)-binding site (KBSc) in *Broad-complex* (*BR-C*) and *Ecdysone-induced protein 93F* (*E93*) of *Bombyx mori*. (**a**) Genomic structures, KBSc and ecdysone response element (EcRE) positions. White boxes and arrows represent the UTRs and transcription start sites, respectively. Black and gray ellipses indicate KBSc and EcRE, respectively, and the numbers below represent distances from the transcription start sites. (**b**) Motif analysis based on Kr-h1 ChIP-seq. The obtained KBSc consensus is aligned with that of *E93* and *BR-C*. Light gray shading indicates highly conserved nucleotides

To identify putative KBSs on a genome-wide basis, we performed Chromatin immunoprecipitation sequencing (ChIP-seq) analysis using *Bombyx* cells overexpressing Kr-h1 with a C-terminal epitope tag. Focusing on the region around the *E93* locus, an enriched peak was observed in the region containing the KBS identified approximately 2 kb upstream (Kayukawa et al. 2017). Additionally, ChIP-qPCR confirmed significant enrichment in the region containing the KBS (Kayukawa et al. 2017). Genome-wide motif analysis of the enriched sequences successfully identified a KBS consensus sequence (KBSc, TGACCTWYSNYAAC), and the KBSs found upstream of *E93* and *BR-C* were highly similar to the KBSc (Kayukawa et al. 2017; Fig. 8.7).

8.6.3 Molecular Mechanism of E93 Repression by Kr-h1

The aforementioned results facilitate exploration of the following mechanism by which Kr-h1 represses *E93* induction: JH-inducible Kr-h1 directly binds to the KBS site upstream of the *E93* locus to repress its transcription in a cell-autonomous manner, thereby preventing larvae from bypassing the pupal stage and progressing to precocious adult development (Kayukawa et al. 2017; Fig. 8.6). This molecular mechanism is similar to that of Kr-h1-mediated repression of *BR-C* (Kayukawa et al. 2016). Because the KBS and EcRE in the *E93* promoter region are separated by only approximately 30 bp, we hypothesized that the binding of Kr-h1 to the KBS sterically inhibits the interaction between EcR/USP and EcRE. However, EMSA analyses clearly indicate that Kr-h1 and EcR/USP independently bind to KBS and EcRE, respectively, and physical interference is unlikely to occur between Kr-h1 and EcR/USP on the *E93* promoter (Kayukawa et al. 2017).

More than 60% of Kr-h1 is composed of zinc-finger domains, and the remaining 40% shares no similarity with other insect Kr-h1s (Minakuchi et al. 2009; Kayukawa et al. 2012). However, we discovered two distinctive motifs (LPPRKR and SVIQFA motifs) in the Kr-h1 C-terminus conserved in four holometabolous insects and termed this region the C-terminal conserved domain (CtC domain) (Kayukawa et al. 2017). To determine whether the CtC domain is involved in the transcriptional inhibitory activity of Kr-h1, reporter assays were performed using the *Bombyx* cell line ectopically expressing wild-type Kr-h1 or Kr-h1 with deletion of the CtC domain (Kr-h1_ΔCtC). Notably, ectopic expression of native Kr-h1 repressed the 20E-dependent activation of a reporter carrying *E93*_P. In contrast, deletion of the CtC domain abolishes the repression by Kr-h1, and ectopic expression of Kr-h1_ΔCtC inhibits the JH-mediated repression through a dominant-negative effect (Kayukawa et al. 2017). Taken together, these findings support the notion that the CtC domain of Kr-h1 contributes to JH-inducible transcriptional repression of target genes.

8.7 Novel Metamorphic Repressor Gene, *chronologically inappropriate morphogenesis* (*Chinmo*)

Recently, chronologically inappropriate morphogenesis (Chinmo) was discovered to function as a "larval specifier" in *D. melanogaster* (Truman and Riddiford 2022). Surprisingly, *Chinmo* RNAi induces precocious puparium formation from the first instar larvae of *D. melanogaster* (Truman and Riddiford 2022). Chinmo is a transcription factor that comprises a BTB domain and a paired C2H2 zinc-finger domain, similar to BR-C, and plays a key role in maintaining the larval stage of

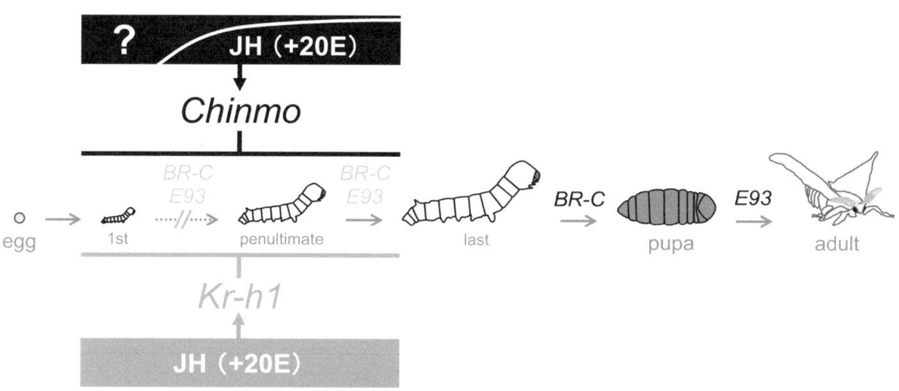

Fig. 8.8 A model for the metamorphic repression of *Broad-Complex* (*BR-C*) and *Ecdysone-induced protein 93F* (*E93*) by collaboration of *Krüppel homolog 1* (*Kr-h1*) and *chronologically inappropriate morphogenesis* (*Chinmo*). JH Juvenile hormone, 20E 20-hydroxyecdysone

D. melanogaster (Truman and Riddiford 2022). *Chinmo* expression continues at high levels until the critical weight of the final larval instar (third) of *D. melanogaster*, which is the growth-related checkpoint that allows preparation for metamorphosis, thereby repressing the expression of genes involved in metamorphosis (*BR-C* and *E93*) (Truman and Riddiford 2022). The action of *Chinmo* in *D. melanogaster* is comparable to that of *Kr-h1*, which is widely conserved among various insects.

In subsequent studies using *S. frugiperda*, *Chinmo* knockout developed normally until the second larval instar, and most of these insects died during the third instar molt without the development of precocious metamorphosis (Chen et al. 2024). *Chinmo* and *Kr-h1* double knockout induces premature metamorphosis in the third larval instar, whereas Kr-h1 knockout only occurs in the fifth instar larvae (penultimate) (Chen et al. 2024). These results suggest that Chinmo may not be involved in repressing metamorphosis during the early larval stage, unlike *D. melanogaster*. Single-cell multiome ATAC analysis revealed that Chinmo participates in chromatin modifications that prevent the promoter accessibility and expression of *BR-C* and *E93* (Chen et al. 2024).

To briefly introduce our results of *Chinmo* analysis in *B. mori* (Kayukawa unpublished data), the developmental expression profile of *Chinmo* is highly correlated with changes in hemolymph JH titer. Notably, tissue and cell culture experiments indicated that *Chinmo* was induced by JH and that superinduction was caused by JH and 20E. These expression patterns were similar to those of *Kr-h1*. Intriguingly, *Chinmo* transcript levels were maintained at high levels in the abdominal epidermis of the neck-ligated larvae during early larval instar, whereas a drastic decrease was observed in the middle larval stage. These results suggest that *Chinmo* is mainly induced by JH during the middle larval stage, but its expression during the early larval stage is regulated by JH and/or unknown factor(s). Reporter assays using the *Bombyx* cell line showed that overexpression of *Chinmo* shut down 20E-dependent activation of *BR-C*_P and *E93*_P. The Chinmo binding site (CBS) differs from the previously identified KBS sequence and is located near the transcription start site in *BR-C* and *E93*. The distance between CBS and KBS was >2 kb in both *BR-C*_P and *E93*_P. Furthermore, two-hybrid assays did not confirm the interaction between Kr-h1 and Chinmo. Therefore, Kr-h1 and Chinmo repress metamorphic genes independently, and precocious metamorphosis is firmly suppressed by this double lock during the larval stage.

The yellow-spotted longicorn beetle, *Psacothea hilaris*, usually metamorphoses to pupa primarily from the fifth or sixth larval instar. In *P. hilaris*, *Kr-h1* and *Chinmo* double knockdown by RNAi causes precocious metamorphosis and induces nearly complete pupae from the second larval instar, whereas it is not observed in *Kr-h1* or *Chinmo* RNAi alone (Kayukawa unpublished data). These results differed among the three species described above, suggesting that metamorphic regulation by *Kr-h1* and *Chinmo* is diverse among insect species. Further analysis of *Chinmo* would provide a deep understanding of the fundamental mechanism underlying metamorphosis, which could contribute to the pest management field, such as developing a novel insect growth regulator that shortens the larval stage to the limit.

References

Antoniewski C, Laval M, Dahan A, Lepesant JA (1994) The ecdysone response enhancer of the *Fbp1* gene of *Drosophila melanogaster* is a direct target for the EcR/USP nuclear receptor. Mol Cell Biol 14:4465–4474. https://doi.org/10.1128/mcb.14.7.4465-4474.1994

Antoniewski C, O'Grady MS, Edmondson RG, Lassieur SM, Benes H (1995) Characterization of an EcR/USP heterodimer target site that mediates ecdysone responsiveness of the *Drosophila Lsp-2* gene. Mol Gen Genet 249:545–556. https://doi.org/10.1007/bf00290580

Ashburner M, Chihara C, Meltzer P, Richards G (1974) Temporal control of puffing activity in polytene chromosomes. Cold Spring Harb Symp Quant Biol 38:655–662. https://doi.org/10.1101/sqb.1974.038.01.070

Ashok M, Turner C, Wilson TG (1998) Insect juvenile hormone resistance gene homology with the bHLH-PAS family of transcriptional regulators. Proc Natl Acad Sci USA 95:2761–2766. https://doi.org/10.1073/pnas.95.6.2761

Baehrecke EH, Thummel CS (1995) The *Drosophila E93* gene from the 93F early puff displays stage- and tissue-specific regulation by 20-hydroxyecdysone. Dev Biol 171:85–97. https://doi.org/10.1006/dbio.1995.1262

Baumann A, Fujiwara Y, Wilson TG (2010) Evolutionary divergence of the paralogs *Methoprene tolerant* (*Met*) and *germ cell expressed* (*gce*) within the genus *Drosophila*. J Insect Physiol 56:1445–1455. https://doi.org/10.1016/j.jinsphys.2010.05.001

Bayer CA, Holley B, Fristrom JW (1996) A switch in *broad-complex* zinc-finger isoform expression is regulated posttranscriptionally during the metamorphosis of *Drosophila* imaginal discs. Dev Biol 177:1–14. https://doi.org/10.1006/dbio.1996.0140

Belles X, Santos CG (2014) The MEKRE93 (Methoprene tolerant-Krüppel homolog 1-E93) pathway in the regulation of insect metamorphosis, and the homology of the pupal stage. Insect Biochem Mol Biol 52:60–68. https://doi.org/10.1016/j.ibmb.2014.06.009

Bowers WS, Fales HM, Thompson MJ, Uebel EC (1966) Juvenile hormone: identification of an active compound from balsam fir. Science 154:1020–1021. https://doi.org/10.1126/science.154.3752.1020

Bowers WS, Ohta T, Cleere JS, Marsella PA (1976) Discovery of insect anti-juvenile hormones in plants: plants yield a potential fourth-generation insecticide. Science 193:542–547. https://doi.org/10.1126/science.986685

Calvez B, Hirn M, De Reggi M (1976) Ecdysone changes in the haemolymph to two silkworms (*Bombyx mori* and *Philosamia cynthia*) during larval and pupal development. FEBS Lett 72:57–61. https://doi.org/10.1016/0014-5793(76)80898-8

Chafino S, López-Escardó D, Benelli G, Kovac H, Casacuberta E, Franch-Marro X et al (2018) Differential expression of the adult specifier E93 in the strepsipteran *Xenos vesparum* Rossi suggests a role in female neoteny. Sci Rep 8:14176. https://doi.org/10.1038/s41598-018-32611-y

Charles J-P, Iwema T, Epa VC, Takaki K, Rynes J, Jindra M (2011) Ligand-binding properties of a juvenile hormone receptor, Methoprene-tolerant. Proc Natl Acad Sci USA 108:21128–21133. https://doi.org/10.1073/pnas.1116123109

Chen X, Koo J, Kumar Arya S, Palli SR (2024) Chronologically inappropriate morphogenesis (*Chinmo*) is required for maintenance of larval stages of fall armyworm. Proc Natl Acad Sci USA 121:e2411286121. https://doi.org/10.1073/pnas.2411286121

Chen X, Palli SR (2022) Identification of species-specific juvenile hormone response elements in the fall armyworm, *Spodoptera frugiperda*. Insect Biochem Mol Biol 151:103860. https://doi.org/10.1016/j.ibmb.2022.103860

Cherbas L, Lee K, Cherbas P (1991) Identification of ecdysone response elements by analysis of the *Drosophila Eip28/29* gene. Genes Dev 5:120–131. https://doi.org/10.1101/gad.5.1.120

Daimon T, Uchibori M, Nakao H, Sezutsu H, Shinoda T (2015) Knockout silkworms reveal a dispensable role for juvenile hormones in holometabolous life cycle. Proc Natl Acad Sci USA 112:E4226–E4235. https://doi.org/10.1073/pnas.1506645112

DiBello PR, Withers DA, Bayer CA, Fristrom JW, Guild GM (1991) The *drosophila broad-complex* encodes a family of related proteins containing zinc fingers. Genetics 129:385–397. https://doi.org/10.1093/genetics/129.2.385
Edwards HE, Gorelick DA (2022) The evolution and structure/function of bHLH-PAS transcription factor family. Biochem Soc Trans 50:1227–1243. https://doi.org/10.1042/BST20211225
Emery IF, Bedian V, Guild GM (1994) Differential expression of *broad-complex* transcription factors may forecast tissue-specific developmental fates during *Drosophila* metamorphosis. Development 120:3275–3287. https://doi.org/10.1242/dev.120.11.3275
Fukuda S (1944) The hormonal mechanism of larval molting and metamorphosis in the silkworm. J Fac Sci Tokyo Imperial Univ 6:447–532
Furness SGB, Lees MJ, Whitelaw ML (2007) The dioxin (aryl hydrocarbon) receptor as a model for adaptive responses of bHLH/PAS transcription factors. FEBS Lett 581:3616–3625. https://doi.org/10.1016/j.febslet.2007.04.011
Furuta K, Yamada N, Kayukawa T (2023) Synthesis of 1,4-benzodioxan derivatives and the evaluation of their biological activity as a novel juvenile hormone signaling inhibitor. Pest Manag Sci 79:5341–5348. https://doi.org/10.1002/ps.7744
Gilbert LI, King DS (1973) Endocrinology. In: Rockstein M (ed) The physiology of insecta Vol. I. Academic Press, London, pp 249–270
Grieneisen ML (1994) Recent advances in our knowledge of ecdysteroid biosynthesis in insects and crustaceans. Insect Biochem Mol Biol 24:115–132. https://doi.org/10.1016/0965-1748(94)90078-7
Hatakoshi M, Agui N, Nakayama I (1986) 2-[1-methyl-2-(4-phenoxyphenoxy) ethoxy] pyridine as a new insect juvenile hormone analogue: induction of supernumerary larvae in *Spodoptera litura* (Lepidoptera: Noctuidae). Appl Entomol Zool 21:351–353. https://doi.org/10.1303/aez.21.351
He Q, Wen D, Jia Q, Cui C, Wang J, Palli SR et al (2014) Heat shock protein 83 (Hsp83) facilitates methoprene-tolerant (Met) nuclear import to modulate juvenile hormone signaling. J Biol Chem 289:27874–27885. https://doi.org/10.1074/jbc.M114.582825
Huang J, Tian L, Peng C, Abdou M, Wen D, Wang Y et al (2011) DPP-mediated TGFβ signaling regulates juvenile hormone biosynthesis by activating the expression of juvenile hormone acid methyltransferase. Development 138:2283–2291. https://doi.org/10.1242/dev.057687
Jindra M, Bellés X, Shinoda T (2015) Molecular basis of juvenile hormone signaling. Curr Opin Insect Sci 11:39–46. https://doi.org/10.1016/j.cois.2015.08.004
Jindra M, McKinstry WJ, Nebl T, Bittova L, Ren B, Shaw J et al (2021a) Purification of an insect juvenile hormone receptor complex enables insights into its post-translational phosphorylation. J Biol Chem 297:101387. https://doi.org/10.1016/j.jbc.2021.101387
Jindra M, Palli SR, Riddiford LM (2013) The juvenile hormone signaling pathway in insect development. Annu Rev Entomol 58:181–204. https://doi.org/10.1146/annurev-ento-120811-153700
Jindra M, Tumova S, Milacek M, Bittova L (2021b) A decade with the juvenile hormone receptor. Adv Insect Physiol 60:37–85. https://doi.org/10.1016/bs.aiip.2021.03.001
Jindra M, Tumova S, Bittova L, Tuma R, Sedlak D (2024) Agonist-dependent action of juvenile hormone receptor. Curr Opin Insect Sci 65:101234. https://doi.org/10.1016/j.cois.2024.101234
Kamimura M, Saito H, Niwa R, Niimi T, Toyoda K, Ueno C et al (2012) Fungal ecdysteroid-22-oxidase, a new tool for manipulating ecdysteroid signaling and insect development. J Biol Chem 287:16488–16498. https://doi.org/10.1074/jbc.M112.341180
Kaneko Y, Takaki K, Iwami M, Sakurai S (2006) Developmental profile of annexin IX and its possible role in programmed cell death of the *Bombyx mori* anterior silk gland. Zool Sci 23:533–542. https://doi.org/10.2108/zsj.23.533
Kayukawa T, Furuta K, Nagamine K, Shinoda T, Yonesu K, Okabe T (2020) Identification of a juvenile-hormone signaling inhibitor via high-throughput screening of a chemical library. Sci Rep 10:18413. https://doi.org/10.1038/s41598-020-75386-x

Kayukawa T, Jouraku A, Ito Y, Shinoda T (2017) Molecular mechanism underlying juvenile hormone-mediated repression of precocious larval–adult metamorphosis. Proc Natl Acad Sci USA 114:1057–1062. https://doi.org/10.1073/pnas.1615423114

Kayukawa T, Minakuchi C, Namiki T, Togawa T, Yoshiyama M, Kamimura M et al (2012) Transcriptional regulation of juvenile hormone-mediated induction of Krüppel homolog 1, a repressor of insect metamorphosis. Proc Natl Acad Sci USA 109:11729–11734. https://doi.org/10.1073/pnas.1204951109

Kayukawa T, Murata M, Kobayashi I, Muramatsu D, Okada C, Uchino K et al (2014) Hormonal regulation and developmental role of Krüppel homolog 1, a repressor of metamorphosis, in the silkworm *Bombyx mori*. Dev Biol 388:48–56. https://doi.org/10.1016/j.ydbio.2014.01.022

Kayukawa T, Nagamine K, Inui T, Yokoi K, Kobayashi I, Nakao H et al (2024) *Dead ringer* acts as a major regulator of juvenile hormone biosynthesis in insects. PNAS Nexus 3:gae435. https://doi.org/10.1093/pnasnexus/pgae435

Kayukawa T, Nagamine K, Ito Y, Nishita Y, Ishikawa Y, Shinoda T (2016) Krüppel homolog 1 inhibits insect metamorphosis via direct transcriptional repression of *broad-complex*, a pupal specifier gene. J Biol Chem 291:1751–1762. https://doi.org/10.1074/jbc.M115.686121

Kayukawa T, Shinoda T (2015) Functional characterization of two paralogous JH receptors, methoprene-tolerant 1 and 2, in the silkworm, *Bombyx mori* (Lepidoptera: Bombycidae). Appl Entomol Zool 50:383–391. https://doi.org/10.1007/s13355-015-0345-8

Kayukawa T, Tateishi K, Shinoda T (2013) Establishment of a versatile cell line for juvenile hormone signaling analysis in *Tribolium castaneum*. Sci Rep 3:1570. https://doi.org/10.1038/srep01570

Kiguchi K, Riddiford LM (1978) A role of juvenile hormone in pupal development of the tobacco hornworm, *Manduca sexta*. J Insect Physiol 24:673–680. https://doi.org/10.1016/0022-1910(78)90063-x

Kiss I, Beaton AH, Tardiff J, Fristrom D, Fristrom JW (1988) Interactions and developmental effects of mutations in the *broad-complex* of *Drosophila melanogaster*. Genetics 118:247–259. https://doi.org/10.1093/genetics/118.2.247

Kiss I, Bencze G, Fodor G, Szabad J, Fristrom JW (1976) Prepupal larval mosaics in *Drosophila melanogaster*. Nature 262:136–138. https://doi.org/10.1038/262136a0

Konopova B, Jindra M (2007) Juvenile hormone resistance gene *Methoprene-tolerant* controls entry into metamorphosis in the beetle *Tribolium castaneum*. Proc Natl Acad Sci USA 104:10488–10493. https://doi.org/10.1073/pnas.0703719104

Konopova B, Jindra M (2008) Broad-complex acts downstream of met in juvenile hormone signaling to coordinate primitive holometabolan metamorphosis. Development 135:559–568. https://doi.org/10.1242/dev.016097

Konopova B, Smykal V, Jindra M (2011) Common and distinct roles of juvenile hormone signaling genes in metamorphosis of holometabolous and hemimetabolous insects. PLoS One 6:e28728. https://doi.org/10.1371/journal.pone.0028728

Lee CY, Baehrecke EH (2001) Steroid regulation of autophagic programmed cell death during development. Development 128:1443–1455. https://doi.org/10.1242/dev.128.8.1443

Lee CY, Wendel DP, Reid P, Lam G, Thummel CS, Baehrecke EH (2000a) *E93* directs steroid-triggered programmed cell death in *Drosophila*. Mol Cell 6:433–443. https://doi.org/10.1016/s1097-2765(00)00042-3

Lee T, Marticke S, Sung C, Robinow S, Luo L (2000b) Cell-autonomous requirement of the USP/EcR-B ecdysone receptor for mushroom body neuronal remodeling in *Drosophila*. Neuron 28:807–818. https://doi.org/10.1016/s0896-6273(00)00155-0

Lehmann M, Korge G (1995) Ecdysone regulation of the *Drosophila Sgs-4* gene is mediated by the synergistic action of ecdysone receptor and SEBP 3. EMBO J 14:716–726. https://doi.org/10.1002/j.1460-2075.1995.tb07050.x

Lehmann M, Wattler F, Korge G (1997) Two new regulatory elements controlling the *Drosophila Sgs-3* gene are potential ecdysone receptor and fork head binding sites. Mech Dev 62:15–27. https://doi.org/10.1016/s0925-4773(96)00644-2

Li M, Liu P, Wiley JD, Ojani R, Bevan DR, Li J et al (2014) A steroid receptor coactivator acts as the DNA-binding partner of the methoprene-tolerant protein in regulating juvenile hormone response genes. Mol Cell Endocrinol 394:47–58. https://doi.org/10.1016/j.mce.2014.06.021

Li M, Mead EA, Zhu J (2011) Heterodimer of two bHLH-PAS proteins mediates juvenile hormone-induced gene expression. Proc Natl Acad Sci USA 108:638–643. https://doi.org/10.1073/pnas.1013914108

Liu H, Wang J, Li S (2014) E93 predominantly transduces 20-hydroxyecdysone signaling to induce autophagy and caspase activity in *Drosophila* fat body. Insect Biochem Mol Biol 45:30–39. https://doi.org/10.1016/j.ibmb.2013.11.005

Liu P, Peng H-J, Zhu J (2015a) Juvenile hormone-activated phospholipase C pathway enhances transcriptional activation by the methoprene-tolerant protein. Proc Natl Acad Sci USA 112:E1871–E1879. https://doi.org/10.1073/pnas.1423204112

Liu X, Dai F, Guo E, Li K, Ma L, Tian L et al (2015b) 20-hydroxyecdysone (20E) primary response gene *E93* modulates 20E signaling to promote *Bombyx* larval-pupal metamorphosis. J Biol Chem 290:27370–27383. https://doi.org/10.1074/jbc.m115.687293

Lozano J, Belles X (2011) Conserved repressive function of Krüppel homolog 1 on insect metamorphosis in hemimetabolous and holometabolous species. Sci Rep 1:163. https://doi.org/10.1038/srep00163

Massari ME, Murre C (2000) Helix-loop-helix proteins: regulators of transcription in eucaryotic organisms. Mol Cell Biol 20:429–440. https://doi.org/10.1128/MCB.20.2.429-440.2000

Minakuchi C, Namiki T, Shinoda T (2009) *Krüppel homolog 1*, an early juvenile hormone-response gene downstream of *Methoprene-tolerant*, mediates its anti-metamorphic action in the red flour beetle *Tribolium castaneum*. Dev Biol 325:341–350. https://doi.org/10.1016/j.ydbio.2008.10.016

Minakuchi C, Namiki T, Yoshiyama M, Shinoda T (2008a) RNAi-mediated knockdown of *juvenile hormone acid O-methyltransferase* gene causes precocious metamorphosis in the red flour beetle *Tribolium castaneum*. FEBS J 275:2919–2931. https://doi.org/10.1111/j.1742-4658.2008.06428.x

Minakuchi C, Zhou X, Riddiford LM (2008b) *Krüppel homolog 1* (*Kr-h1*) mediates juvenile hormone action during metamorphosis of *Drosophila melanogaster*. Mech Dev 125:91–105. https://doi.org/10.1016/j.mod.2007.10.002

Miura K, Oda M, Makita S, Chinzei Y (2005) Characterization of the *Drosophila Methoprene* -tolerant gene product. Juvenile hormone binding and ligand-dependent gene regulation. FEBS J 272:1169–1178. https://doi.org/10.1111/j.1742-4658.2005.04552.x

Muramatsu D, Kinjoh T, Shinoda T, Hiruma K (2008) The role of 20-hydroxyecdysone and juvenile hormone in pupal commitment of the epidermis of the silkworm, *Bombyx mori*. Mech Dev 125:411–420. https://doi.org/10.1016/j.mod.2008.02.001

Nagamine K, Kayukawa T, Hoshizaki S, Matsuo T, Shinoda T, Ishikawa Y (2014) Cloning, phylogeny, and expression analysis of the broad-complex gene in the longicorn beetle *Psacothea hilaris*. Springerplus 3:539. https://doi.org/10.1186/2193-1801-3-539

Niimi S, Sakurai S (1997) Development changes in juvenile hormone and juvenile hormone acid titers in the hemolymph and *in vitro* juvenile hormone synthesis by corpora allata of the silkworm, *Bombyx mori*. J Insect Physiol 43:875–884. https://doi.org/10.1016/s0022-1910(97)00021-8

Nishita Y (2014) Ecdysone response elements in the distal promoter of the *Bombyx Broad-complex* gene, *BmBR-C*. Insect Mol Biol 23:341–356. https://doi.org/10.1111/imb.12085

Nishita Y, Takiya S (2009) Differential responsiveness of *BmBR-C* promoters to ecdysone signals. J Insect Biotech Sericol 78:3_127-3_138. https://doi.org/10.11416/jibs.78.3_127

Okamoto N, Viswanatha R, Bittar R, Li Z, Haga-Yamanaka S, Perrimon N et al (2018) A membrane transporter is required for steroid hormone uptake in *Drosophila*. Dev Cell 47:294–305.e7. https://doi.org/10.1016/j.devcel.2018.09.012

Partch CL, Gardner KH (2010) Coactivator recruitment: a new role for PAS domains in transcriptional regulation by the bHLH-PAS family. J Cell Physiol 223:553–557. https://doi.org/10.1002/jcp.22067

Parthasarathy R, Tan A, Palli SR (2008) bHLH-PAS family transcription factor methoprene-tolerant plays a key role in JH action in preventing the premature development of adult structures during larval-pupal metamorphosis. Mech Dev 125:601–616. https://doi.org/10.1016/j.mod.2008.03.004

Pecasse F, Beck Y, Ruiz C, Richards G (2000) *Krüppel-homolog*, a stage-specific modulator of the prepupal ecdysone response, is essential for *Drosophila* metamorphosis. Dev Biol 221:53–67. https://doi.org/10.1006/dbio.2000.9687

Piulachs M-D, Pagone V, Bellés X (2010) Key roles of the *broad-complex* gene in insect embryogenesis. Insect Biochem Mol Biol 40:468–475. https://doi.org/10.1016/j.ibmb.2010.04.006

Postlethwait JH (1974) Juvenile hormone and the adult development of *Drosophila*. Biol Bull 147:119–135. https://doi.org/10.2307/1540573

Reza AMS, Kanamori Y, Shinoda T, Shimura S, Mita K, Nakahara Y et al (2004) Hormonal control of a metamorphosis-specific transcriptional factor broad-complex in silkworm. Comp Biochem Physiol B 139:753–761. https://doi.org/10.1016/j.cbpc.2004.09.009

Riddiford LM (1994) Cellular and molecular actions of juvenile hormone I. General considerations and premetamorphic actions. Adv Insect Physiol 24:213–274. https://doi.org/10.1016/S0065-2806(08)60084-3

Riddiford LM, Ajami AM (1973) Juvenile hormone: its assay and effects on pupae of *Manduca sexta*. J Insect Physiol 19:749–762. https://doi.org/10.1016/0022-1910(73)90147-9

Riddiford LM, Roseland CR, Thalberg S, Curtis AT (1983) Action of two juvenile hormone antagonists on lepidopteran epidermis. J Insect Physiol 29:281–286. https://doi.org/10.1016/0022-1910(83)90096-3

Riddihough G, Pelham HR (1987) An ecdysone response element in the *Drosophila* hsp27 promoter. EMBO J 6:3729–3734. https://doi.org/10.1002/j.1460-2075.1987.tb02707.x

Sailsbery JK, Atchley WR, Dean RA (2012) Phylogenetic analysis and classification of the fungal bHLH domain. Mol Biol Evol 29:1301–1318. https://doi.org/10.1093/molbev/msr288

Satake S, Kaya M, Sakurai S (1998) Hemolymph ecdysteroid titer and ecdysteroid-dependent developmental events in the last-larval stadium of the silkworm, *Bombyx mori*: role of low ecdysteroid titer in larval-pupal metamorphosis and a reappraisal of the head critical period. J Insect Physiol 44:867–881. https://doi.org/10.1016/s0022-1910(98)00075-4

Sehnal F, Sláma K (1966) The effect of corpus allatum hormone on respiratory metabolism during larval development and metamorphosis of *Galleria mellonella* L. J Insect Physiol 12:1333–1342. https://doi.org/10.1016/0022-1910(66)90022-9

Sláma K, Williams CM (1965) Juvenile hormone activity for the bug *Pyrrhocoris apterus*. Proc Natl Acad Sci USA 54:411–414. https://doi.org/10.1073/pnas.54.2.411

Spokony RF, Restifo LL (2007) Anciently duplicated *broad complex* exons have distinct temporal functions during tissue morphogenesis. Dev Genes Evol 217:499–513. https://doi.org/10.1007/s00427-007-0159-y

Suzuki Y, Truman JW, Riddiford LM (2008) The role of broad in the development of *Tribolium castaneum*: implications for the evolution of the holometabolous insect pupa. Development 135:569–577. https://doi.org/10.1242/dev.015263

Touhara K, Lerro KA, Bonning BC, Hammock BD, Prestwich GD (1993) Ligand binding by a recombinant insect juvenile hormone binding protein. Biochemistry 32:2068–2075. https://doi.org/10.1021/bi00059a026

Truman JW, Riddiford LM (2022) *Chinmo* is the larval member of the molecular trinity that directs *Drosophila* metamorphosis. Proc Natl Acad Sci USA 119:e2201071119. https://doi.org/10.1073/pnas.2201071119

Tumova S, Dolezel D, Jindra M (2024) Conserved and unique roles of bHLH-PAS transcription factors in insects - from clock to hormone reception. J Mol Biol 436:168332. https://doi.org/10.1016/j.jmb.2023.168332

Unnithan GC, Andersen JF, Hisano T, Kuwano E, Feyereisen R (1995) Inhibition of juvenile hormone biosynthesis and methyl farnesoate epoxidase activity by 1,5-disubstituted imidazoles in the cockroach, *Diploptera punctata*. Pestic Sci 43:13–19. https://doi.org/10.1002/ps.2780430103

Ureña E, Chafino S, Manjón C, Franch-Marro X, Martín D (2016) The occurrence of the holometabolous pupal stage requires the interaction between E93, Krüppel-homolog 1 and broad-complex. PLoS Genet 12:e1006020. https://doi.org/10.1371/journal.pgen.1006020

Ureña E, Manjón C, Franch-Marro X, Martín D (2014) Transcription factor E93 specifies adult metamorphosis in hemimetabolous and holometabolous insects. Proc Natl Acad Sci USA 111:7024–7029. https://doi.org/10.1073/pnas.1401478111

Wigglesworth VB (1936) The function of the corpus allatum in the growth and reproduction of *Rhodnius prolixus* (Hemiptera). J Cell Sci S2-79:91–121. https://doi.org/10.1242/jcs.s2-79.313.91

Wigglesworth VB (1958) Some methods for assaying extracts of the juvenile hormone in insects. J Insect Physiol 2:73–84. https://doi.org/10.1016/0022-1910(58)90031-3

Wilson TG, Fabian J (1986) A *Drosophila melanogaster* mutant resistant to a chemical analog of juvenile hormone. Dev Biol 118:190–201. https://doi.org/10.1016/0012-1606(86)90087-4

Yamanaka N (2021) Ecdysteroid signalling in insects—from biosynthesis to gene expression regulation. Adv Insect Physiol 60:1–36. https://doi.org/10.1016/bs.aiip.2021.03.002

Zhou B, Hiruma K, Shinoda T, Riddiford LM (1998) Juvenile hormone prevents ecdysteroid-induced expression of broad complex RNAs in the epidermis of the tobacco hornworm, *Manduca sexta*. Dev Biol 203:233–244. https://doi.org/10.1006/dbio.1998.9059

Zhou B, Riddiford LM (2001) Hormonal regulation and patterning of the broad-complex in the epidermis and wing discs of the tobacco hornworm, *Manduca sexta*. Dev Biol 231:125–137. https://doi.org/10.1006/dbio.2000.0143

Zhou X, Riddiford LM (2002) Broad specifies pupal development and mediates the 'status quo' action of juvenile hormone on the pupal-adult transformation in *Drosophila* and *Manduca*. Development 129:2259–2269. https://doi.org/10.1242/dev.129.9.2259

Zhou X, Zhou B, Truman JW, Riddiford LM (2004) Overexpression of broad: a new insight into its role in the *Drosophila* prothoracic gland cells. J Exp Biol 207:1151–1161. https://doi.org/10.1242/jeb.00855

Chapter 9
Hormonal Regulation of Development and Sexual Dimorphism of Hemimetabolous Insects

Chieka Minakuchi

Abstract Hemimetabolous and holometabolous insects have evolved morphologically distinct larval (or nymphal) and adult forms, enabling them to disperse and occupy different habitats. Hemimetaboly comprises nymphal and adult stages, whereas holometaboly comprises larval, pupal, and adult stages. Similar to holometabolous insects, hemimetabolous development is regulated by ecdysteroids and juvenile hormones (JHs). In the embryonic development of hemimetabolous insects, JH biosynthesis starts in the middle of the embryonic stage, and events such as nymphal cuticle formation are regulated by JH. In postembryonic development, JH is secreted from the corpora allata into the hemolymph during the nymphal stage but disappears in the final nymphal instar as they prepare for metamorphosis to adulthood, indicating that JH has anti-metamorphic action to prevent precocious adult metamorphosis. The signaling pathway of JH, including the JH receptor Methoprene-tolerant (Met)-Taiman, the early JH-response gene Krüppel homolog 1 (Kr-h1), and the adult specifier ecdysone-induced protein 93F (E93), is conserved in hemimetabolous insects. In contrast, the function of Broad is different from that in Holometabola; Broad is essential for progressive wing development in Hemimetabola. A comparison of the expression profiles of transcription factors in hormonal signaling among insect species has provided insights into the evolution of metamorphosis in thrips, which develop from larvae to adults through nonfeeding and quiescent stages. In addition to regulating embryonic and postembryonic development, JH is involved in polyphenism in aphids and sexual dimorphism in scale insects.

Keywords Embryonic development · Hemimetaboly · Polyphenism · Postembryonic development · Sexual dimorphism

---

C. Minakuchi (✉)
Graduate School of Bioagricultural Sciences, Nagoya University, Nagoya, Japan
e-mail: c_mina@agr.nagoya-u.ac.jp

T. Shinoda et al. (eds.), *Juvenile Hormone*, Entomology Monographs,
https://doi.org/10.1007/978-981-95-8605-9_9

9.1 Hemimetaboly

Insect metamorphosis is a successful biological adaptation. Ancestral insects did not undergo metamorphosis, and their juveniles closely resembled adults in terms of external morphology. This type of development is called ametaboly. Through evolution, insects have obtained morphologically different larval and adult forms, which have enabled them to disperse and occupy different habitats. In addition to ametaboly, there are two types of insect metamorphosis: hemimetaboly and holometaboly. Hemimetaboly, also called "incomplete metamorphosis," comprises nymphal and adult stages in postembryonic development. Unlike holometabolous insects, there is no pupal stage in hemimetabolous insects. In contrast, holometabolous insects develop through larval, pupal, and adult stages. Adult morphogenesis in holometabolous insects proceeds rapidly during the pupal stage (Belles 2020; Truman and Riddiford 1999). In this chapter, we describe the hormonal regulation of the development and metamorphosis of hemimetabolous insects.

Hemimetabolous insects are classified into three superorders: Palaeoptera, Polyneoptera, and Paraneoptera. Palaeoptera includes primitive winged insects, such as Odonata (dragonflies and damselflies) and Ephemeroptera (mayflies). Polyneoptera includes the orders Orthoptera (grasshoppers, bush-crickets, crickets, and locusts), Phasmida (also known as Phasmatoptera or Phasmatodea: stick insects), Plecoptera (stoneflies), Dermaptera (earwigs), Dictyoptera (cockroaches, termites, and mantids), Embioptera (webspinners), Grylloblattaria (also known as Grylloblattodea: rock crawlers), Mantophasmatodea (heelwalkers), and Zoraptera (angel insects). Paraneoptera comprises the orders Hemiptera (true bugs), Psocodea (lice, booklice, and barklice), and Thysanoptera (thrips) (Belles 2020).

9.2 Juvenile Hormones and Their Secretory Glands

The presence of a humoral factor that inhibits metamorphosis to adulthood was first described by V. B. Wigglesworth in 1934 in the hemimetabolous blood-sucking bug *Rhodnius prolixus* (Wigglesworth 1934). Subsequently, the chemical structure of JH was first elucidated in the cecropia moth, *Hyalophora cecropia*, in 1967 (Röller et al. 1967). Since then, the chemical structure of JH has been determined in a variety of hemimetabolous and holometabolous insect species. JH III has been identified in several hemimetabolous species, such as Orthoptera and Dictyoptera. In contrast, in the hemipteran stink bug *Plautia stali*, the structure of JH has been identified as JH III skipped bisepoxide (Kotaki et al. 2009).

The corpora allata (CA) are glands responsible for the biosynthesis and secretion of JH. The morphology of the CA has been well studied in several species, including the German cockroach, *Blattella germanica*. Generally, there are nervous connections from the CA to the corpora cardiaca (CC) and from the CA to the subesophageal ganglion. In *B. germanica*, there are major nerves from the CA:

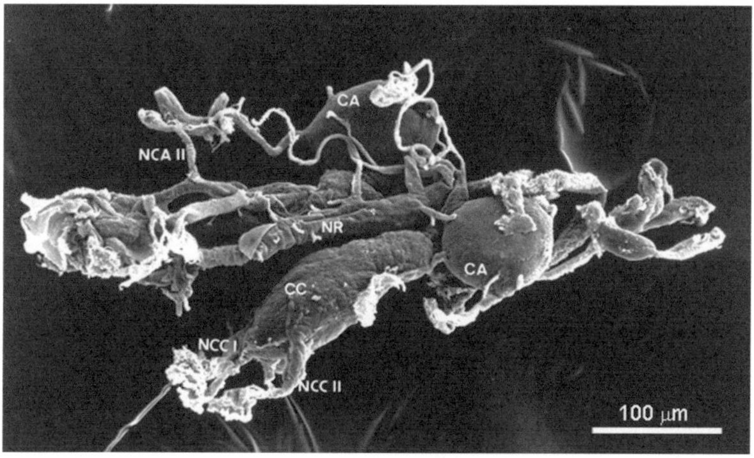

Fig. 9.1 Ultrastructural image showing the main nervous connections of the corpora allata in the cockroach *Blattella germanica*. CA, corpora allata; CC, corpora cardiaca; NCAII, nervi corporis allati II; NCCI, nervi corporis cardiaci I; NCCII, nervi corporis cardiaci II; NR, recurrent nerve and its connections. (Photo courtesy of Maria-Dolors Piulachs and from Belles (2020), with permission)

nervi corporis allati I (NCA I), which connect the CA to the CC, whereas nervi corporis allati II (NCA II) connect the CA to the suboesophageal ganglion. Regarding the CC, there are nerves (nervi corporis cardiaci, NCC) that connect to the brain (Fig. 9.1). These nervous connections are diverse in insect species (Belles 2020).

9.3 Role of JH in Embryonic Development

Embryonic development in hemimetabolous insects proceeds in a manner similar to that in holometabolous insects, with the formation of the blastoderm followed by germband formation (Belles 2020). In mid-embryogenesis of some hemimetabolous insects, a process called blastokinesis occurs, which is the movement of the embryo into the yolk mass involving the amniotic and serosal membranes and usually results in partial revolution of the embryonic body. Blastokinesis is divided into anatrepsis and katatrepsis (Belles 2020; Panfilio 2008). The events during embryonic development are summarized in Fig. 9.2.

In hemimetabolous species in Orthoptera, Odonata, and Plecoptera, the formation and deposition of a couple of cuticular layers were observed during embryonic development (Belles 2020; Konopova and Zrzavy 2005; Sbrenna 1990). In the embryonic development of the orthopteran *Locusta migratoria*, there are three peaks of ecdysteroids, which likely coincide with the formation of three cuticle layers (Lagueux et al. 1979). In general, JH synthesis starts in the middle of embryonic development (Belles 2020): JH is detected at 50% of development in *L. migratoria*

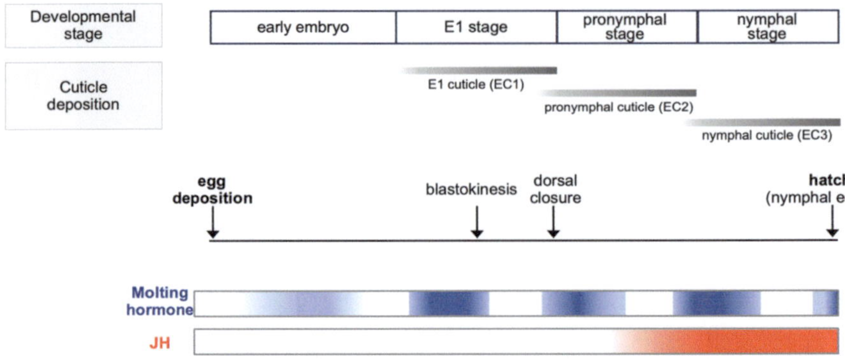

Fig. 9.2 Diagram of events during embryonic development of hemimetabolous insects. Deposition of cuticles at the E1 stage (EC1), pronymphal stage (EC2), and nymphal stage (EC3) is indicated by gray lines. Events such as egg deposition, blastokinesis, dorsal closure, and hatching are shown with arrows. Estimated changes in hormonal titers are illustrated below. (This figure is drawn based on reports in crickets (Erezyilmaz et al. 2004) and grasshoppers (Lagueux et al. 1979; Temin et al. 1986))

(Temin et al. 1986) and *B. germanica* (Maestro et al. 2010). Thus, JH appears in late embryonic development and is present during the formation of the first-instar nymphal cuticle (Fig. 9.2) (Temin et al. 1986).

The role of JH in embryonic development has been examined in several hemimetabolous species. Inhibition of JH synthesis by treatment with a precocene analog in short germband embryos of *L. migratoria* resulted in lethality at hatching or in the nymphal stage, as well as retarded nymphal growth (Aboulafia-Baginsky et al. 1984; Konopova 2025). In the ovoviviparous cockroach *Nauphoeta cinerea*, the levels of JH III and methyl farnesoate are high in the embryos (Brüning et al. 1985): treatment of the embryos with a precocene analog caused suppression of JH synthesis, resulting in retarded growth and malformation of organs such as eyes, gut, and nymphal cuticle (Brüning and Lanzrein 1987; Konopova 2025). In contrast, treatment of desert locust *Schistocerca gregaria* embryos with JH mimics before endogenous JH secretion caused morphological and growth defects (Novák 1969). Moreover, treatment of *S. gregaria* embryos with JH mimics redirected precocious nymphal features on the second embryonic cuticle (EC2: pronymphal cuticle) (Sbrenna-Micciarelli 1977). In the cricket *Acheta domestica*, JH mimic treatment of embryos prior to the first embryonic cuticle (EC1) formation produced small embryos that failed to complete normal katatrepsis, while JH mimic treatment during the period from formation of the first embryonic cuticle (EC1) to that of the second embryonic cuticle (pronymphal cuticle, EC2) caused the formation of a precocious nymphal cuticle as well as defects in dorsal closure (Erezyilmaz et al. 2004). Thus, treatment of JH mimics had different effects on embryonic development, depending on the timing of treatment. Specifically, JH mimic treatment from EC1 to EC2 caused precocious nymphal cuticle formation. These effects are different from the "status quo" action of JH in the postembryonic development of holometabolous insects, in which JH prevents metamorphosis and differentiation of imaginal tissues.

9.4 Role of JH in Postembryonic Development

The existence of a humoral factor that prevents metamorphosis was first demonstrated in *R. prolixus* by Wigglesworth (1934, 1936). Through experiments using classical techniques such as extirpation and replantation of organs and parabiosis, he showed that humoral factors regulate postembryonic development and metamorphosis (Belles 2020; Wigglesworth 1936).

In the postembryonic development of hemimetabolous insects, JH is present in the hemolymph during the nymphal stage but disappears in the last nymphal instar as they prepare for metamorphosis to adulthood (Truman and Riddiford 2019). Topical application of JH to the final instar nymphs of *P. stali* inhibited their transition to adulthood (Kotaki et al. 2009). This indicates that endogenous JH in pre-final nymphal instars has an anti-metamorphic action that prevents precocious adult metamorphosis.

9.5 Signaling Pathway of JH

9.5.1 Signaling in the Postembryonic Development

The signaling pathways of JH and ecdysteroids have been elucidated in holometabolous model insects, such as the fruit fly *Drosophila melanogaster*, the silkworm *Bombyx mori*, and the tobacco hornworm *Manduca sexta* (Jindra et al. 2015). In *D. melanogaster*, Methoprene-tolerant (Met) acts as a JH receptor by forming a heterodimer with Taiman (Tai), also known as steroid receptor coactivator (SRC) (Li et al. 2011; Zhang et al. 2011). In addition to these holometabolous species, it has also been studied in hemimetabolous species such as the cockroach *B. germanica*. In *B. germanica*, RNA interference (RNAi)-mediated knockdown of *Met* in the pre-final nymphal instar caused precocious adult metamorphosis and morphogenesis, suggesting that the role of *Met* as the JH receptor is conserved in this species (Belles 2020; Lozano and Belles 2014). In *B. germanica*, four Tai variants are generated from different alternative splicing events, each with a specific function, and RNAi-mediated depletion of all variants causes precocious adult metamorphosis (Lozano et al. 2014).

Krüppel homolog 1 (Kr-h1) is a transcription factor that is directly induced by the JH-Met/Tai receptor complex. Krüppel homolog 1 was originally identified in the fruit fly *D. melanogaster* as a transcription factor whose expression is induced by ecdysteroids and JH (Minakuchi et al. 2008; Pecasse et al. 2000). Subsequently, its role as a key factor in JH signaling was proven in the red flour beetle, *Tribolium castaneum*: RNAi-mediated knockdown of *Kr-h1* caused precocious metamorphosis (Minakuchi et al. 2009). The action of *Kr-h1* has also been examined in various insect species. In hemimetabolous species such as the cockroach *B. germanica* and the bugs *Pyrrhocoris apterus* and *R. prolixus*, knockdown of *Kr-h1* in the pre-final nymphal stage induced precocious metamorphosis to adults (Konopova et al. 2011; Lozano and Belles 2011), suggesting that the role of *Kr-h1* is conserved in these species. Moreover, *Kr-h1* has been identified in other hemimetabolous species, including the common bed bug *Cimex lectularius* (Gujar and Palli 2016), the Japanese mealybug *Planococcus kraunhiae* (Vea et al. 2016), the brown planthopper *Nilaparvata lugens* (Li et al. 2018), the bean bug *Riptortus pedestris* (Dong et al. 2022), and *P. stali* (Oishi et al. 2023).

The ecdysone-induced protein 93F gene (*E93*) is an early response gene for ecdysteroids in *D. melanogaster*. E93 was found to be involved in the degeneration of salivary glands during the adult development of *D. melanogaster* (Baehrecke and Thummel 1995). RNAi-mediated knockdown of *E93* in the nymphal stage prevented nymph-adult transition in the hemimetabolous *B. germanica*, whereas *E93* knockdown prevented pupal-adult transition in holometabolous species, including *D. melanogaster* and *T. castaneum* (Ureña et al. 2014). E93 is indispensable for female reproduction by promoting previtellogenic ovarian development and vitellogenesis in *B. germanica* (Bai et al. 2024). These reports have elucidated that E93 is a key transcription factor that induces adult development. E93 has been identified in other hemimetabolous species, such as *C. lectularius* (Gujar and Palli 2016), *N. lugens* (Li et al. 2018), *P. kraunhiae* (Vea et al. 2019), and *P. stali* (Oishi et al. 2023).

Broad is a transcription factor that belongs to the Broad-Tramtrack-Bric-a-brac (BTB) family (Bayer et al. 1996; DiBello et al. 1991) and induces larva-pupa transition in holometabolous insects. The role of Broad in hemimetabolous insects is distinct from that in holometabolous insects. In the milkweed bug, *Oncopeltus fasciatus*, *broad* is expressed from the embryonic stage to the onset of the final nymphal instar. Knockdown of *broad* in the pre-final nymphal stage suppressed the progressive morphogenesis of wing pads (Erezyilmaz et al. 2006). In the fire bug *P. apterus*, knockdown of *broad* caused retarded wing growth (Konopova et al. 2011). In *B. germanica*, Broad is involved in wing pad growth, and *broad* expression is enhanced by JH (Huang et al. 2013). These observations suggest that Broad is essential for progressive wing development in hemimetabolous insects.

The signaling pathways, including Met, Kr-h1, and E93, have been well investigated in *B. germanica*. Knockdown of *Kr-h1* resulted in the upregulation of *E93* (Belles and Santos 2014), suggesting that the pathway through Kr-h1 and E93 is present (Belles 2020; Belles and Santos 2014). The signaling pathway has also been examined in Odonata; in the damselfly *Ischnura senegalensis*, metamorphosis is regulated by the Met-Kr-h1-E93 (MEKRE93) pathway, whereas Broad regulates a substantial number of nymph-specific and adult-specific genes independently of Kr-h1 and E93 (Okude et al. 2022). Thus, it was suggested that the hormonal signaling pathway (MEKRE93) is present in hemimetabolous species. The proposed

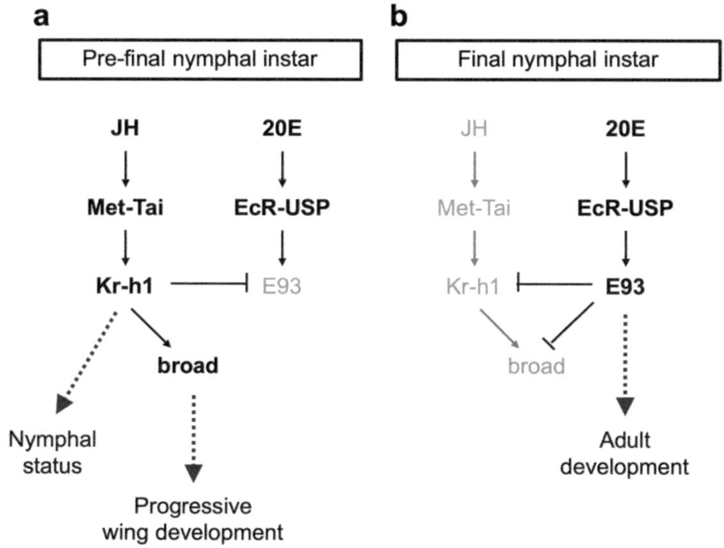

Fig. 9.3 Model of hormonal signaling pathways in hemimetabolous insects. (**a**) In the pre-final nymphal stage, juvenile hormone (JH) induces the expression of *Krüppel homolog 1* (*Kr-h1*) through its receptor methoprene tolerant (Met)-Taiman (Tai), while Kr-h1 suppresses the expression of *ecdysone-induced protein 93F* (*E93*); *broad* is upregulated by JH and promotes progressive wing development. (**b**) In the final nymphal instar, under extremely low JH titer, *Kr-h1* expression level decreases, which allows the upregulation of *E93* as well as the adult development. (Figures are based on the reports by Urena et al. (2016) and Belles (2020))

model is shown in Fig. 9.3. In the pre-final nymphal stage of hemimetabolous species, JH induces the expression of Kr-h1 through its receptor Met-Tai, whereas Kr-h1 suppresses the expression of E93. In contrast, JH biosynthesis becomes inactive in the final nymphal instar. Under extremely low JH levels in the hemolymph, *Kr-h1* expression decreases, which allows the upregulation of *E93* as well as adult development. In the pre-final nymphal stage, *broad* is upregulated by JH and is involved in progressive wing development (Fig. 9.3).

9.5.2 *Signaling in the Embryonic Development*

As described in the previous section, JH synthesis begins in the middle of embryonic development. JH signaling during the embryonic stage has been examined using RNAi techniques. In *B. germanica*, maternal RNAi of *Met* and *Kr-h1* revealed that JH is necessary during mid- and late embryogenesis to control proper cuticle tanning and promote embryo fitness (Fernandez-Nicolas and Belles 2017).

In hemimetabolous insects, such as *O. fasciatus* and *B. germanica*, *broad* expression is observed during embryonic development (Erezyilmaz et al. 2009; Piulachs et al. 2010). This is in contrast to the observations in holometabolous insects: in holometabolous species such as *D. melanogaster* and *M. sexta*, immunocytochemical experiments revealed that the Broad protein is exclusively present in the central nervous system during late embryonic development (Zhou et al. 2009), while *broad* is expressed more highly during larva-pupa metamorphosis than in the embryonic and larval stages (Konopova and Jindra 2008). RNAi-mediated knockdown of *broad* in *O. fasciatus* results in morphological defects during embryonic development (Erezyilmaz et al. 2009). Maternal RNAi of *broad* in *B. germanica* resulted in various defects in the formation of the germ band, in the progress beyond 25–30% of total development when the amnion and serosa rupture and the secondary dorsal organ is formed, and in the hatching process (Piulachs et al. 2010). These results suggest that *broad* is essential for normal embryonic development in hemimetabolous insects. In contrast, it seems that *broad* is not involved in morphogenesis in the embryonic stage in Holometabola: *broad* null mutants of *D. melanogaster* survive through embryogenesis but finally die at larval pupal metamorphosis (Kiss et al. 1988): knockout of *broad* in *B. mori* resulted in a slight reduction in viability during embryogenesis, but many larvae survived until larval-pupal transition (Daimon et al. 2015). Thus, the function of Broad differs between hemimetabolous and holometabolous species.

9.6 Hormonal Regulation of Postembryonic Development of Thrips (Thysanoptera)

Thrips (Thysanoptera) are members of Hemimetabola and are considered the sister group of Hemiptera. The postembryonic development of thrips is unusual because there are nonfeeding and quiescent stages between the larval and adult stages. These stages are called "pupa" and "propupa," which are reminiscent of the holometabolous pupal stage. The external morphology of thrips larvae resembles that of adults, except for the absence of adult-type wings.

In the western flower thrips, *Frankliniella occidentalis*, and the predatory thrips *Haplothrips brevitubus*, hormonal regulation of metamorphosis was examined by analyzing the developmental expression profiles of the transcription factors *Kr-h1*, *broad*, and *E93* (Fig. 9.4). In both species, the transcript levels of *Kr-h1* were high during the embryonic stage, remained at a moderate level in the larval and propupal stages, and were low in the pupal stage (Minakuchi et al. 2011). *broad* is highly expressed in the embryonic stage as well as during the larval-propupal transition (Minakuchi et al. 2011). In contrast, *E93* transcript levels increase in the propupal and pupal stages, whereas *Kr-h1* and *broad* transcript levels decrease (Suzuki et al. 2021). These observations suggest that the MEKRE93 signaling pathway is also present in thrips. Furthermore, based on a comparison of the expression profiles of these transcription factors among insect species, it was proposed that the propupal stage of thrips is analogous to the holometabolous prepupal stage, whereas the

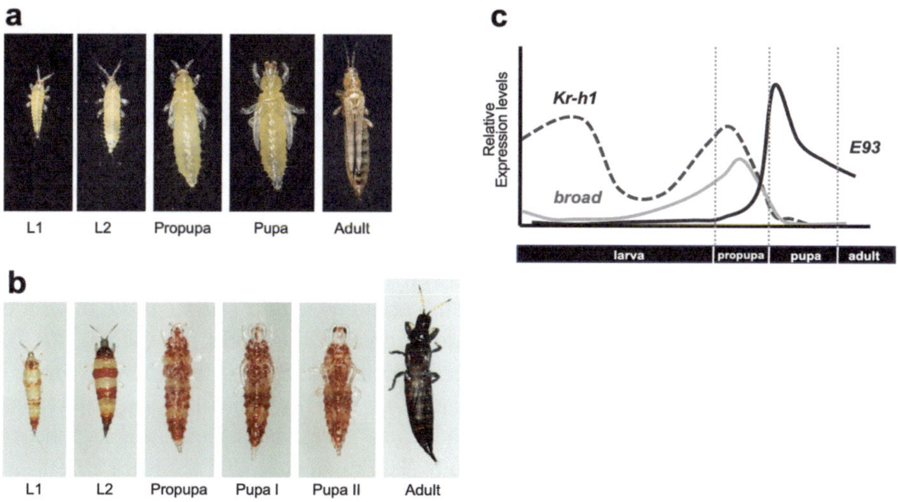

Fig. 9.4 Development of thrips and the diagram of expression profiles of transcription factors. (**a**) Postembryonic development of the western flower thrips, *Frankliniella occidentalis*. (**b**) Postembryonic development of the predatory thrips *Haplothrips brevitubus*. L1, first-instar larva; L2, second-instar larva. (**c**) Diagram of the expression profiles of key transcription factors in hormonal signaling in thrips. The expression profiles of *Krüppel homolog 1* (*Kr-h1*), *broad*, and *ecdysone-induced protein 93F* (*E93*) are based on previous studies (Minakuchi et al. 2011; Suzuki et al. 2021)

pupal stage of thrips is similar to the pupal stage in holometabolous species (Suzuki et al. 2021).

9.7 Polyphenism and Sexual Dimorphism

9.7.1 Aphid Polyphenism

Aphids exhibit polyphenisms, such as wing and reproductive polyphenisms, during their annual life cycle (Miura 2019; Ogawa and Miura 2014). In autumn, males and sexual oviparous females mate and produce overwintering eggs, from which parthenogenetic female aphids hatch the following spring and produce clonal offspring through parthenogenetic viviparous reproduction from spring to summer.

Winged and wingless adult females are produced in response to environmental conditions. This phenomenon is called wing polyphenism (Miura 2019; Ogawa and Miura 2014). Winged aphids are produced in response to unfavorable environmental conditions, such as crowding, poor food quality, or the presence of predators (Müller et al. 2001; Vellichirammal et al. 2017). Winged morphs include flightless morphs (brachypterous) with reduced wings and morphs capable of flight with large functional wings (macropterous) (Müller et al. 2001). Wingless morphs have shorter developmental times and increased fecundity (Müller et al. 2001).

The roles of ecdysteroids and JH in wing development and wing polyphenism have been examined in many insect species, including aphids (Miura 2019; Zera and Denno 1997). It has been hypothesized that an elevated level of JH during some critical period of development blocks the appropriate morphogenesis of wings, flight muscles, and associated structures, resulting in a brachypterous or wingless morph (Nijhout 1994; Zera and Denno 1997). For example, in the cricket *Gryllus rubens*, JH titer and the activity of JH-degrading enzymes (JH esterase) are involved in wing polyphenism between brachypterous and macropterous adults, although other factors seem to be responsible as well (Zera and Denno 1997). Molecular studies have been conducted using the vetch aphid *Megoura crassicauda*, in which winged aphids can easily be induced under high-density conditions (Miura 2019). In this species, the JH III titer was higher in presumptive wingless than in winged aphids (Ishikawa et al. 2013). The application of JH III to third-instar nymphs disturbs appropriate wing development, leading to winged/wingless intermorphs or juvenized aphids with supernumerary molts (Ishikawa et al. 2013; Miura 2019).

Juvenile hormones are also involved in reproductive polyphenism. For example, the CA volume differed between long-day and short-day reared aphids, *Megoura viciae* (Hardie 1987). When JH analogs were topically applied to aphids under short-day conditions, parthenogenetic reproduction was induced. In the pea aphid, *Acyrthosiphon pisum*, the JH titer in aphids reared under short-day conditions was lower than that under long-day conditions, and this suppression of JH titer under short-day conditions was suggested to be caused by the upregulation of JH-degrading enzymes, such as JH esterases and JH epoxide hydrolase (Ishikawa et al. 2012; Miura 2019).

9.7.2 Sexual Dimorphism of Scale Insects

Scale insects and mealybugs (Hemiptera: Coccoidea) exhibit extreme sexual dimorphism. Males generally undergo two nymphal instars and develop into one or two quiescent stages called "prepupa" and "pupa." They then emerge as winged adults. In contrast, females undergo three or four nymphal instars and emerge as wingless adults with juvenile-like external morphology (Belles 2020). In the case of the Japanese mealybug, *Planococcus kraunhiae*, the external morphology of males and females is similar until the first few days of the second nymphal instar, and the sexes start to differentiate at the end of the second nymphal instar (Fig. 9.5a) (Vea et al. 2016).

Growth hormones may be involved in the establishment of differential pathways. An anatomical study of the soft scale insect *Parthenolecanium corni* (Coccidae) showed that the female CA appears to be larger than that of males (Pflugfelder 1936). This finding suggests that more JH is secreted in females than in males,

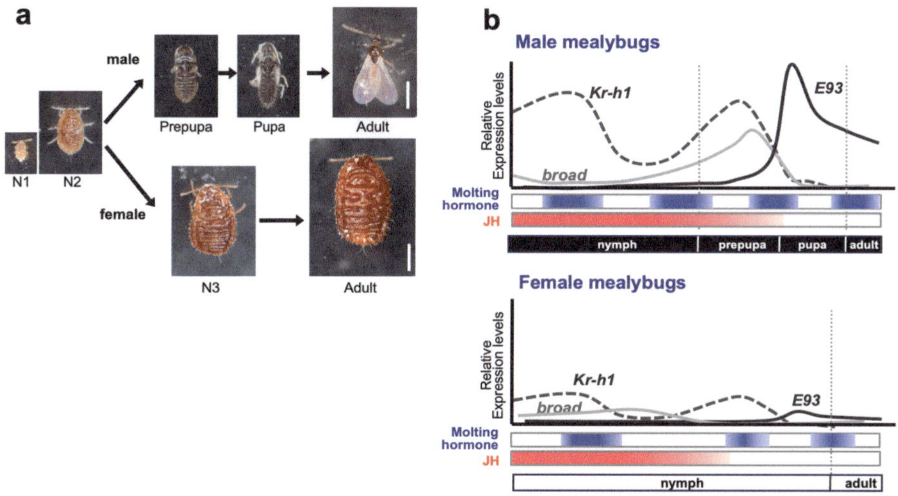

Fig. 9.5 Development of the Japanese mealybugs and the diagram of expression profiles of transcription factors. (**a**) Postembryonic development of the Japanese mealybug, *Planococcus kraunhiae*. Photos are taken after removing the white secretion from the body surface. N1, first--instar nymph; N2, second-instar nymph; N3, female third-instar nymph. Scale bars in the pictures are 500 μm. (**b**) Diagram of the expression profiles of key transcription factors in hormonal signaling and the estimated titer of molting hormone and juvenile hormone (JH) in *P. kraunhiae*. The expression profiles of *Krüppel homolog 1* (*Kr-h1*), *broad*, and *ecdysone-induced protein 93F* (*E93*) are based on previous studies (Vea et al. 2016, 2019). The hormonal titer is estimated from the expression profiles of biosynthetic enzymes of the hormones (Muramatsu et al. 2020; Vea et al. 2016)

resulting in the suppression of imaginal structures and reproductive maturation in females (Matsuda 1987). A recent study analyzing the developmental expression profiles of the JH biosynthetic enzyme *JH acid O-methyltransferase* (*JHAMT*) and JH-response gene *Kr-h1* in *P. kraunhiae* showed that both *JHAMT* and *Kr-h1* expression levels are higher in males than in females when sexes are differentiated from the end of the second nymphal instar, indicating that JH titer is higher in males than in females during this period (Vea et al. 2016).

In *P. kraunhiae*, the hormonal regulation of metamorphosis was examined by analyzing the developmental expression profiles of the transcription factors *Kr-h1*, *broad*, and *E93* (Fig. 9.5b). Based on the expression levels of hormonal biosynthetic enzymes, it has been suggested that both ecdysteroid and JH titers are higher in males than in females (Muramatsu et al. 2020; Vea et al. 2016). The transcript levels of *Kr-h1*, *broad*, and *E93* were higher in males, suggesting that the adult male development of mealybugs is directed by hormones through these transcription factors (Vea et al. 2016; Vea et al. 2019). Thus, ecdysteroids and JH are involved in establishing sexually dimorphic metamorphosis in mealybugs.

9.8 Conclusions and Perspectives

As stated in this chapter, the role of JH and its signaling pathways has been elucidated in hemimetabolous insects. In addition, the involvement of JH in the polyphenism of aphids and sexual dimorphism in scale insects has been reported.

To clarify JH action and its signaling pathway, functional analysis of genes involved in JH action has been performed, mainly using RNAi. It has been reported that RNAi-mediated knockdown by microinjection of double-stranded RNA, which was originally developed with the beetle *T. castaneum* (Tomoyasu and Denell 2004), is also effective against hemimetabolous species such as true bugs (Dong et al. 2022; Erezyilmaz et al. 2009; Konopova et al. 2011) and cockroaches (Lozano and Belles 2014). However, RNAi is not always sufficiently effective against some insect species, which hampers the functional analysis of the genes of interest. To date, gene-editing techniques such as CRISPR/Cas9 have been developed and applied to insects. The first successful examples of gene editing were limited to a few species of insect models. Recently, direct parental CRISPR (DIPA-CRISPR) was developed, in which gene editing can be performed by microinjection of Cas9 ribonucleoproteins (RNPs) into the hemocoel of adult females, followed by ovary uptake of injected Cas9 RNPs (Shirai et al. 2022). Using the DIPA-CRISPR technique, highly efficient gene editing has been conducted in a variety of species, including *B. germanica* (Shirai et al. 2022), the white-backed planthopper, *Sogatella furcifera* (Zhang et al. 2024), and the predatory flower bug, *Orius strigicollis* (Matsuda et al. 2024). DIPA-CRISPR is applicable to a wide variety of insect species, including non-model insects. This technology will be helpful in elucidating the functions of genes involved in hormonal signaling in hemimetabolous insects.

References

Aboulafia-Baginsky N, Pener MP, Staal GB (1984) Chemical allatectomy of late *Locusta* embryos by a synthetic precocene and its effect on hopper morphogenesis. J Insect Physiol 30:839–852. https://doi.org/10.1016/0022-1910(84)90057-X

Baehrecke EH, Thummel CS (1995) The *Drosophila E93* gene from the 93F early puff displays stage- and tissue-specific regulation by 20-hydroxyecdysone. Dev Biol 171:85–97. https://doi.org/10.1006/dbio.1995.1262

Bai Y, Lv YN, Zeng M, Yan ZY, Huang DY, Wen JZ et al (2024) E93 is indispensable for reproduction in ametabolous and hemimetabolous insects. Development 151:dev202518. https://doi.org/10.1242/dev.202518

Bayer CA, Holley B, Fristrom JW (1996) A switch in *broad-complex* zinc-finger isoform expression is regulated posttranscriptionally during the metamorphosis of *drosophila* imaginal discs. Dev Biol 177:1–14. https://doi.org/10.1006/dbio.1996.0140

Belles X (2020) Insect metamorphosis– from natural history to regulation of development and evolution. Academic, Cambridge https://doi.org/10.1016/C2016-0-04530-8

Belles X, Santos CG (2014) The MEKRE93 (Methoprene tolerant-Krüppel homolog 1-E93) pathway in the regulation of insect metamorphosis, and the homology of the pupal stage. Insect Biochem Mol Biol 52:60–68. https://doi.org/10.1016/j.ibmb.2014.06.009

Brüning E, Lanzrein B (1987) Function of juvenile hormone III in embryonic development of the cockroach, *Nauphoeta cinerea*. Int J Invert Reprod Dev 12:29–44. https://doi.org/10.1080/01688170.1987.10510300

Brüning E, Saxer A, Lanzrein B (1985) Methyl farnesoate and juvenile hormone III in normal and precocene treated embryos of the ovoviviparous cockroach *Nauphoeta cinerea*. Int J Invert Reprod Dev 8:269–278. https://doi.org/10.1080/01688170.1985.10510155

Daimon T, Uchibori M, Nakao H, Sezutsu H, Shinoda T (2015) Knockout silkworms reveal a dispensable role for juvenile hormones in holometabolous life cycle. Proc Natl Acad Sci USA 112:E4226–E4235. https://doi.org/10.1073/pnas.1506645112

DiBello PR, Withers DA, Bayer CA, Fristrom JW, Guild GM (1991) The *Drosophila Broad-Complex* encodes a family of related proteins containing zinc fingers. Genetics 129:385–397. https://doi.org/10.1093/genetics/129.2.385

Dong L, Muramatsu N, Numata H, Ito C (2022) Functional analysis of a juvenile hormone inducible transcription factor, Krüppel homolog 1, in the bean bug, *Riptortus pedestris*. Zool Sci 39:562–569. https://doi.org/10.2108/zs220025

Erezyilmaz DF, Riddiford LM, Truman JW (2004) Juvenile hormone acts at embryonic molts and induces the nymphal cuticle in the direct-developing cricket. Dev Genes Evol 214:313–323. https://doi.org/10.1007/s00427-004-0408-2

Erezyilmaz DF, Riddiford LM, Truman JW (2006) The pupal specifier *broad* directs progressive morphogenesis in a direct-developing insect. Proc Natl Acad Sci USA 103:6925–6930. https://doi.org/10.1073/pnas.0509983103

Erezyilmaz DF, Rynerson MR, Truman JW, Riddiford LM (2009) The role of the pupal determinant *broad* during embryonic development of a direct-developing insect. Dev Genes Evol 219:535–544. https://doi.org/10.1007/s00427-009-0315-7

Fernandez-Nicolas A, Belles X (2017) Juvenile hormone signaling in short germ-band hemimetabolan embryos. Development 144:4637–4644. https://doi.org/10.1242/dev.152827

Gujar H, Palli SR (2016) Krüppel homolog 1 and E93 mediate juvenile hormone regulation of metamorphosis in the common bed bug, *Cimex lectularius*. Sci Rep 6:26092. https://doi.org/10.1038/srep26092

Hardie J (1987) The corpus allatum, neurosecretion and photoperiodically controlled polymorphism in an aphid. J Insect Physiol 33:201–205. https://doi.org/10.1016/0022-1910(87)90147-8

Huang JH, Lozano J, Belles X (2013) Broad-complex functions in postembryonic development of the cockroach *Blattella germanica* shed new light on the evolution of insect metamorphosis. Biochim Biophys Acta 1830:2178–2187. https://doi.org/10.1016/j.bbagen.2012.09.025

Ishikawa A, Ogawa K, Gotoh H, Walsh TK, Tagu D, Brisson JA et al (2012) Juvenile hormone titre and related gene expression during the change of reproductive modes in the pea aphid. Insect Mol Biol 21:49–60. https://doi.org/10.1111/j.1365-2583.2011.01111.x

Ishikawa A, Gotoh H, Abe T, Miura T (2013) Juvenile hormone titer and wing-morph differentiation in the vetch aphid *Megoura crassicauda*. J Insect Physiol 59:444–449. https://doi.org/10.1016/j.jinsphys.2013.02.004

Jindra M, Belles X, Shinoda T (2015) Molecular basis of juvenile hormone signaling. Curr Opin Insect Sci 11:39–46. https://doi.org/10.1016/j.cois.2015.08.004

Kiss I, Beaton AH, Tardiff J, Fristrom D, Fristrom JW (1988) Interactions and developmental effects of mutations in the broad-complex of *Drosophila melanogaster*. Genetics 118:247–259. https://doi.org/10.1093/genetics/118.2.247

Konopova B (2025) Evolution of insect metamorphosis – an update. Curr Opin Insect Sci 67:101289. https://doi.org/10.1016/j.cois.2024.101289

Konopova B, Jindra M (2008) Broad-Complex acts downstream of Met in juvenile hormone signaling to coordinate primitive holometabolan metamorphosis. Development 135:559–568. https://doi.org/10.1242/dev.016097

Konopova B, Zrzavy J (2005) Ultrastructure, development, and homology of insect embryonic cuticles. J Morphol 264:339–362. https://doi.org/10.1002/jmor.10338

Konopova B, Smykal V, Jindra M (2011) Common and distinct roles of juvenile hormone signaling genes in metamorphosis of holometabolous and hemimetabolous insects. PLoS One 6:e28728. https://doi.org/10.1371/journal.pone.0028728

Kotaki T, Shinada T, Kaihara K, Ohfune Y, Numata H (2009) Structure determination of a new juvenile hormone from a heteropteran insect. Org Lett 11:5234–5237. https://doi.org/10.1021/ol902161x

Lagueux M, Hetru C, Goltzene F, Kappler C, Hoffmann JA (1979) Ecdysone titer and metabolism in relation to cuticulogenesis in embryos of *Locusta migratoria*. J Insect Physiol 25:709–723. https://doi.org/10.1016/0022-1910(79)90123-9

Li M, Mead EA, Zhu J (2011) Heterodimer of two bHLH-PAS proteins mediates juvenile hormone-induced gene expression. Proc Natl Acad Sci USA 108:638–643. https://doi.org/10.1073/pnas.1013914108

Li KL, Yuan SY, Nanda S, Wang WX, Lai FX, Fu Q et al (2018) The roles of E93 and Kr-h1 in metamorphosis of *Nilaparvata lugens*. Front Physiol 9:1677. https://doi.org/10.3389/fphys.2018.01677

Lozano J, Belles X (2011) Conserved repressive function of Krüppel homolog 1 on insect metamorphosis in hemimetabolous and holometabolous species. Sci Rep 1:163. https://doi.org/10.1038/srep00163

Lozano J, Belles X (2014) Role of Methoprene-tolerant (Met) in adult morphogenesis and in adult ecdysis of *Blattella germanica*. PLoS One 9:e103614. https://doi.org/10.1371/journal.pone.0103614

Lozano J, Kayukawa T, Shinoda T, Belles X (2014) A role for Taiman in insect metamorphosis. PLoS Genet 10:e1004769. https://doi.org/10.1371/journal.pgen.1004769

Maestro JL, Pascual N, Treiblmayr K, Lozano J, Belles X (2010) Juvenile hormone and allatostatins in the German cockroach embryo. Insect Biochem Mol Biol 40:660–665. https://doi.org/10.1016/j.ibmb.2010.06.006

Matsuda R (1987) Animal evolution in changing environments with special reference to abnormal metamorphosis. Wiley, New York

Matsuda N, Takahashi M, Shirai Y, Hinomoto N, Daimon T (2024) Direct parental CRISPR gene editing in the predatory bug *Orius strigicollis*, a biocontrol agent against small arthropods. Pest Manag Sci 80:5465–5472. https://doi.org/10.1002/ps.8275

Minakuchi C, Zhou X, Riddiford LM (2008) *Krüppel homolog 1* (*Kr-h1*) mediates juvenile hormone action during metamorphosis of *Drosophila melanogaster*. Mech Dev 125:91–105. https://doi.org/10.1016/j.mod.2007.10.002

Minakuchi C, Namiki T, Shinoda T (2009) *Krüppel homolog 1*, an early juvenile hormone-response gene downstream of *Methoprene-tolerant*, mediates its anti-metamorphic action in the red flour beetle *Tribolium castaneum*. Dev Biol 325:341–350. https://doi.org/10.1016/j.ydbio.2008.10.016

Minakuchi C, Tanaka M, Miura K, Tanaka T (2011) Developmental profile and hormonal regulation of the transcription factors *broad* and *Krüppel homolog 1* in hemimetabolous thrips. Insect Biochem Mol Biol 41:125–134. https://doi.org/10.1016/j.ibmb.2010.11.004

Miura T (2019) Juvenile hormone as a physiological regulator mediating phenotypic plasticity in pancrustaceans. Develop Growth Differ 61:85–96. https://doi.org/10.1111/dgd.12572

Müller CB, Williams IS, Hardie J (2001) The role of nutrition, crowding and interspecific interactions in the development of winged aphids. Ecol Entomol 26:330–340. https://doi.org/10.1046/j.1365-2311.2001.00321.x

Muramatsu M, Tsuji T, Tanaka S, Shiotsuki T, Jouraku A, Miura K et al (2020) Sex-specific expression profiles of ecdysteroid biosynthesis and ecdysone response genes in extreme sexual dimorphism of the mealybug *Planococcus kraunhiae* (Kuwana). PLoS One 15:e0231451. https://doi.org/10.1371/journal.pone.0231451

Nijhout HF (1994) Insect hormones, Princeton University Press, Princeton

Novák VJ (1969) Morphogenetic analysis of the effects of juvenile hormone analogues and other morphogenetically active substances on embryos of *Schistocerca gregaria* (Forskål). J Embryol Exp Morphol 21:1–21. https://doi.org/10.1242/dev.21.1.1

Ogawa K, Miura T (2014) Aphid polyphenisms: trans-generational developmental regulation through viviparity. Front Physiol 5:1. https://doi.org/10.3389/fphys.2014.00001

Oishi S, Moriyama M, Mizutani M, Futahashi R, Fukatsu T (2023) Regulation and remodeling of microbial symbiosis in insect metamorphosis. Proc Natl Acad Sci USA 120: e2304879120. https://doi.org/10.1073/pnas.2304879120

Okude G, Moriyama M, Kawahara-Miki R, Yajima S, Fukatsu T, Futahashi R (2022) Molecular mechanisms underlying metamorphosis in the most-ancestral winged insect. Proc Natl Acad Sci USA 119:e2114773119. https://doi.org/10.1073/pnas.2114773119

Panfilio KA (2008) Extraembryonic development in insects and the acrobatics of blastokinesis. Dev Biol 313:471–491. https://doi.org/10.1016/j.ydbio.2007.11.004

Pecasse F, Beck Y, Ruiz C, Richards G (2000) *Krüppel-homolog*, a stage-specific modulator of the prepupal ecdysone response, is essential for *drosophila* metamorphosis. Dev Biol 221:53–67. https://doi.org/10.1006/dbio.2000.9687

Pflugfelder O (1936) Bau und morphologische bedeutung der sog. Ocellen der Schildlaus-Männchen (Lecanium corni March). Zool Anz 114:49–55

Piulachs MD, Pagone V, Belles X (2010) Key roles of the *Broad-Complex* gene in insect embryogenesis. Insect Biochem Mol Biol 40:468–475. https://doi.org/10.1016/j.ibmb.2010.04.006

Röller H, Dahm KH, Sweely CC, Trost BM (1967) The structure of the juvenile hormone. Angew Chem Int Ed Engl 6:179–180. https://doi.org/10.1002/anie.196701792

Sbrenna G (1990) Roles of morphogenetic hormones in embryonic cuticle deposition in arthropods. In: Gupta AP (ed) Morphogenetic hormones of arthropods, vol 3. Rutgers University Press, New Brunswick, pp 44–82

Sbrenna-Micciarelli A (1977) Effects of farnesyl methyl ether on embryos of *Schistocerca gregaria* (Orthoptera). Acta Embryol Exp 3:295–303

Shirai Y, Piulachs MD, Belles X, Daimon T (2022) DIPA-CRISPR is a simple and accessible method for insect gene editing. Cell Rep Methods 2:100215. https://doi.org/10.1016/j.crmeth.2022.100215

Suzuki Y, Shiotsuki T, Jouraku A, Miura K, Minakuchi C (2021) Characterization of E93 in neometabolous thrips *Frankliniella occidentalis* and *Haplothrips brevitubus*. PLoS One 16:e0254963. https://doi.org/10.1371/journal.pone.0254963

Temin G, Zander M, Roussel JP (1986) Physicochemical (GC-MS) measurements of juvenile hormone-III titers during embryogenesis of *Locusta migratoria*. Int J Invert Reprod Dev 9:105–112. https://doi.org/10.1080/01688170.1986.10510184

Tomoyasu Y, Denell RE (2004) Larval RNAi in *Tribolium* (Coleoptera) for analyzing adult development. Dev Genes Evol 214:575–578. https://doi.org/10.1007/s00427-004-0434-0

Truman JW, Riddiford LM (1999) The origins of insect metamorphosis. Nature 401:447–452. https://doi.org/10.1038/46737

Truman JW, Riddiford LM (2019) The evolution of insect metamorphosis: a developmental and endocrine view. Philos T R Soc B 374:20190070. https://doi.org/10.1098/rstb.2019.0070

Ureña E, Manjon C, Franch-Marro X, Martin D (2014) Transcription factor E93 specifies adult metamorphosis in hemimetabolous and holometabolous insects. Proc Natl Acad Sci USA 111:7024–7029. https://doi.org/10.1073/pnas.1401478111

Ureña E, Chafino S, Manjon C, Franch-Marro X, Martin D (2016) The occurrence of the holometabolous pupal stage requires the interaction between E93, Krüppel-homolog 1 and broad-complex. PLoS Genet 12:e1006020. https://doi.org/10.1371/journal.pgen.1006020

Vea IM, Tanaka S, Shiotsuki T, Jouraku A, Tanaka T, Minakuchi C (2016) Differential juvenile hormone variations in scale insect extreme sexual dimorphism. PLoS One 11:e0149459. https://doi.org/10.1371/journal.pone.0149459

Vea IM, Tanaka S, Tsuji T, Shiotsuki T, Jouraku A, Minakuchi C (2019) *E93* expression and links to the juvenile hormone in hemipteran mealybugs with insights on female neoteny. Insect Biochem Mol Biol 104:65–72. https://doi.org/10.1016/j.ibmb.2018.11.008

Vellichirammal NN, Gupta P, Hall TA, Brisson JA (2017) Ecdysone signaling underlies the pea aphid transgenerational wing polyphenism. Proc Natl Acad Sci USA 114:1419–1423. https://doi.org/10.1073/pnas.1617640114

Wigglesworth VB (1934) The physiology of ecdysis in *Rhodnius prolixus* (Hemiptera). II. Factors controlling moulting and 'metamorphosis'. J Cell Sci s2-77:191–222. https://doi.org/10.1242/jcs.s2-77.306.191

Wigglesworth VB (1936) The function of the corpus allatum in the growth and reproduction of *Rhodnius prolixus* (Hemiptera). J Cell Sci s2-79:91–121. https://doi.org/10.1242/jcs.s2-79.313.91

Zera AJ, Denno RF (1997) Physiology and ecology of dispersal polymorphism in insects. Annu Rev Entomol 42:207–230. https://doi.org/10.1146/annurev.ento.42.1.207

Zhang Z, Xu J, Sheng Z, Sui Y, Palli SR (2011) Steroid receptor co-activator is required for juvenile hormone signal transduction through a bHLH-PAS transcription factor, methoprene tolerant. J Biol Chem 286:8437–8447. https://doi.org/10.1074/jbc.M110.191684

Zhang MQ, Gong LL, Zhao YQ, Ma YF, Long GJ, Guo H et al (2024) Efficient DIPA-CRISPR-mediated knockout of an eye pigment gene in the white-backed planthopper, *Sogatella furcifera*. Insect Sci 31:1015–1025. https://doi.org/10.1111/1744-7917.13286

Zhou B, Williams DW, Altman J, Riddiford LM, Truman JW (2009) Temporal patterns of broad isoform expression during the development of neuronal lineages in Drosophila. Neural Dev 4:39. https://doi.org/10.1186/1749-8104-4-39

Chapter 10
The Molecular Makeup of Juvenile Hormone Receptor Signaling

Sarka Tumova and Marek Jindra

Abstract The action of juvenile hormone (JH), vitally important for regulating physiological processes in insect development and reproduction, relies on the functionality of the juvenile hormone receptor (JHR). The essential ligand-binding component of JHR has been identified as the product of *Methoprene-tolerant* (*Met*) genes. Upon binding JH, the basic helix-loop-helix-PER-ARNT-SIM (bHLH-PAS) protein MET forms a transcriptionally active complex with another member of the same family to execute specific gene-expression programs. We will outline the basic molecular attributes of the signaling before discussing recent developments on the mode of JHR functioning at both the gene-regulatory and nongenomic levels. Structural and mechanistic features of JHR and an analogous ligand-regulated bHLH-PAS protein, aryl hydrocarbon receptor, will be compared with emphasis on agonist-induced protein interactions during the receptor activation and on ligand specificity and selectivity. While this chapter reflects substantial advances in JH research over the last two decades, undoubtedly, many additional complexities of JHR signaling are yet awaiting their discovery.

Keywords Juvenile hormone · Sesquiterpenoids · Hormone receptor · bHLH-PAS proteins · Agonists · Ligand binding · Dimerization · Transcriptional regulation · Insect development

---

S. Tumova · M. Jindra (✉)
Institute of Entomology, Biology Center, Czech Academy of Sciences,
Ceske Budejovice, Czech Republic
e-mail: jindra@entu.cas.cz

T. Shinoda et al. (eds.), *Juvenile Hormone*, Entomology Monographs,
https://doi.org/10.1007/978-981-95-8605-9_10

10.1 Introduction

Juvenile hormone (JH) signaling is unique to arthropods and best studied in insects. JH provides critical stimuli during insect development as well as in adulthood, where it regulates reproduction and other physiological processes. JH signaling relies on receptors belonging to the basic helix-loop-helix-PER-ARNT-SIM (bHLH-PAS) family. In most insects, a single gene named *Methoprene-tolerant* (*Met*) encodes the JH-binding receptor protein. Brachyceran flies, including *Drosophila* species, additionally possess a paralog called *germ cell-expressed* (*gce*), and two copies of *Met* also occur in the Lepidoptera. The functional distinction between the duplicated paralogs remains unclear. In the absence of JH, the MET or GCE proteins reside within a cytoplasmic chaperone assembly with heat shock protein 83 (HSP83), but the interaction with JH induces conformational changes that lead to the receptor release and promote heterodimerization with another bHLH--PAS protein, taiman (TAI). The resulting complex functions as a transcription factor regulating the expression of JH-responsive genes through binding their DNA response elements. Recent years have seen major progress in understanding the molecular mechanisms that govern JH signaling, yet many aspects still remain elusive. This chapter will first set the stage by briefly outlining the critical JH attributes and relevant specifics of JH production, decay, and transport before proceeding with a short historical perspective on the JH receptor discovery. Some of these issues are specifically dealt with by other parts of this volume and have been comprehensively covered in a recent book chapter (Jindra and Noriega 2026). Finally, we will concentrate on the molecular details of JH signaling, with the emphasis on new findings, including nongenomic JH actions.

10.2 Juvenile Hormone: What is in a Name

JH was named for its most distinguishing feature, the ability to maintain a larval, or juvenile, character of the insect body, although it has turned out to be much more versatile in its physiological activities. Those can range from regulating fertility, reproduction, or metabolic balance to mating behavior or caste specification in eusocial insects. Most of these activities depend on the presence of the intracellular JH receptor (JHR) protein MET and require de novo gene expression. In some cases, a separate branch of nongenomic signaling can also be involved.

JH signaling was originally described decades before the identification of the signaling molecule itself, in the seminal work of Vincent Wigglesworth (1934, 1936, 1940, 1958). His experiments used the extracts from the cecropia silkmoth, *Hyalophora cecropia,* to interfere with the epidermal development of the kissing bug, *Rhodnius prolixus*, and gave the first indication of the cell-autonomous character of JH signaling, showing that its presence or absence was sufficient to confer the respective larval or adult character on the organism or its parts.

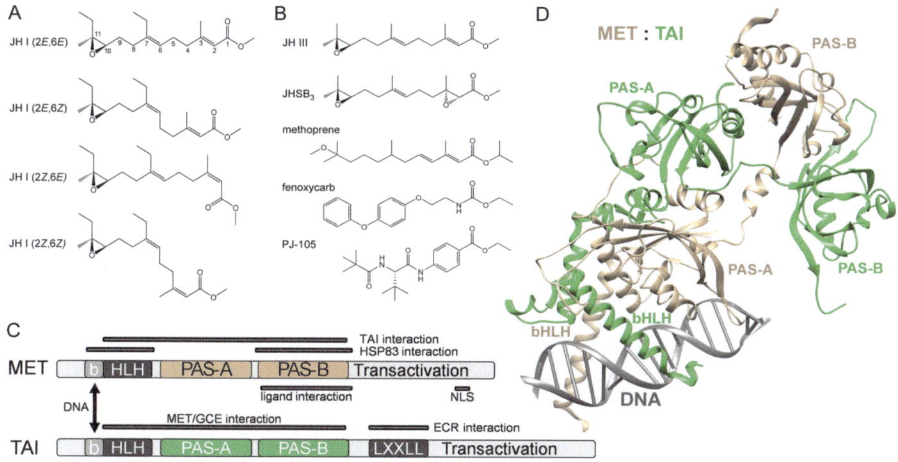

Fig. 10.1 Juvenile hormone (JH) and JH receptor (JHR) structures. (**a**) The native lepidopteran JH form, JH I (2*E*,6*E*), and its artificial geometric stereoisomers. (**b**) The structures of ubiquitous JH III, heteropteran $JHSB_3$, and the synthetic JHR agonists methoprene, fenoxycarb, and peptidic juvenoid PJ-105. (**c**) Domain architecture of the JHR bHLH-PAS proteins, the JH binder methoprene-tolerant (MET) and its dimerizing partner taiman (TAI), with the critical interacting regions outlined. HSP83 heat shock protein 83, ECR ecdysone receptor, NLS nuclear localization signal. Schematic not drawn to scale. (**d**) An AlphaFold-predicted structure of the MET:TAI heterodimer bound to the DNA response element. TAI is depicted in green, MET in tan, and the DNA double helix in gray. The model was provided by Roman Tuma

The active substances were later identified as sesquiterpenoids with a methyl ester at the C1 position and an epoxide group between C10 and C11 (Fig. 10.1a, b). Following this initial characterization of lepidopteran JH I and JH II (Meyer et al. 1968; Röller et al. 1967), JH III was discovered (Judy et al. 1973) as the most widespread JH form in insects (Helvig et al. 2004; Minakuchi et al. 2008; Niwa et al. 2008) and possibly the ancestral one (Baker et al. 1984; Truman et al. 2024). The JH III structure is similar to the lepidopteran JHs, except for methyl groups at the C7 and C11 positions, which are ethylated in JH I. The Diptera and Hemiptera possess bisepoxide forms of JH III with additional epoxide rings either between C6 and C7, as in JHB_3 of the fruit fly *Drosophila melanogaster* (Richard et al. 1989), or between C2 and C3 in JH III skipped bisepoxide ($JHSB_3$), which has been detected as the unique JH in heteropterans such as the stink bug *Plautia stali* (Kotaki et al. 2009), the linden bug *Pyrrhocorris apterus* (Hejnikova et al. 2022), and *R. prolixus* (Villalobos-Sambucaro et al. 2020) (Fig. 10.1b). Methyl farnesoate (MF), which functions as an active sesquiterpenoid hormone in crustaceans, is a detectable JH precursor in insects that have evolved the ability to epoxidize MF to produce JH III (Smykal and Dolezel 2023; Tsang et al. 2020). MF itself displays JH signaling activity, albeit weaker than the epoxidized JHs (Bittova et al. 2019; Kayukawa et al. 2012).

Structural features critical for JH (and MF) activity include the presence of the methyl ester group at C1 and the geometry of the double bonds at positions C2–3 and C6–7 (Fig. 10.1a). The natural configuration 2*E*,6*E*, common to MF and JHs synthesized by the insect endocrine glands (Helvig et al. 2004), is the most active one. The mixed isomers 2*E*,6*Z* or 2*Z*,6*E* are less potent, and the 2*Z*,6*Z* variant is completely inert, as was originally shown in vivo (Dahm et al. 1968; Wigglesworth 1969) and later confirmed for the JH receptor at the molecular level (Bittova et al. 2019; Tumova and Jindra 2023). Matching stereoselectivity was displayed in JH interaction with the JH-binding protein (JHBP) from larval hemolymph of the tobacco hornworm, *Manduca sexta* (Goodman et al. 1978).

Compounds with JH-like activity (JH mimics or juvenoids) were synthesized early on after the initial discovery of JH. While some juvenoids, such as methoprene, resemble the JH structurally, others, such as fenoxycarb or pyriproxyfen, have different chemistries (Fig. 10.1b). Juvenoids are agonists of the JHR (Bittova et al. 2019; Charles et al. 2011; Jindra and Bittova 2020; Tumova et al. 2022) and can typically replicate most JH effects both in vitro and in vivo, making them excellent tools for JH research. Several juvenoids have been used as insecticides of the Insect Growth Regulator (IGR) class.

10.3 How JH is Synthesized and Degraded: From Cradle to Grave

As has already been apparent in the original Wigglesworth's studies, controlling the hormone titer is a crucial aspect of the JH signaling mechanism. There are intricate ways in which JH levels can be regulated through biosynthesis, transport, storage, and degradation. Although alternative sites of synthesis have been suggested (Hejnikova et al. 2022; Rahman et al. 2017), JHs are generally produced and secreted into the hemolymph by the endocrine glands named corpora allata (Goodman and Cusson 2012; Jindra and Noriega 2026; Rivera-Pérez et al. 2020). The JH synthesis and degradation are described in detail in Chaps. 3, 4, 5, and 7. Therefore, we will just briefly summarize some of the main principles and mechanisms critical for the JH signaling potential.

The early steps of the JH synthesis (Bellés et al. 2005; Rivera-Pérez et al. 2020) are conserved in eukaryotes, as they overlap with the mevalonate pathway essential for isoprenoid synthesis. However, the follow-up reactions are specific to JH production. The terminal steps, methyl esterification by JH acid methyltransferase (JHAMT) (Shinoda and Itoyama 2003) and epoxidation by P450 enzymes of the CYP15 family (Helvig et al. 2004) (hereafter referred to as EPOX), procure the JH-defining modifications to generate the high-affinity signaling molecule. JHAMT activity and regulation are instrumental in shaping the JH titers, and its removal leads to complete JH deficiency (Minakuchi et al. 2008; Nouzova et al. 2021). EPOX encoding genes have not been detected in arthropods other than insects, and the epoxidation they supply may provide an evolutionary advantage by improving reproductive fitness (Nouzova et al. 2021).

Several types of enzymes have been implicated in inactivating surplus JH (Goodman and Cusson 2012; Rivera-Pérez et al. 2020), of which JH esterase (JHE) and JH epoxide hydrolase (JHEH) are best studied. The intracellular JHEHs may have other functions unrelated to JH processing, such as detoxification or lipid messenger modifications, but the JHEs have been directly linked to regulating JH titers by removing JH either from the circulation or within insect tissues. A fine example of the latter has been provided recently in the Florida carpenter ant, *Camponotus floridanus*, where a JHE isoform lacking the N-terminal signal peptide for secretion localizes at the blood-brain barrier and reduces JH levels to reprogram the caste-associated behavior (Ju et al. 2023). Abundant JHE-like proteins have also been detected in the trophallactic fluid of the same species, raising the possibility that the amount of JH present and shared via nutritive fluid exchange between individuals or in larval feeding could be further regulated by JH degradation (LeBouef et al. 2018).

10.4 Transport and JH-Binding Proteins: The Go-Between

The lipophilic nature of JH prevents it from remaining free during transport in the hemolymph. Four distinct groups of proteins have been described with the ability to transport, store, and protect JH in different species (Goodman and Cusson 2012; Rivera-Pérez et al. 2020; Trowell 1992). Some of the earliest identified high-affinity JH-binding proteins (JHBPs) secreted into the hemolymph were found in lepidopterans. These low-molecular-weight JHBPs harbor a central lipophilic cavity (Suzuki et al. 2011), and a pH-dependent collisional mechanism of ligand release (Dupas et al. 2020) has been proposed for JH delivery to target cells. Structurally distinct hemolymph JHBPs, related to odorant-binding proteins, have been discovered in pupae and adults of the yellow fever mosquito, *Aedes aegypti* (Kim et al. 2017). Their ability to bind JH appears to be essential for mounting a normal immune response to bacterial infection (Kim et al. 2020).

High-molecular-weight multimeric transport JHBPs known as hexamerins and lipophorins have been identified and studied in orthopterans (locusts and crickets) and cockroaches, respectively. They typically bind JH with affinity in the nanomolar (<100 nM) or even subnanomolar range (Rivera-Pérez et al. 2020; Trowell 1992). Lipophorins function as reusable lipoprotein shuttles capable of delivering pheromones and lipids such as diacylglycerol, phospholipids, or cholesterol to storage and peripheral tissues, and they have a high capacity for JH transport as well (Canavoso et al. 2001). In hexamerins, each of the six subunits (Burmester 2002) can bind one molecule of JH, providing extra storage capacity. In addition to the transport function, they can serve as reservoirs available for rapid JH release (Goodman and Cusson 2012) or sequestration (Zhou et al. 2007). On the other hand, JH activity itself has been shown to regulate hexamerin expression (Hawkings et al. 2019), suggesting a complex feedback mechanism.

The identity of JH transport carriers in other insect species or the exact mode of JH delivery into the cell is still unknown. Owing to its hydrophobicity, JH is thought to traverse the plasma membrane via simple diffusion. However, by analogy to the *Drosophila* ecdysone importer (EcI) for the steroid hormone (Okamoto et al. 2018), the existence of a dedicated JH importer protein cannot be ruled out. Yet, if an essential cellular JH importer exists, it must also be present in mammalian cells engineered to express insect JHR proteins, which can instantaneously respond to added JH (Bittova et al. 2019; Kayukawa et al. 2012; Tumova and Jindra 2023). One might imagine that once JH is synthesized and released from the corpora allata, an orchestrated system would be in place to guide its delivery to the target tissues and the intracellular receptor, but, with the exception of the proteins mentioned above, further details or other likely components in such a pathway remain unknown.

10.5 Finding the Signaling Receptor for JH: The Arduous Path

Once the signaling hormone molecule had been identified, the search for a specific cellular receptor mediating its developmental effects was on. The initial efforts were paradoxically derailed by using the otherwise excellent genetic model of *Drosophila*, where, in a promising start, a candidate gene named *Methoprene-tolerant* (*Met*) was identified in a genetic screen (Wilson and Fabian 1986). However, the mutant flies suffered only subtle anomalies inconsistent with complete removal of JH signaling (for reviews, see Dubrovsky and Bernardo 2014; Jindra and Noriega 2026). Given the nature of known receptors for other lipophilic signals such as steroids or retinoic acid, it seemed rather surprising that MET was a bHLH-PAS protein (Ashok et al. 1998). Although MET bound JH with high affinity (Miura et al. 2005), the apparent phenotype inconsistency paused the pursuit of MET as a JHR. It turned out that, as already anticipated (Wilson and Ashok 1998), the mild *Met* mutant phenotype had been caused by genetic redundancy (Abdou et al. 2011). In brachyceran flies such as *D. melanogaster*, *Met* is a result of gene duplication of the ancestral gene *germ cell-expressed* (*gce*), and it only becomes indispensable when *gce* is deleted. Loss of both *Met* and *gce* is unconditionally lethal during pupation, a phenotype also caused by JH deficiency (Riddiford et al. 2010). Since MET and GCE can independently act as JHRs and substitute for each other's function (Abdou et al. 2011; Jindra et al. 2015), flies lacking only one of the paralogs are viable.

The identity of JHR was eventually established with the help of the red flour beetle, *Tribolium castaneum*, which only possesses one *Met* gene and provides a better representation of "canonical" JH signaling. Here, JH application during the last larval or the pupal stages causes the respective supernumerary juvenile molts. Conversely, precocious metamorphosis occurs when JH is withdrawn prematurely. Removing MET through systemic RNAi knockdown had the same effect, confirming the indispensability of MET in JH signaling (Konopova and Jindra 2007).

T. castaneum MET was fully established as the JHR by JH binding studies coupled with mutagenesis of the ligand-binding domain (Charles et al. 2011). Cell-based reporter assays characterized the JH-dependent transcriptional activity of MET/GCE proteins from other insects (Jindra et al. 2015; Kayukawa et al. 2012, 2013; Li et al. 2011), and the requirement for MET/GCE to bind JH for its role in metamorphosis was demonstrated in vivo using the *D. melanogaster* genetic model (Jindra et al. 2015). The essential function of MET in JH signaling has since been validated in numerous species, as reviewed in Jindra et al. (2021b, 2024) and Palli (2023).

As noted above, MET proteins are members of the bHLH-PAS family of transcription factors (Fig. 10.1c). The genes encoding MET orthologs are conserved in arthropods but have been lost early in the vertebrate lineage (Tumova et al. 2024), where JH itself is also lacking. The bHLH-PAS proteins typically act as heterodimers, with the subunits providing specific functions categorized as class I, whereas their more promiscuous and versatile partners are designated as class II. MET has not been formally categorized, but it can form complexes with a class II protein cycle (CYC), a partner of the circadian protein clock (CLK), to regulate circadian rhythmic expression of JH-response genes in *A. aegypti* (Shin et al. 2012). JH-independent interactions of *D. melanogaster* MET with CLK (He et al. 2023) or of *A. aegypti* MET with the versatile class II bHLH-PAS protein tango (TGO) (Shin et al. 2012), orthologous to the vertebrate aryl hydrocarbon receptor nuclear translocator (ARNT), were reported, but their physiological significance is unclear. As we shall see below, currently the best characterized signaling JHR complexes mediating JH action are bHLH-PAS heterodimers comprising either MET or GCE and their partner TAI.

But first, we will take a brief detour to mention another type of MET complexes with unresolved function. The MET and GCE proteins from *D. melanogaster* were long ago reported to form homo- and heteromeric complexes (presumably dimers) that would dissociate upon JH addition (Godlewski et al. 2006). Later, it was demonstrated for *T. castaneum* MET that the ability to bind JH is required for the MET:MET complex disassembly, since mutations preventing the interaction with JH caused MET to stay within the complex when JH was added (Charles et al. 2011). As long as at least one of the MET monomers could bind JH, the ligand addition led to the rapid complex dissociation (Charles et al. 2011; Tumova and Jindra 2023). However, what relationships might exist between the MET:MET or MET:GCE complexes and the liganded, transcriptionally active MET/GCE:TAI heterodimer, or what their function could be, is unclear. Within the bHLH-PAS family, a homodimer has been reported for ARNT and shown to activate transcription in a cell-based assay (Sogawa et al. 1995). Whether complexes containing MET and/or GCE without a heterodimerizing partner exert any effects on transcription and what might be the role of the JH-dependent complex dissociation are purely speculative at this point. Despite attempts, no DNA sequence that could bind to the presumed *A. aegypti* MET homodimer has been identified (Li et al. 2014). Uncovering the physiological role of these homomeric complexes requires further study.

As alluded to above, MET/GCE partners with TAI to mediate JH signaling (Kayukawa et al. 2012; Li et al. 2011, 2014; Zhang et al. 2011). TAI is an atypical bHLH-PAS family member and an insect ortholog of the vertebrate nuclear receptor coactivators (NCoAs 1–3), also known as steroid receptor coactivators (SRCs 1–3), which regulate nuclear hormone receptors and histone acetylases (Tumova et al. 2024). TAI had been reported under different names (such as SRC-1 or FISC) and in alternative complexes, for example, as a coactivator to the nuclear receptors ECR and FTZ-F1 in *D. melanogaster* (Bai et al. 2000) and *A. aegypti* (Zhu et al. 2006), respectively. Like the vertebrate NCoAs, TAI binds the insect nuclear receptors through its C-terminal LXXLL motifs without direct interaction with DNA. A distinct mode of engagement is observed in the JHR complex, which follows the general principles described for other bHLH-PAS heterodimers (Jindra et al. 2024; Wu and Rastinejad 2017), where the N-termini of both monomers participate in the DNA binding (Fig. 10.1c, d). The bHLH-PAS region is both necessary and sufficient for the MET:TAI interaction (Charles et al. 2011; Jindra et al. 2021a; Li et al. 2011). The 1:1 stoichiometry had been anticipated based on the two-hybrid assays (Li et al. 2011) and was confirmed when the functional native-like *T. castaneum* MET:TAI complex was purified (Fig. 10.2a) and characterized in vitro (Jindra et al.

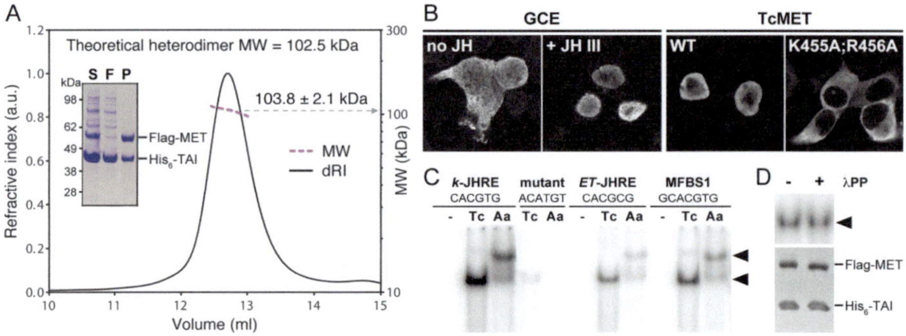

Fig. 10.2 Juvenile hormone receptor (JHR) protein complex, nuclear localization, and interaction with DNA. (**a**) Size-exclusion chromatography/multiangle laser-light scattering of the purified JHR complex of tagged *Tribolium castaneum* methoprene-tolerant (MET) and taiman (TAI) co-expressed in insect *Sf9* cells determined 1:1 stoichiometry. The protein concentration is indicated in refractive index units (dRI), with the molecular weight (MW) denoted by a purple dashed line. The inset shows electrophoretic analysis of the starting material following enrichment on a Ni-NTA column (S), flow-through (F), and Flag-affinity purification (P). (**b**) Myc-tagged *Drosophila melanogaster* GCE and *T. castaneum* (TcMET) proteins were expressed in human HEK293 cells. GCE is evenly distributed throughout the cells in JH absence and localized to the nuclei upon JH III addition. Nuclear localization of wild-type (WT) TcMET was blocked by mutating the conserved NLS sequence (K455A; R456A). (**c**) Purified recombinant *T. castaneum* (Tc) and *Aedes aegypti* (Aa) MET:TAI complexes interacted with DNA containing E-box-like JH response element (JHRE) motifs but not with a mutated sequence in an electrophoretic mobility shift assay. The sequences are from the *Krüppel-homolog 1* (*k*-JHRE) and *A. aegypti early trypsin* (*ET*-JHRE) genes; MFBS1 (MET-FISC binding site 1) was determined by random oligonucleotide selection (Li et al. 2014). (**d**) Dephosphorylation of the purified MET:TAI complex from (**a**) with λPP phosphatase reduced the molecular weight of MET (bottom) but did not prevent binding of the MET:TAI dimer to the *k*-JHRE DNA (top). DNA:JHR complexes in (**c**, **d**) are indicated by arrowheads. (Data are from Jindra et al. (2021a))

2021a). Detailed structural information on either of the partners is still lacking, but molecular modeling aided by AlphaFold (Jumper et al. 2021) suggests a tight association between the PAS domains of *D. melanogaster* GCE and TAI (Sedlak et al. 2025) or *T. castaneum* MET and TAI (Fig. 10.1d). Upon agonist-stimulated nuclear import of GCE/MET (Fig. 10.2b), the dimerization with TAI enables the JHR complex to bind to DNA (Fig. 10.2c). The JHR models broadly resemble the structurally resolved heterodimers formed by ARNT with the aryl hydrocarbon receptor (AHR) (Dai et al. 2022; Diao et al. 2025; Schulte et al. 2017). The analogies between JHR and the AHR:ARNT signaling will be further discussed in Sect. 10.7.

10.6 JH-Binding and JHR-Signaling Complexes: Seeing is Believing

The primary function of MET/GCE as a JH-binding protein has been validated in direct ligand-binding assays with radioactive [10-^{3}H]JH III (Fig. 10.3a) and in vitro translated (Bittova et al. 2019; Charles et al. 2011; Li et al. 2014; Milacek et al. 2021) or exogenously expressed and purified (Jindra et al. 2021a) recombinant proteins. The measured affinity for JH III generally falls within the low nanomolar range, with examples of determined dissociation constants (K_d) shown in Table 10.1. The binding of unlabeled compounds can be tested through competition against the radiolabeled JH III, providing the inhibitory constant (K_i) as a good estimate of affinity. Indeed, the K_i for unlabeled JH III competing against [^{3}H]JH III closely matched the respective K_d (Bittova et al. 2019; Milacek et al. 2021), and also the interaction between lepidopteran JH I and *D. melanogaster* GCE protein yielded similar K_i (Table 10.1). Lower affinity (higher K_i) was observed for JHB_3 and MF binding to GCE (Bittova et al. 2019). The specificity of JH binding has been demonstrated in two ways: (1) by mutating critical residues within the ligand-binding cavity of the MET or GCE PAS-B domain (Charles et al. 2011; Jindra et al. 2015; Li et al. 2014), and (2) by using inactive stereoisomers such as JH I (2*Z*,6*Z*) (Fig. 10.1a), which fail to compete against [^{3}H]JH III in the binding assay (Bittova et al. 2019).

The JH-induced formation of MET:TAI heterodimeric complexes has been observed for diverse species in numerous studies using co-immunoprecipitation, two-hybrid (Fig. 10.3b), or split luciferase (Fig. 10.3c) techniques. The effect of JH on MET:TAI interaction is common to distant holometabolous and hemimetabolous orders, underscoring the broadly conserved role of this specific JHR heterodimer in insect metamorphosis and reproduction (Rivera-Pérez et al. 2020). *Met* or *Met*-like genes have been detected throughout the Protostomia (Tumova et al. 2024). However, apart from insects, the functionality of MET in sesquiterpenoid signaling has only been demonstrated for crustaceans, where two-hybrid assays revealed interaction between the MET and TAI orthologs of the water fleas *Daphnia pulex* and *D. magna* in the presence of the native crustacean MF or insect JHs (Miyakawa

Table 10.1 Binding affinities and agonist potencies of selected juvenile hormone receptor (JHR) hormonal ligands

Ligand	Receptor	Species[a]	Potency (nM)	Reference
Direct binding				
[^{3}H]-JH III	MET	*D. melanogaster*	$K_d = 5.3$	Miura et al. (2005)
[^{3}H]-JH III	GCE	*D. melanogaster*	$K_d = 19.3$	Jindra et al. (2015)
[^{3}H]-JH III	MET	*T. castaneum*	$K_d = 2.9$	Charles et al. (2011)
[^{3}H]-JH III	MET PAS-B	*T. castaneum*	$K_d = 12.3$	Charles et al. (2011)
[^{3}H]-JH III	MET	*A. aegypti*	$K_d = 4.4$	Li et al. (2014)
[^{3}H]-JH III	MET	*P. simplex*	$K_d = 5.7$	Milacek et al. (2021)
[^{3}H]-JH III	MET PAS-B	*B. germanica*	$K_d = 8.5$	Milacek et al. (2021)
Competition				
JH III	MET	*P. simplex*	$K_i = 4.9$	Milacek et al. (2021)
JH III	MET	*B. germanica*	$K_i = 7.3$	Milacek et al. (2021)
JH III	GCE	*D. melanogaster*	$K_i = 11.0$	Bittova et al. (2019)
JH I	GCE	*D. melanogaster*	$K_i = 13.8$	Bittova et al. (2019)
JHB_3	GCE	*D. melanogaster*	$K_i = 83.3$	Bittova et al. (2019)
MF	GCE	*D. melanogaster*	$K_i = 89.8$	Bittova et al. (2019)
Two-hybrid				
JH III	GCE:TAI	*D. melanogaster*	EC_{50}[b] = 58.4	Bittova et al. (2019)
JH I	GCE:TAI	*D. melanogaster*	$EC_{50} = 53.4$	Bittova et al. (2019)
JHB_3	GCE:TAI	*D. melanogaster*	$EC_{50} = 310$	Bittova et al. (2019)
MF	GCE:TAI	*D. melanogaster*	$EC_{50} = 300$	Bittova et al. (2019)
$JHSB_3$	MET:TAI	*P. apterus*	$EC_{50} = 67.8$	Tumova et al. (2022)
NanoBiT (split luciferase)				
JH III	MET:TAI	*T. castaneum*	$EC_{50} = 6.1$	Tumova, unpublished
$JHSB_3$	MET:TAI	*P. apterus*	$EC_{50} = 3.3$	Tumova, unpublished

[a]*D. melanogaster*, *Drosophila melanogaster* (Diptera); *T. castaneum*, *Tribolium castaneum* (Coleoptera); *A. aegypti*, *Aedes aegypti* (Diptera); *P. simplex*, *Prorhinotermes simplex* (Blattodea); *B. germanica*, *Blattella germanica* (Blattodea); *P. apterus*, *Pyrrhocoris apterus* (Hemiptera, Heteroptera)
[b]Half-maximal effective concentration

Fig. 10.3 (continued) induced dimerization of MET/germ cell-expresse (GCE) and taiman (TAI). Natural JHs and methyl farnesoate (MF) induced GCE:TAI interaction in a dose-dependent manner. (**c**) Split luciferase assay with heat shock protein 83 (HSP83) and the JHR proteins tagged with complementing fragments of the NanoLuc luciferase. Interactions of *T. castaneum* MET with HSP83 or TAI were followed in real time, showing that JH binding to wild type (WT) MET triggered disassembly of the HSP83:MET complex, while the MET T254Y and V297F mutants that cannot bind JH did not dissociate in the agonist presence (left). On a similar time scale, the assay detected MET:TAI dimerization induced by JH binding to WT MET, but the association was prevented by the loss of JH binding in the MET mutants (right). (**d**) The JH signaling mechanism inferred from the above and other published studies (see the main text). DBD, DNA-binding domain; VP16, transactivation domain; LB and SB, large-bit and small-bit NanoLuc fragments; Luc, luciferase. (Data are from Charles et al. (2011) (**a**), Bittova et al. (2019) (**b**), and Tumova and Jindra (2023) (**c**))

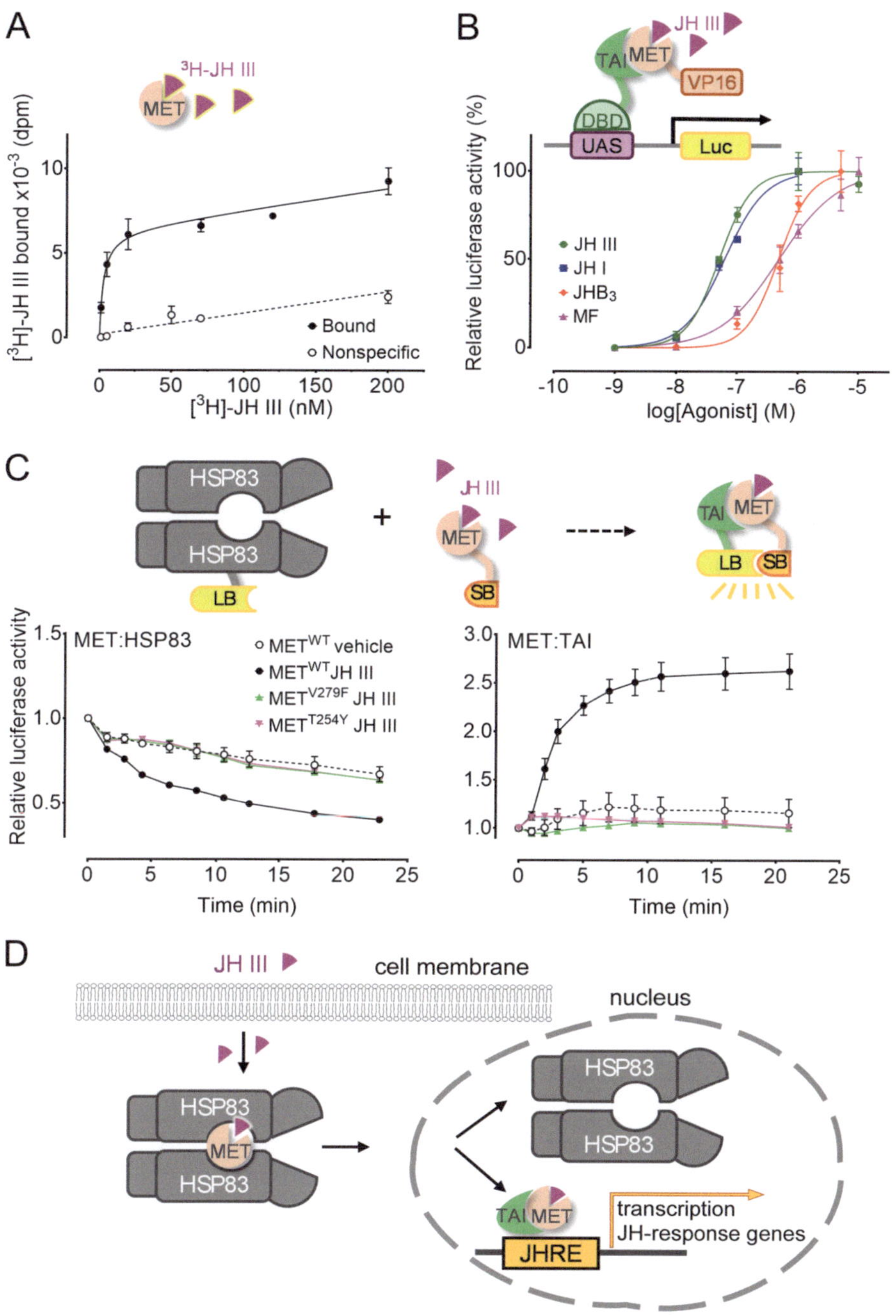

Fig. 10.3 Juvenile hormone receptor (JHR) activity and signaling complex dynamics. (**a**) Direct in vitro binding assay determined the affinity of *Tribolium castaneum* methoprene-tolerant (MET) for radiolabeled JH III (K_d = 2.9 nM). (**b**) Two-hybrid luciferase reporter assay based on ligand-

and Iguchi 2017; Miyakawa et al. 2013). The *Daphnia* MET:TAI complex also engages in agonist-dependent transcriptional activation (Tanaka et al. 2019). While the MET:TAI (or GCE:TAI) complex has been firmly established as the JH signaling assembly, MET can interact with other partners, both in JH presence (CYC) or absence (TGO or MET/GCE itself). Therefore, the existence of additional MET signaling complexes, with or without JH involvement, could be relevant in some physiological contexts.

Besides revealing MET/GCE:TAI interaction, the cell-based two-hybrid system also provides a quantitative measure of JHR activation (Fig. 10.3b and Table 10.1). As demonstrated through mutagenesis within the PAS-B domain, the ligand-binding function of MET/GCE is required for dimerization with TAI (Milacek et al. 2021; Tumova et al. 2022). Therefore, the two-hybrid method can partly substitute for a direct ligand-binding assay.

Studies on JH cellular signaling have predominantly relied on endpoint assays detecting co-immunoprecipitated protein complexes or transcriptional reporter activity. By their nature, these methods are unable to address events ensuing immediately after JH exposure. The availability of a split luciferase assay for monitoring cellular protein-protein interactions in real time (Dixon et al. 2016) has allowed for a more dynamic approach that could follow the JH-induced assembly or disassembly of the protein partners tagged with complementing fragments of the NanoLuc luciferase (Fig. 10.3c). Thus, interactions of JHR proteins, detected as changes in luminescence, could be captured in live cells right after the agonist application and followed in time (Tumova and Jindra 2023). The assay can yield quantitative data for JHR agonist potency (Table 10.1). As will be discussed in the next section, this new technique has enabled addressing some outstanding questions regarding the early steps during JHR activation.

10.7 The JHR Signaling Mechanism: Aryl Hydrocarbon Receptor Parallels

Although low levels of MET:TAI complexes occur even in the absence of JH, the heterodimerization is dramatically enhanced once JH becomes available. The formation of a stable signaling heterodimer requires MET that is competent for JH binding, as has been documented for MET mutants that are unable to dimerize with TAI when their ability to bind JH has been compromised (Charles et al. 2011; Li et al. 2014; Milacek et al. 2021; Tumova and Jindra 2023; Tumova et al. 2022). Prior to the ligand engagement, MET is retained in a cytosolic chaperone complex containing HSP83 (Tumova and Jindra 2023), the insect ortholog of HSP90, in a manner resembling the ligand-activated vertebrate bHLH-PAS transcription factor AHR (Bonati et al. 2024; Gruszczyk et al. 2022; Jindra et al. 2024). The interaction with HSPs ensures proper folding and protection from degradation, keeping the client receptors in a ligand-binding-competent state. Indeed, HSP90/83 inhibition leads to rapid degradation of AHR or MET (Chen et al. 1997; Tumova and Jindra 2023), likely by targeting the stalled complex for proteasomal degradation.

Structural analysis of the AHR:HSP90 assembly suggests a partially unfolded AHR with one α-helix of the PAS-B domain threaded through the HSP90 dimer (Gruszczyk et al. 2022). AlphaFold-aided modeling has suggested a comparable arrangement for HSP83-bound GCE (Sedlak et al. 2025). For AHR, interaction with ligands such as indirubin induces a conformational change that leads to nuclear import, release from the chaperone, and dimerization with the class II partner ARNT. The molecular details of the process are still subject to investigation (Bonati et al. 2024; Gruszczyk et al. 2022).

A similar mechanism could be envisioned for the JHR. Studies on *D. melanogaster* MET have suggested that upon ligand binding, the MET-HSP83 complex migrates to the nucleus (He et al. 2014, 2017). The nuclear transport has also been shown for liganded GCE or *T. castaneum* MET (Fig. 10.2b). Indeed, both nuclear export and nuclear localization signals (NLS) have been identified in MET/GCE proteins (Greb-Markiewicz et al. 2011, 2015; Jindra et al. 2021a). TAI is considered a nuclear resident, and so it would be expected that is where the signaling MET:TAI heterodimer forms, but the exact course of events is poorly understood. As has been inferred from the split luciferase assay in mammalian cells expressing MET and HSP83 from *T. castaneum* (Tumova and Jindra 2023), the dissociation of MET from the chaperone occurs within minutes following ligand application (Fig. 10.3c). The ligand interaction with MET is a prerequisite to the complex disassembly, since mutated MET variants that are unable to bind JH do not dissociate from HSP83 when JH is supplied. On a similar time scale, interaction between the NanoLuc-tagged MET and TAI could be monitored upon JH addition (Fig. 10.3c). As expected, this interaction is blocked by mutations preventing MET from binding JH or when an inactive JH stereoisomer is used (Tumova and Jindra 2023). These results have afforded a model for JH signaling analogous to the AHR activation pathway, in which ligand binding to the chaperone-bound MET triggers a conformational change leading to MET dissociation from HSP83 and dimerization with TAI (Fig. 10.3d). In an analogy with AHR, structural changes in MET/GCE induced by ligand binding could lead to the NLS exposure and importin-aided nuclear translocation (He et al. 2014).

Once assembled, the nuclear JHR complex puts on its transcription factor hat and uses the combined bHLH domains to bind DNA at JH response elements (JHREs) (Fig. 10.2c) to activate the transcription of target genes in a manner analogous to other bHLH-PAS transcription factors (Wu and Rastinejad 2017) (Figs. 10.1d and 10.3d). The JHRE sequences containing an E-box-like hexanucleotide core, CACGTG or CACGCG (Fig. 10.2c), were identified in regulatory regions of JH-response genes such as *Krüppel-homolog 1* (*Kr-h1*) or *early trypsin* (*ET*) from *A. aegypti* (Li et al. 2011; Shin et al. 2012), the silkworm *Bombyx mori* (Kayukawa et al. 2012), and *T. castaneum* (Kayukawa et al. 2013). An unbiased search for DNA motifs binding to the recombinant *A. aegypti* MET:TAI complex predominantly yielded sequences sharing the CACGTG palindrome, to which the heterodimer bound with affinities reaching K_d in the nanomolar range (Li et al. 2014). Other JH response elements could be pertinent depending on the target genes.

10.8 Ligand-Binding Pocket Structure and Selectivity: Getting to the Core of It

In general, the activity of most class I bHLH-PAS proteins can be influenced by interactions with low-molecular-weight compounds (Wu et al. 2016). However, MET/GCE and the vertebrate AHR thus far represent the only family members known to require agonist ligands for activation. Their ligand-binding function lies in the PAS-B domain (Figs. 10.1c and 10.4a). Homology modeling for GCE, supported by AlphaFold, predicted a binding cavity formed by a cluster of β strands interspersed with several helices (Figs. 10.1d and 10.4b) (Sedlak et al. 2025), largely reminiscent of the human AHR PAS-B domain structure (Diao et al. 2025; Gruszczyk et al. 2022). In both cases, the ligand-binding pockets are lined by about an equal mix of hydrophobic and polar residues, of which the latter may engage in hydrogen bonding that stabilizes the cavity architecture or the respective ligand. Reassuringly, some of the JHR residues implicated in contacts with the ligand groups have been previously shown as critical for JH binding by the GCE and MET proteins (Charles et al. 2011; Jindra et al. 2015; Li et al. 2014; Sedlak et al. 2025). In support of the analogy, the same positions are shared by several ligand-binding residues of AHR (Fig. 10.4a). One of the best described mutations that preclude JH binding is T254Y in *T. castaneum* MET (Charles et al. 2011) and the homologous mutations of the conserved threonine in MET/GCE proteins from other species (Jindra et al. 2015; Li et al. 2014; Milacek et al. 2021; Tumova et al. 2022). Mutation of the corresponding T283 in murine AHR also abolishes ligand binding (Pandini et al. 2009) (Fig. 10.4a). Interestingly, molecular dynamics simulations of a GCE variant with the homologous T542Y mutation suggest that rather than making an indispensable contact with the ligand, the residue may be critical for the PAS-B domain stability, since replacing it makes the cavity prone to collapse during the simulations (Sedlak et al. 2025).

A comparison between the JH III-engaged GCE and the indirubin-bound human AHR shows one notable difference. There are five phenylalanine residues lining the AHR binding pocket, of which only F324 aligns with F576 in GCE (Fig. 10.4a, b). Another two positions (F287 and F351) are occupied by the more polar tyrosine residues (Y540, Y604) that make hydrogen bonds to JH III; accordingly, Y540F and Y604F mutations reduce JH III binding to GCE (Sedlak et al. 2025). Three of the AHR phenylalanines make contact with indirubin. This could be one reason why JH III and indirubin show no cross-reactivity toward AHR and GCE, respectively. However, the ligand pocket malleability in both receptors may allow for other shared ligands, as was demonstrated in a recent high-throughput screen, where about 25% of compounds identified as JHR agonists also showed some activity toward human AHR (Sedlak et al. 2025).

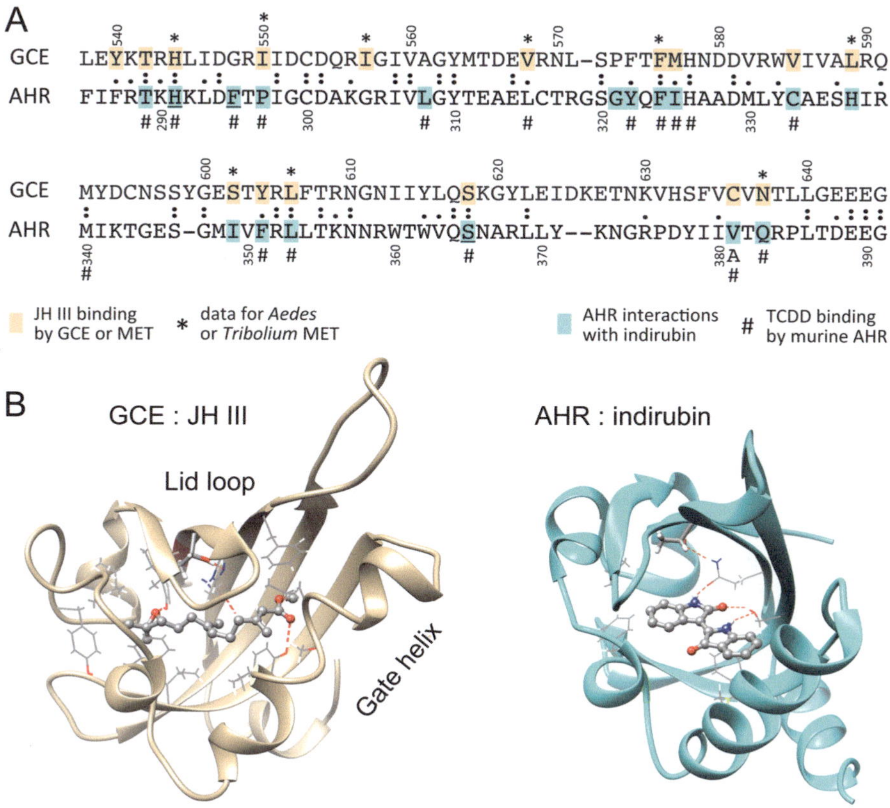

Fig. 10.4 Similarity of ligand engagement between juvenile hormone receptor (JHR) and aryl hydrocarbon receptor (AHR). (**a**) PAS-B domain residues critical for germ cell-expressed (GCE)/methoprene-tolerant (MET) and AHR ligand binding occupy corresponding positions. *Drosophila melanogaster* GCE and human AHR sequences are shown. TCDD, 2,3,7,8-tetrachlorodibenzo-*p*-dioxin. (**b**) Structural comparison between GCE (tan) and human AHR (cyan) ligand-binding cavities within the PAS-B domain occupied by their respective ligands, JH III and indirubin. The representations were generated by AlphaFold2 modeling and molecular dynamics simulations (Sedlak et al. 2025)

While the natural JHR agonists occur in several forms, with JH III being the most prevalent in insects, the different JHs, including the lepidopteran-specific JH I, generally activate MET/GCE receptors across insect orders (Bittova et al. 2019; Kayukawa and Shinoda 2015; Kayukawa et al. 2012, 2013; Tumova et al. 2022). At the same time, the binding of the native hormones is highly stereoselective, as demonstrated in various assays for JH I stereoisomers (Bittova et al. 2019; Tumova and Jindra 2023). Yet the MET/GCE ligand-binding cavity can accommodate synthetic compounds with JH activity that greatly differ from the natural JH in their structure, like the broad-spectrum IGR insecticide fenoxycarb with two benzene rings (Fig. 10.1b), which interacts with MET/GCE proteins from different insects and daphnid crustaceans rather indiscriminately and with potency superior to that of the natural hormones (Bittova et al. 2019; Miyakawa and Iguchi 2023). Intriguingly, there are synthetic JHR agonists with a remarkable level of selectivity for MET proteins from certain insects. A class of small molecules containing peptide bonds (Fig. 10.1b), originally inspired by a balsam fir product, juvabione (Zaoral et al. 1970), exclusively affects true bugs of the Pyrrhocoridae family. These "peptidic juvenoids" indeed selectively activate MET from the pyrrhocorid bugs but no other tested species (Fig. 10.5a). Chemical modification of the peptidic juvenoids has further improved agonist potency to reach picomolar concentrations, thus surpassing the natural $JHSB_3$ by three orders of magnitude (Tumova et al. 2022). The improved potency may result from an increased number of contacts with the residues in the binding cavity of *P. apterus* MET adjusting to the synthetic ligand (Fig. 10.5b).

The features underlying the remarkable ligand selectivity illustrated above likely reside within the PAS-B domain of MET/GCE proteins. Homology modeling and molecular dynamics have identified structural features important for the general mechanism (Tumova et al. 2022). In good agreement with the structural predictions of the GCE model, the PAS-B domain of *P. apterus* MET contains an elongated binding cavity lined by β-sheets on one side and on the opposite side by α-helices, including one that may function as a gate, opening and closing upon the ligand entry into the binding space (Fig. 10.5b). A flexible loop at the N-terminus of PAS-B appears to act like a lid, switching between open (apo) and closed (ligand-bound) positions. The dynamic arrangement apparently enables partial opening for the ligand entry, followed by a slight expansion allowing even larger molecules to be accommodated. The relevance of this model has been further validated by computed ligand interaction energies, which correlate well with the agonist potencies determined for various ligands by in vitro and in vivo assays (Tumova et al. 2022). Both JHR models share similarity with the flexible ligand-binding pocket of AHR (Fig. 10.4b) that likewise accommodates ligands of highly variable structures and chemistries.

Swapping the PAS-B domains between MET proteins from different species has been used to address ligand selectivity. MET2 from *B. mori* displayed a slight (3.6-fold) preference for the lepidopteran native hormone JH I over JH III, and this preference was shifted toward JH III when the *B. mori* PAS-B domain was replaced with that of *T. castaneum* MET (Kayukawa and Shinoda 2015). A much more dramatic effect was achieved by exchanging PAS-B domains between MET proteins from two bug species (Heteroptera), of which only *P. apterus* is sensitive to the peptidic juvenoids such as PJ-105 (Fig. 10.1b) (Tumova et al. 2022). As described above, these compounds activate MET from *P. apterus* but not from the bed bug *Cimex lectularius* (Fig. 10.5a). Our unpublished data show that introducing the PAS-B domain from *P. apterus* into *C. lectularius* MET imparts responsiveness to PJ-105 (Fig. 10.5c), further reinforcing the concept that ligand specificity is determined within the PAS-B region. While being evolutionarily the best conserved of MET domains, PAS-B still presents a considerable diversity, with only 58% amino acid identity between the two heteropteran species.

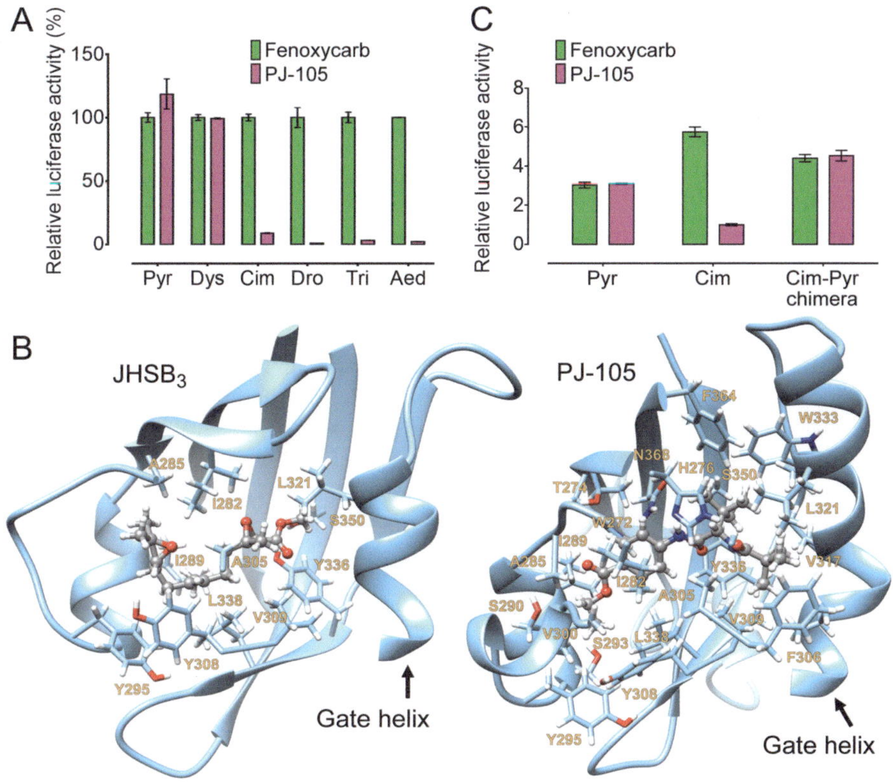

Fig. 10.5 Ligand-specific interactions of juvenile hormone receptor (JHR) from diverse insects. (**a**) The two-hybrid assay showed that while fenoxycarb activated JHRs from all tested species, peptidic juvenoid (PJ-105) selectively induced interaction of methoprene-tolerant (MET) and taiman (TAI) from the Pyrrhocoridae family, exemplified by *Pyrrhocoris apterus* (Pyr) and *Dysdercus fasciatus* (Dys), but not from *Cimex lectularius* (Cim), *Drosophila melanogaster* (Dro), *Tribolium castaneum* (Tri), or *Aedes aegypti* (Aed). (**b**) The *P. apterus* MET ligand-binding pocket occupied either by the natural ligand $JHSB_3$ (left) or PJ-105 (right) indicates an increased number of amino acid residues in contact with the latter, potentially explaining its superior affinity. The ligands are shown in ball-and-stick representation. Parts of the backbone, denoted by the cyan ribbon, were removed to reveal the contacts. (**c**) Replacing the PAS-B domain in MET from *C. lectularius*, nonresponsive to PJ-105, with the PAS-B of *P. apterus* was sufficient to confer the PJ-105 responsiveness as demonstrated using the two-hybrid assay with TAI and the chimeric (Cim-Pyr) MET protein. The values indicate average fold induction relative to the signal obtained in the absence of ligand (= 1). (Data in (**a**) and models in (**b**) are from Tumova et al. (2022))

10.9 Nongenomic JH Signaling: Another Way Around the Same Corner?

Ample evidence suggests that JH signaling is not limited to regulating gene expression via the MET:TAI-mediated transcriptional activation. JH additionally elicits more rapid cellular responses involving second messenger signaling and protein phosphorylation cascades originating at the cell membrane. Efforts to identify postulated cell membrane receptors for JH have been largely unsuccessful. A recent study provided some evidence of interaction between JH III and either of two receptor tyrosine kinases (RTKs), cadherin 96ca (CAD96CA) or fibroblast growth factor receptor (FGFR1), in the cotton bollworm, *Helicoverpa armigera* (Li et al. 2025). Their engagement by JH apparently could induce Ca^{2+}-dependent signaling, leading to MET1 and TAI phosphorylation and JHR transcriptional activity. However, many aspects of the proposed model remain questionable, such as how the sesquiterpenoid hormone binds to and activates the RTKs or what the role of the intracellular JHR is in the RTK signaling.

A membrane receptor has also been implicated in JH signaling in *A. aegypti* (Liu et al. 2015; Ojani et al. 2016). Here, JH III triggers Ca^{2+} and phospholipase C (PLC)-mediated response via the calcium/calmodulin-dependent protein kinase type II (CaMKII) and PKC protein kinases, inducing phosphorylation of MET and TAI, which in turn increases their interaction with JHRE DNA and enhances the transcriptional activation of JH-responsive genes. Thus, rather paradoxically, the membrane-initiated JH signaling events converge back at the gene-regulatory action of JHR.

The work on *A. aegypti* has been followed up in a study (Zhao et al. 2025), suggesting that the elusive cell-membrane JH receptor is the MET protein itself. The authors show that a subpopulation of MET residing at the cell membrane interacts with the RTK known as PVR, an insect counterpart of the vertebrate platelet-derived growth factor receptor (PDGFR) and vascular endothelial growth factor receptor (VEGFR), in a manner requiring the MET PAS-B domain, whereas the DNA-binding function of MET is dispensable. In response to JH, MET triggers phosphorylation of the intracellular region of PVR, which then relays the signal via PLC to downstream targets and genes regulated by PVR and JHR signaling, the latter including the established JH-responsive *Kr-h1*. Thus, again, the membrane-initiated branch of JH signaling partly synergizes with the gene-regulatory branch. The PVR-MET signaling is important for oogenesis in the *A. aegypti* mosquito. The scenario in which MET serves both the nuclear and the membrane receptor functions offers an elegant explanation for the conundrum of JH dual action. Interestingly, the human AHR also operates in a nongenomic mode involving Ca^{2+} and protein kinase-dependent signaling (Matsumura 2009; Sondermann et al. 2023), further extending the functional analogy between the two receptors (Jindra and Tumova 2026).

Although phosphorylation of MET has been reported in several studies, its impact on JH signaling is not well established. Multiple serine and threonine residues, some differentially phosphorylated in the presence of JH agonists, have been identified in MET/GCE proteins from *A. aegypti* (Kim et al. 2021), *T. castaneum* (Jindra et al. 2021a), and *D. melanogaster* (Gao et al. 2022). However, the functional relevance of these posttranslational modifications has not been ascertained. Mutation of multiple phosphosites in the PAS-A domain and the disordered C-terminal region of *T. castaneum* MET affected neither JH III binding nor nuclear translocation of MET after JH III treatment, although the non-phosphorylatable mutants had slightly higher retention in the cytoplasm than the control protein (Jindra et al. 2021a). Dephosphorylation of the purified recombinant MET:TAI heterodimer did not prevent its interaction with the JHRE DNA (Fig. 10.2d). In a *Drosophila* cell line, phosphorylation of four MET residues located between the PAS domains and at the C terminus, detected after methoprene application, had no major impact on MET activity and localization (Gao et al. 2022). Phosphorylation of two amino acids in the bHLH region apparently contributed to the transcriptional activity of MET (Gao et al. 2022).

A single specific threonine residue (T393) in a predicted loop of the PAS-B domain of MET1 was shown to be phosphorylated in *H. armigera* larvae experiencing high JH III titers (Li et al. 2021). This modification promoted MET1 interaction with TAI and the subsequent transcriptional activity, as well as the dissociation of MET1 homomers, both processes known to depend on JH binding. The effects were lost in the non-phosphorylatable T393A MET1 mutant. The pathway responsible for the T393 phosphorylation has not been identified.

10.10 Conclusions and Further Directions

Recent years have seen a steady increase in the number of publications related to JH research, with over 2200 articles churned out during the last decade. However, our understanding of the signaling process is still rudimentary. A large segment of the research expands the field laterally, addressing the roles of JH signaling in a multitude of different species rather than pursuing the mechanisms in depth. The number and, mainly, the enormous diversity of insect models clearly make exploring the details of JH signaling in a focused manner challenging.

On the positive side, the assembly comprising MET (or GCE) and TAI has been consistently designated as the JH signaling effector regulating gene expression during insect development and reproduction. While well supported by genetic and molecular-level evidence, many details of the MET/GCE:TAI signal transduction remain poorly defined. In the absence of empirical structural information, the JH-binding function of JHR and the molecular mechanism of its activation are now being explored with the help of molecular modeling, using parallels with other bHLH-PAS proteins. Thus, it has become clear that the ligand-binding capability that lies within the PAS-B domain of MET/GCE requires structural flexibility and relies on intra- and interdomain interactions in order to transmit the agonist activity. This approach has benefited enormously from the ongoing progress in the AlphaFold modeling, and we expect it to yield further specific structural information in the near future. Regardless of the increasing quality of the models, however, one must keep in mind that any computational data are but predictions that necessitate experimental validation through mutagenesis and functional assays.

The steps that precede and follow the formation of the MET:TAI complex, namely the interaction with HSP83 or nuclear import/export and its dynamics, are integral to the signaling process. It would also be important to know how the enigmatic MET:MET and MET:GCE presumed dimers fit in. Phosphorylation and other posttranslational modifications of the JHR likely bear major significance, which, for the time being, remains unclear. All these processes will need to be thoroughly investigated to discern the full mechanism. We anticipate that many presently unknown molecular components of the pathway will be uncovered. High-throughput cell-based and genetic approaches should facilitate the quest.

There is growing evidence of the versatile roles JH plays in diverse aspects of insect physiology that are mediated by the intracellular JHR at the gene-regulatory level. It is highly likely that part of the remarkable JH multitasking relies on MET/GCE interactions with partners in addition to TAI. Although alternative MET heterodimers have been described, little is known about their actual functions. Part of the JH pleiotropy is undoubtedly attained through crosstalk with other signaling pathways, likely in a combination of gene-regulatory and nongenomic JH actions. As we have seen, the outstanding recent findings indicate that the two modes of JH action are neither mutually exclusive nor independent, as MET can relay the JH signal to the receptor tyrosine kinase PVR acting at the cell membrane.

Deciphering the complex JHR signaling mechanisms will continue to present challenges requiring creative thinking. We have high hopes for upcoming exciting studies that will not only fill in the above knowledge gaps but may also surprise us with unexpected discoveries on the unique hormone-receptor system.

Acknowledgments We would like to thank Roman Tuma for his molecular modeling in Fig. 10.1d and Lenka Bittova for the immunostaining of GCE shown in Fig. 10.2b. Our work was supported by projects 20-05151X and 25-17423 from the Czech Science Foundation.

References

Abdou MA, He Q, Wen D, Zyaan O, Wang J, Xu J et al (2011) *Drosophila* Met and Gce are partially redundant in transducing juvenile hormone action. Insect Biochem Mol Biol 41:938–945. https://doi.org/10.1016/j.ibmb.2011.09.003

Ashok M, Turner C, Wilson TG (1998) Insect juvenile hormone resistance gene homology with the bHLH-PAS family of transcriptional regulators. Proc Natl Acad Sci USA 95:2761–2766. https://doi.org/10.1073/pnas.95.6.2761

Bai J, Uehara Y, Montel DJ (2000) Regulation of invasive cell behavior by taiman, a *Drosophila* protein related to AIB1, a steroid receptor coactivator amplified in breast cancer. Cell 103:1047–1058. https://doi.org/10.1016/S0092-8674(00)00208-7

Baker FC, Lanzrein B, Miller CA, Tsai LW, Jamieson GC, Schooley DA (1984) Detection of only JH III in several life-stages of *Nauphoeta cinerea* and *Thermobia domestica*. Life Sci 35:1553–1560. https://doi.org/10.1016/0024-3205(84)90353-9

Bellés X, Martin D, Piulachs M-D (2005) The mevalonate pathway and the synthesis of juvenile hormone in insects. Annu Rev Entomol 50:181–199. https://doi.org/10.1146/annurev.ento.50.071803.130356

Bittova L, Jedlicka P, Dracinsky M, Kirubakaran P, Vondrasek J, Hanus R et al (2019) Exquisite ligand stereoselectivity of a *Drosophila* juvenile hormone receptor contrasts with its broad agonist repertoire. J Biol Chem 294:410–423. https://doi.org/10.1074/jbc.RA118.005992

Bonati L, Motta S, Callea L (2024) The AhR signaling mechanism: a structural point of view. J Mol Biol 436:168296. https://doi.org/10.1016/j.jmb.2023.168296

Burmester T (2002) Origin and evolution of arthropod hemocyanins and related proteins. J Comp Physiol B 172:95–107. https://doi.org/10.1007/s00360-001-0247-7

Canavoso LE, Jouni ZE, Karnas KJ, Pennington JE, Wells MA (2001) Fat metabolism in insects. Annu Rev Nutr 21:23–46. https://doi.org/10.1146/annurev.nutr.21.1.23

Charles J-P, Iwema T, Epa VC, Takaki K, Rynes J, Jindra M (2011) Ligand-binding properties of a juvenile hormone receptor, Methoprene-tolerant. Proc Natl Acad Sci USA 108:21128–21133. https://doi.org/10.1073/pnas.1116123109

Chen HS, Singh SS, Perdew GH (1997) The Ah receptor is a sensitive target of geldanamycin-induced protein turnover. Arch Biochem Biophys 348:190–198. https://doi.org/10.1006/abbi.1997.0398

Dahm K, Röller H, Trost BM (1968) The juvenile hormone - IV. Stereochemistry of juvenile hormone and biological activity of some of its isomers and related compounds. Life Sci 7:129–137. https://doi.org/10.1016/0024-3205(68)90296-8

Dai S, Qu L, Li J, Zhang Y, Jiang L, Wei H et al (2022) Structural insight into the ligand binding mechanism of aryl hydrocarbon receptor. Nat Commun 13:6234. https://doi.org/10.1038/s41467-022-33858-w

Diao X, Shang Q, Guo M, Huang Y, Zhang M, Chen X et al (2025) Structural basis for the ligand-dependent activation of heterodimeric AHR-ARNT complex. Nat Commun 16:1282. https://doi.org/10.1038/s41467-025-56574-7

Dixon AS, Schwinn MK, Hall MP, Zimmerman K, Otto P, Lubben TH et al (2016) NanoLuc complementation reporter optimized for accurate measurement of protein interactions in cells. ACS Chem Biol 11:400–408. https://doi.org/10.1021/acschembio.5b00753

Dubrovsky EB, Bernardo TJ (2014) The juvenile hormone receptor and molecular mechanisms of juvenile hormone action. Adv Insect Physiol 46:305–388. https://doi.org/10.1016/B978-0-12-417010-0.00004-5

Dupas S, Neiers F, Granon E, Rougeux E, Dupont S, Beney L et al (2020) Collisional mechanism of ligand release by *Bombyx mori* JHBP, a member of the TULIP/Takeout family of lipid transporters. Insect Biochem Mol Biol 117:103293. https://doi.org/10.1016/j.ibmb.2019.103293

Gao Y, Chen N, Zhang X, Li EY, Luo W, Zhang J et al (2022) Juvenile hormone membrane signaling enhances its intracellular signaling through phosphorylation of Met and Hsp83. Front Physiol 13:872889. https://doi.org/10.3389/fphys.2022.872889

Godlewski J, Wang S, Wilson TG (2006) Interaction of bHLH-PAS proteins involved in juvenile hormone reception in *Drosophila*. Biochem Biophys Res Commun 342:1305–1311. https://doi.org/10.1016/j.bbrc.2006.02.097

Goodman WG, Cusson M (2012) The juvenile hormones. In: Gilbert LI (ed) Insect endocrinology. Elsevier, Amsterdam, pp 310–365. https://doi.org/10.1016/B978-0-12-384749-2.10008-1

Goodman W, Schooley DA, Gilbert LI (1978) Specificity of the juvenile hormone binding protein: the geometrical isomers of juvenile hormone I. Proc Natl Acad Sci USA 75:185–189. https://doi.org/10.1073/pnas.75.1.185

Greb-Markiewicz B, Orłowski M, Dobrucki J, Ożyhar A (2011) Sequences that direct subcellular traffic of the *Drosophila* methoprene-tolerant protein (MET) are located predominantly in the PAS domains. Mol Cell Endocrinol 345:16–26. https://doi.org/10.1016/j.mce.2011.06.035

Greb-Markiewicz B, Sadowska D, Surgut N, Godlewski J, Zarębski M, Ożyhar A (2015) Mapping of the sequences directing localization of the *Drosophila* germ cell-expressed protein (GCE). PLoS One 10:e0133307. https://doi.org/10.1371/journal.pone.0133307

Gruszczyk J, Grandvuillemin L, Lai-Kee-Him J, Paloni M, Savva CG, Germain P et al (2022) Cryo-EM structure of the agonist-bound Hsp90-XAP2-AHR cytosolic complex. Nat Commun 13:7010. https://doi.org/10.1038/s41467-022-34773-w

Hawkings C, Calkins TL, Pietrantonio PV, Tamborindeguy C (2019) Caste-based differential transcriptional expression of hexamerins in response to a juvenile hormone analog in the red imported fire ant (*Solenopsis invicta*). PLoS One 14:e0216800. https://doi.org/10.1371/journal.pone.0216800

He Q, Wen D, Jia Q, Cui C, Wang J, Palli SR et al (2014) Heat shock protein 83 (Hsp83) facilitates Methoprene-tolerant (Met) nuclear import to modulate juvenile hormone signaling. J Biol Chem 289:27874–27885. https://doi.org/10.1074/jbc.m114.582825
He L, Wu B, Shi J, Du J, Zhao Z (2023) Regulation of feeding and energy homeostasis by clock-mediated Gart in *Drosophila*. Cell Rep 42:112912. https://doi.org/10.1016/j.celrep.2023.112912
He Q, Zhang Y, Zhang X, Xu D, Dong W, Li S et al (2017) Nucleoporin Nup358 facilitates nuclear import of Methoprene-tolerant (Met) in an importin β- and Hsp83-dependent manner. Insect Biochem Mol Biol 81:10–18. https://doi.org/10.1016/j.ibmb.2016.12.005
Hejnikova M, Nouzova M, Ramirez CE, Fernandez-Lima F, Noriega FG, Dolezel D (2022) Sexual dimorphism of diapause regulation in the hemipteran bug *Pyrrhocoris apterus*. Insect Biochem Mol Biol 142:103721. https://doi.org/10.1016/j.ibmb.2022.103721
Helvig C, Koener JF, Unnithan GC, Feyereisen R (2004) CYP15A1, the cytochrome P450 that catalyzes epoxidation of methyl farnesoate to juvenile hormone III in cockroach *corpora allata*. Proc Natl Acad Sci USA 101:4024–4029. https://doi.org/10.1073/pnas.0306980101
Jindra M, Bittova L (2020) The juvenile hormone receptor as a target of juvenoid "insect growth regulators.". Arch Insect Biochem Physiol 103:e21615. https://doi.org/10.1002/arch.21615
Jindra M, McKinstry WJ, Nebl T, Bittova L, Ren B, Shaw J et al (2021a) Purification of an insect juvenile hormone receptor complex enables insights into its post-translational phosphorylation. J Biol Chem 297:101387. https://doi.org/10.1016/j.jbc.2021.101387
Jindra M, Noriega FG (2026) Juvenile hormone signaling. In: Atkinson PW, Yamanaka N (eds) Comprehensive molecular insect science, 2nd edn. Elsevier, Amsterdam, pp 138–165. https://doi.org/10.1016/B978-0-323-95424-2.00036-4
Jindra M, Tumova S (2026) MET goes to the membrane—a new place for the "old" juvenile hormone receptor. Proc Natl Acad Sci USA 123:e2530588122. https://doi.org/10.1073/pnas.2530588122
Jindra M, Tumova S, Bittova L, Tuma R, Sedlak D (2024) Agonist-dependent action of the juvenile hormone receptor. Curr Opin Insect Sci 65:101234. https://doi.org/10.1016/j.cois.2024.101234
Jindra M, Tumova S, Milacek M, Bittova L (2021b) A decade with the juvenile hormone receptor. Adv Insect Physiol 60:37–85. https://doi.org/10.1016/bs.aiip.2021.03.001
Jindra M, Uhlirova M, Charles J-P, Smykal V, Hill RJ (2015) Genetic evidence for function of the bHLH-PAS protein Gce/Met as a juvenile hormone receptor. PLoS Genet 11:e1005394. https://doi.org/10.1371/journal.pgen.1005394
Ju L, Glastad KM, Sheng L, Gospocic J, Kingwell CJ, Davidson SM et al (2023) Hormonal gatekeeping via the blood-brain barrier governs caste-specific behavior in ants. Cell 186:4289–4309.e23. https://doi.org/10.1016/j.cell.2023.08.002
Judy KJ, Schooley DA, Dunham LL, Hall MS, Bergot BJ, Siddall JB (1973) Isolation, structure, and absolute configuration of a new natural insect juvenile hormone from *Manduca sexta*. Proc Natl Acad Sci USA 70:1509–1513. https://doi.org/10.1073/pnas.70.5.1509
Jumper J, Evans R, Pritzel A, Green T, Figurnov M, Ronneberger O et al (2021) Highly accurate protein structure prediction with AlphaFold. Nature 596:583–589. https://doi.org/10.1038/s41586-021-03819-2
Kayukawa T, Minakuchi C, Namiki T, Togawa T, Yoshiyama M, Kamimura M et al (2012) Transcriptional regulation of juvenile hormone-mediated induction of Krüppel homolog 1, a repressor of insect metamorphosis. Proc Natl Acad Sci USA 109:11729–11734. https://doi.org/10.1073/pnas.1204951109
Kayukawa T, Shinoda T (2015) Functional characterization of two paralogous JH receptors, methoprene-tolerant 1 and 2, in the silkworm, *Bombyx mori* (Lepidoptera: Bombycidae). Appl Entomol Zool 50:383–391. https://doi.org/10.1007/s13355-015-0345-8
Kayukawa T, Tateishi K, Shinoda T (2013) Establishment of a versatile cell line for juvenile hormone signaling analysis in *Tribolium castaneum*. Sci Rep 3:1570. https://doi.org/10.1038/srep01570

Kim K, Albishi NM, Palli SR (2021) Identification of juvenile hormone-induced posttranslational modifications of methoprene tolerant and Krüppel homolog 1 in the yellow fever mosquito, *Aedes aegypti*. J Proteomics 242:104257. https://doi.org/10.1016/j.jprot.2021.104257

Kim IH, Castillo JC, Aryan A, Martin-Martin I, Nouzova M, Noriega FG et al (2020) A mosquito juvenile hormone binding protein (mJHBP) regulates the activation of innate immune defenses and hemocyte development. PLoS Pathog 16:e1008288. https://doi.org/10.1371/journal.ppat.1008288

Kim IH, Pham V, Jablonka W, Goodman WG, Ribeiro JMC, Andersen JF (2017) A mosquito hemolymph odorant-binding protein family member specifically binds juvenile hormone. J Biol Chem 292:15329–15339. https://doi.org/10.1074/jbc.m117.802009

Konopova B, Jindra M (2007) Juvenile hormone resistance gene Methoprene-tolerant controls entry into metamorphosis in the beetle *Tribolium castaneum*. Proc Natl Acad Sci USA 104:10488–10493. https://doi.org/10.1073/pnas.0703719104

Kotaki T, Shinada T, Kaihara K, Ohfune Y, Numata H (2009) Structure determination of a new juvenile hormone from a heteropteran insect. Org Lett 11:5234–5237. https://doi.org/10.1021/ol902161x

LeBoeuf AC, Cohanim AB, Stoffel C, Brent CS, Waridel P, Privman E et al (2018) Molecular evolution of juvenile hormone esterase-like proteins in a socially exchanged fluid. Sci Rep 8:17830. https://doi.org/10.1038/s41598-018-36048-1

Li YX, Kang XL, Li YL, Wang XP, Yan Q, Wang JX et al (2025) Receptor tyrosine kinases CAD96CA and FGFR1 function as the cell membrane receptors of insect juvenile hormone. elife 13:RP97189. https://doi.org/10.7554/elife.97189

Li M, Liu P, Wiley JD, Ojani R, Bevan DR, Li J et al (2014) A steroid receptor coactivator acts as the DNA-binding partner of the methoprene-tolerant protein in regulating juvenile hormone response genes. Mol Cell Endocrinol 394:47–58. https://doi.org/10.1016/j.mce.2014.06.021

Li M, Mead EA, Zhu J (2011) Heterodimer of two bHLH-PAS proteins mediates juvenile hormone-induced gene expression. Proc Natl Acad Sci USA 108:638–643. https://doi.org/10.1073/pnas.1013914108

Li YX, Wang D, Zhao WL, Zhang JY, Kang XL, Li YL et al (2021) Juvenile hormone induces methoprene-tolerant 1 phosphorylation to increase interaction with Taiman in *Helicoverpa armigera*. Insect Biochem Mol Biol 130:103519. https://doi.org/10.1016/j.ibmb.2021.103519

Liu P, Peng H-J, Zhu J (2015) Juvenile hormone-activated phospholipase C pathway enhances transcriptional activation by the methoprene-tolerant protein. Proc Natl Acad Sci USA 112:E1871–E1879. https://doi.org/10.1073/pnas.1423204112

Matsumura F (2009) The significance of the nongenomic pathway in mediating inflammatory signaling of the dioxin-activated Ah receptor to cause toxic effects. Biochem Pharmacol 77:608–626. https://doi.org/10.1016/j.bcp.2008.10.013

Meyer AS, Schneiderman HA, Hanzmann E, Ko J (1968) The two juvenile hormones from the *Cecropia* silk moth. Proc Natl Acad Sci USA 60:853–860. https://doi.org/10.1073/pnas.60.3.853

Milacek M, Bittova L, Tumova S, Luksan O, Hanus R, Kyjakova P et al (2021) Binding of *de novo* synthesized radiolabeled juvenile hormone (JH III) by JH receptors from the Cuban subterranean termite *Prorhinotermes simplex* and the German cockroach *Blattella germanica*. Insect Biochem Mol Biol 139:103671. https://doi.org/10.1016/j.ibmb.2021.103671

Minakuchi C, Namiki T, Yoshiyama M, Shinoda T (2008) RNAi-mediated knockdown of juvenile hormone acid *O*-methyltransferase gene causes precocious metamorphosis in the red flour beetle *Tribolium castaneum*. FEBS J 275:2919–2931. https://doi.org/10.1111/j.1742-4658.2008.06428.x

Miura K, Oda M, Makita S, Chinzei Y (2005) Characterization of the *Drosophila* Methoprene-tolerant gene product. FEBS J 272:1169–1178. https://doi.org/10.1111/j.1742-4658.2005.04552.x

Miyakawa H, Iguchi T (2017) Comparative luciferase assay for establishing reliable *in vitro* screening system of juvenile hormone agonists. J Appl Toxicol 37:1082–1090. https://doi.org/10.1002/jat.3459

Miyakawa H, Iguchi T (2023) Establishment of a testing system for juvenoids that disturb the sex determination system of *Daphnia*, based on a molecular interaction of ligands and the JH receptor. Curr Opin Toxicol 36:100425. https://doi.org/10.1016/j.cotox.2023.100425

Miyakawa H, Toyota K, Hirakawa I, Ogino Y, Miyagawa S, Oda S et al (2013) A mutation in the receptor Methoprene-tolerant alters juvenile hormone response in insects and crustaceans. Nat Commun 4:1856. https://doi.org/10.1038/ncomms2868

Niwa R, Niimi T, Honda N, Yoshiyama M, Itoyama K, Kataoka H et al (2008) Juvenile hormone acid *O*-methyltransferase in *Drosophila melanogaster*. Insect Biochem Mol Biol 38:714–720. https://doi.org/10.1016/j.ibmb.2008.04.003

Nouzova M, Edwards MJ, Michalkova V, Ramirez CE, Ruiz M, Areiza M et al (2021) Epoxidation of juvenile hormone was a key innovation improving insect reproductive fitness. Proc Natl Acad Sci USA 118:E2109381118. https://doi.org/10.1073/pnas.2109381118

Ojani R, Liu P, Fu X, Zhu J (2016) Protein kinase C modulates transcriptional activation by the juvenile hormone receptor methoprene-tolerant. Insect Biochem Mol Biol 70:44–52. https://doi.org/10.1016/j.ibmb.2015.12.001

Okamoto N, Viswanatha R, Bittar R, Li Z, Haga-Yamanaka S, Perrimon N et al (2018) A membrane transporter is required for steroid hormone uptake in *Drosophila*. Dev Cell 47:294–305.e7. https://doi.org/10.1016/j.devcel.2018.09.012

Palli SR (2023) Juvenile hormone receptor methoprene tolerant: functions and applications. Vitam Horm 123:619–644. https://doi.org/10.1016/bs.vh.2023.03.002

Pandini A, Soshilov AA, Song Y, Zhao J, Bonati L, Denison MS (2009) Detection of the TCDD binding-fingerprint within the Ah receptor ligand binding domain by structurally driven mutagenesis and functional analysis. Biochemistry 48:5972–5983. https://doi.org/10.1021/bi900259z

Rahman MM, Franch-Marro X, Maestro JL, Martin D, Casali A (2017) Local juvenile hormone activity regulates gut homeostasis and tumor growth in adult *Drosophila*. Sci Rep 7:11677. https://doi.org/10.1038/s41598-017-11199-9

Richard DS, Applebaum SW, Sliter TJ, Baker FC, Schooley DA, Reuter CC et al (1989) Juvenile hormone bisepoxide biosynthesis in vitro by the ring gland of *Drosophila melanogaster*: a putative juvenile hormone in the higher Diptera. Proc Natl Acad Sci USA 86:1421–1425. https://doi.org/10.1073/pnas.86.4.1421

Riddiford LM, Truman JW, Mirth CK, Shen YC (2010) A role for juvenile hormone in the prepupal development of *Drosophila melanogaster*. Development 137:1117–1126. https://doi.org/10.1242/dev.037218

Rivera-Pérez C, Clifton ME, Noriega FG, Jindra M (2020) Juvenile hormone regulation and function. In: Saleuddin S, Lange A, Orchard I (eds) Advances in invertebrate (neuro)endocrinology. Apple Academic Press, Burlington, pp 1–76. https://doi.org/10.1201/9781003029861

Röller H, Dahm KH, Sweeley CC, Trost BM (1967) Die Struktur des Juvenilhormons. Angew Chemie 4:190–191. https://doi.org/10.14411/eje.2015.073

Schulte KW, Green E, Wilz A, Platten M, Daumke O (2017) Structural basis for aryl hydrocarbon receptor-mediated gene activation. Structure 25:1025–1033. https://doi.org/10.1016/j.str.2017.05.008

Sedlak D, Tuma R, Kolla JN, Mokhamatam RB, Bahrova L, Lisova M et al (2025) Unique and common agonists activate the insect juvenile hormone receptor and the human AHR. J Mol Biol 437:168883. https://doi.org/10.1016/j.jmb.2024.168883

Shin SW, Zou Z, Saha TT, Raikhel AS (2012) bHLH-PAS heterodimer of methoprene-tolerant and cycle mediates circadian expression of juvenile hormone-induced mosquito genes. Proc Natl Acad Sci USA 109:16576–16581. https://doi.org/10.1073/pnas.1214209109

Shinoda T, Itoyama K (2003) Juvenile hormone acid methyltransferase: a key regulatory enzyme for insect metamorphosis. Proc Natl Acad Sci USA 100:11986–11991. https://doi.org/10.1073/pnas.2134232100

Smykal V, Dolezel D (2023) Evolution of proteins involved in the final steps of juvenile hormone synthesis. J Insect Physiol 145:104487. https://doi.org/10.1016/j.jinsphys.2023.104487

Sogawa K, Nakano R, Kobayashi A, Kikuchi Y, Ohe N, Matsushita N et al (1995) Possible function of Ah receptor nuclear translocator (Arnt) homodimer in transcriptional regulation. Proc Natl Acad Sci USA 92:1936–1940. https://doi.org/10.1073/pnas.92.6.1936
Sondermann NC, Faßbender S, Hartung F, Hätälä AM, Rolfes KM, Vogel CFA et al (2023) Functions of the aryl hydrocarbon receptor (AHR) beyond the canonical AHR/ARNT signaling pathway. Biochem Pharmacol 208:115371. https://doi.org/10.1016/j.bcp.2022.115371
Suzuki R, Fujimoto Z, Shiotsuki T, Tsuchiya W, Momma M, Tase A et al (2011) Structural mechanism of JH delivery in hemolymph by JHBP of silkworm, *Bombyx mori*. Sci Rep 1:133. https://doi.org/10.1038/srep00133
Tanaka T, Iguchi T, Miyakawa H (2019) Establishment of a high-sensitivity reporter system in mammalian cells for detecting juvenoids using juvenile hormone receptors of *Daphnia pulex*. J Appl Toxicol 39:241–246. https://doi.org/10.1002/jat.3713
Trowell SC (1992) High affinity juvenile hormone carrier proteins in the haemolymph of insects. Comp Biochem Physiol 103B:795–807. https://doi.org/10.1016/0305-0491(92)90196-X
Truman JW, Riddiford LM, Konopova B, Nouzova M, Noriega FG, Herko M (2024) The embryonic role of juvenile hormone in the firebrat, *Thermobia domestica*, reveals its function before its involvement in metamorphosis. elife 12:RP92643. https://doi.org/10.7554/elife.92643
Tsang SSK, Law STS, Li C, Qu Z, Bendena WG, Tobe SS et al (2020) Diversity of insect sesquiterpenoid regulation. Front Genet 11:1027. https://doi.org/10.3389/fgene.2020.01027
Tumova S, Dolezel D, Jindra M (2024) Conserved and unique roles of bHLH-PAS transcription factors in insects—from clock to hormone reception. J Mol Biol 436:168332. https://doi.org/10.1016/j.jmb.2023.168332
Tumova S, Jindra M (2023) Ligand-dependent protein interactions of the juvenile hormone receptor captured in real time. FEBS J 290:2881–2894. https://doi.org/10.1111/febs.16719
Tumova S, Milacek M, Snajdr I, Muthu M, Tuma R, Reha D et al (2022) Unique peptidic agonists of a juvenile hormone receptor with species-specific effects on insect development and reproduction. Proc Natl Acad Sci USA 119:e2215541119. https://doi.org/10.1073/pnas.2215541119
Villalobos-Sambucaro MJ, Nouzova M, Ramirez CE, Eugenia Alzugaray M, Fernandez-Lima F, Ronderos JR et al (2020) The juvenile hormone described in *Rhodnius prolixus* by Wigglesworth is juvenile hormone III skipped bisepoxide. Sci Rep 10:3091. https://doi.org/10.1038/s41598-020-59495-1
Wigglesworth VB (1934) The physiology of ecdysis in *Rhodnius prolixus* (Hemiptera). II. Factors controlling moulting and "metamorphosis". Quart J Micr Sci 77:191–222
Wigglesworth VB (1936) The function of the *corpus allatum* in the growth and reproduction of *Rhodnius prolixus* (Hemiptera). Quart J Micr Sci 79:91–121
Wigglesworth VB (1940) The determination of characters at metamorphosis of *Rhodnius prolixus* (Hemiptera). J Exp Biol 17:201–223. https://doi.org/10.1242/jeb.17.2.201
Wigglesworth VB (1958) Some methods for assaying extracts of the juvenile hormone in insects. J Insect Physiol 2:73–84. https://doi.org/10.1016/0022-1910(58)90031-3
Wigglesworth VB (1969) Chemical structure and juvenile hormone activity: comparative tests on *Rhodnius prolixus*. J Insect Physiol 15:73–94. https://doi.org/10.1016/0022-1910(69)90213-3
Wilson TG, Ashok M (1998) Insecticide resistance resulting from an absence of target-site gene product. Proc Natl Acad Sci USA 95:14040–14044. https://doi.org/10.1073/pnas.95.24.14040
Wilson TG, Fabian J (1986) A *Drosophila melanogaster* mutant resistant to a chemical analog of juvenile hormone. Dev Biol 118:190–201. https://doi.org/10.1016/0012-1606(86)90087-4
Wu D, Rastinejad F (2017) Structural characterization of mammalian bHLH-PAS transcription factors. Curr Opin Struct Biol 43:1–9. https://doi.org/10.1016/j.sbi.2016.09.011
Wu D, Su X, Potluri N, Kim Y, Rastinejad F (2016) NPAS1-ARNT and NPAS3-ARNT crystal structures implicate the bHLH-PAS family as multi-ligand binding transcription factors. elife 5:e18790. https://doi.org/10.7554/elife.18790
Zaoral M, Sláma K (1970) Peptides with juvenile hormone activity. Science 170:92–93. https://doi.org/10.1126/science.170.3953.92

Zhang Z, Xu J, Sheng Z, Sui Y, Palli SR (2011) Steroid receptor co-activator is required for juvenile hormone signal transduction through a bHLH-PAS transcription factor, methoprene tolerant. J Biol Chem 286:8437–8447. https://doi.org/10.1074/jbc.m110.191684

Zhao W, Griffith K, Liang J, Dorodnitsyna M, Liu P, Saunders TR et al (2025) Membrane-localized MET engages PVR to mediate extranuclear juvenile hormone signaling in *Aedes aegypti*. Proc Natl Acad Sci USA 122:e2516796122. https://doi.org/10.1073/pnas.2516796122

Zhou X, Tarver MR, Scharf ME (2007) Hexamerin-based regulation of juvenile hormone-dependent gene expression underlies phenotypic plasticity in a social insect. Development 134:601–610. https://doi.org/10.1242/dev.02755

Zhu J, Chen L, Sun G, Raikhel AS (2006) The competence factor beta Ftz-F1 potentiates ecdysone receptor activity via recruiting a p160/SRC coactivator. Mol Cell Biol 26:9402–9412. https://doi.org/10.1128/mcb.01318-06

Chapter 11
The Role of Juvenile Hormone in Insect Ovarian Development

Yoshitomo Kurogi and Ryusuke Niwa

Abstract Juvenile hormone (JH) plays a crucial role in the regulation of insect reproduction. In female insects, JH regulates oogenesis by promoting the production and uptake of vitellogenin (Vg). This function has been confirmed since the beginning of JH research more than 80 years ago by transplantation of the corpus allatum, the JH-producing organ. However, it is only in the past two decades that the details of the molecular mechanisms of JH-dependent oogenesis regulation have become clear. In this chapter, we summarize the molecular mechanism of JH action in oogenesis, focusing on the fruit fly, *Drosophila melanogaster*, which has been extensively studied using powerful genetic tools.

Keywords Juvenile hormone · Vitellogenesis · Yolk protein · Germline stem cell · Ovarian muscle · Ovulation

11.1 The Discovery of the Gonadotropic Function of Juvenile Hormone

In 1936, using the kissing bug, *Rhodnius prolixus*, Sir Vincent B. Wigglesworth discovered that the corpus allatum (plural: corpora allata; CA) cells shrink toward the end of the final-instar nymph and enlarge their cytoplasm after the last-instar nymph-adult eclosion (Wigglesworth 1936). These changes in the CA cells were also reported to be regulated by feeding. The size of the CA cells decreases when adult females are starved but increases when they are refed. In any case, the timing of CA development coincides with that of the development of the reproductive

Y. Kurogi
Institute for Protein Research, The University of Osaka, Osaka, Japan

R. Niwa (✉)
Life Science Center for Survival Dynamics, Tsukuba Advanced Research Alliance (TARA), University of Tsukuba, Tsukuba, Japan
e-mail: ryusuke-niwa@tara.tsukuba.ac.jp

T. Shinoda et al. (eds.), *Juvenile Hormone*, Entomology Monographs,
https://doi.org/10.1007/978-981-95-8605-9_11

organs, such as the ovary and the male accessory gland. Experiments with *R. prolixus* showed that ovarian development does not occur when the adult head containing the CA is ablated, while it does occur when only the brain but not the CA is ablated (Wigglesworth 1936) (Fig. 11.1). Furthermore, ovarian development was induced even in the CA-less animals when they were parabiosed to animals with CA (Fig. 11.1).

Further experimental studies suggested that the factor(s) secreted from the CA, possibly JH, regulate ovarian development via yolk accumulation. In *R. prolixus* ovaries, oocytes are closely associated with nurse cells to obtain nutrients and other biomolecules until oocytes reach a specific size (Nunes-da-Fonseca et al. 2017). This process is known as previtellogenesis. After the previtellogenesis stage, oocytes

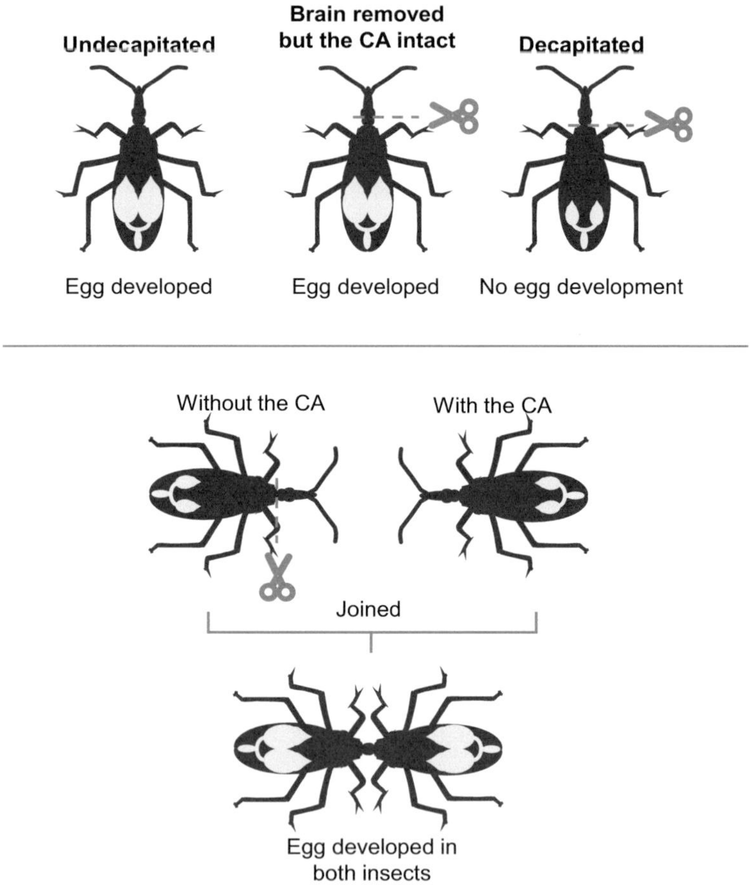

Fig. 11.1 Experiments of Sir Vincent B. Wigglesworth with *Rhodnius prolixus*. Schematic diagram of Wigglesworth's experiments with *R. prolixus*. (Top) Eggs develop in undecapitated animals and in animals in which only the brain has been removed. On the other hand, eggs do not develop in decapitated animals in which the corpus allatum (CA) has been removed. (Bottom) By pairing the decapitated animal with the undecapitated one, eggs develop in both animals

are disconnected from nurse cells, followed by encasing of oocytes by follicle cells and the start of yolk accumulation, known as vitellogenesis. In the CA-ablated animals, while previtellogenesis normally occurs, vitellogenesis, which is yolk accumulation, does not occur, leading to oocyte degeneration. Therefore, in the past, JH research pioneers had already recognized the role of JH in regulating oogenesis. Nowadays, it is widely recognized across various insect species that JH primarily regulates ovarian development during the yolk accumulation stage (Leyria 2024; Raikhel et al. 2005; Santos et al. 2019; Wyatt and Davey 1996).

11.2 The Relationship Between Juvenile Hormone and Reproduction in Various Insect Species

11.2.1 General Function of Juvenile Hormone in Vitellogenesis

Reproduction is a highly energy-demanding physiological process that requires precise regulation. JH is a key hormone known to play an important role in regulating reproduction in insects (Leyria 2024; Raikhel et al. 2005; Santos et al. 2019; Wyatt and Davey 1996). As mentioned above, JH is involved in the yolk production and uptake in adult females. A major constituent of yolk is vitellogenin (Vg), which serves as the predominant precursor of yolk proteins (Yps) (Tufail and Takeda 2008). Vg is primarily produced in the fat body and then transported through the hemolymph to the ovary (Raikhel et al. 2005; Raikhel and Dhadialla 1992; Wu et al. 2020b; Wyatt and Davey 1996). JH induces the synthesis of Vg in the fat body, promotes the opening of cell junctions in the follicle cells, and facilitates the selective uptake of Vg from the hemolymph via receptor-mediated endocytosis (Leyria 2024; Raikhel et al. 2005; Santos et al. 2019; Wu et al. 2020b). In the vitellogenic process, the intracellular receptor methoprene-tolerant (Met), a member of the basic helix–loop–helix Per-Arnt-Sim (bHLH-PAS) transcription factor family, coordinates JH-dependent oogenesis regulation (Jindra et al. 2013, 2015a). When JH binds, Met forms a heterodimer with another bHLH-PAS protein, Taiman (Tai), to act as a transcription factor (Charles et al. 2011; Dubrovsky and Bernardo 2014; Jindra et al. 2015b; Konopova and Jindra 2007). RNA interference (RNAi)-mediated knockdown of *Met* in many adult females results in reduced *Vg* expression and ovarian development (Hu et al. 2019; Lin et al. 2015; Ma et al. 2018; Marchal et al. 2014; Parthasarathy et al. 2010; Shpigler et al. 2014; Smykal et al. 2014; Song et al. 2014; Yue et al. 2018). Downstream of the JH–Met–Tai complex, the expression of several target genes is induced, including the zinc-finger transcription factor gene *Krüppel homolog 1* (*Kr-h1*) (Guo et al. 2014; Jindra et al. 2015a; Kayukawa et al. 2012; Li et al. 2011; Wu et al. 2018, 2016). While *Kr-h1* is crucial for controlling metamorphosis, it also promotes vitellogenesis and controls ovarian development via the JH signaling pathway in reproductive regulation (Kayukawa et al. 2012; Lin et al. 2015; Minakuchi et al. 2008; Parthasarathy et al. 2010; Shin et al. 2012; Yue et al. 2018).

This pathway plays a role in adjusting reproductive activity according to environmental conditions and developmental stages, ensuring the proper timing of ovarian maturation.

JH is crucial for the control of reproduction in a variety of insect species. For more details, we strongly recommend reading several excellent review articles (Leyria 2024; Raikhel et al. 2005; Roy et al. 2018). Here, the function of JH in some insects in different orders is briefly outlined.

11.2.2 Orthoptera

Studies on the migratory locust, *Locusta migratoria,* have advanced our understanding of the regulation of reproduction by JH (Leyria 2024; Roy et al. 2018; Wyatt and Davey 1996). In this species, elevated JH titers induce a Met-dependent upregulation of *Kr-h1* expression, which subsequently induces *Vg* expression downstream of Kr-h1, thereby promoting vitellogenesis (Song et al. 2014). Furthermore, Met and Kr-h1 are required for lipid accumulation in the oocytes (Song et al. 2014). Additionally, JH-Met-dependent polyploidy cell formation promotes rapid and extensive Vg synthesis, leading to the synchronous development of numerous oocytes. In the JH-Met-dependent polyploidization, minichromosome maintenance 3 (Mcm3), Mcm4, Mcm7, cell division cycle 6 (Cdc6), cyclin-dependent kinase 6 (Cdk6), and E2F transcription factor 1 (E2f1) play critical roles (Guo et al. 2014; Wu et al. 2020a, 2018, 2016). Studies in locusts also suggest that JH may facilitate Vg uptake through phosphorylation of G protein–coupled receptors (GPCRs) or receptor tyrosine kinases (RTKs), activating the inositol triphosphate pathway (Jing et al. 2018). This pathway is thought to create intercellular gaps through which Vg is taken up by the oocytes. Additionally, phospholipase C signaling downstream of JH promotes Vg uptake by enhancing Vg receptor (VgR) phosphorylation (Jing et al. 2021).

11.2.3 Blattodea

Cockroaches, belonging to the order Blattodea, are insects that have been classically used to study the role of JH in yolk formation in female insects (Korb and Belles 2017). In cockroaches, JH plays a crucial role in the reproductive regulation of females (Korb and Belles 2017). In oviparous species such as the American cockroach, *Periplaneta americana*, it has been shown that removal of the CA suppresses eggshell development. This suppression can be reversed by treatment with the JH analog fenoxycarb. Furthermore, treatment with this agent accelerates ovarian development, suggesting that JH level is a key factor in regulating the pace of ovarian maturation (Weaver and Edwards 1990). In addition, JH effects on *Vg* expression and Vg titers have been reported in several species, and JH-mediated mechanisms of reproduction have been observed (Bradfield et al. 1990; Comas et al. 2001, 1999; Keeley et al. 1988; Marchal et al. 2014).

The viviparous Pacific beetle cockroach, *Diploptera punctata,* represents another example of JH-regulated reproduction in Blattodea (Marchal et al. 2013). In *D. punctata*, JH titers fluctuate cyclically throughout the gonotrophic cycle and reach their highest levels during mid-cycle, when oocyte growth is most rapid (Stay and Tobe 1978; Tobe et al. 1985). Allatectomy arrests oocyte development, confirming that JH is required for initiating vitellogenesis (Engelmann 1959). In addition, ovariectomy early in adulthood reduces JH biosynthesis throughout the first gonotrophic cycle, whereas transplantation of vitellogenic ovarian tissue restores the normal cyclic pattern, indicating that the ovary releases stage-dependent signals that modulate CA activity (Rankin and Stay 1984; Stay et al. 1983; Stay and Tobe 1978). Young, vitellogenic ovaries produce stimulatory factors that enhance JH synthesis and suppress JH-catabolic pathways. In contrast, post-vitellogenic ovaries release inhibitory peptides that reduce the CA function. These data suggest that multiple ovarian factors act sequentially to regulate JH output (Chang et al. 2005; Elliott et al. 2006; Sutherland et al. 2000).

11.2.4 Isoptera

Termites, belonging to the order Isoptera, are social insects, and a single colony contains animals of various castes. JH is a regulator of termite caste differentiation, and JH signaling promotes worker differentiation into soldiers in several species (Howard and Haverty 1979; Hrdý 1976; Korb and Belles 2017; Maekawa et al. 2014; Okot-kotber 1980; Tsuchiya et al. 2008). Within a caste, there are primary reproductives that form new colonies and neotenic reproductives that replace primary reproductives and are capable of reproduction (Thorne 1996). In the subterranean termite, *Reticulitermes speratus,* the JH level of primary reproductives increases prior to the increase in *Vg* expression, and neotenic reproductives show higher JH titer (Maekawa et al. 2010). In nymphoids, one of the neotenic reproductives, *Met* knockdown reduces *Vg* expression (Saiki et al. 2015). These results indicate that JH signaling affects not only termite caste differentiation but also reproduction regulation.

11.2.5 Hemiptera

In adult females of Hemiptera, including *R. prolixus* described above, JH has been shown to be a major regulator of various aspects of female reproduction in multiple ways. In the kissing bug, *Dipetalogaster maxima*, oogenesis is mediated by Met in the ovaries and fat body (Ramos et al. 2022). *D. maxima* starts vitellogenesis after blood feeding. During this process, the expression of *Met* in the ovary and fat body increases and peaks at day 4 after feeding (Ramos et al. 2022). In *Met* knockdown and blood-fed animals, the expression of *Vg* and *lipophorin* (*Lp*) in the fat body, as well as *vitellogenin receptor* (*VgR*) and *lipophorin receptor* (*LpR*) in the ovary, is reduced (Ramos et al. 2022). Additionally, in the ovaries, the size of the terminal follicle is decreased, and immunostaining signals for Vg and Lp in the perioocytic space and oocyte membrane are diminished due to *Met* knockdown, suggesting an inhibition of uptake (Ramos et al. 2022). In addition to *R. prolixus* and *D. maxima*, recent studies on the brown-winged green bug, *Plautia stali*, identified *P. stali* myo-inhibitory peptide (Plast-MIP) as a neuropeptide that inhibits JH biosynthesis under diapause-inducing conditions and suppresses oogenesis (Hasebe and Shiga 2021; Hasegawa et al. 2020; Matsumoto et al. 2017).

11.2.6 Coleoptera

In the red flour beetle, *Tribolium castaneum*, a coleopteran model insect, JH affects *Vg* expression by inducing the expression of *insulin-like peptide* genes (Parthasarathy et al. 2010; Sheng et al. 2011). In addition, the dopamine D2-like receptor likely regulates Vg uptake in response to JH (Bai and Palli 2016). RNAi of this receptor inhibits Vg uptake into oocytes (Bai and Palli 2016). Furthermore, in human embryonic kidney–derived HEK293 culture cells expressing *T. castaneum dopamine D2-like receptor*, JH triggers a dose-dependent inhibition of intracellular cyclic adenosine monophosphate levels (Bai and Palli 2016). These results raise the possibility that JH is received by the dopamine D2-like receptor, although no physical interaction between JH and the receptor has been demonstrated.

11.2.7 Hymenoptera

The importance of JH in regulating reproduction in hymenopteran insects varies depending on the degree of sociality exhibited by each species. In the bumblebee, *Bombus terrestris*, which exhibits a relatively primitive form of sociality, JH functions as a gonadotropic hormone, and its levels correspond with reproductive activity in queens and dominant workers (Bloch et al. 2000; Bloch and Hefetz 1999). Furthermore, JH treatment accelerates oogenesis in *B. terrestris* (Röseler 1977). However, in adult queens of the highly eusocial honey bee, *Apis mellifera*, reproductive status does not appear to be linked to JH levels (Robinson et al. 1992, 1991; Robinson and Vargo 1997). JH treatment also has no visible activity on workers in terms of enhancing reproduction (Robinson et al. 1992). In the Indian jumping ant, *Harpegnathos saltator*, which shows caste differentiation plasticity between workers and gamergates depending on the outcome of social conflict, JH is essential for determining caste status but not fertility (Gospocic et al. 2021; Peeters and

Hölldobler 1995; Penick et al. 2011). Therefore, some eusocial hymenopterans appear to have secondarily reduced the gonadotropic role of JH during evolution.

11.2.8 Lepidoptera

In many lepidopteran species, vitellogenesis is initiated after adult eclosion and depends on JH. In the fall armyworm, *Spodoptera frugiperda*, Vg first appears in the hemolymph several hours after eclosion, and both JH and ecdysteroid participate in Vg induction (Ramaswamy et al. 1997; Raikhel et al. 2005; Telfer 2009). Decapitation shortly after eclosion suppresses Vg accumulation, whereas treatment with a JH analog restores Vg production, indicating that JH signaling is required for the onset of vitellogenesis (Sorge et al. 2000).

In contrast to *S. frugiperda*, *Vg* expression in the silkworm, *Bombyx mori*, begins in the early pupal stage and is driven primarily by ecdysone rather than JH, and 20-hydroxyecdysone (20E) titer in early pupae triggers previtellogenic and vitellogenic processes in *B. mori* (Tsuchida et al. 1987). 20E induces the expression of regulatory genes such as nuclear receptors *E75*, *HR3*, and *Br-C* (Swevers and Iatrou 2003; Zhang et al. 2017).

The tobacco hawk moth, *Manduca sexta*, occupies a position between species that initiate vitellogenesis after adult eclosion and those in which Vg synthesis begins during the pupal stage. In *M. sexta*, the JH analog (methoprene) treatment induced Vg synthesis and *Vg* mRNA in prepupae and freshly molted pupae (Satyanarayana et al. 1994). Consistent with this timing, Vg is first detectable in the female hemolymph about 4 days before adult eclosion, and Vg levels rise sharply after eclosion (Imboden and Law 1983). Paradoxically, the CA-ablated pupae still eclose with yolk-containing oocytes, and their hemolymph vitellogenin concentrations at eclosion are not reduced relative to intact females (Nijhout and Riddiford 1974). These observations indicate that JH is not required for Vg synthesis or uptake. Currently, the involvement of JH in pupal vitellogenesis in *M. sexta* is still unclear. This situation, however, does not mean that JH is unnecessary for *M. sexta* oogenesis, as JH is essential for the completion of oocyte growth and chorion formation from the adult eclosion (Nijhout and Riddiford 1974).

Together, these examples illustrate that the contribution of JH to vitellogenesis varies widely among Lepidoptera. Therefore, in lepidopterans, it seems that reproductive control appears to have evolved independently in different lineages.

11.2.9 Diptera

In Diptera, one group that has long been the subject of research is disease-transmitting mosquitoes. Ovarian development in female mosquitoes is closely linked to their blood-feeding behavior, and various studies have investigated the interplay between blood feeding, hormonal changes, and ovarian regulation (Klowden 1997; Leyria 2024; Raikhel et al. 2005; Santos et al. 2019; Zhu and Noriega 2016). The most well-studied example is the yellow fever mosquito, *Aedes aegypti*. In this species, JH plays a crucial role in priming the fat body for nutrient storage before blood feeding and in promoting the growth of previtellogenic oocytes (Clifton and Noriega 2012; Feinsod and Spielman 1980; Gwadz and Spielman 1973; Raikhel and Lea 1985; Shapiro et al. 1986; Wang et al. 2017). The JH titer increases from the end of the pupal stage and reaches its peak dozens of hours after eclosion in females (Fig. 11.2) (Hernández-Martínez et al. 2015; Shapiro et al. 1986; Zhao et al. 2016). Expression and protein levels of *Kr-h1* also peak at 48 h after eclosion and knockdown of *Kr-h1*-inhibited oogenesis. Following a blood meal, JH and ecdysone titers decrease and conversely increase, respectively, promoting Vg production in the fat body (Fig. 11.2) (Browne 2001; Hagedorn et al. 1975; Hernández-Martínez et al. 2015; Raikhel et al. 2005; Shapiro et al. 1986; Van Handel and Lea 1984). This process cross talks with the wingless/Wnt and mTOR pathways (Attardo et al. 2005; Hansen et al. 2005; Weng and Shiao 2015). Recent studies have also shown that JH mediates a quality check of the pre-yolk-forming follicles. The formation of the JH–Met–Tai complex regulates the expression of receptors critical for yolk uptake, such as Vg receptors, ovary-specific lipophorin receptor isoforms, ovary-specific clathrin heavy chain, and other genes involved in vitellogenesis (Clifton and Noriega 2012). This regulation increases the lipid content in oocytes, ensuring high-quality eggs after blood feeding. Recently, a mutant lacking *CYP15* (*epox*), the JH biosynthetic P450 epoxidase gene, was established, enabling the study of the role of the JH precursor methyl farnesoate (MF) in reproduction (Nouzova et al. 2021). *epox* mutants are unable to synthesize epoxidized JH because of their inability to epoxidize MF. Compared to wild-type females, *epox*

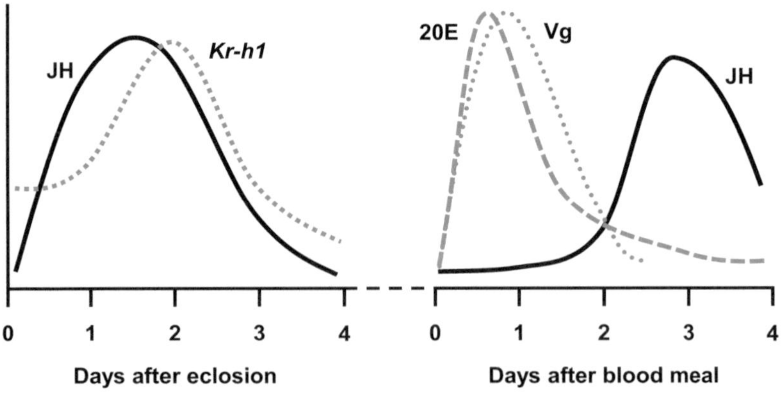

Fig. 11.2 Changes in juvenile hormone (JH) titer, 20-hydroxyecdysone (20E) titer, *Krüppel homolog 1* (*Kr-h1*) expression level, and vitellogenin (Vg) titer in *Aedes aegypti*. Schematic diagram of the time-dependent changes in JH titer and 20E titer in *A. aegypti*. After eclosion, JH peaks for several tens of hours and then declines. The expression of *Kr-h1* also fluctuates according to this change in JH titer. After blood feeding, JH titer is suppressed while 20E titer and *Vg* expression increase. Then, JH rises to prepare for the next reproductive cycle

mutants exhibit smaller previtellogenic follicles with reduced lipid storage and produce approximately 40% fewer eggs (Nouzova et al. 2021). These findings suggest that although MF is sufficient to maintain baseline reproductive functions in female mosquitoes, epoxidized JH is required to achieve the full reproductive potential observed in wild-type females. Thus, in *A. aegypti*, JH primarily supports the early stages of egg formation, ensuring proper oocyte development before the influence of ecdysone after feeding. Similar ecdysone-mediated modification of JH functions are observed in Diptera (Raikhel et al. 2005; Roy et al. 2018).

The representative model of dipterans other than mosquitoes is the fruit fly, *Drosophila melanogaster*. The JH-dependent oogenesis process in this species is described in detail in the next section.

11.3 Mechanisms Regulating Ovarian Development Via Juvenile Hormone in the Fruit Fly, *Drosophila melanogaster*

11.3.1 JH Signaling in D. melanogaster

The role of JH in the development and physiology of *D. melanogaster* has also been studied by many researchers (Ashburner 1970; Flatt et al. 2005; Madhavan 1973; Riddiford and Ashburner 1991; Zhang et al. 2021). Similar to the hormonal regulation of oogenesis in mosquitoes, ecdysone, JH, and their cross talk are involved in oogenesis in *D. melanogaster* (Leyria 2024; Santos et al. 2019). The advanced genetic tools available for this model organism have significantly contributed to a deeper understanding of its molecular mechanisms (Zhang et al. 2021).

Drosophila melanogaster has two genes encoding intracellular JH receptors: *Met* and its paralog, *germ cell-expressed* (*gce*) (Abdou et al. 2011; Jindra et al. 2015a, b; Wilson 1996). Met and Gce bind to specific JH response elements on target genes and induce their transcription in the presence of JH (He et al. 2014; Jindra et al. 2015b; Kayukawa et al. 2012; Li et al. 2014, 2011; Zhang et al. 2011). Using the transgenic flies, it was reported that both Met and Gce are widely expressed in many tissues in adults (Baumann et al. 2017). In addition, the development of reporter strains visualizing JH signaling intensity has enabled the exploration of tissues affected by JH (Barton et al. 2024; He et al. 2014; Kurogi et al. 2024). Furthermore, tissue-specific knockdown of JH receptor genes allowed researchers to analyze the role of JH signaling in different tissues, revealing that JH plays a role not only in vitellogenesis but also in various stages of oogenesis. This section summarizes the role of JH signaling in *D. melanogaster* oogenesis processes, such as yolk formation, ovulation, and germline stem cell (GSC) regulation.

11.3.2 Regulation of Yolk Accumulation by Juvenile Hormone

As mentioned previously, the classical function of JH in oogenesis is the regulation of vitellogenesis (Raikhel et al. 2005; Santos et al. 2019). In female *D. melanogaster*, three yolk protein genes (*Yolk protein 1*, *Yolk protein 2*, and *Yolk protein 3*) are expressed predominantly in the fat body and, to a lesser extent, in the ovary (Brennan et al. 1982; Isaac and Bownes 1982). Methoprene treatment increases the expression of yolk protein genes (Bownes and Blair 1986; Soller et al. 1999). Additionally, a specific loss-of-function allele of the *apterous* gene, which exhibits reduced JH levels, suppresses vitellogenesis (Dubrovsky et al. 2002; Postlethwait and Weiser 1973). Moreover, CA-ablated flies exhibit the reduced yolk protein gene expression, leading to the suppression of vitellogenesis and the reduction of mature egg formation, which is rescued by methoprene treatment (Yamamoto et al. 2013). Reduced mature egg formation is also observed in the loss-of-function mutant of *juvenile hormone acid O-methyltransferase* (*jhamt*), a rate-limiting JH biosynthetic enzyme responsible for the methylation of JH precursors (Niwa et al. 2008; Wen et al. 2015). These findings suggest the importance of JH signaling in egg production via vitellogenesis (Fig. 11.3). However, it remains unclear whether JH directly regulates yolk protein gene expression. A study showed that a potential promoter of yolk protein genes is responsive to 20E but not methoprene in the fat body (Bownes et al. 1996). Therefore, ecdysteroid but not JH is likely primarily responsible for regulating yolk protein gene expression.

The involvement of JH in the uptake of yolk proteins into the ovaries is also unclear. Yolk proteins produced in the fat body are released into the hemolymph. Methoprene treatment reduces yolk protein levels in the hemolymph, indicating that JH signaling facilitates yolk protein uptake (Soller et al. 1999). Methoprene treatment in *jhamt* mutants induces follicular patency in maturing oocytes (Luo et al. 2021). The opening of intercellular spaces between follicle cells allows hemolymph proteins to access the oocyte during the vitellogenic stages of oogenesis (Fig. 11.3) (Row et al. 2021; Zimowska et al. 1994). However, the exact mechanism by which JH regulates yolk protein uptake remains to be elucidated. In some insects, it has been suggested that an uncharacterized membrane receptor for JH may be involved in yolk protein uptake (Bai and Palli 2016; Jindra et al. 2015a; Jing et al. 2018; Liu et al. 2015).

The promotion of yolk accumulation by JH is influenced by various environmental conditions, such as reproductive dormancy and mating stimuli. When *D. melanogaster* is exposed to a combination of low temperatures and short-day photoperiods, reproductive slowdown, known as reproductive dormancy, is induced (Kurogi et al. 2021). Reproductive dormancy is closely linked to the suppression of JH biosynthesis in the CA (Saunders et al. 1990). Under dormancy-inducing conditions, JH biosynthesis is reduced due to decreased insulin signaling, increased dopamine signaling, and neural regulation of the CA through the action of diuretic

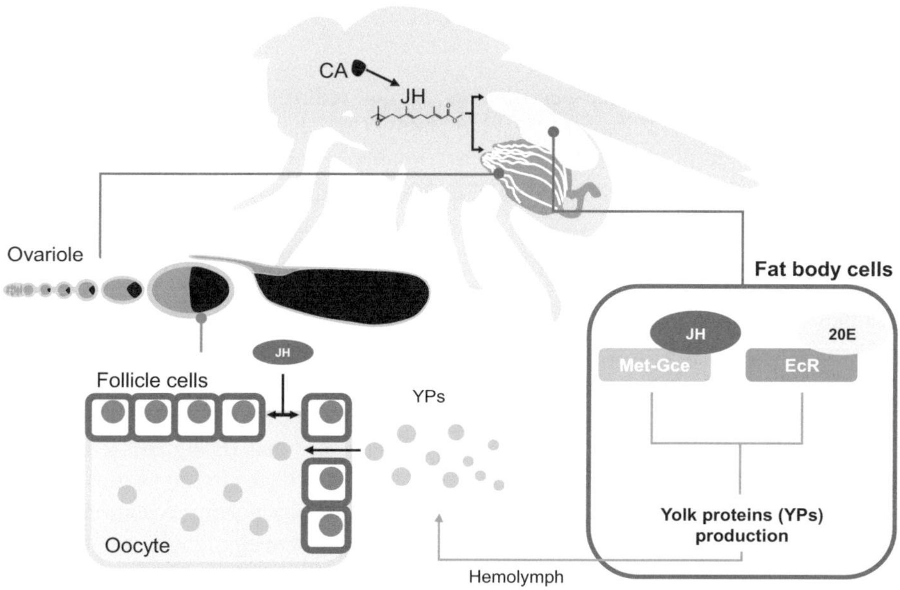

Fig. 11.3 Juvenile hormone (JH)-mediated regulation of vitellogenesis in *Drosophila melanogaster*. Schematic representation of the function of JH in vitellogenesis in *D. melanogaster*. JH and 20-hydroxyecdysone (20E) induce the *yolk protein* (*yp*) gene expression in the fat body. YPs produced in the fat body are transported to the ovary via the hemolymph and taken up by the ovary. The uptake into the ovary is promoted by JH

hormone 31 (Andreatta et al. 2018; Kurogi et al. 2023; Ojima et al. 2018). This JH biosynthesis inhibition suppresses ovarian development, leading to reproductive dormancy. In current studies on reproductive dormancy in *D. melanogaster*, the presence or absence of yolk accumulation is widely used as an indicator of whether reproductive dormancy has been induced (Kurogi et al. 2021).

JH also facilitates yolk accumulation after mating. Mated females can enhance JH biosynthesis in response to the reception of sex peptide (SP) found in the male seminal fluid (Bontonou et al. 2015; Moshitzky et al. 1996; Reiff et al. 2015; Schwenke and Lazzaro 2017; Soller et al. 1997; Sugime et al. 2017). This mating-dependent regulation of JH levels is thought to be controlled either by nerves projecting from the ovary to the ventral nerve cord (VNC) or by direct reception of the sex peptide in the CA (Moshitzky et al. 1996; Soller et al. 1997; Zhang et al. 2022). Mating-induced elevation of JH is involved not only in yolk accumulation but also in midgut remodeling, which ensures that the energy required for laying eggs is efficiently absorbed, thereby contributing to reproductive success (Reiff et al. 2015). Environment-dependent JH regulation allows *D. melanogaster* to promote ovarian development under appropriate reproductive conditions.

11.3.3 Regulation of Ovulation

In *D. melanogaster*, *Met* and *gce* mutants display reduced fecundity (Abdou et al. 2011; Luo et al. 2021; Wilson and Fabian 1986). This phenotype is similar to that of the *jhamt* mutant and CA-ablated flies (Gruntenko et al. 2010; Wen et al. 2015). However, *jhamt* mutant and CA-ablated flies have smaller ovaries, whereas *Met* and *gce* mutant flies have larger ovaries. These observations suggest that JH signaling has functions other than yolk protein accumulation and production. Consistent with this expectation, an experiment using a JH reporter strain indicated that JH signaling may also function in the muscles surrounding the ovary (Luo et al. 2021).

Knockdown of the muscle-specific *Met*, *gce*, and *Kr-h1* reduces the frequency of muscle contraction and induces larger ovaries (Luo et al. 2021). A reduction in muscle contraction frequency was also observed in *jhamt* mutants and was restored by methoprene feeding (Luo et al. 2021). This indicates that JH signaling regulates the muscles, which, in turn, influences the number of oocytes in the ovaries.

Furthermore, JH-mediated ovarian muscle contraction during ovulation affects egg morphology. In ovaries with reduced JH signaling, the egg length was found to be reduced, although the egg width remained unchanged (Luo et al. 2021). The depletion of laminin in muscle cells has previously been reported to reduce ovarian muscle contraction and egg length (Andersen and Horne-Badovinac 2016). Inhibition of JH signaling results in reduced expression of *Laminin A*, *Laminin B1,* and *Laminin B2* (Luo et al. 2021). Moreover, knockdown of laminin genes reduced muscle contraction and egg laying. In addition to laminins, another extracellular matrix component, collagen IV, is involved in this phenotype. Collagen IV is not produced in the ovary but in the fat body (Van De Bor et al. 2015; Zang et al. 2015). Collagen IV is released into the hemolymph and subsequently deposited in the ovarian muscle to form the extracellular matrix (ECM) (Pastor-Pareja and Xu 2011; Van De Bor et al. 2015; Zang et al. 2015). Fat body–specific knockdown of *Met-gce* and *Kr-h1* reduces the expression of collagen IV subunit genes, *Viking* and *Col4a1* (Luo et al. 2021). Furthermore, reduced muscle contraction, reduced fecundity, and egg deformation were observed in these flies. In other words, JH acts on the ovarian muscle and fat body to promote ECM production and regulate ovulation (Fig. 11.4).

11.3.4 Germline Stem Cell Regulation

Another aspect of JH-mediated reproductive regulation in *D. melanogaster* is the regulation of germ stem cells (GSCs). The *D. melanogaster* ovary is composed of multiple ovarioles, with GSCs located at the tips of each ovariole (Kirilly and Xie 2007; Xie 2008). GSCs attach to the cap cells that form the stem cell niche. GSCs divide asymmetrically to produce one daughter cell that remains undifferentiated and adjacent to the cap cells, thus maintaining its status as a stem cell (Fuchs et al. 2004; Spradling et al. 2001). The other daughter cell, which is separated from the

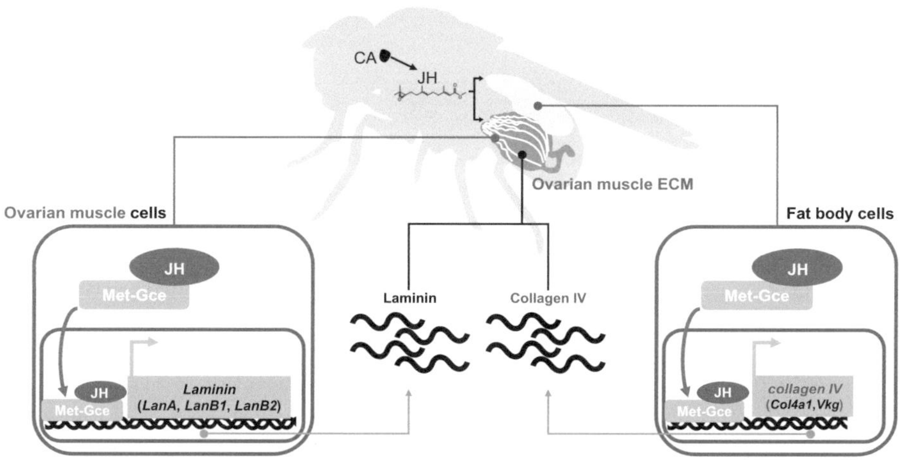

Fig. 11.4 Juvenile hormone (JH)-mediated regulation of the ovarian muscle in *Drosophila melanogaster*. Schematic representation of the function of JH in ovulation in *D. melanogaster*. JH acts on ovarian muscle cells to induce the expression of *laminin*. JH also acts on the fat body to induce the expression of *collagen IV*. These inductions allow the construction of ovarian muscle, contributing to normal ovulation

cap cells, differentiates into a cystoblast and subsequently develops into an oocyte and a nurse cell (Xie and Spradling 2000).

Cap cells release a signaling molecule called decapentaplegic (Dpp), maintaining their self-renewal and differentiation potential. GSCs require the stem cell niche, which is primarily composed of cap cells (Kirilly and Xie 2007; Spradling et al. 2001; Xie 2008; Xie and Spradling 2000) and regulates the stemness of GSCs (Xie and Spradling 1998). Dpp is received by neighboring GSCs, activating the signaling pathway. This pathway helps maintain GSCs in an undifferentiated state by suppressing the expression of *Bag-of-Marble* (*Bam*), a protein required for differentiation during oogenesis (Chen and McKearin 2003). Conversely, when cells are generated by mitosis and physically separate from the cap cells, they become unresponsive to Dpp, leading to *Bam* expression and differentiation into cystoblasts. The regulation of GSCs and their niches is the foundation for controlling fertility in female *D. melanogaster*.

JH regulates GSCs in *D. melanogaster*. *Met* and *gce* mutants show a gradual decline in germline stem cells with each passing day after eclosion (Luo et al. 2020). This reduction in GSC numbers was not observed in the wild-type flies. A similar decline in GSC numbers was also observed in *jhamt* mutant flies (Luo et al. 2020). GSCs are regulated by JH through the reception of JH by GSCs themselves and by cap cells constituting the stem cell niche. Niche-specific *Met-gce* knockdown reduced GSC numbers, whereas cap cell numbers remained unchanged. In contrast, GSC-specific knockdown of *Met-gce* leads to an increase in GSC numbers (Luo et al. 2020). In addition, phosphorylated Mothers against Dpp (p-Mad), which is downstream of Dpp, is increased in flies in which JH signaling is inhibited in a

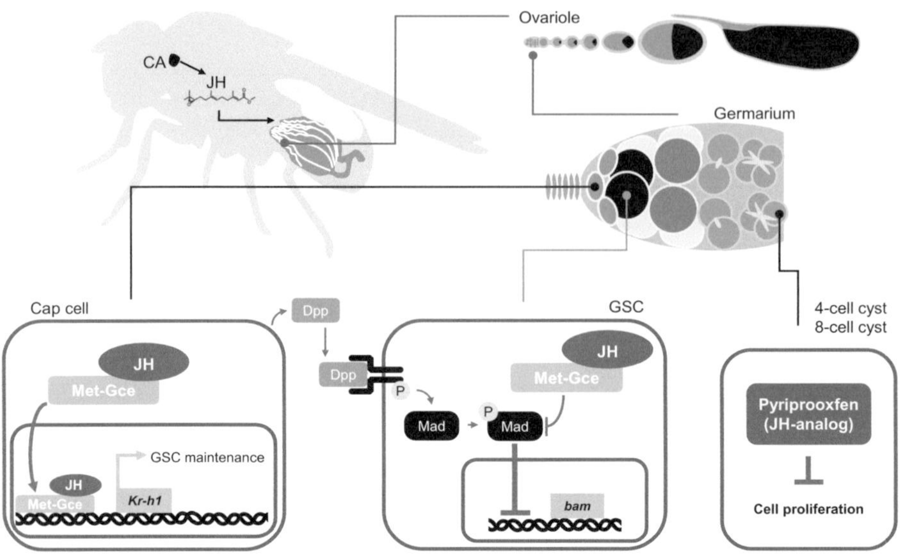

Fig. 11.5 Juvenile hormone (JH)-mediated regulation of germline stem cell (GSC) in *Drosophila melanogaster*. Schematic representation of the function of JH in GSC regulation in *D. melanogaster*. JH maintains the number of GSCs by acting on cap cells, while JH inhibits decapentaplegic (Dpp) signaling by acting on GSCs themselves, thereby reducing their number. It has also been reported that pyriproxyfen, a JH analog, promotes cystoblast division

GSC-specific manner (Luo et al. 2020), suggesting that JH signaling in GSCs influences their number via Dpp signaling (Fig. 11.5).

In another study, feeding on a JH analog, pyriproxyfen, reduced fecundity. In pyriproxyfen-fed flies, the number of GSCs increased, whereas the number of cystoblasts decreased (Khalid et al. 2024). In these flies, cell proliferation is suppressed in cystoblasts at the four- and eight-cell stages (Khalid et al. 2024). Thus, JH signaling may be involved in the regulation of cystoblast differentiation from GSCs.

It has been shown that various environmental cues, such as aging, mating, nutrition, and dormancy-inducing conditions, can regulate GSCs (Hoshino and Niwa 2021; Yoshinari et al. 2019). For example, JH levels in *D. melanogaster* have been reported to increase after mating, and GSC numbers also increase with mating (Ameku and Niwa 2016; Reiff et al. 2015). However, it has been observed that CA-specific knockdown of *jhamt* and knockdown of *Met* and *gce* in ovarian somatic cells still show normal postmating GSC proliferation (Ameku and Niwa 2016). These data suggest that JH is not required for mating-dependent GSC proliferation in *D. melanogaster*. Additionally, a recent study demonstrated that GSC numbers are reduced in a JH-dependent manner in dormancy-inducing conditions (Easwaran et al. 2022).

11.4 Conclusion

The first half of the chapter focuses on the facilitative effects of JH on ovarian development in several insect species. Based on the meticulous observations of many entomologists, the most classically known function of JH, which is mostly common across many insect groups, is to promote the production of Vg and related yolk proteins and their uptake into the ovary (Leyria 2024; Raikhel et al. 2005; Santos et al. 2019; Wyatt and Davey 1996). In contrast, the degree of involvement of JH in reproduction varies among the different insect species, which is interesting from an evolutionary standpoint.

The second half of the chapter focuses on the fruit fly, *D. melanogaster*, which excels as a genetic tool among a variety of interesting insect species. In particular, the identification and functional characterization of the "bona fide" JH receptors, Met and Gce, in *D. melanogaster* have contributed significantly to our understanding of JH-dependent biological phenomena, including reproduction. Recent studies in *D. melanogaster* have revealed that JH regulates various aspects of reproduction, including ovulation control and GSC maintenance.

One of the remaining questions to be addressed is whether and how JH orchestrates and synchronizes gene expression in different types of ovarian cells, such as GSCs, more differentiated female germline cells, somatic niche cells, somatic follicle cells, and even ovarian muscle cells, to achieve proper ovarian development in insects. To address this question, tissue-specific manipulation of JH signaling in *D. melanogaster* will facilitate further investigation into the various functions of JH in multiple ovarian cell types. Another important issue in understanding the relationship between JH and reproduction is how JH-dependent regulation of reproduction is linked to and regulated by other JH-dependent biological events, such as development, immunity, lifespan, and behavior (Flatt et al. 2005; Zhang et al. 2021). Future studies on insect physiology should position reproductive regulation within the systemic regulation exerted by a single hormone, JH.

References

Abdou MA, He Q, Wen D, Zyaan O, Wang Jing XJ et al (2011) *Drosophila* Met and Gce are partially redundant in transducing juvenile hormone action. Insect Biochem Mol Biol 41:938–945. https://doi.org/10.1016/J.IBMB.2011.09.003

Ameku T, Niwa R (2016) Mating-induced increase in germline stem cells via the neuroendocrine system in female *Drosophila*. PLoS Genet 12:e1006123. https://doi.org/10.1371/journal.pgen.1006123

Andersen D, Horne-Badovinac S (2016) Influence of ovarian muscle contraction and oocyte growth on egg chamber elongation in *Drosophila*. Development 143:1375–1387. https://doi.org/10.1242/dev.131276

Andreatta G, Kyriacou CP, Flatt T, Costa R (2018) Aminergic signaling controls ovarian dormancy in *Drosophila*. Sci Rep 8:2030. https://doi.org/10.1038/s41598-018-20407-z

Ashburner M (1970) Effects of juvenile hormone on adult differentiation of *Drosophila melanogaster*. Nature 227:187–189. https://doi.org/10.1038/227187a0

Attardo GM, Hansen IA, Raikhel AS (2005) Nutritional regulation of vitellogenesis in mosquitoes: implications for anautogeny. Insect Biochem Mol Biol 35:661–675. https://doi.org/10.1016/j.ibmb.2005.02.013

Bai H, Palli SR (2016) Identification of G protein-coupled receptors required for vitellogenin uptake into the oocytes of the red flour beetle, *Tribolium castaneum*. Sci Rep 6:27648. https://doi.org/10.1038/srep27648

Barton LJ, Sanny J, Packard Dawson E, Nouzova M, Noriega FG, Stadtfeld M et al (2024) Juvenile hormones direct primordial germ cell migration to the embryonic gonad. Curr Biol 34:505–518.e6. https://doi.org/10.1016/j.cub.2023.12.033

Baumann AA, Texada MJ, Chen HM, Etheredge JN, Miller DL, Picard S et al (2017) Genetic tools to study juvenile hormone action in *Drosophila*. Sci Rep 7:2132. https://doi.org/10.1038/s41598-017-02264-4

Bloch G, Borst DW, Huang Z-Y, Robinson GE, Cnaani J, Hefetz A (2000) Juvenile hormone titers, juvenile hormone biosynthesis, ovarian development and social environment in *Bombus terrestris*. J Insect Physiol 46:47–57. https://doi.org/10.1016/s0022-1910(99)00101-8

Bloch G, Hefetz A (1999) Regulation of reproduction by dominant workers in bumblebee (*Bombus terrestris*) queenright colonies. Behav Ecol Sociobiol 45:125–135. https://doi.org/10.1007/s002650050546

Bontonou G, Shaik HA, Denis B, Wicker-Thomas C (2015) Acp70A regulates *Drosophila* pheromones through juvenile hormone induction. Insect Biochem Mol Biol 56:36–49. https://doi.org/10.1016/j.ibmb.2014.11.008

Bownes M, Blair M (1986) The effects of a sugar diet and hormones on the expression of the *drosophila* yolk-protein genes. J Insect Physiol 32:493–501. https://doi.org/10.1016/0022-1910(86)90011-9

Bownes M, Ronaldson E, Mauchline D (1996) 20-Hydroxyecdysone, but not juvenile hormone, regulation of *yolk protein* gene expression can be mapped to *cis*-acting DNA sequences. Dev Biol 173:475–489. https://doi.org/10.1006/dbio.1996.0041

Bradfield JY, Berlin RL, Keeley LL (1990) Contrasting modulations of gene expression by a juvenile hormone analog. Insect Biochem 20:105–111. https://doi.org/10.1016/0020-1790(90)90002-c

Brennan MD, Weiner AJ, Goralski TJ, Mahowald AP (1982) The follicle cells are a major site of vitellogenin synthesis in *Drosophila melanogaster*. Dev Biol 89:225–236. https://doi.org/10.1016/0012-1606(82)90309-8

Browne LB (2001) Quantitative aspects of the regulation of ovarian development in selected anautogenous Diptera: integration of endocrinology and nutrition. Entomol Exp Appl 100:137–149. https://doi.org/10.1046/j.1570-7458.2001.00859.x

Chang L-W, Tsai C-M, Yang D-M, Chiang A-S (2005) Cell size control by ovarian factors regulates juvenile hormone synthesis in corpora allata of the cockroach, *Diploptera punctata*. Insect Biochem Mol Biol 35:41–50. https://doi.org/10.1016/j.ibmb.2004.10.006

Charles J-P, Iwema T, Epa VC, Takaki K, Rynes J, Jindra M (2011) Ligand-binding properties of a juvenile hormone receptor, Methoprene-tolerant. Proc Natl Acad Sci USA 108:21128–21133. https://doi.org/10.1073/pnas.1116123109

Chen D, McKearin D (2003) Dpp signaling silences *bam* transcription directly to establish asymmetric divisions of germline stem cells. Curr Biol 13:1786–1791. https://doi.org/10.1016/j.cub.2003.09.033

Clifton ME, Noriega FG (2012) The fate of follicles after a blood meal is dependent on previtellogenic nutrition and juvenile hormone in *Aedes aegypti*. J Insect Physiol 58:1007–1019. https://doi.org/10.1016/j.jinsphys.2012.05.005

Comas D, Piulachs MD, Bellés X (2001) Induction of vitellogenin gene transcription in vitro by juvenile hormone in *Blattella germanica*. Mol Cell Endocrinol 183:93–100. https://doi.org/10.1016/s0303-7207(01)00589-5

Comas D, Piulachs MD, Bellés X (1999) Fast induction of vitellogenin gene expression by juvenile hormone III in the cockroach *Blattella germanica* (L.) (Dictyoptera, Blattellidae). Insect Biochem Mol Biol 29:821–827. https://doi.org/10.1016/s0965-1748(99)00058-2

Dubrovsky EB, Bernardo TJ (2014) The juvenile hormone receptor and molecular mechanisms of juvenile hormone action. Adv Insect Physiol 46:305–388. https://doi.org/10.1016/b978-0-12-417010-0.00004-5

Dubrovsky EB, Dubrovskaya VA, Berger EM (2002) Juvenile hormone signaling during oogenesis in *Drosophila melanogaster*. Insect Biochem Mol Biol 32:1555–1565. https://doi.org/10.1016/s0965-1748(02)00076-0

Easwaran S, Van Ligten M, Kui M, Montell DJ (2022) Enhanced germline stem cell longevity in *Drosophila* diapause. Nat Commun 13:711. https://doi.org/10.1038/s41467-022-28347-z

Elliott KL, Woodhead AP, Stay B (2006) A stage-specific ovarian factor with stable stimulation of juvenile hormone synthesis in corpora allata of the cockroach *Diploptera punctata*. J Insect Physiol 52:929–935. https://doi.org/10.1016/j.jinsphys.2006.05.012

Engelmann F (1959) The control of reproduction in *Diploptera punctata* (Blattaria). Biol Bull 116:406–419. https://doi.org/10.2307/1538949

Feinsod FM, Spielman A (1980) Nutrient-mediated juvenile hormone secretion in mosquitoes. J Insect Physiol 26:113–117. https://doi.org/10.1016/0022-1910(80)90050-5

Flatt T, Tu MP, Tatar M (2005) Hormonal pleiotropy and the juvenile hormone regulation of *Drosophila* development and life history. BioEssays 27:999–1010. https://doi.org/10.1002/bies.20290

Fuchs E, Tumbar T, Guasch G (2004) Socializing with the neighbors: stem cells and their niche. Cell 116:769–778. https://doi.org/10.1016/s0092-8674(04)00255-7

Gospocic J, Glastad KM, Sheng L, Shields EJ, Berger SL, Bonasio R (2021) Kr-h1 maintains distinct caste-specific neurotranscriptomes in response to socially regulated hormones. Cell 184:5807–5823.e14. https://doi.org/10.1016/j.cell.2021.10.006

Gruntenko NE, Wen D, Karpova EK, Adonyeva NV, Liu Y, He Q et al (2010) Altered juvenile hormone metabolism, reproduction and stress response in *Drosophila* adults with genetic ablation of the *corpus allatum* cells. Insect Biochem Mol Biol 40:891–897. https://doi.org/10.1016/j.ibmb.2010.09.001

Guo W, Wu Z, Song J, Jiang F, Wang Z, Deng S et al (2014) Juvenile hormone-receptor complex acts on *Mcm4* and *Mcm7* to promote polyploidy and vitellogenesis in the migratory locust. PLoS Genet 10:e1004702. https://doi.org/10.1371/journal.pgen.1004702

Gwadz RW, Spielman A (1973) Corpus allatum control of ovarian development in *Aedes aegypti*. J Insect Physiol 19:1441–1448. https://doi.org/10.1016/0022-1910(73)90174-1

Hagedorn HH, O'Connor JD, Fuchs MS, Sage B, Schlaeger DA, Bohm MK (1975) The ovary as a source of alpha-ecdysone in an adult mosquito. Proc Natl Acad Sci USA 72:3255–3259. https://doi.org/10.1073/pnas.72.8.3255

Hansen IA, Attardo GM, Roy SG, Raikhel AS (2005) Target of rapamycin-dependent activation of S6 kinase is a central step in the transduction of nutritional signals during egg development in a mosquito. J Biol Chem 280:20565–20572. https://doi.org/10.1074/jbc.M500712200

Hasebe M, Shiga S (2021) Photoperiodic response in the pars intercerebralis neurons, including Plast-MIP neurons, in the brown-winged green bug, *Plautia stali*. Zool Sci 38:317–325. https://doi.org/10.2108/zs210005

Hasegawa T, Hasebe M, Shiga S (2020) Immunohistochemical and direct mass spectral analyses of *Plautia stali* myoinhibitory peptides in the cephalic ganglia of the brown-winged green bug *Plautia stali*. Zool Sci 37:42–49. https://doi.org/10.2108/zs190092

He Q, Wen D, Jia Q, Cui C, Wang J, Pallii SR et al (2014) Heat shock protein 83 (Hsp83) facilitates Methoprene-tolerant (Met) nuclear import to modulate juvenile hormone signaling. J Biol Chem 289:27874–27885. https://doi.org/10.1074/JBC.M114.582825

Hernández-Martínez S, Rivera-Perez C, Nouzova M, Noriega FG (2015) Coordinated changes in JH biosynthesis and JH hemolymph titers in *Aedes aegypti* mosquitoes. J Insect Physiol 72:22–27. https://doi.org/10.1016/j.jinsphys.2014.11.003

Hoshino R, Niwa R (2021) Regulation of mating-induced increase in female germline stem cells in the fruit fly *Drosophila melanogaster*. Front Physiol 12:785435. https://doi.org/10.3389/fphys.2021.785435

Howard RW, Haverty MI (1979) Termites and juvenile hormone analogues: a review of methodology and observed effects. Sociobiology 4:277–285

Hrdý I (1976) The influence of juvenile hormone analogues on caste development in termites. In: Lüscher M (ed) Phase and caste determination in insects. Elsevier, Amsterdam, p 71. https://doi.org/10.1016/b978-0-08-021256-2.50011-x

Hu K, Tian P, Tang Y, Yang L, Qiu L, He H et al (2019) Molecular characterization of vitellogenin and its receptor in *Sogatella furcifera*, and their function in oocyte maturation. Front Physiol 10:1532. https://doi.org/10.3389/fphys.2019.01532

Imboden H, Law JH (1983) Heterogeneity of vitellins and vitellogenins of the tobacco hornworm, *Manduca sexta* L. time course of vitellogenin appearance in the haemolymph of the adult female. Insect Biochem 13:151–162. https://doi.org/10.1016/0020-1790(83)90078-1

Isaac PG, Bownes M (1982) Ovarian and fat-body vitellogenin synthesis in *Drosophila melanogaster*. Eur J Biochem 123:527–534. https://doi.org/10.1111/j.1432-1033.1982.tb06563.x

Jindra M, Bellés X, Shinoda T (2015a) Molecular basis of juvenile hormone signaling. Curr Opin Insect Sci 11:39–46. https://doi.org/10.1016/j.cois.2015.08.004

Jindra M, Palli SR, Riddiford LM (2013) The juvenile hormone signaling pathway in insect development. Annu Rev Entomol 58:181–204. https://doi.org/10.1146/annurev-ento-120811-153700

Jindra M, Uhlirova M, Charles JP, Smykal V, Hill RJ (2015b) Genetic evidence for function of the bHLH-PAS protein Gce/Met as a juvenile hormone receptor. PLoS Genet 11:e1005394. https://doi.org/10.1371/JOURNAL.PGEN.1005394

Jing Y-P, An H, Zhang S, Wang N, Zhou S (2018) Protein kinase C mediates juvenile hormone-dependent phosphorylation of Na^+/K^+-ATPase to induce ovarian follicular patency for yolk protein uptake. J Biol Chem 293:20112–20122. https://doi.org/10.1074/jbc.RA118.005692

Jing Y-P, Wen X, Li L, Zhang S, Zhang C, Zhou S (2021) The vitellogenin receptor functionality of the migratory locust depends on its phosphorylation by juvenile hormone. Proc Natl Acad Sci USA 118:e2106908118. https://doi.org/10.1073/pnas.2106908118

Kayukawa T, Minakuchi C, Namiki T, Togawa T, Yoshiyama M, Kamimura M et al (2012) Transcriptional regulation of juvenile hormone-mediated induction of Krüppel homolog 1, a repressor of insect metamorphosis. Proc Natl Acad Sci USA 109:11729–11734. https://doi.org/10.1073/pnas.1204951109

Keeley LL, Sowa SM, Hayes TK, Bradfield JY (1988) Neuroendocrine and juvenile hormone effects on fat body protein synthesis during the reproductive cycle in female *Blaberus discoidalis* cockroaches. Gen Comp Endocrinol 72:364–373. https://doi.org/10.1016/0016-6480(88)90158-x

Khalid MZ, Liu J, Zhang J, Yang L, Sun Z, Zhong G (2024) Pyriproxyfen enhances germline stem cell proliferation and reduces reproduction in *Drosophila* by up-regulating juvenile hormone signaling. Pest Manag Sci 80:5099–5111. https://doi.org/10.1002/ps.8234

Kirilly D, Xie T (2007) The *Drosophila* ovary: an active stem cell community. Cell Res 17:15–25. https://doi.org/10.1038/sj.cr.7310123

Klowden MJ (1997) Endocrine aspects of mosquito reproduction. Arch Insect Biochem Physiol 35:491–512. https://doi.org/10.1002/(SICI)1520-6327(1997)35:4<491::AID-ARCH10>3.0.CO;2-5

Konopova B, Jindra M (2007) Juvenile hormone resistance gene *Methoprene-tolerant* controls entry into metamorphosis in the beetle *Tribolium castaneum*. Proc Natl Acad Sci USA 104:10488–10493. https://doi.org/10.1073/pnas.0703719104

Korb J, Belles X (2017) Juvenile hormone and hemimetabolan eusociality: a comparison of cockroaches with termites. Curr Opin Insect Sci 22:109–116. https://doi.org/10.1016/j.cois.2017.06.002

Kurogi Y, Imura E, Mizuno Y, Hoshino R, Nouzova M, Matsuyama S et al (2023) Female reproductive dormancy in *Drosophila* is regulated by DH31-producing neurons projecting into the corpus allatum. Development 150:dev201186. https://doi.org/10.1242/dev.201186

Kurogi Y, Mizuno Y, Hayashi R, Goyins K, Okamoto N, Barton L et al (2024) The seminal vesicle is a juvenile hormone-responsive tissue in adult male *Drosophila melanogaster*. Open Biol 14:240315. https://doi.org/10.1098/rsob.240315

Kurogi Y, Mizuno Y, Imura E, Niwa R (2021) Neuroendocrine regulation of reproductive dormancy in the fruit fly *Drosophila melanogaster*: a review of juvenile hormone-dependent regulation. Front Ecol Evol 9:715029. https://doi.org/10.3389/FEVO.2021.715029

Leyria J (2024) Endocrine factors modulating vitellogenesis and oogenesis in insects: an update. Mol Cell Endocrinol 587:112211. https://doi.org/10.1016/j.mce.2024.112211

Li M, Liu P, Wiley JD, Ojani R, Bevan DR, Li J et al (2014) A steroid receptor coactivator acts as the DNA-binding partner of the methoprene-tolerant protein in regulating juvenile hormone response genes. Mol Cell Endocrinol 394:47–58. https://doi.org/10.1016/J.MCE.2014.06.021

Li M, Mead EA, Zhu J (2011) Heterodimer of two bHLH-PAS proteins mediates juvenile hormone-induced gene expression. Proc Natl Acad Sci USA 108:638–643. https://doi.org/10.1073/pnas.1013914108

Lin X, Yao Y, Wang B (2015) *Methoprene-tolerant* (*Met*) and *Krüpple-homologue 1* (*Kr-h1*) are required for ovariole development and egg maturation in the brown plant hopper. Sci Rep 5:18064. https://doi.org/10.1038/srep18064

Liu P, Peng H-J, Zhu J (2015) Juvenile hormone-activated phospholipase C pathway enhances transcriptional activation by the methoprene-tolerant protein. Proc Natl Acad Sci USA 112:E1871–E1879. https://doi.org/10.1073/pnas.1423204112

Luo W, Liu S, Zhang W, Yang L, Huang J, Zhou S et al (2021) Juvenile hormone signaling promotes ovulation and maintains egg shape by inducing expression of extracellular matrix genes. Proc Natl Acad Sci USA 118:e2104461118. https://doi.org/10.1073/PNAS.2104461118

Luo W, Veeran S, Wang J, Li S, Li K, Liu S (2020) Dual roles of juvenile hormone signaling during early oogenesis in *Drosophila*. Insect Sci 27:665–674. https://doi.org/10.1111/1744-7917.12698

Ma L, Zhang W, Liu C, Chen L, Xu Y, Xiao H et al (2018) Methoprene-tolerant (Met) is indispensable for larval metamorphosis and female reproduction in the cotton bollworm *Helicoverpa armigera*. Front Physiol 9:1601. https://doi.org/10.3389/fphys.2018.01601

Madhavan K (1973) Morphogenetic effects of juvenile hormone and juvenile hormone mimics on adult development of *Drosophila*. J Insect Physiol 19:441–453. https://doi.org/10.1016/0022-1910(73)90119-4

Maekawa K, Hayashi Y, Lee T, Lo N (2014) Presoldier differentiation of Australian termite species induced by juvenile hormone analogues: termite presoldier differentiation. Aust Entomol 53:138–143. https://doi.org/10.1111/aen.12060

Maekawa K, Ishitani K, Gotoh H, Cornette R, Miura T (2010) Juvenile hormone titre and vitellogenin gene expression related to ovarian development in primary reproductives compared with nymphs and nymphoid reproductives of the termite *Reticulitermes speratus*. Physiol Entomol 35:52–58. https://doi.org/10.1111/j.1365-3032.2009.00711.x

Marchal E, Hult EF, Huang J, Pang Z, Stay B, Tobe SS (2014) Methoprene-tolerant (Met) knockdown in the adult female cockroach, *Diploptera punctata* completely inhibits ovarian development. PLoS One 9:e106737. https://doi.org/10.1371/journal.pone.0106737

Marchal E, Hult EF, Huang J, Stay B, Tobe SS (2013) *Diploptera punctata* as a model for studying the endocrinology of arthropod reproduction and development. Gen Comp Endocrinol 188:85–93. https://doi.org/10.1016/j.ygcen.2013.04.018

Matsumoto K, Suetsugu Y, Tanaka Y, Kotaki T, Goto SG, Shinoda T et al (2017) Identification of allatostatins in the brown-winged green bug *Plautia stali*. J Insect Physiol 96:21–28. https://doi.org/10.1016/j.jinsphys.2016.10.005

Minakuchi C, Zhou X, Riddiford LM (2008) *Krüppel homolog 1* (*Kr-h1*) mediates juvenile hormone action during metamorphosis of *Drosophila melanogaster*. Mech Dev 125:91–105. https://doi.org/10.1016/j.mod.2007.10.002

Moshitzky P, Fleischmann I, Chaimov N, Saudan P, Klauser S, Kubli E et al (1996) Sex-peptide activates juvenile hormone biosynthesis in the *Drosophila melanogaster* corpus allatum. Arch

Insect Biochem Physiol 32:363–374. https://doi.org/10.1002/(SICI)1520-6327(1996)32:3/4 <363::AID-ARCH9>3.0.CO;2-T

Nijhout MM, Riddiford LM (1974) The control of egg maturation by juvenile hormone in the tobacco hornworm moth, *Manduca sexta*. Biol Bull 146:377–392. https://doi.org/10.2307/1540412

Niwa R, Niimi T, Honda N, Yoshiyama M, Itoyama K, Kataoka H et al (2008) Juvenile hormone acid *O*-methyltransferase in *Drosophila melanogaster*. Insect Biochem Mol Biol 38:714–720. https://doi.org/10.1016/j.ibmb.2008.04.003

Nouzova M, Edwards MJ, Michalkova V, Ramirez CE, Ruiz M, Areiza M et al (2021) Epoxidation of juvenile hormone was a key innovation improving insect reproductive fitness. Proc Natl Acad Sci USA 118:e2109381118. https://doi.org/10.1073/pnas.2109381118

Nunes-da-Fonseca R, Berni M, Tobias-Santos V, Pane A, Araujo HM (2017) *Rhodnius prolixus*: from classical physiology to modern developmental biology. Genesis 55:e22995. https://doi.org/10.1002/dvg.22995

Ojima N, Hara Y, Ito H, Yamamoto D (2018) Genetic dissection of stress-induced reproductive arrest in *Drosophila melanogaster* females. PLoS Genet 14:e1007434. https://doi.org/10.1371/journal.pgen.100434

Okot-kotber BM (1980) The influence of juvenile hormone analogue on soldier differentiation in the higher termite, *Macrotermes michaelseni*. Physiol Entomol 5:407–416. https://doi.org/10.1111/j.1365-3032.1980.tb00250.x

Parthasarathy R, Sun Z, Bai H, Palli SR (2010) Juvenile hormone regulation of vitellogenin synthesis in the red flour beetle, *Tribolium castaneum*. Insect Biochem Mol Biol 40:405–414. https://doi.org/10.1016/j.ibmb.2010.03.006

Pastor-Pareja JC, Xu T (2011) Shaping cells and organs in *Drosophila* by opposing roles of fat body-secreted collagen IV and perlecan. Dev Cell 21:245–256. https://doi.org/10.1016/j.devcel.2011.06.026

Peeters C, Hölldobler B (1995) Reproductive cooperation between queens and their mated workers: the complex life history of an ant with a valuable nest. Proc Natl Acad Sci USA 92:10977–10979. https://doi.org/10.1073/pnas.92.24.10977

Penick CA, Liebig J, Brent CS (2011) Reproduction, dominance, and caste: endocrine profiles of queens and workers of the ant *Harpegnathos saltator*. J Comp Physiol A Neuroethol Sens Neural Behav Physiol 197:1063–1071. https://doi.org/10.1007/s00359-011-0667-0

Postlethwait JH, Weiser K (1973) Vitellogenesis induced by juvenile hormone in the female sterile mutant *apterous-four* in *Drosophila melanogaster*. Nat New Biol 244:284–285. https://doi.org/10.1038/newbio244284a0

Raikhel AS, Brown MR, Belles X (2005) Hormonal control of reproductive processes. In: Gilbert LI (ed) Comprehensive molecular insect science. Elsevier, Amsterdam, pp 433–491. https://doi.org/10.1016/B0-44-451924-6/00040-5

Raikhel AS, Dhadialla TS (1992) Accumulation of yolk proteins in insect oocytes. Annu Rev Entomol 37:217–251. https://doi.org/10.1146/annurev.en.37.010192.001245

Raikhel AS, Lea AO (1985) Hormone-mediated formation of the endocytic complex in mosquito oocytes. Gen Comp Endocrinol 57:422–433. https://doi.org/10.1016/0016-6480(85)90224-2

Ramos FO, Nouzova M, Fruttero LL, Leyria J, Ligabue-Braun R, Noriega FG et al (2022) Role of Methoprene-tolerant in the regulation of oogenesis in *Dipetalogaster maxima*. Sci Rep 12:14195. https://doi.org/10.1038/s41598-022-18384-5

Rankin SM, Stay B (1984) The changing effect of the ovary on rates of juvenile hormone synthesis in *Diploptera punctata*. Gen Comp Endocrinol 54:382–388. https://doi.org/10.1016/0016-6480(84)90151-5

Reiff T, Jacobson J, Cognigni P, Antonello Z, Ballesta E, Tan KJ et al (2015) Endocrine remodelling of the adult intestine sustains reproduction in *Drosophila*. elife 4:e06930. https://doi.org/10.7554/eLife.06930

Riddiford LM, Ashburner M (1991) Effects of juvenile hormone mimics on larval development and metamorphosis of *Drosophila melanogaster*. Gen Comp Endocrinol 82:172–183. https://doi.org/10.1016/0016-6480(91)90181-5

Robinson GE, Strambi C, Strambi A, Feldlaufer MF (1991) Comparison of juvenile hormone and ecdysteroid haemolymph titres in adult worker and queen honey bees (*Apis mellifera*). J Insect Physiol 37:929–935. https://doi.org/10.1016/0022-1910(91)90008-n

Robinson GE, Strambi C, Strambi A, Huang ZY (1992) Reproduction in worker honey bees is associated with low juvenile hormone titers and rates of biosynthesis. Gen Comp Endocrinol 87:471–480. https://doi.org/10.1016/0016-6480(92)90055-o

Robinson GE, Vargo EL (1997) Juvenile hormone in adult eusocial Hymenoptera: gonadotropin and behavioral pacemaker. Arch Insect Biochem Physiol 35:559–583. https://doi.org/10.1002/(SICI)1520-6327(1997)35:4<559::AID-ARCH13>3.0.CO;2-9

Röseler P-F (1977) Juvenile hormone control of oögenesis in bumblebee workers, *Bombus terrestris*. J Insect Physiol 23:985–992. https://doi.org/10.1016/0022-1910(77)90126-3

Row S, Huang Y-C, Deng W-M (2021) Developmental regulation of oocyte lipid intake through "patent" follicular epithelium in *Drosophila melanogaster*. iScience 24:102275. https://doi.org/10.1016/j.isci.2021.102275

Roy S, Saha TT, Zou Z, Raikhel AS (2018) Regulatory pathways controlling female insect reproduction. Annu Rev Entomol 63:489–511. https://doi.org/10.1146/annurev-ento-020117-043258

Saiki R, Gotoh H, Toga K, Miura T, Maekawa K (2015) High juvenile hormone titre and abdominal activation of JH signalling may induce reproduction of termite neotenics: JH titre and signalling pathways in termites. Insect Mol Biol 24:432–441. https://doi.org/10.1111/imb.12169

Santos CG, Humann FC, Hartfelder K (2019) Juvenile hormone signaling in insect oogenesis. Curr Opin Insect Sci 31:43–48. https://doi.org/10.1016/J.COIS.2018.07.010

Satyanarayana K, Bradfield JY, Bhaskaran G, Dahm KH (1994) Stimulation of vitellogenin production by methoprene in prepupae and pupae of *Manduca sexta*. Arch Insect Biochem Physiol 25:21–37. https://doi.org/10.1002/arch.940250104

Saunders DS, Richard DS, Applebaum SW, Ma M, Gilbert LI (1990) Photoperiodic diapause in *Drosophila melanogaster* involves a block to the juvenile hormone regulation of ovarian maturation. Gen Comp Endocrinol 79:174–184. https://doi.org/10.1016/0016-6480(90)90102-R

Schwenke RA, Lazzaro BP (2017) Juvenile hormone suppresses resistance to infection in mated female *Drosophila melanogaster*. Curr Biol 27:596–601. https://doi.org/10.1016/j.cub.2017.01.004

Shapiro AB, Wheelock GD, Hagedorn HH, Baker FC, Tsai LW, Schooley DA (1986) Juvenile hormone and juvenile hormone esterase in adult females of the mosquito *Aedes aegypti*. J Insect Physiol 32:867–877. https://doi.org/10.1016/0022-1910(86)90102-2

Sheng Z, Xu J, Bai H, Zhu F, Palli SR (2011) Juvenile hormone regulates vitellogenin gene expression through insulin-like peptide signaling pathway in the red flour beetle, *Tribolium castaneum*. J Biol Chem 286:41924–41936. https://doi.org/10.1074/jbc.M111.269845

Shin SW, Zou Z, Saha TT, Raikhel AS (2012) bHLH-PAS heterodimer of methoprene-tolerant and cycle mediates circadian expression of juvenile hormone-induced mosquito genes. Proc Natl Acad Sci USA 109:16576–16581. https://doi.org/10.1073/pnas.1214209109

Shpigler H, Amsalem E, Huang ZY, Cohen M, Siegel AJ, Hefetz A et al (2014) Gonadotropic and physiological functions of juvenile hormone in Bumblebee (*Bombus terrestris*) workers. PLoS One 9:e100650. https://doi.org/10.1371/journal.pone.0100650

Smykal V, Bajgar A, Provaznik J, Fexova S, Buricova M, Takaki K et al (2014) Juvenile hormone signaling during reproduction and development of the linden bug, *Pyrrhocoris apterus*. Insect Biochem Mol Biol 45:69–76. https://doi.org/10.1016/j.ibmb.2013.12.003

Soller M, Bownes M, Kubli E (1999) Control of oocyte maturation in sexually mature *drosophila* females. Dev Biol 208:337–351. https://doi.org/10.1006/dbio.1999.9210

Soller M, Bownes M, Kubli E (1997) Mating and sex peptide stimulate the accumulation of yolk in oocytes of *Drosophila melanogaster*. Eur J Biochem 243:732–738. https://doi.org/10.1111/j.1432-1033.1997.00732.x

Song J, Wu Z, Wang Z, Deng S, Zhou S (2014) Krüppel-homolog 1 mediates juvenile hormone action to promote vitellogenesis and oocyte maturation in the migratory locust. Insect Biochem Mol Biol 52:94–101. https://doi.org/10.1016/j.ibmb.2014.07.001

Spradling A, Drummond-Barbosa D, Kai T (2001) Stem cells find their niche. Nature 414:98–104. https://doi.org/10.1038/35102160

Stay B, Tobe SS (1978) Control of juvenile hormone biosynthesis during the reproductive cycle of a viviparous cockroach. II. Effects of unilateral allatectomy, implantation of supernumerary corpora allata, and ovariectomy. Gen Comp Endocrinol 34:276–286. https://doi.org/10.1016/0016-6480(78)90249-6

Stay B, Tobe SS, Mundall EC, Rankin S (1983) Ovarian stimulation of juvenile hormone biosynthesis in the viviparous cockroach, *Diploptera punctata*. Gen Comp Endocrinol 52:341–349. https://doi.org/10.1016/0016-6480(83)90171-5

Sugime Y, Watanabe D, Yasuno Y, Shinada T, Miura T, Tanaka NK (2017) Upregulation of juvenile hormone titers in female *Drosophila melanogaster* through mating experiences and host food occupied by eggs and larvae. Zool Sci 34:52–57. https://doi.org/10.2108/ZS160150

Sutherland TD, Unnithan GC, Feyereisen R (2000) Terpenoid omega-hydroxylase (CYP4C7) messenger RNA levels in the corpora allata: a marker for ovarian control of juvenile hormone synthesis in *Diploptera punctata*. J Insect Physiol 46:1219–1227. https://doi.org/10.1016/s0022-1910(00)00042-1

Swevers L, Iatrou K (2003) The ecdysone regulatory cascade and ovarian development in lepidopteran insects: insights from the silkmoth paradigm. Insect Biochem Mol Biol 33:1285–1297. https://doi.org/10.1016/j.ibmb.2003.06.012

Thorne BL (1996) Termite terminology. Sociobiology 28:253–263

Tobe SS, Ruegg RP, Stay BA, Baker FC, Miller CA, Schooley DA (1985) Juvenile hormone titre and regulation in the cockroach *Diploptera punctata*. Experientia 41:1028–1034. https://doi.org/10.1007/bf01952127

Tsuchida K, Nagata M, Suzuki A (1987) Hormonal control of ovarian development in the silkworm, *Bombyx mori*. Arch Insect Biochem Physiol 5:167–177. https://doi.org/10.1002/arch.940050303

Tsuchiya M, Watanabe D, Maekawa K (2008) Effect on mandibular length of juvenile hormones and regulation of soldier differentiation in the termite *Reticulitermes speratus* (Isoptera: Rhinotermitidae). Appl Entomol Zool 43:307–314. https://doi.org/10.1303/aez.2008.307

Tufail M, Takeda M (2008) Molecular characteristics of insect vitellogenins. J Insect Physiol 54:1447–1458. https://doi.org/10.1016/j.jinsphys.2008.08.007

Van De Bor V, Zimniak G, Papone L, Cerezo D, Malbouyres M, Juan T et al (2015) Companion blood cells control ovarian stem cell niche microenvironment and homeostasis. Cell Rep 13:546–560. https://doi.org/10.1016/j.celrep.2015.09.008

Van Handel E, Lea AO (1984) Vitellogenin synthesis in blood-fed *Aedes aegypti* in the absence of the head thorax and ovaries. J Insect Physiol 30:871–875. https://doi.org/10.1016/0022-1910(84)90061-1

Wang J-L, Saha TT, Zhang Y, Zhang C, Raikhel AS (2017) Juvenile hormone and its receptor methoprene-tolerant promote ribosomal biogenesis and vitellogenesis in the *Aedes aegypti* mosquito. J Biol Chem 292:10306–10315. https://doi.org/10.1074/jbc.M116.761387

Weaver RJ, Edwards JP (1990) The role of the corpora allata and associated nerves in the regulation of ovarian cycles in the oviparous cockroach *Periplaneta americana*. J Insect Physiol 36:51–59. https://doi.org/10.1016/0022-1910(90)90150-e

Wen D, Rivera-Perez C, Abdou M, Jia Q, He Q, Liu X et al (2015) Methyl farnesoate plays a dual role in regulating *Drosophila* metamorphosis. PLoS Genet 11:e1005038. https://doi.org/10.1371/journal.pgen.1005038

Weng S-C, Shiao S-H (2015) Frizzled 2 is a key component in the regulation of TOR signaling-mediated egg production in the mosquito *Aedes aegypti*. Insect Biochem Mol Biol 61:17–24. https://doi.org/10.1016/j.ibmb.2015.03.010

Wigglesworth VB (1936) Memoirs: the function of the corpus allatum in the growth and reproduction of *Rhodnius prolixus* (Hemiptera). J Cell Sci s2-79:91–121. https://doi.org/10.1242/JCS.S2-79.313.91

Wilson TG (1996) Genetic evidence that mutants of the *Methoprene-tolerant* gene of *Drosophila melanogaster* are null mutants. Arch Insect Biochem Physiol 32:641–649. https://doi.org/10.1002/(SICI)1520-6327(1996)32:3/4<641::AID-ARCH35>3.0.CO;2-A

Wilson TG, Fabian J (1986) A *Drosophila melanogaster* mutant resistant to a chemical analog of juvenile hormone. Dev Biol 118:190–201. https://doi.org/10.1016/0012-1606(86)90087-4

Wu Z, Guo W, Xie Y, Zhou S (2016) Juvenile hormone activates the transcription of cell-division-cycle 6 (*Cdc6*) for polyploidy-dependent insect vitellogenesis and oogenesis. J Biol Chem 291:5418–5427. https://doi.org/10.1074/jbc.M115.698936

Wu Z, Guo W, Yang L, He Q, Zhou S (2018) Juvenile hormone promotes locust fat body cell polyploidization and vitellogenesis by activating the transcription of *Cdk6* and *E2f1*. Insect Biochem Mol Biol 102:1–10. https://doi.org/10.1016/j.ibmb.2018.09.002

Wu Z, He Q, Zeng B, Zhou H, Zhou S (2020a) Juvenile hormone acts through FoxO to promote *Cdc2* and *Orc5* transcription for polyploidy-dependent vitellogenesis. Development 147:dev.188813. https://doi.org/10.1242/dev.188813

Wu Z, Yang L, He Q, Zhou S (2020b) Regulatory mechanisms of vitellogenesis in insects. Front Cell Dev Biol 8:593613. https://doi.org/10.3389/fcell.2020.593613

Wyatt GR, Davey KG (1996) Cellular and molecular actions of juvenile hormone. II. Roles of juvenile hormone in adult insects. Adv Insect Physiol 26:1–155. https://doi.org/10.1016/s0065-2806(08)60030-2

Xie T (2008) Germline stem cell niches. In: Stembook [Internet]. Harvard Stem Cell Institute, Cambridge. https://doi.org/10.3824/stembook.1.23.1

Xie T, Spradling AC (2000) A niche maintaining germ line stem cells in the *Drosophila* ovary. Science 290:328–330. https://doi.org/10.1126/science.290.5490.328

Xie T, Spradling AC (1998) *Decapentaplegic* is essential for the maintenance and division of germline stem cells in the *Drosophila* ovary. Cell 94:251–260. https://doi.org/10.1016/S0092-8674(00)81424-5

Yamamoto R, Bai H, Dolezal AG, Amdam G, Tatar M (2013) Juvenile hormone regulation of *Drosophila* aging. BMC Biol 11:85. https://doi.org/10.1186/1741-7007-11-85

Yoshinari Y, Kurogi Y, Ameku T, Niwa R (2019) Endocrine regulation of female germline stem cells in the fruit fly *Drosophila melanogaster*. Curr Opin Insect Sci 31:14–19. https://doi.org/10.1016/j.cois.2018.07.001

Yue Y, Yang R-L, Wang W-P, Zhou Q-H, Chen E-H, Yuan G-R et al (2018) Involvement of *Met* and *Kr-h1* in JH-mediated reproduction of female *Bactrocera dorsalis* (Hendel). Front Physiol 9:482. https://doi.org/10.3389/fphys.2018.00482

Zang Y, Wan M, Liu M, Ke H, Ma S, Liu L-P et al (2015) Plasma membrane overgrowth causes fibrotic collagen accumulation and immune activation in *Drosophila* adipocytes. elife 4:e07187. https://doi.org/10.7554/eLife.07187

Zhang C, Kim AJ, Rivera-Perez C, Noriega FG, Kim Y-J (2022) The insect somatostatin pathway gates vitellogenesis progression during reproductive maturation and the post-mating response. Nat Commun 13:969. https://doi.org/10.1038/s41467-022-28592-2

Zhang Q, Sun W, Sun B-Y, Xiao Y, Zhang Z (2017) The dynamic landscape of gene regulation during *Bombyx mori* oogenesis. BMC Genomics 18:714. https://doi.org/10.1186/s12864-017-4123-6

Zhang X, Li S, Liu S (2021) Juvenile hormone studies in *Drosophila melanogaster*. Front Physiol 12:785320. https://doi.org/10.3389/fphys.2021.785320

Zhang Z, Xu J, Sheng Z, Sui Y, Palli SR (2011) Steroid receptor co-activator is required for juvenile hormone signal transduction through a bHLH-PAS transcription factor, methoprene tolerant. J Biol Chem 286:8437–8447. https://doi.org/10.1074/JBC.M110.191684

Zhao B, Hou Y, Wang J, Kokoza VA, Saha TT, Wang X-L et al (2016) Determination of juvenile hormone titers by means of LC-MS/MS/MS and a juvenile hormone-responsive Gal4/UAS system in *Aedes aegypti* mosquitoes. Insect Biochem Mol Biol 77:69–77. https://doi.org/10.1016/j.ibmb.2016.08.003

Zhu J, Noriega FG (2016) The role of juvenile hormone in mosquito development and reproduction. Adv Insect Physiol 51:93–113. https://doi.org/10.1016/bs.aiip.2016.04.005

Zimowska G, Shirk PD, Silhacek DL, Shaaya E (1994) Yolk sphere formation is initiated in oocytes before development of patency in follicles of the moth, *Plodia interpunctella*. Rouxs Arch Dev Biol 203:215–226. https://doi.org/10.1007/BF00636337

Chapter 12
Roles of Juvenile Hormone in Insect Diapause

Shin G. Goto and Sakiko Shiga

Abstract Juvenile hormone (JH) plays a crucial role in regulating insect diapause, a physiological state of developmental or reproductive arrest that allows insects to synchronize their life cycles with favorable environmental conditions. This chapter provides an overview of the current knowledge on the role of JH in insect diapause at various developmental stages. In adult diapause, a low JH titer in the hemolymph is the main cause of diapause induction and maintenance. JH controls adult diapause through canonical signaling pathways involving Methoprene-tolerant, Taiman, and Krüppel homolog 1 as well as noncanonical pathways that may be species-specific. In contrast, the roles of JH in embryonic, larval, and pupal diapause are more variable. Although JH may induce embryonic diapause in some species, it plays little or no role in other species. In larval and pupal diapauses, a shutdown of the brain–prothoracic gland axis and the resulting lack of ecdysteroids is a general feature; however, JH can also induce diapause in some species by maintaining high titers. This chapter further discusses the role of JH in the maternal effects on diapause induction in the next generation.

Keywords Adult diapause · Dormancy · Embryonic diapause · Larval diapause · Maternal effect · Photoperiodism · Pupal diapause

12.1 Introduction: Diapause

Most temperate insects enter diapause, a physiological state in which development or reproduction is suppressed or arrested, to synchronize their activities (development and reproduction) with favorable periods during the year (Denlinger 2022).

S. G. Goto (✉)
Graduate School of Science, Osaka Metropolitan University, Osaka, Japan
e-mail: shingoto@omu.ac.jp

S. Shiga
Graduate School of Science, The University of Osaka, Toyonaka, Japan

T. Shinoda et al. (eds.), *Juvenile Hormone*, Entomology Monographs,
https://doi.org/10.1007/978-981-95-8605-9_12

This is accompanied by a significant shutdown of endocrine signaling and metabolic activity. Many insects destined for diapause accumulate additional energy reserves before the onset of diapause, ensuring a high probability of survival during and after adverse seasons. Unlike simple quiescence, which is an immediate response to unfavorable environmental conditions at any stage, diapause is anticipatory and occurs much earlier than the onset of adverse seasons (Koštál 2006). Some insects enter diapause at a particular stage in each generation, regardless of environmental conditions; this is known as obligate diapause. In other insects, entry into diapause can be triggered by environmental factors such as day length (photoperiod), temperature, and food quality. This environmentally controlled diapause is referred to as facultative diapause (Goto 2022). Photoperiod is a cue that numerous organisms utilize, and the capacity of these organisms to react to photoperiod is known as photoperiodism. Most insects that enter diapause for overwintering (winter diapause) react to a shortening of the day length or short photoperiod. Summer diapause is also widespread, induced by a long photoperiod, and is the inverse of winter diapause in many ways.

Diapause occurs at a specific developmental stage (or sometimes at multiple stages), unlike quiescence (Koštál 2006; Izadi 2025). Embryonic diapause is common in Lepidoptera, Hemiptera, Orthoptera, and some Diptera species, such as mosquitoes and midges (Denlinger 2022). Although each species has a fixed embryonic stage at which its development is arrested, diapause is not restricted to a specific stage of embryonic development when we consider many insect species. Larval diapause is especially common in Lepidoptera, certain groups of Diptera, Hymenoptera, Coleoptera, Neuroptera, Odonata, Orthoptera, Hemiptera, and Plecoptera (Denlinger 2022). It frequently occurs during the final instar stage but is also known to occur during early instars. Pupal diapause is particularly common in Lepidoptera and Diptera and frequently occurs in the early pupal stage (Denlinger 2022). In some cases, it appears in pharate adults, in which adult development has been completed, but the insect remains within the pupal cuticle. Adult diapause is widespread among Coleoptera, Hemiptera, Hymenoptera, Orthoptera, Neuroptera, Diptera, and Lepidoptera (Denlinger 2022). Embryonic, larval, and pupal diapauses are characterized by developmental arrest, whereas adult diapause is characterized by reproductive arrest (also called reproductive diapause) and the suppression of reproductive organ development and mating behavior. Although juvenile hormone (JH) may regulate diapause at any stage, its significance has been repeatedly emphasized in the context of adult diapause.

Insect endocrinology is intricately associated with diapause. Early researchers have leveraged diapause as an effective tool to identify the endocrine glands and hormones that are suppressed during developmental suspension. Their keen observations have been instrumental in understanding diapause and identifying the hormones that orchestrate the intricate processes of insect development and reproduction (Denlinger 2022). This overview presents the current knowledge on the role of JH in insect diapause. It should be noted that this chapter is interconnected with the other chapters in this volume. More comprehensive information can be found in the literature.

12.2 Roles of JH in Adult Diapause

As JH acts as the principal gonadotropic hormone to initiate vitellogenesis in many insects, the suppression of JH titers in the hemolymph is the main cause of the induction and maintenance of adult diapause.

Pioneering studies arguing for a role of JH in adult diapause have been conducted in the Colorado potato beetle, *Leptinotarsa decemlineata,* in which JH titers measured by the *Galleria* assay are low in diapausing beetles under short days, application of JH analogs (JHAs) terminates diapause, and removal of the corpus allatum (CA), allatectomy, from reproductive insects causes diapause-like characteristics (Table 12.1). The CA is the primary organ for JH production in insects. The JH production rates by the CA were measured, showing that they were high under reproductive conditions but suppressed under diapause conditions. The importance of the CA in the control of adult diapause has been demonstrated in many species, including the Asian comma butterfly, *Polygonia c-aureum* (Endo 1972, 1973), and the brown-winged green bug, *Plautia stali* (Kotaki and Yagi 1989). Following earlier studies, the importance of JH in controlling adult diapause has been verified in many species and has become certain (Table 12.1).

Following the use of gas chromatography/mass spectrometry (GC/MS) in *L. decemlineata* and the oriental migratory locust, *Locusta migratoria* (de Kort et al. 1982; Okuda et al. 1996), liquid chromatography/tandem mass spectrometry (LC/MS/MS) has been used to precisely measure JH titers in the hemolymph, confirming that JH titers decrease during adult diapause. This was supported by the decreased expression of JH-responsive genes. In the Asian lady beetle, *Harmonia axyridis*, JH titers and mRNA levels of *Krüppel homolog 1* (*Kr-h1*), a JH-responsive transcription factor, are significantly lower under diapause-inducing short-day conditions than under diapause-avoiding long-day conditions (Gao et al. 2022a) (Fig. 12.1). The application of an exogenous JHA to diapause females with immature ovaries causes the upregulation of *vitellogenin* and *Kr-h1* mRNA levels with ovarian development, and RNAi of JH receptor *Methoprene-tolerant* (*Met*) disrupts these JHA effects (Gao et al. 2022a). Males of *H. axyridis* also enter diapause under short days with slow development of the internal reproductive system, and this photoperiodic control is mediated by a canonical Met-dependent JH signaling pathway (Gao et al. 2022b).

Among processes controlling JH titer, such as the synthesis, degradation, sequestration, and secretion of JH, JH synthesis seems to be critical. In the cabbage beetle, *Colaphellus bowringi*, which enters adult diapause induced by larval long days, 18 of 22 genes encoding enzymes in the mevalonate pathway and in the JH branch of JH biosynthesis exhibited a significant decrease in expression in diapausing females compared to reproductive females. RNAi of most genes in reproductive females reduced the expression of *Kr-h1*, suggesting a decrease in JH titer (Tian et al. 2021). Knockdown of JH biosynthetic enzymes, *3-hydroxy-3-methylglutaryl-CoA reductase 2* (*HMGR2*), *farnesyl-pyrophosphate synthase 1* (*Fpps1*), and *juvenile hormone acid methyltransferase 1* (*jhamt1*), induced diapause-associated immature

Table 12.1 List of major papers examining the roles of juvenile hormones in reproductive diapause

Species	Comparison between diapause and nondiapause insects			Effects on diapause or nondiapause phenotypes			References
	JH titer in hemolymph	JH production rates in CA	JH-related gene expression	CA removal, transplantation	JH or JH analog application	Knockdown of JH-related genes	
Coleoptera							
Leptinotarsa decemlineata	√	√			√		deWilde et al. (1968), Schooneveld et al. (1977), deKort et al. (1982, 1987), Lefevere and de Wilde (1984)
Colaphellus bowringi			√		√	√	Liu et al. (2016), Zhu et al. (2017), Tian et al. (2021), Wang et al. (2025)
Galeruca daurica			√		√		Ma et al. (2021a, b)
Gastrophysa atrocyanea					√		Ojima et al. (2015)
Harmonia axyridis	√		√		√	√	Gao et al. (2022a, b)
Coccinella septempunctata			√			√	Li et al. (2022)
Conotrachelus nenuphar					√		Hoffman et al. (2007)
Diptera							
Drosophila suzukii	√		√				Zhai et al. (2019)
Drosophila melanogaster		√					Saunders et al. (1990)
Protophormia terraenovae		√		√	√		Matsuo et al. (1997), Shiga et al. (2003)

Culex pipiens		√		√	√		Spielman (1974), Readio et al. (1999), Kim and Denlinger (2010)
Hymenoptera							
Osmia rufa					√		Wasielewski et al. (2011)
Osmia bicornis					√		Giejdasz et al. (2023)
Bombus terrestris	√	√	√				Larrere et al. (1993), Chen et al. (2021a, b), Liu et al. (2024)
Lepidoptera							
Polygonia c-aureum				√	√		Endo (1972, 1973), Hiroyoshi et al. (2017)
Speyeria idalia	√						Kopper et al. (2001)
Caloptilia fraxinella					√		Evenden et al. (2007)
Hemiptera							
Plautia stali		√		√			Kotaki and Yagi (1989), Matsumoto et al. (2013)
Aspongopus chinensis	√		√		√		Zhou et al. (2022)
Riptortus pedestris	√		√		√	√	Morita and Numata (1997), Morita et al. (1999) Ando et al. (2020), Mano and Goto (2022), Hafeez et al. (2025)

(continued)

Table 12.1 (continued)

Species	Comparison between diapause and nondiapause insects			Effects on diapause or nondiapause phenotypes			References
	JH titer in hemolymph	JH production rates in CA	JH-related gene expression	CA removal, transplantation	JH or JH analog application	Knockdown of JH-related genes	
Pyrrhocoris apterus	√		√	√	√	√	Hodkova (2008), Smykal et al. (2014), Hejníková et al. (2022)
Orthoptera							
Nomadacris succincta		√					Okuda et al. (1996)
Locusta migratoria	√	√			√		Okuda et al. (1996)
Anacridium aegyptium				√			Greenfield and Pener (1992)

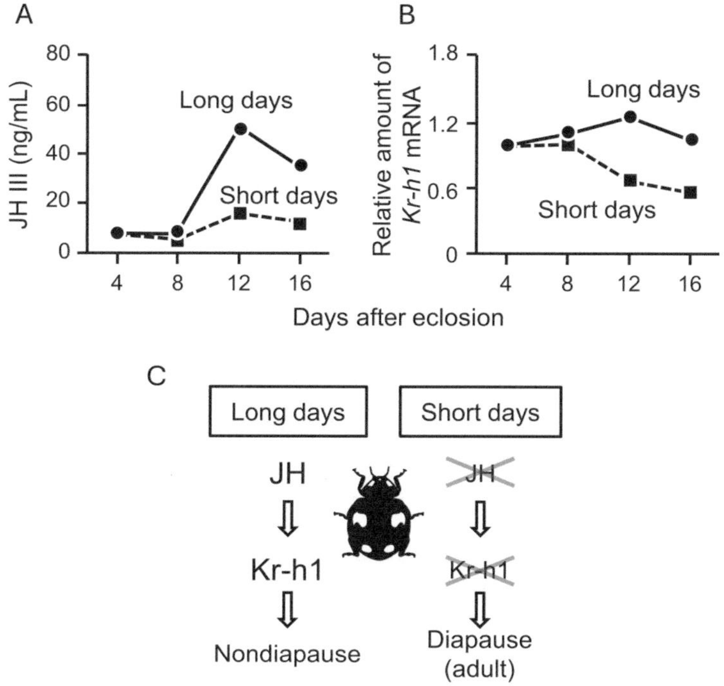

Fig. 12.1 Hemolymph juvenile hormone (JH) III (**A**) and *Krüppel homolog 1* (*Kr-h1*) expression (**B**) under diapause-averting long-day and diapause-inducing short-day conditions and the schematic drawing of the mechanisms underlying adult diapause (**C**) in the Asian lady beetle, *Harmonia axyridis*. (Adapted from Gao et al. (2022a), with permission from John Wiley and Sons)

ovaries, large lipid accumulations, and burrowing behavior (Tian et al. 2021). Transcriptome analysis in *H. axyridis* showed that the expression levels of genes involved in JH biosynthesis were decreased, and that most genes involved in JH degradation were significantly increased, in diapausing females compared to reproductive females (Gao et al. 2022a). RNAi of *Fpps1*, *Fpps2*, *farnesol dehydrogenase 1* (*Fold1*), and *jhamt1* decreased *Kr-h1* mRNA levels and caused diapause-like phenotypes, including inhibition of ovarian development and increased nutrient storage. However, the knockdown of *juvenile hormone diol kinase* (*Jhdk*) and *juvenile hormone epoxide hydrolase* (*Jheh*), which are involved in JH inactivation, caused milder effects in females during diapause. It significantly increased the expression level of *Kr-h1* but did not affect ovarian development (Gao et al. 2022a). Similar results have been reported for *C. bowringi* (Guo et al. 2019). These findings suggest that JH biosynthesis, rather than the JH degradation pathway, determines the reduction in JH titer in diapausing females of *C. bowringi* and *H. axyridis* (Guo et al. 2019; Tian et al. 2021; Gao et al. 2022a). In contrast, in the seven-spotted ladybeetle, *Coccinella septempunctata*, knockdown of JHEH and JH esterase (JHE) reduced diapause-related characteristics and promoted reproduction with ovarian growth, with increased *Kr-h1* and *vitellogenin* expression (Li et al. 2022). The importance of

JH degradation during diapause may be species-specific. Transcriptome analysis has revealed that *jhamt* is downregulated during adult diapause in the buff-tailed bumblebee, *Bombus terrestris*, but its functional significance in adult diapause remains unclear (Zhai et al. 2019; Chen et al. 2021b). Although skepticism about the existence of true diapause has been expressed, downregulation of *jhamt* has also been observed in the spotted wing drosophila, *Drosophila suzukii*, during winter reproductive dormancy (Toxopeus et al. 2016; Zhai et al. 2019; Colinet and Kustre 2025).

The role of JH in adult diapause has also been extensively studied in hemipteran insects. After the identification of juvenile hormone III skipped bisepoxide ($JHSB_3$) as an inherent JH in *P. stali* (Kotaki et al. 2011), JH biosynthetic rates and JH titers have been investigated, and the importance of JH signaling in adult diapause has been verified in several species, including the bean bug, *Riptortus pedestris*; the linden bug, *Pyrrhocoris apterus*; and *P. stali* (Table 12.1). In diapausing females, JH biosynthesis or titers are suppressed (Matsumoto et al. 2013; Smykal et al. 2014; Mano and Goto 2022). In *P. apterus,* RNAi of *Met* or its partner *Taiman* (*Tai*) blocked ovarian development and suppressed *vitellogenin* expression in the fat body under reproductive conditions. Promotion of vitellogenesis by JHA treatment in diapausing females also requires both *Met* and *Tai* in the fat body, suggesting that the Met–Tai complex likely functions in vitellogenesis under diapause-averting conditions (Smykal et al. 2014). In *R. pedestris*, the importance of JH biosynthetic enzymes, cytochrome P450 15 (CYP15) and JHAMT, has been shown in controlling both the arrest of ovarian development and lipid accumulation in adult diapause phenotypes (Mano and Goto 2022; Hafeez et al. 2025). However, the roles of Met and Kr-h1 are different between ovary and lipid accumulation in this species. Silencing either *Met* or *Kr-h1* expression under long-day reproductive conditions solely affected ovarian development, with no change in lipid accumulation. In addition, the knockdown did not alter the effects of JHA on lipid reduction under short-day diapause conditions. Collectively, these results suggest that JH controls ovarian development through the canonical Met–Kr-h1 pathway while modulating lipid accumulation through an alternative noncanonical pathway during adult diapause in female *R. pedestris* (Hafeez et al. 2025).

A noncanonical pathway for JH signaling has also been suggested in the adult diapause regulation of *P. apterus* males (Hejníková et al. 2022). They enter adult diapause, in which the male accessory gland (MAG) growth slows down, and mating behavior occurs less frequently under short-day conditions. JH titers are low under short days but are upregulated under long days (Hejníková et al. 2022). Ectopic application of JHA effectively terminated male diapause through Met and Tai. However, diapause termination by exposing males to long days occurs even under *Met* or *Tai* silencing, suggesting a different JH signaling pathway for the photoperiodic termination of male diapause. Furthermore, JH hemolymph titers under long days after allatectomy did not differ from those of intact or sham-operated males, suggesting that a source of JH might exist outside the CA in *P. apterus* males (Hejníková et al. 2022). $JHSB_3$ and *jhamt* mRNA were detected in the isolated MAG under long days but were almost undetectable under short days (Hejníková et al. 2022). *Met* or *Tai* silencing in diapause males prior to their transfer to a long photoperiod did not prevent mating, as suggested in other studies on *P. apterus* males (Urbanová et al. 2016; Hejnikova et al. 2016). The synthesis of JH by the MAG has been reported in the cecropia moth, *Hyalophora cecropia,* and the yellow fever mosquito, *Aedes aegypti* (Shirk et al. 1976; Borovsky et al. 1994), both of which do not undergo adult diapause. More species may employ JH signaling originating from the MAG, possibly employing *Met*- or *Tai*-independent (noncanonical) pathways for diapause control. The sexual dimorphism of diapause regulation related to JH signaling might be explained by the different reproductive costs for each sex (Hejníková et al. 2022).

JH signaling appears to be the primary factor controlling adult diapause phenotypes, including different target organs and behaviors, in both males and females, with some exceptions in *R. pedestris* (Morita et al. 1999). However, the downstream signaling of JH may vary. One is canonical signaling mediated by Met/Tai and Kr-h1, and another pathway controlling fat body development or mating behavior is found in hemipteran species (Fig. 12.2). As the canonical pathway is operative in JH control of the fat body in *H. axyridis* (Gao et al. 2022b), this might be unique to some orders of insects. An interesting question for the future is how the noncanonical JH signal controls adult diapause.

12.3 Roles of JH in Diapause at Stages Other than Adult

In contrast to adult diapause, JH may play little or no role in embryonic, larval, nymphal, and pupal diapauses in many species. However, only a few species use JH as a critical endocrinological regulator. In this section, we review the general endocrine mechanisms regulating these diapauses and then focus on the role of JH in some cases.

12.3.1 Embryonic Diapause

The hormonal basis of embryonic diapause remains largely unexplored, likely because of the challenges posed by the small size of embryos, which hinders gland removal, transplantation, and hormone administration. Embryonic diapause may begin as early as the initial stages of blastoderm formation and can extend until the completion of pharate larval development (Masaki 1956; Tanigawa et al. 2009). Numerous checkpoints must exist during embryonic development, a feature underscored by the diverse stages of embryogenesis in which diapause manifests. This implies that diverse mechanisms arrest embryonic development at various stages (Denlinger 2022).

Sorbitol, ecdysteroids, and JH are critical regulators of embryonic diapause. The silkworm, *Bombyx mori,* enters diapause at an early embryonic stage, immediately

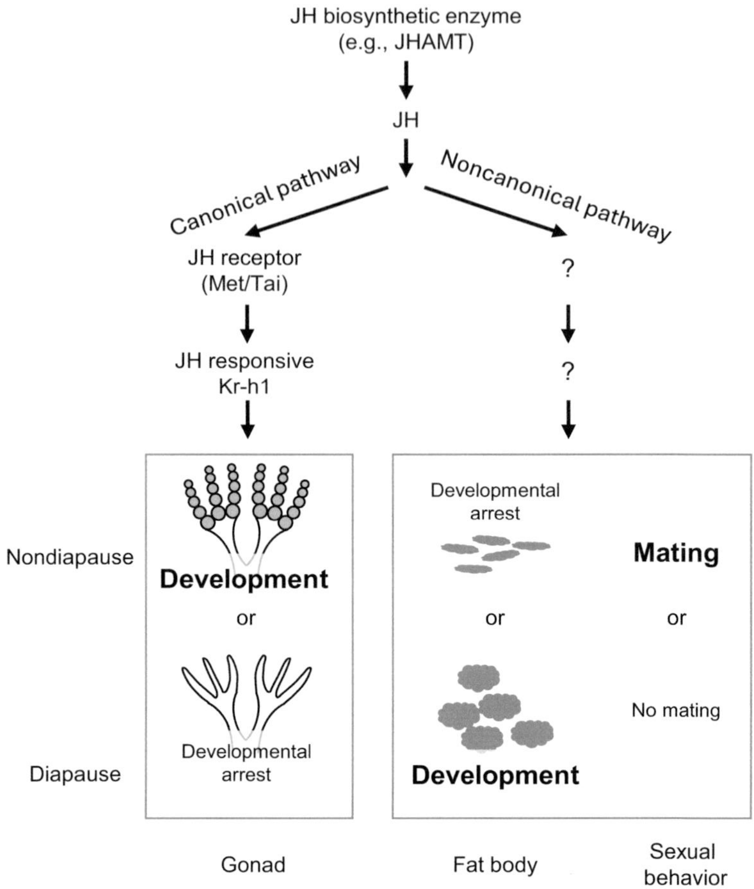

Fig. 12.2 Canonical and noncanonical juvenile hormone (JH) signaling pathways in the control of adult diapause in some hemipteran insects. JHAMT Juvenile hormone acid *O*-methyltransferase, Met Methoprene-tolerant, Tai Taiman

after the formation of the cephalic lobe and mesoderm segmentation. Diapausing eggs contain a high concentration of sorbitol (0.2 M), which strongly inhibits embryonic development. Diapausing eggs cultured in a sorbitol-free medium can complete postembryonic development normally, whereas adding sorbitol to the medium surrounding embryos causes a developmental halt (Horie et al. 2000; Iwata et al. 2005). These results highlight sorbitol as a strong arresting factor in embryonic development, but the story is not so simple. Diapause can be induced in artificially glycogen-reduced (less than 3% of the control) eggs using the trehalase inhibitor trehazolin (Katagiri et al. 1998). In these eggs, glycogen-sorbitol conversion does not occur. In addition, sorbitol does not induce prediapause embryos to initiate diapause (Iwata et al. 2005). These results indicate that two distinct processes underlie the regulation of embryonic diapause in *B. mori*. Sorbitol can stabilize embryonic diapause but is not involved in its induction (Katagiri et al. 1998; Iwata et al. 2005).

A low ecdysteroid titer in embryos elicits diapause in the Australian plague locust, *Chortoicetes terminifera,* and *L. migratoria* (Gregg et al. 1987; Tawfik et al. 2002). In *L. migratoria*, a drastic increase in ecdysteroids is detected at diapause termination, and exogenous ecdysteroids can also terminate diapause (Tawfik et al. 2002; Kidokoro et al. 2006). In contrast to these examples, several lines of evidence indicate that elevated ecdysteroid levels induce obligate diapause at the pharate first-instar larval stage in the spongy moth, *Lymantria dispar* (Lee and Denlinger 1997; Lee et al. 1997). Ecdysteroids may also direct entry into embryonic diapause in the southern ground cricket, *Allonemobius socius*, as suggested by the prominent mRNA signatures of ecdysteroidogenesis noted during prediapause (Reynolds and Hand 2009).

A low titer of JH appears to induce embryonic diapause in the Asian tiger mosquito, *Aedes albopictus*; the titer of 20-hydroxyecdysone (20E) did not differ between diapausing and nondiapausing eggs, whereas that of JH III was lower in diapausing eggs (Batz et al. 2019) (Fig. 12.3). JH titers were corroborated by transcriptional evidence. In addition, the application of JHA to diapausing eggs terminates diapause in a dose-dependent manner (Suman et al. 2015). Although transcriptional differences in JH-related genes have also been observed between diapause and nondiapause eggs in *B. mori* and *L. migratoria*, their causal involvement in diapause induction has not been investigated (Sasibhushan et al. 2013; Tu et al. 2015; Cui et al. 2019). Notably, applying JHA to diapausing eggs terminates diapause but shows severe teratological effects, i.e., developmental malformations, in *L. migratoria* and the big-headed grasshopper, *Aulocara elliotti* (Neumann-Visscher 1976; Kidokoro et al. 2006). These results suggest that the diapause-terminating effects of JH application in these species may be pharmacological artifacts.

In contrast, sorbitol, ecdysteroids, and JH do not appear to be involved in embryonic diapause in the band-legged ground cricket, *Dianemobius nigrofasciatus*. Sorbitol levels gradually increased in nondiapausing eggs toward egg hatching, but no difference was detected between diapausing and nondiapausing eggs at the stage where diapause occurs (Goto et al. 2008). No differences were observed in the amount of 20E between these eggs, and JH was not detected in them (Shimizu and Goto 2024). The factors responsible for arresting embryonic development have not been identified in this species.

12.3.2 Larval and Pupal Diapauses

Developmental arrest at the larval (nymphal) or pupal stage implies that the hormonal signal for advancing to the next stage has not yet been released. This suggests the failure of the prothoracic gland (PG) in producing the ecdysteroids required to initiate the next molt. Indeed, a shutdown of the brain–PG axis is a general and

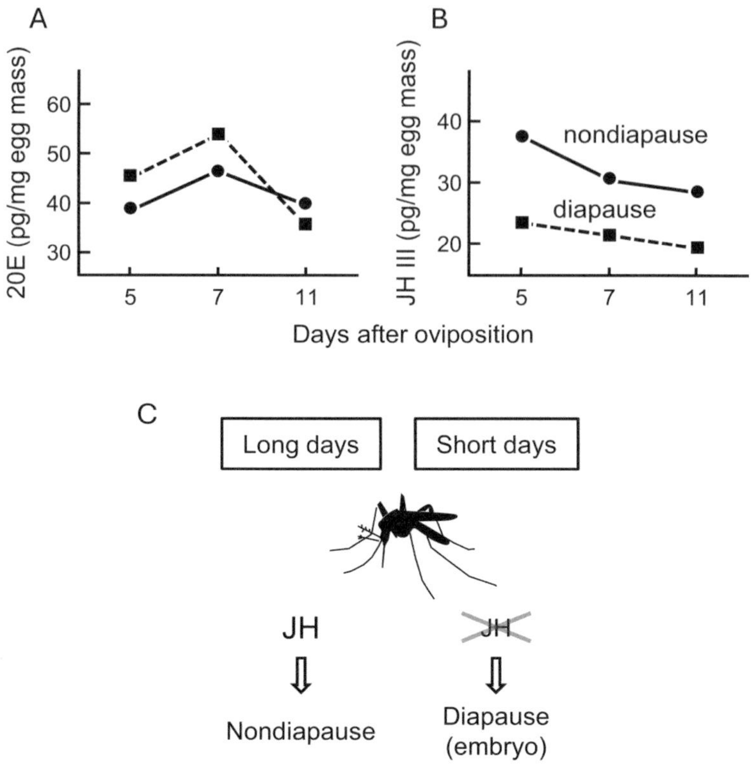

Fig. 12.3 Abundance of 20-hydroxyecdysone (20E) (**A**) and juvenile hormone (JH) III (**B**) in diapause and nondiapause eggs and the schematic drawing of the role of JH in embryonic diapause (**C**) in the Asian tiger mosquito, *Aedes albopictus*. (Adapted from Batz et al. (2019). Copyright 2019 Batz, Brent, Marias, Sugijanto, and Armbruster. Licensed under CC BY 4.0)

conspicuous feature of larval and pupal diapauses (Denlinger 2022). However, in some species, JH plays an essential role.

Various insect species have reported a shutdown of the brain–PG axis during larval and pupal diapauses. A pair of neurosecretory cells in the pars lateralis of each brain hemisphere produces prothoracicotropic hormone (PTTH), stimulating ecdysone secretion from the PG (Smith and Rybczynski 2012). Intracellular electrical recordings from PTTH-expressing cells in the tobacco hornworm *Manduca sexta* have shown that these cells are silent during pupal diapause (Tomioka et al. 1995). PTTH levels are low during diapause in some species (Yin and Chippendale 1979; Gelman et al. 1992; Yoshinaga and Goto 2025), or PTTH is present even during diapause but not released in other species (Bowen et al. 1984; Richard and Saunders 1987; Mizoguchi et al. 2013). In the cabbage armyworm, *Mamestra brassicae*, PTTH is abundantly released into the hemolymph of nondiapause individuals, followed by a significant surge in ecdysteroids, whereas very little PTTH or ecdysteroids are detected in the hemolymph of individuals destined for pupal diapause (Mizoguchi et al. 2013; Ogihara et al. 2017). Low PTTH and ecdysteroid titers during diapause have been observed in various insect species (Hao et al. 2013; Wang et al. 2013; Ahmadi et al. 2021; Süess et al. 2022; Yoshinaga and Goto 2025).

The lack of ecdysteroids is a hallmark of larval diapause in the bamboo borer, *Omphisa fuscidentalis* (Singtripop et al. 1999). Interestingly, JHA application is highly effective for diapause termination in this species, in contrast to 20E injections (Singtripop et al. 2000). Ecdysteroid titers increased after JHA application, suggesting that JH may be involved in activating the PG at diapause termination. When JHA was applied to late-instar larvae of *M. brassicae* destined for pupal diapause, some pupae failed to enter diapause (Hiruma 1979). A series of experiments revealed that an increase in PTTH activity in the brain is elicited by ecdysone from the PG activated by JH. However, it remains unknown whether JH plays a role in averting or breaking diapause in nature.

The profound significance of CA in larval diapause was first reported in the rice stem borer, *Chilo suppressalis* (Fukaya and Mitsuhashi 1957). The implantation of the PG did not terminate diapause but caused excessive molting into the larval–pupal intermorph. Conversely, implanting the PG into the isolated abdomen of diapause larvae resulted in normal pupation. Ligation between the head and thorax of the diapause larvae, with the CA and PG located in the anterior and posterior body compartments, respectively, induced pupation only in the posterior parts of the larvae. Additionally, implanting the PG into allatectomized diapause larvae successfully triggered pupation (Fukaya and Mitsuhashi 1958). The histological appearance and volume changes suggest high CA activity during diapause (Mitsuhashi and Fukaya 1960). Later, it was confirmed that JH secreted from the CA is the key factor in diapause induction and maintenance (Yagi and Fukaya 1974). JH administration induced larval diapause even under long-day photoperiods, which typically prevent diapause. The JH titer is high in diapause larvae but decreases shortly after the final larval molt in the nondiapause counterparts. These results indicate that high JH titers trigger larval diapause in *C. suppressalis*. The same applies to the yellow-spotted longicorn beetle, *Psacothea hilaris* (Munyiri and Ishikawa 2004). JH titers of the larvae destined for pupation decreased continuously, whereas those of the prediapause larvae increased gradually (Fig. 12.4). The significance of JH in larval diapause applies to the southwestern corn borer, *Diatraea grandiosella* (Yin and Chippendale 1973, 1979); the Mediterranean corn borer, *Sesamia nonagrioides* (Eizaguirre et al. 1998, 2005); the beet webworm, *Loxostege sticticalis* (Jiang et al. 2011); the oriental moth, *Monema flavescens* (Takeda 1978); and the brown planthopper, *Laodelphax striatellus* (Yin et al. 2018).

The cricket *Svercacheta siamensis* (previously known as *Modicogryllus siamensis*) appears twice a year and shows clear photoperiodic responses during nymphal development (Taniguchi and Tomioka 2013). At 25 °C, most nymphs become adults at approximately 60 d after hatching from eggs, undergoing seven or eight molts under long-day conditions. In contrast, under short-day conditions, the nymphal period elongates with an increased number of molts for overwintering. Thus, the nymphal diapause of this species is not merely a developmental arrest but involves the suppression of adult emergence, along with an extension of the nymphal period

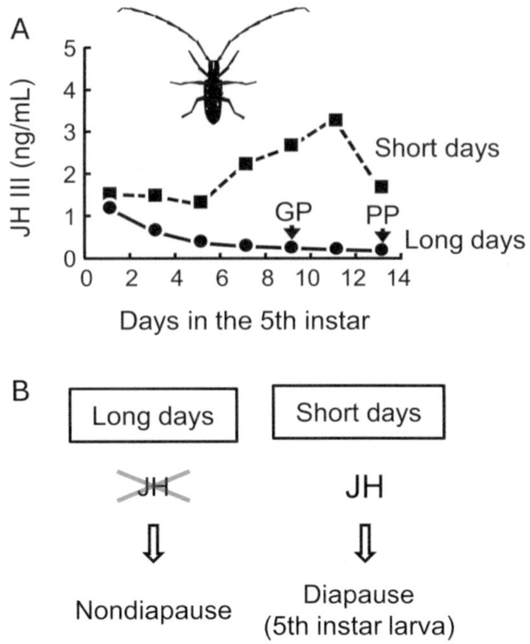

Fig. 12.4 Hemolymph juvenile hormone (JH) III titer during the fifth-instar larval development under diapause-averting long-day and diapause-inducing short-day conditions (**A**) and the schematic drawing of the role of JH in larval diapause (**B**) in the yellow-spotted longicorn beetle, *Psacothea hilaris*. GP gut purge and PP prepupa. (Adapted from Munyiri and Ishikawa (2004), with permission from Elsevier Ltd.)

and an increase in the number of molts. It is well-known that the transforming growth factor-β (TGF-β) ligand Myoglianin (Myo) negatively regulates *jhamt* expression in various insect species (Kamsoi and Belles 2019; Chen et al. 2022; Kawamoto et al. 2025). In *S. siamensis*, *myo* expression increases under long-day conditions, leading to the suppression of JH synthesis (Miki et al. 2020). When *myo* expression is suppressed, *jhamt* is upregulated, and molting is increased to prolong the nymphal period, even under long-day conditions. These results suggest that photoperiod influences JH biosynthesis, thereby controlling the nymphal diapause.

The metabolic rate is periodically elevated with a four-to-six-day periodicity in *Sarcophaga* flesh flies during pupal diapause (Denlinger et al. 1972). The regular cyclic patterns of substantial metabolic alternations resemble periodic arousal in hibernating mammals. JH drives this cycle in the flesh fly. The hemolymph titer progressively increases during the cycle trough, reaches a critical point to trigger a bout of high oxygen consumption, and then drops precipitously (Denlinger et al. 1984). The rapid decline in JH titer is aided by an increase in JH esterase (Denlinger and Tanaka 1989). Flesh fly pupae in diapause depend on anaerobic glycolysis during metabolic depression and switch to aerobic respiration during periodic oxygen consumption. Reactive oxygen species (ROS) levels regulate the timing of oxygen consumption by influencing the activity of key metabolic enzymes. These findings suggest that ROS signaling plays a central role in controlling metabolic cycles during insect diapause, similar to the mechanisms of mammalian hibernation (Chen et al. 2021a). However, it remains unclear whether JH is regulated by ROS or whether JH signaling controls ROS and downstream metabolism.

12.4 The Maternal Effect

In species that enter photoperiodic facultative diapause, the photoperiod-sensitive stage, in which the species receives photoperiodic cues, occurs before the diapause stage. The most extreme example of advanced photoperiodic perception is when this sensitive stage occurs in one generation and developmental arrest occurs in the next generation. This is referred to as the maternal effect (Mousseau and Dingle 1991). This can be found in Hymenoptera, Diptera, Orthoptera, Lepidoptera, Hemiptera, and Psocodea, particularly in many parasitic wasps, higher Diptera, and some species of Lepidoptera. The complex processes behind the maternal effect are attractive, and many scientists have sought to elucidate them. Although these processes remain largely unknown, they may differ between species.

In *B. mori*, females exposed to short days and cool temperatures during embryonic development or those exposed to long days and warm temperatures during the larval stage become nondiapause producers, laying eggs that do not enter diapause. Conversely, females that experience long days and high temperatures as embryos or short days and low temperatures as larvae become diapause producers, laying eggs destined for embryonic diapause (Shimizu 2024). Diapause producers release diapause hormones (DHs) from the subesophageal ganglion during the pupal stage. DH, a 24-amino acid peptide belonging to the FXPRLamide family, targets the ovaries, alters carbohydrate metabolism, and elevates sorbitol levels in the egg yolk (Tsuchiya et al. 2021). The involvement of DH in the maternal induction of diapause has been demonstrated in the tussock moth, *Orgyia thyellina* (Uehara et al. 2011).

Because of the apparent effect of DH on maternal diapause induction in *B. mori*, the involvement of other endocrine factors has been neglected. However, a recent study on *B. mori* revealed that the brains (with no corpora cardiaca–CA complex) of fifth-instar female larvae destined to become diapause producers accumulated higher amounts of *jhamt* mRNA than those of their counterparts. Although the role of JHAMT in the brain, but not in the CA, is unknown, silencing maternal *jhamt* mRNA significantly decreased the incidence of diapause in the next generation (Egi and Sakamoto 2022). Interestingly, older literature has also reported that JH injection into larvae destined to be nondiapause producers alters their developmental trajectory into diapause producers (Morohoshi et al. 1972). Another recent study in *B. mori* revealed that the mRNA encoding mevalonate kinase (MK), a key enzyme in the JH biosynthetic pathway, is highly methylated in the ovaries of pupae destined to be nondiapause producers compared to their counterparts. RNAi of *MK* in pupae changes the developmental trajectory of nondiapause producers to diapause producers (Chen et al. 2023). Although the authors emphasized the significance of the JH signaling pathway in maternal induction of diapause, it is largely unknown whether MK in the pupal ovary affects the production of JH. A previous study reported that JH injection into pupae did not affect the maternal status (Morohoshi et al. 1972). The relationship between JH and DH in the maternal induction of diapause in *B. mori* is yet to be elucidated.

A recent study on the jewel wasp, *Nasonia vitripennis,* clearly revealed that JH is the causal factor regulating the maternal effect (Mukai et al. 2022). Adult females of this species under long days become nondiapause producers, laying eggs that develop into pupae with no interruption. In contrast, adult females under short days become diapause producers, laying eggs destined to enter larval diapause (Saunders 1966). Diapause producers typically downregulate *jhamt* expression, thereby reducing hemolymph JH concentrations. Silencing *jhamt* transforms nondiapause producers into diapause producers, and the topical application of JH temporarily transforms diapause producers into nondiapause producers (Mukai et al. 2022). These results indicate that *N. vitripennis* females photoperiodically regulate the developmental trajectory of their offspring through JH (Fig. 12.5). In this case, JH in the mother's hemolymph did not regulate ovarian development and fecundity (Mukai et al. 2022). Although DNA methylation has been verified as the key to the maternal effect in this species (Pegoraro et al. 2016), the mechanism by which JH affects DNA methylation to regulate the maternal effect remains unknown.

In contrast to *N. vitripennis*, the involvement of JH in the maternal effect has been denied in several insect species. In adult females of *D. nigrofasciatus*, hemolymph JH concentrations were higher under long-day conditions than under short-day conditions; however, the topical application of JH did not affect the maternal status (diapause or nondiapause producers) (Shimizu and Goto 2024). The flesh fly, *Sarcophaga bullata,* exhibits a unique maternal effect. This species shuts down the capacity for diapause in the following generation if the mother receives short-day photoperiodic cues (Henrich and Denlinger 1982). None of the common insect hormonal agents, including JHA and ecdysone, can mimic maternal effects (Rockey et al. 1989).

12.5 Future Perspectives

Despite significant advancements in our understanding of the role of JH in insect diapause, numerous questions remain. In adult diapause, the canonical JH signaling pathway, which includes Met, Tai, and Kr-h1, has been extensively studied in various species. Nonetheless, new findings suggest the presence of noncanonical pathways that influence specific diapause characteristics, such as lipid storage and mating behavior, especially in hemipteran insects. Investigating the molecular elements and regulatory processes of these alternative pathways presents a promising avenue for future studies. Moreover, the role of JH in diapause during non-adult stages, such as embryonic, larval, and pupal stages, remains poorly understood and seems to be highly species-specific. Additional comparative studies across different taxa are necessary to determine whether JH acts as a universal or lineage-specific regulator during these developmental stages. The maternal effect is another fascinating area where JH has been linked to the transgenerational regulation of diapause in some species but not in others. Exploring how JH interacts with other hormonal and epigenetic factors in this context could provide new insights into the evolution

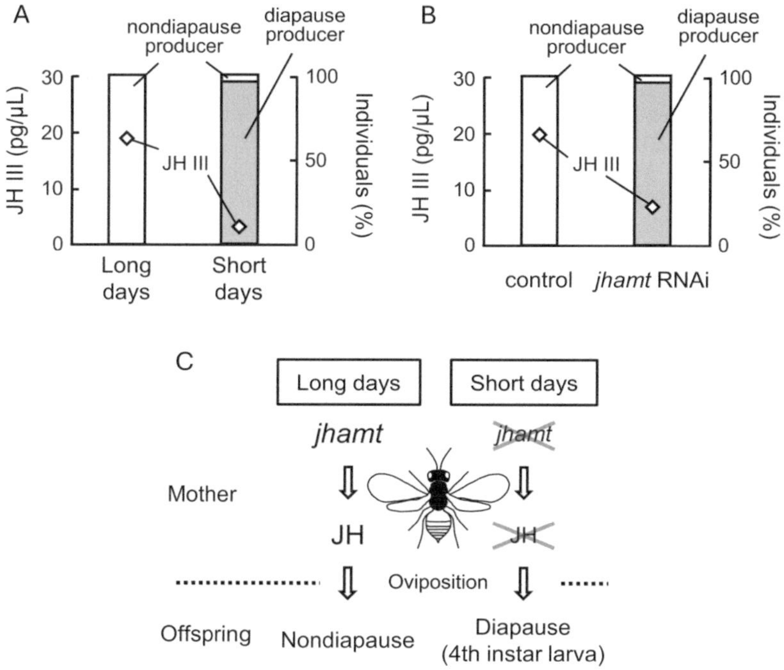

Fig. 12.5 Juvenile hormone (JH) III titer and maternal status (diapause producer or nondiapause producer) under long-day and short-day conditions (**A**) and after silencing of *jhamt* under long-day conditions (**B**) and the schematic drawing of the mechanisms underlying the maternal induction of larval diapause (**C**) in the jewel wasp, *Nasonia vitripennis*. (Adapted from Mukai et al. (2022), with permission from Elsevier Ltd.)

of diapause strategies. To fully understand the complex roles of JH in diapause regulation, it is crucial to integrate molecular, physiological, and phylogenetic approaches.

References

Ahmadi F, Mikani A, Moharramipour S (2021) Induction of diapause by clock proteins period and timeless via changes in PTTH and ecdysteroid titer in the sugar beet moth, *Scrobipalpa ocellatella* (Lepidoptera: Gelechiidae). Arch Insect Biochem Physiol 107:e21790. https://doi.org/10.1002/arch.21790

Ando Y, Matsumoto K, Misaki K, Mano G, Shiga S, Numata H, Kotaki T, Shinada T, Goto, SG (2020) Juvenile hormone III skipped bisepoxide not its stereoisomers as a juvenile hormone of the bean bug *Riptortus pedestris*. Gen Comp Endocrinol 289:113394. https://doi.org/10.1016/j.ygcen.2020.113394

Batz ZA, Brent CS, Marias MR, Sugijanto J, Armbruster PA (2019) Juvenile hormone III but not 20-hydroxyecdysone regulates the embryonic diapause of *Aedes albopictus*. Front Physiol 10:1352. https://doi.org/10.3389/fphys.2019.01352

Borovsky D, Carlson DA, Hancock RG, Rembold H, Van Handel E (1994) *De novo* biosynthesis of juvenile hormone III and I by the accessory glands of the male mosquito. Insect Biochem Mol Biol 24:437–444. https://doi.org/10.1016/0965-1748(94)90038-8

Bowen MF, Bollenbacher WE, Gilbert LI (1984) *In vitro* studies on the role of the brain and prothoracic glands in the pupal diapause of *Manduca sexta*. J Exp Biol 108:9–24. https://doi.org/10.1242/jeb.108.1.9

Chen C, Mahar R, Merritt ME, Denlinger DL, Hahn DA (2021a) ROS and hypoxia signaling regulate periodic metabolic arousal during insect dormancy to coordinate glucose, amino acid, and lipid metabolism. Proc Natl Acad Sci USA 118:e2017603118. https://doi.org/10.1073/pnas.2017603118

Chen H, Wu G, Zhou H, Dai X, Steeghs NWF, Dong X et al (2021b) Hormonal regulation of reproductive diapause that occurs in the year-round mass rearing of *Bombus terrestris* queens. J Proteome Res 20:2240–2250. https://doi.org/10.1021/acs.jproteome.0c00776

Chen F, Zhang X-Q, Wu J-J, Jin L, Li G-Q (2022) Requirement of Myoglianin for metamorphosis in the beetle *Henosepilachna vigintioctopunctata*. Insect Mol Biol 31:671–685. https://doi.org/10.1111/imb.12795

Chen Y-H, Jiang T, Yasen A, Fan B-Y, Zhu J, Wang M-X et al (2023) m^6A-dependent mevalonate kinase in juvenile hormone synthesis pathway regulates the diapause process of bivoltine silkworm (*Bombyx mori*). Mol Biol Rep 50:5295–5306. https://doi.org/10.1007/s11033-023-08489-z

Colinet H, Kustre A (2025) The apparent seasonal biphenism in *Drosophila suzukii* stems in reality from continuous reaction norms. Pest Manag Sci 81:507–517. https://doi.org/10.1002/ps.8452

Cui D, Tu X, Hao K, Raza A, Chen J, McNeill M et al (2019) Identification of diapause-associated proteins in migratory locust, *Locusta migratoria* L. (Orthoptera: Acridoidea) by label-free quantification analysis. J Integr Agric 18:2579–2588. https://doi.org/10.1016/S2095-3119(19)62607-9

de Kort CAD, Bergot BJ, Schooley DA (1982) The nature and titre of juvenile hormone in the Colorado potato beetle, *Leptinotarsa decemlineata*. J Insect Physiol 28:471–474. https://doi.org/10.1016/0022-1910(82)90077-4

de Kort CAD, Khan MA, Koopmanschap AB (1987) Juvenile hormone and the control of adult diapause in the Colorado potato beetle, *Leptinotarsa decemlineata*. Insect Biochem 17:985–988. https://doi.org/10.1016/0020-1790(87)90107-7

de Wilde J, Staal GB, de Kort CAD, de Loon A, Baard G (1968) Juvenile hormone titer in the haemolymph as a function of photoperiodic treatment in the adult Colorado beetle (*Leptinotarsa decemlineata* Say). Proc K Ned Akad C 71:321–326

Denlinger DL (2022) Insect diapause. Cambridge University Press, Cambridge

Denlinger DL, Shukla M, Faustini DL (1984) Juvenile hormone involvement in pupal diapause of the flesh fly *Sarcophaga crassipalpis*: regulation of infradian cycles of O_2 consumption. J Exp Biol 109:191–199. https://doi.org/10.1242/jeb.109.1.191

Denlinger DL, Tanaka S (1989) Cycles of juvenile hormone esterase activity during the juvenile hormone-driven cycles of oxygen consumption in pupal diapause of flesh flies. Experientia 45:474–476. https://doi.org/10.1007/BF01952035

Denlinger DL, Willis JH, Fraenkel G (1972) Rates and cycles of oxygen consumption during pupal diapause in *Sarcophaga* flesh flies. J Insect Physiol 18:871–882. https://doi.org/10.1016/0022-1910(72)90026-1

Egi Y, Sakamoto K (2022) Genome-wide screening of genes involved in programming diapause in the next generation in silkworm, *Bombyx mori* (Lepidoptera: Bombycidae). Eur J Entomol 119:405–412. https://doi.org/10.14411/eje.2022.042

Eizaguirre M, Prats J, Abellana M, Carmen López, Llovera M, Canela R (1998) Juvenile hormone and diapause in the Mediterranean corn borer, *Sesamia nonagrioides*. J Insect Physiol 44:419–425. https://doi.org/10.1016/S0022-1910(98)00019-5

Eizaguirre M, Schafellner C, López C, Sehnal F (2005) Relationship between an increase of juvenile hormone titer in early instars and the induction of diapause in fully grown larvae of *Sesamia nonagrioides*. J Insect Physiol 51:1127–1134. https://doi.org/10.1016/j.jinsphys.2005.05.010

Endo K (1972) Activation of the corpora allata in relation to ovarian maturation in the seasonal forms of the butterfly, *Polygonia c-aureum* L. Dev Growth Differ 14:263–274. https://doi.org/10.1111/j.1440-169X.1972.00263.x

Endo K (1973) Hormonal regulation of mating in the butterfly, *Polygonia c-aureum* L. Dev Growth Differ 15:1–10. https://doi.org/10.1111/j.1440-169X.1973.00001.x

Evenden ML, Armitage G, Lau R (2007) Effects of nutrition and methoprene treatment upon reproductive diapause in *Caloptilia fraxinella* (Lepidoptera: Gracillariidae). Physiol Entomol 32:275–282. https://doi.org/10.1111/j.1365-3032.2007.00581.x

Fukaya M, Mitsuhashi J (1957) The hormonal control of larval diapause in the rice stem borer, *Chilo suppressalis*. I. Some factors in the head maintaining larval diapause. Jpn J Appl Entomol Zool 1:145–154. https://doi.org/10.1303/jjaez.1.145

Fukaya M, Mitsuhashi J (1958) The hormonal control of larval diapause in the rice stem borer, *Chilo suppressalis*. II. The activity of the corpora allata during the diapausing period. Jpn J Appl Entomol Zool 2:223–227. https://doi.org/10.1303/jjaez.2.223

Gao Q, Li B, Tian Z, De Loof A, Wang J-L, Wang X-P et al (2022a) Key role of juvenile hormone in controlling reproductive diapause in females of the Asian lady beetle. Pest Manag Sci 78:193–204. https://doi.org/10.1002/ps.6619

Gao Q, Li B, Wei B-X, Liu W, Wang P, Wang J-L et al (2022b) Juvenile hormone regulates photoperiod-mediated male reproductive diapause via the methoprene-tolerant gene in the ladybeetle *Harmonia axyridis*. Insect Sci 29:139–150. https://doi.org/10.1111/1744-7917.12918

Gelman DB, Thyagaraja BS, Kelly TJ, Masler EP, Bell RA, Borkovec AB (1992) Prothoracicotropic hormone levels in brains of the European corn borer, *Ostrinia nubilalis*: Diapause vs the non-diapause state. J Insect Physiol 38:383–395. https://doi.org/10.1016/0022-1910(92)90063-J

Giejdasz K, Fliszkiewicz M, Wasielewski O (2023) Methoprene, a juvenile hormone analog, causes winter diapause elimination in univoltine bee species *Osmia bicornis* L. Animals 13:3344. https://doi.org/10.3390/ani13213344

Goto SG (2022) Photoperiodic time measurement, photoreception, and circadian clocks in insect photoperiodism. Appl Entomol Zool 57:193–212. https://doi.org/10.1007/s13355-022-00785-7

Goto SG, Doi K, Nakayama S, Numata H (2008) Maternal control of cold and desiccation tolerance in eggs of the band-legged ground cricket *Dianemobius nigrofasciatus* in relation to embryonic diapause. Entomol Res 38:17–23. https://doi.org/10.1111/j.1748-5967.2008.00140.x

Greenfield MD, Pener MP (1992) Alternative schedules of male reproductive diapause in the grasshopper *Anacridium aegyptium* (L.): effects of the corpora allata on sexual behavior (Orthoptera: Acrididae). J Insect Behav 5:245–261. https://doi.org/10.1007/BF01049292

Gregg PC, Roberts B, Wentworth SL (1987) Levels of ecdysteroids in diapause and non-diapause eggs of the Australian plague locust, *Chortoicetes terminifera* (Walker). J Insect Physiol 33:237–242. https://doi.org/10.1016/0022-1910(87)90043-6

Guo S, Sun D, Tian Z, Liu W, Zhu F, Wang X-P (2019) The limited regulatory roles of juvenile hormone degradation pathways in reproductive diapause preparation of the cabbage beetle, *Colaphellus bowringi*. J Insect Physiol 119:103967. https://doi.org/10.1016/j.jinsphys.2019.103967

Hafeez A, Wang K, Liu W, Wang X-P (2025) Juvenile hormone regulates reproductive diapause through both canonical and noncanonical pathways in the bean bug *Riptortus pedestris*. Insect Biochem Mol Biol 177:104233. https://doi.org/10.1016/j.ibmb.2024.104233

Hao Z-P, Zhao Y-Q, Yuan Z-Q, Shi Z-H (2013) Effects of photoperiod on body size and hormone titres relating to diapause regulation in *Cotesia vestalis* (Hymenoptera: Braconidae). Can Entomol 145:369–386. https://doi.org/10.4039/tce.2013.6

Hejníková M, Nouzova M, Ramirez CE, Fernandez-Lima F, Noriega FG, Doležel D (2022) Sexual dimorphism of diapause regulation in the hemipteran bug *Pyrrhocoris apterus*. Insect Biochem Mol Biol 142:103721. https://doi.org/10.1016/j.ibmb.2022.103721

Hejnikova M, Paroulek M, Hodkova M (2016) Decrease in *Methoprene tolerant* and *Taiman* expression reduces juvenile hormone effects and enhances the levels of juvenile hormone circulating in males of the linden bug *Pyrrhocoris apterus*. J Insect Physiol 93–94:72–80. https://doi.org/10.1016/j.jinsphys.2016.08.009
Henrich VC, Denlinger DL (1982) Selection for late pupariation affects diapause incidence and duration in the flesh fly, *Sarcophaga bullata*. Physiol Entomol 7:407–411. https://doi.org/10.1111/j.1365-3032.1982.tb00316.x
Hiroyoshi S, Reddy GVP, Mitsuhashi J (2017) Effects of juvenile hormone analogue (methoprene) and 20-hydroxyecdysone on reproduction in *Polygonia c-aureum* (Lepidoptera: Nymphalidae) in relation to adult diapause. J Comp Physiol A 203:635–647. https://doi.org/10.1007/s00359-017-1179-3
Hiruma K (1979) Prevention of pupal diapause by the application of juvenile hormone analogue to the last instar larvae of *Mamestra brassicae*. Appl Entomol Zool 14:76–82. https://doi.org/10.1303/aez.14.76
Hodkova M (2008) Tissue signaling pathways in the regulation of life-span and reproduction in females of the linden bug, *Pyrrhocoris apterus*. J Insect Physiol 54:508–517. https://doi.org/10.1016/j.jinsphys.2007.11.011
Hoffmann EJ, VanderJagt J, Whalon ME (2007) Pyriproxyfen activates reproduction in prediapause northern strain plum curculio (*Conotrachelus nenuphar* Herbst). Pest Manag Sci 63:835–840. https://doi.org/10.1002/ps.1397
Horie Y, Kanda T, Mochida Y (2000) Sorbitol as an arrester of embryonic development in diapausing eggs of the silkworm, *Bombyx mori*. J Insect Physiol 46:1009–1016. https://doi.org/10.1016/S0022-1910(99)00212-7
Iwata K-I, Fujiwara Y, Takeda M (2005) Effects of temperature, sorbitol, alanine and diapause hormone on the embryonic development in *Bombyx mori*: *in vitro* tests of old hypotheses. Physiol Entomol 30:317–323. https://doi.org/10.1111/j.1365-3032.2005.00463.x
Izadi H (2025) Endocrine and enzymatic shifts during insect diapause: a review of regulatory mechanisms. Front Physiol 16:1544198. https://doi.org/10.3389/fphys.2025.1544198
Jiang X, Huang S, Luo L (2011) Juvenile hormone changes associated with diapause induction, maintenance, and termination in the beet webworm, *Loxostege sticticalis* (Lepidoptera: Pyralidae). Arch Insect Biochem Physiol 77:134–144. https://doi.org/10.1002/arch.20429
Kamsoi O, Belles X (2019) Myoglianin triggers the premetamorphosis stage in hemimetabolan insects. FASEB J 33:3659–3669. https://doi.org/10.1096/fj.201801511R
Katagiri N, Ando O, Yamashita O (1998) Reduction of glycogen in eggs of the silkworm, *Bombyx mori*, by use of a trehalase inhibitor, trehazolin, and diapause induction in glycogen-reduced eggs. J Insect Physiol 44:1205–1212. https://doi.org/10.1016/S0022-1910(98)00088-2
Kawamoto K, Ishimaru Y, Tomonari S, Watanabe T, Noji S, Mito T (2025) Myoglianin is a crucial factor for the transition to the juvenile hormone-dependent phase during hemimetabolous nymphal development. Insect Biochem Mol Biol 178:104274. https://doi.org/10.1016/j.ibmb.2025.104274
Kidokoro K, Iwata K, Fujiwara Y, Takeda M (2006) Effects of juvenile hormone analogs and 20-hydroxyecdysone on diapause termination in eggs of *Locusta migratoria* and *Oxya yezoensis*. J Insect Physiol 52:473–479. https://doi.org/10.1016/j.jinsphys.2006.01.001
Kim M, Denlinger DL (2010) A potential role for ribosomal protein S2 in the gene network regulating reproductive diapause in the mosquito *Culex pipiens*. J Comp Physiol B 180:171–178. https://doi.org/10.1007/s00360-009-0406-9
Kopper BJ, Shu S, Charlton RE, Ramaswamy SB (2001) Evidence for reproductive diapause in the fritillary *Speyeria idalia* (Lepidoptera: Nymphalidae). Ann Entomol Soc Am 94:427–432. https://doi.org/10.1603/0013-8746(2001)094[0427:EFRDIT]2.0.CO;2
Košťál V (2006) Eco-physiological phases of insect diapause. J Insect Physiol 52:113–127. https://doi.org/10.1016/j.jinsphys.2005.09.008

Kotaki T, Shinada T, Kaihara K, Ohfune Y, Numata H (2011) Biological activities of juvenile hormone III skipped bisepoxide in last instar nymphs and adults of a stink bug, *Plautia stali*. J Insect Physiol 57:147–152. https://doi.org/10.1016/j.jinsphys.2010.10.003

Kotaki T, Yagi S (1989) Hormonal control of adult diapause in the brown-winged green bug, *Plautia stali* SCOTT (Heteroptera: Pentatomidae). Appl Entomol Zool 24:42–51. https://doi.org/10.1303/aez.24.42

Larrere M, Lavenseau L, Tasei JN, Couillaud F (1993) Juvenile hormone biosynthesis and diapause termination in *Bombus terrestris*. Invertebr Reprod Dev 23:7–14. https://doi.org/10.1080/07924259.1993.9672288

Lee K-Y, Denlinger DL (1997) A role for ecdysteroids in the induction and maintenance of the pharate first instar diapause of the gypsy moth, *Lymantria dispar*. J Insect Physiol 43:289–296. https://doi.org/10.1016/S0022-1910(96)00082-0

Lee K-Y, Valaitis AP, Denlinger DL (1997) Further evidence that diapause in the gypsy moth, *Lymantria dispar*, is regulated by ecdysteroids: a comparison of diapause and nondiapause strains. J Insect Physiol 43:897–903. https://doi.org/10.1016/S0022-1910(97)00054-1

Lefevere KS, de Wilde J (1984) Effect of photoperiod and allatectomy on reproduction and (second) diapause induction in post-diapause adult females of the Colorado potato beetle, *Leptinotarsa decemlineata* Say. Int J Invertebr Reprod 7:69–72. https://doi.org/10.1080/01688170.1984.10510073

Li Y-Y, Chen J-J, Liu M-Y, He W-W, Reynolds JA, Wang Y-N et al (2022) Enhanced degradation of juvenile hormone promotes reproductive diapause in the predatory ladybeetle *Coccinella septempunctata*. Front Physiol 13:877153. https://doi.org/10.3389/fphys.2022.877153

Liu W, Li Y, Zhu L, Zhu F, Lei C-L, Wang X-P (2016) Juvenile hormone facilitates the antagonism between adult reproduction and diapause through the *methoprene-tolerant* gene in the female *Colaphellus bowringi*. Insect Biochem Mol Biol 74:50–60. https://doi.org/10.1016/j.ibmb.2016.05.004

Liu Y, Su L, Wang R, Dai X, Li X, Chang Y et al (2024) Comparative 4D label-free quantitative proteomic analysis of *Bombus terrestris* provides insights into proteins and processes associated with diapause. Int J Mol Sci 25:326. https://doi.org/10.3390/ijms25010326

Ma H-Y, Li Y-Y, Li L, Tan Y, Pang B-P (2021a) Regulation of juvenile hormone on summer diapause of *Geleruca daurica* and its pathway analysis. Insects 12:237. https://doi.org/10.3390/insects12030237

Ma H-Y, Li Y-Y, Li L, Tan Y, Pang B-P (2021b) Juvenile hormone regulates the reproductive diapause through Methoprene-tolerant gene in *Galeruca daurica*. Insect Mol Biol 30:446–458. https://doi.org/10.1111/imb.12710

Mano G, Goto SG (2022) Photoperiod controls insulin and juvenile hormone signaling pathways via the circadian clock in the bean bug *Riptortus pedestris* (Hemiptera: Alydidae). Appl Entomol Zool 57:363–377. https://doi.org/10.1007/s13355-022-00795-5

Masaki S (1956) The effect of temperature on the termination of diapause in the egg of *Lymantria dispar* Linné (Lepidoptera: Lymantriidae). Jpn J Appl Zool 21:148–157

Matsumoto K, Numata H, Shiga S (2013) Role of the brain in photoperiodic regulation of juvenile hormone biosynthesis in the brown-winged green bug *Plautia stali*. J Insect Physiol 59:387–393. https://doi.org/10.1016/j.jinsphys.2013.01.007

Matsuo J, Nakayama S, Numata H (1997) Role of the corpus allatum in the control of adult diapause in the blow fly, *Protophormia terraenovae*. J Insect Physiol 43:211–216. https://doi.org/10.1016/S0022-1910(96)00103-5

Miki T, Shinohara T, Chafino S, Noji S, Tomioka K (2020) Photoperiod and temperature separately regulate nymphal development through JH and insulin/TOR signaling pathways in an insect. Proc Natl Acad Sci USA 117:5525–5531. https://doi.org/10.1073/pnas.1922747117

Mitsuhashi J, Fukaya M (1960) The hormonal control of larval diapause in the rice stem borer, *Chilo suppressalis*. III. Histological studies on the neurosecretory cells of the brain and the secretory cells of the corpora allata during diapause and post diapause. Jpn J Appl Entomol Zool 4:127–136. https://doi.org/10.1303/jjaez.4.127

Mizoguchi A, Ohsumi S, Kobayashi K, Okamoto N, Yamada N, Tateishi K et al (2013) Prothoracicotropic hormone acts as a neuroendocrine switch between pupal diapause and adult development. PLOS ONE 8:e60824. https://doi.org/10.1371/journal.pone.0060824

Morita A, Numata H (1997) Role of the neuroendocrine complex in the control of adult diapause in the bean bug, *Riptortus clavatus*. Arch Insect Biochem Physiol 35:347–355. https://doi.org/10.1002/(SICI)1520-6327(199705)35:3<347::AID-ARCH8>3.0.CO;2-Q

Morita A, Soga K, Hoson T, Kamisaka S, Numata H (1999) Changes in mechanical properties of the cuticle and lipid accumulation in relation to adult diapause in the bean bug, *Riptortus clavatus*. J Insect Physiol 45:241–247. https://doi.org/10.1016/S0022-1910(98)00119-X

Morohoshi S, Ishida S, Fugo H (1972) The control of growth and development in *Bombyx mori*. XVIII. Production of non-diapause eggs by injection of the juvenile hormone. Proc Jpn Acad 48:736–741. https://doi.org/10.2183/pjab1945.48.736

Mousseau TA, Dingle H (1991) Maternal effects in insect life histories. Annu Rev Entomol 36:511–534. https://doi.org/10.1146/annurev.en.36.010191.002455

Mukai A, Mano G, Des Marteaux L, Shinada T, Goto SG (2022) Juvenile hormone as a causal factor for maternal regulation of diapause in a wasp. Insect Biochem Mol Biol 144:103758. https://doi.org/10.1016/j.ibmb.2022.103758

Munyiri FN, Ishikawa Y (2004) Endocrine changes associated with metamorphosis and diapause induction in the yellow-spotted longicorn beetle, *Psacothea hilaris*. J Insect Physiol 50:1075–1081. https://doi.org/10.1016/j.jinsphys.2004.09.009

Neumann-Visscher S (1976) The embryonic diapause of *Aulocara elliotti* (Orthoptera, Acrididiae). Cell Tissue Res 174:433–452. https://doi.org/10.1007/BF00232831

Ogihara MH, Ikeda H, Yamada N, Hikiba J, Nakaoka T, Fujimoto Y et al (2017) Identification of ecdysteroidogenic enzyme genes and their expression during pupal diapause in the cabbage armyworm, *Mamestra brassicae*. Insect Mol Biol 26:286–297. https://doi.org/10.1111/imb.12291

Ojima N, Ishiguro S, An Y, Kadosawa T, Suzuki K (2015) Male reproductive maturity and diapause termination in the leaf beetle *Gastrophysa atrocyanea*. Physiol Entomol 40:277–283. https://doi.org/10.1111/phen.12112

Okuda T, Tanaka S, Kotaki T, Ferenz H-J (1996) Role of the corpora allata and juvenile hormone in the control of imaginal diapause and reproduction in three species of locusts. J Insect Physiol 42:943–951. https://doi.org/10.1016/0022-1910(96)00055-8

Pegoraro M, Bafna A, Davies NJ, Shuker DM, Tauber E (2016) DNA methylation changes induced by long and short photoperiods in *Nasonia*. Genome Res 26:203–210. https://doi.org/10.1101/gr.196204.115

Readio J, Chen M-H, Meola R (1999) Juvenile hormone biosynthesis in diapausing and nondiapausing *Culex pipiens* (Diptera: Culicidae). J Med Entomol 36:355–360. https://doi.org/10.1093/jmedent/36.3.355

Reynolds JA, Hand SC (2009) Embryonic diapause highlighted by differential expression of mRNAs for ecdysteroidogenesis, transcription and lipid sparing in the cricket *Allonemobius socius*. J Exp Biol 212:2075–2084. https://doi.org/10.1242/jeb.027367

Richard DS, Saunders DS (1987) Prothoracic gland function in diapause and non-diapause *Sarcophaga argyrostoma* and *Calliphora vicina*. J Insect Physiol 33:385–392. https://doi.org/10.1016/0022-1910(87)90016-3

Rockey SJ, Miller BB, Denlinger DL (1989) A diapause maternal effect in the flesh fly, *Sarcophaga bullata*: transfer of information from mother to progeny. J Insect Physiol 35:553–558. https://doi.org/10.1016/0022-1910(89)90143-1

Sasibhushan S, Rao CGP, Ponnuvel KM (2013) Genome wide microarray based expression profiles during early embryogenesis in diapause induced and non-diapause eggs of polyvoltine silkworm *Bombyx mori*. Genomics 102:379–387. https://doi.org/10.1016/j.ygeno.2013.07.007

Saunders DS (1966) Larval diapause of maternal origin—II. The effect of photoperiod and temperature on *Nasonia vitripennis*. J Insect Physiol 12:569–581. https://doi.org/10.1016/0022-1910(66)90095-3

Saunders DS, Richard DS, Applebaum SW, Ma M, Gilbert LI (1990) Photoperiodic diapause in *Drosophila melanogaster* involves a block to the juvenile hormone regulation of ovarian maturation. Gen Comp Endocrinol 79:174–184. https://doi.org/10.1016/0016-6480(90)90102-R

Schooneveld H, Otazo-Sanchez A, de Wilde J (1977) Juvenile hormone-induced break and termination of diapause in the Colorado potato beetle. J Insect Physiol 23:689–696. https://doi.org/10.1016/0022-1910(77)90085-3

Shiga S, Hamanaka Y, Tatsu Y, Okuda T, Numata H (2003) Juvenile hormone biosynthesis in diapause and nondiapause females of the adult blow fly *Protophormia terraenovae*. Zoolog Sci 20:1199–1206. https://doi.org/10.2108/zsj.20.1199

Shimizu I (2024) Photoperiodism of diapause induction in the silkworm, *Bombyx mori*. Zoolog Sci 41:141–158. https://doi.org/10.2108/zs230036

Shimizu Y, Goto SG (2024) No evidence for the involvement of juvenile hormone III and 20-hydroxyecdysone in maternal decisions for embryonic diapause and diapause entry in the band-legged ground cricket *Dianemobius nigrofasciatus* (Orthoptera: Trigonidiidae). Appl Entomol Zool 59:51–59. https://doi.org/10.1007/s13355-023-00853-6

Shirk PD, Dahm KH, Röller H (1976) The accessory sex glands as the repository for juvenile hormone in male cecropia moths. Z Naturforsch C 31:199–200. https://doi.org/10.1515/znc-1976-3-421

Singtripop T, Wanichacheewa S, Sakurai S (2000) Juvenile hormone-mediated termination of larval diapause in the bamboo borer, *Omphisa fuscidentalis*. Insect Biochem Mol Biol 30:847–854. https://doi.org/10.1016/S0965-1748(00)00057-6

Singtripop T, Wanichacheewa S, Tsuzuki S, Sakurai S (1999) Larval growth and diapause in a tropical moth, *Omphisa fuscidentalis* Hampson. Zool Sci 16:725–733. https://doi.org/10.2108/zsj.16.725

Smith W, Rybczynski R (2012) Prothoracicotropic hormone. In: Gilbert LI (ed) Insect endocrinology. Academic Press, Cambridge, pp 1–62

Smykal V, Bajgar A, Provaznik J, Fexova S, Buricova M, Takaki K et al (2014) Juvenile hormone signaling during reproduction and development of the linden bug, *Pyrrhocoris apterus*. Insect Biochem Mol Biol 45:69–76. https://doi.org/10.1016/j.ibmb.2013.12.003

Spielman A (1974) Effect of synthetic juvenile hormone on ovarian diapause of *Culex pipiens* mosquitoes. J Med Entomol 11:223–225. https://doi.org/10.1093/jmedent/11.2.223

Süess P, Dircksen H, Roberts KT, Gotthard K, Nässel DR, Wheat CW et al (2022) Time- and temperature-dependent dynamics of prothoracicotropic hormone and ecdysone sensitivity co-regulate pupal diapause in the green-veined white butterfly *Pieris napi*. Insect Biochem Mol Biol 149:103833. https://doi.org/10.1016/j.ibmb.2022.103833

Suman DS, Wang Y, Gaugler R (2015) The insect growth regulator pyriproxyfen terminates egg diapause in the Asian tiger mosquito, *Aedes albopictus*. PLOS ONE 10:e0130499. https://doi.org/10.1371/journal.pone.0130499

Takeda N (1978) Hormonal control of prepupal diapause in *Monema flavescens* (Lepidoptera). Gen Comp Endocrinol 34:123–131. https://doi.org/10.1016/0016-6480(78)90202-2

Tanigawa N, Matsumoto K, Yasuyama K, Numata H, Shiga S (2009) Early embryonic development and diapause stage in the band-legged ground cricket *Dianemobius nigrofasciatus*. Dev Genes Evol 219:589–596. https://doi.org/10.1007/s00427-010-0320-x

Taniguchi N, Tomioka K (2013) Duration of development and number of nymphal instars are differentially regulated by photoperiod in the cricket *Modicogryllus siamensis* (Orthoptera: Gryllidae). Eur J Entomol 100:275–281. https://doi.org/10.14411/eje.2003.043

Tawfik AI, Tanaka Y, Tanaka S (2002) Possible involvement of ecdysteroids in embryonic diapause of *Locusta migratoria*. J Insect Physiol 48:743–749. https://doi.org/10.1016/S0022-1910(02)00099-9

Tian Z, Guo S, Li J-X, Zhu F, Liu W, Wang X-P (2021) Juvenile hormone biosynthetic genes are critical for regulating reproductive diapause in the cabbage beetle. Insect Biochem Mol Biol 139:103654. https://doi.org/10.1016/j.ibmb.2021.103654

Tomioka K, Agui N, Bollenbacher WE (1995) Electrical properties of the cerebral prothoracicotropic hormone cells in diapausing and non-diapausing pupae of the tobacco hornworm, *Manduca sexta*. Zoolog Sci 12:165–173. https://doi.org/10.2108/zsj.12.165

Toxopeus J, Jakobs R, Ferguson LV, Gariepy TD, Sinclair BJ (2016) Reproductive arrest and stress resistance in winter-acclimated *Drosophila suzukii*. J Insect Physiol 89:37–51. https://doi.org/10.1016/j.jinsphys.2016.03.006

Tsuchiya R, Kaneshima A, Kobayashi M, Yamazaki M, Takasu Y, Sezutsu H et al (2021) Maternal GABAergic and GnRH/corazonin pathway modulates egg diapause phenotype of the silkworm *Bombyx mori*. Proc Natl Acad Sci USA 118:e2020028118. https://doi.org/10.1073/pnas.2020028118

Tu X, Wang J, Hao K, Whitman DW, Fan Y, Cao G et al (2015) Transcriptomic and proteomic analysis of pre-diapause and non-diapause eggs of migratory locust, *Locusta migratoria* L. (Orthoptera: Acridoidea). Sci Rep 5:11402. https://doi.org/10.1038/srep11402

Uehara H, Senoh Y, Yoneda K, Kato Y, Shiomi K (2011) An FXPRLamide neuropeptide induces seasonal reproductive polyphenism underlying a life-history tradeoff in the tussock moth. PLOS ONE 6:e24213. https://doi.org/10.1371/journal.pone.0024213

Urbanová V, Bazalová O, Vaněčková H, Dolezel D (2016) Photoperiod regulates growth of male accessory glands through juvenile hormone signaling in the linden bug, *Pyrrhocoris apterus*. Insect Biochem Mol Biol 70:184–190. https://doi.org/10.1016/j.ibmb.2016.01.003

Wang K, Zhao Y-L, Jiang Y-Z, Liu W, Wang X-P (2025) *Seven up* regulates reproductive diapause initiation via juvenile hormone biosynthesis in the cabbage beetle *Colaphellus bowringi*. Insect Sci 33:85-98. https://doi.org/10.1111/1744-7917.13497

Wang Q, Mohamed AAM, Takeda M (2013) Serotonin receptor B may lock the gate of PTTH release/synthesis in the Chinese silk moth, *Antheraea pernyi*; A diapause initiation/maintenance mechanism? PLOS ONE 8:e79381. https://doi.org/10.1371/journal.pone.0079381

Wasielewski O, Wojciechowicz T, Giejdasz K, Krishnan N (2011) Influence of methoprene and temperature on diapause termination in adult females of the over-wintering solitary bee, *Osmia rufa* L. J Insect Physiol 57:1682–1688. https://doi.org/10.1016/j.jinsphys.2011.09.002

Yagi S, Fukaya M (1974) Juvenile hormone as a key factor regulating larval diapause of the rice stem borer, *Chilo suppressalis* (Lepidoptera: Pyralidae). Appl Entomol Zool 9:247–255. https://doi.org/10.1303/aez.9.247

Yin C-M, Chippendale GM (1979) Diapause of the southwestern corn borer, *Diatraea grandiosella*: further evidence showing juvenile hormone to be the regulator. J Insect Physiol 25:513–523. https://doi.org/10.1016/S0022-1910(79)80010-4

Yin C-M, Chippendale GM (1973) Juvenile hormone regulation of the larval diapause of the Southwestern corn borer, *Diatraea grandiosella*. J Insect Physiol 19:2403–2420. https://doi.org/10.1016/0022-1910(73)90244-8

Yin Z-J, Dong X-L, Kang K, Chen H, Dai X-Y, Wu G-A et al (2018) FoxO transcription factor regulate hormone mediated signaling on nymphal diapause. Front Physiol 9:1654. https://doi.org/10.3389/fphys.2018.01654

Yoshinaga N, Goto SG (2025) Expression of short neuropeptide F and prothoracicotropic hormone in relation to photoperiodic regulation of pupariation and pupal diapause in the flesh fly, *Sarcophaga similis* (Diptera: Sarcophagidae). Appl Entomol Zool 60:127–134

Zhai Y, Dong X, Gao H, Chen H, Yang P, Li P et al (2019) Quantitative proteomic and transcriptomic analyses of metabolic regulation of adult reproductive diapause in *Drosophila suzukii* (Diptera: Drosophilidae) females. Front Physiol 10:344. https://doi.org/10.3389/fphys.2019.00344

Zhou W-Z, Wu Y-F, Yin Z-Y, Guo J-J, Li H-Y (2022) Juvenile hormone is an important factor in regulating *Aspongopus chinensis* Dallas diapause. Front Physiol 13:873580. https://doi.org/10.3389/fphys.2022.873580

Zhu L, Yin T-Y, Sun D, Liu W, Zhu F, Lei C-L et al (2017) Juvenile hormone regulates the differential expression of putative *juvenile hormone esterases* via *methoprene-tolerant* in non-diapause-destined and diapause-destined adult female beetle. Gene 627:373–378. https://doi.org/10.1016/j.gene.2017.06.061

Chapter 13
Role of Juvenile Hormone in Polyphenism

Yudai Masuoka

Abstract Some types of polyphenism are known to be under the control of juvenile hormone (JH). These polyphenisms include morphological changes, such as weaponry and body coloration; changes in internal organs, such as the gonads; and behavioral changes. As a notable example of polyphenism involving many changes, this chapter summarizes recent advances in research on the regulatory mechanisms of caste differentiation in termites by JH and other factors and signal transduction pathways. In addition to caste differentiation in hymenopteran social insects, the role of JH in other plastic phenomena, such as male weapon polyphenism in beetles, changes in reproductive modes in aphids, and phase polyphenism in locusts, will be presented.

Keywords Phenotypic plasticity · Termite · Caste differentiation · 20-Hydroxyecdysone(20E) · Methoprene-tolerant (Met)

13.1 Polyphenism

The phenotypes of organisms can be plastically altered not only by genetic factors but also by environmental factors. This phenomenon is called phenotypic plasticity and is an important mechanism for flexible responses to environmental changes. In particular, when environmental factors induce discrete alternative phenotypes from the same genotype, this form of phenotypic plasticity is known as polyphenism. It is also known that such plasticity can be enhanced or suppressed by selective pressures (West-Eberhard 2003). For example, it has been reported that when plastic changes are adaptive to environmental changes, plasticity itself is under selection pressure, and the range of plasticity is amplified (genetic accommodation) (Gilbert and Epel 2009). However, when there is a change in the environment or a bias

Y. Masuoka (✉)
National Agriculture and Food Research Organization, Tsukuba, Japan
e-mail: masuoka.yudai003@naro.go.jp

T. Shinoda et al. (eds.), *Juvenile Hormone*, Entomology Monographs,
https://doi.org/10.1007/978-981-95-8605-9_13

toward environmental stimuli, the range of plasticity is narrowed, and the trait is fixed, that is, genetic assimilation (Gilbert and Epel 2009). Thus, phenotypic plasticity is also attracting attention as a driving force in evolution. Polyphenism with plastic changes in traits has been observed in many insect species. In many cases, external factors, such as seasonal changes and nutritional conditions during the larval stage, provide environmental cues that alter the postembryonic developmental pathway, resulting in different phenotypes. An example of a more complex polyphenism is the intraspecific polymorphism (caste) exhibited by social insects, which display specialized traits for each task they perform. The environmental factors that promote the emergence of such castes are unique in that, in addition to environmental factors outside the nest, interactions between castes play an important role. The most important polyphenism exhibited by social insects is the presence of reproductive and sterile castes. Individuals differentiated into sterile castes do not produce their own offspring throughout their lives but instead engage in altruistic labor, such as caring for other individuals in the nest. In termites, the sterile caste is further subdivided into workers and soldiers. The latter exhibit specialized traits for nest defense. The reproductive caste includes not only the queen (founder female) and king (founder male) but also supplementary reproductive castes that appear secondarily. Many cases have been reported in which juvenile hormones (JHs) play an important role in the emergence of polyphenism in such insects (Simpson et al. 2011). In these cases, JH induces morphogenesis and changes in individual developmental phases as a reaction to environmental factors. This chapter introduces the role of JH in insect polyphenism, focusing on the relationship between JH and the diverse caste differentiation exhibited by termites.

13.2 Termite Caste

Termites are representative social insects, with approximately 2600 species known worldwide. All extant species have a high degree of sociality (eusociality), with a clear division of labor among individuals. The division of labor among individuals is determined by morphologically and behaviorally distinct polymorphisms (castes) within a colony. Castes include workers who care for and forage for blood relatives, soldiers who defend the colony, and reproductive insects responsible for breeding activities (Thorne 1996; Roisin 2000). Differentiation into each caste is essentially determined during postembryonic development (Roisin 2000), and juvenile hormone (JH) has long been known to be an important factor (Nijhout and Wheeler 1982; Korb 2015). In particular, soldier differentiation is caused by an increase in the amount of JH in the bodies of workers, and it has been shown that soldier differentiation can be induced by implanting the corpora allatum, a JH-synthesizing organ (Fig. 13.1), into the abdomen of workers (Lüscher 1958). It has also been reported in several species that soldier differentiation can be artificially induced by the external administration of JH or its analogs (JHA) (Hrdý and Krecek 1972; Howard and Haverty 1979; Scharf et al. 2003; Cornette et al. 2008; Tsuchiya et al. 2008; Toga et al. 2009; Maekawa et al. 2014). Soldiers specialized for defense are a

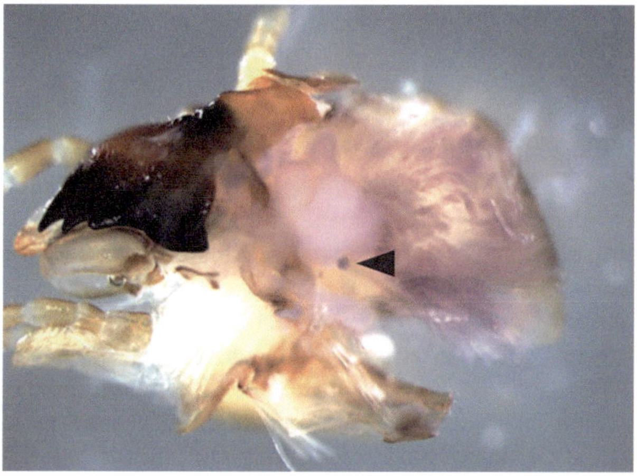

Fig. 13.1 Corpora allata of worker in *Zootermopsis nevadensis*. Stained by in situ hybridization (ISH) targeting the juvenile hormone synthesis gene (juvenile hormone acid *O*-methyltransferase; JHAMT)

caste not found in ants and bees and are one of the key features of termite sociality. Therefore, much attention has been paid to the differentiation mechanism of soldiers, and morphological and molecular embryological analyses have been performed in several species using artificial induction of soldier differentiation (Miura and Scharf 2011; Watanabe et al. 2014). However, the initial mechanism by which soldier differentiation occurs (e.g., what factors cause the increase in JH levels and what signals trigger characteristic morphogenesis) remains unclear. The main reason for this is that it is extremely difficult to observe the process of soldier differentiation under natural conditions. This is because the number of soldiers in a normal mature colony is overwhelmingly small compared to the number of workers, and it is almost impossible to identify in advance which individuals will differentiate into soldiers. However, we recently identified individuals destined to differentiate into soldiers in a colony of the Nevada dampwood termite, *Zootermopsis nevadensis*, shortly after its establishment (Maekawa et al. 2012). This is the first time that we have been able to follow the entire process of soldier differentiation under natural conditions. In addition, with the recent development of omics data obtained by bioinformatics (Terrapon et al. 2014) and gene function analysis technology using RNA interference (RNAi), insights into the relationship between caste differentiation and JH, including soldiers, have been accumulating.

13.3 The Mechanism of Determining Soldier Differentiation by JH Titer Variation

Among the castes found in social insects, termite soldiers are highly specialized morphologically and behaviorally (Deligne et al. 1981). Their sole role is colony defense, with characteristic weaponry. All species possess soldiers, except for some species that lost them secondarily, and soldiers are thought to have been acquired only once in the common ancestor of termites (Lo and Eggleton 2011). Termites are also considered to be the first group to evolve soldiers as a sterile caste, in contrast to the sociality of ants and bees ("soldier-first eusociality"; Tian and Zhou 2014). Thus, elucidating the differentiation mechanism of termite soldiers provides molecular evolutionary insight into the morphological diversification that occurs and is a crucial task for understanding the initial conditions for the evolution of termite sociality. Soldiers differentiate through an intermediate stage, the presoldier stage, which is caused by an increase in the amount of JH in workers (Cornette et al. 2008). The hexamerin gene, known as a JH-binding protein, is upregulated in JH-treated workers of the eastern subterranean termite, *Reticulitermes flavipes* (Scharf et al. 2005), and knockdown of its expression in workers promotes soldier differentiation (Zhou et al. 2006). Therefore, hexamerin may be involved in the regulation of JH levels during soldier differentiation in this species. In the aforementioned early colonies of *Z. nevadensis*, JH synthase genes (*JH acid methyltransferase* (*JHAMT*) and *CYP15A1*) were highly expressed in workers who would become soldiers in the future (Yaguchi et al. 2015). Furthermore, these individuals frequently exchanged nutrients with the queen, and RNA-seq results indicated that lipocalin genes were highly expressed in them. Knockdown of the lipocalin gene also suppressed the frequency of nutrient exchange with the queen and soldier differentiation (Yaguchi et al. 2018). Thus, fluctuations in JH levels due to social communication within the colony may be involved in the early stages of soldier differentiation in termites.

It is also possible that biogenic amines play a role in social behavior by regulating fluctuations in JH levels. Biogenic amines are important neurotransmitters, neuromodulators, and neurohormones that function in the peripheral and central nervous systems of insects (Evans 1980). In the American cockroach, *Periplaneta americana*, the administration of dopamine (DA) in body fluids reduces the frequency of self-grooming (Weisel-Eichler et al. 1999). The role of DA in the generation of phenotypic polymorphism has also been demonstrated in grasshoppers, bees, and ants (Ma et al. 2011; Matsuyama et al. 2015; Okada et al. 2015). Individuals that will become soldiers in the future have elevated brain DA levels after molting, and antagonist treatment of DA receptors has been shown to reduce the frequency of nutrient exchange with the reproductive caste and the rate of soldier differentiation (Yaguchi et al. 2015).

13.4 The Role of JH Downstream Genes During Soldier Differentiation

Recently, our understanding of the JH signaling pathway has significantly advanced. In particular, the insect JH receptor, which was long thought to be unknown, has been identified as a transcription factor called methoprene-tolerant (Met) (Charles et al. 2011; Jindra et al. 2015). Furthermore, the presence of steroid receptor coactivator (SRC), which acts as a Met coactivator, and its downstream transcription factor, Krüppel homolog 1 (Kr-h1), has been revealed (Minakuchi et al. 2008; Kayukawa et al. 2012). Analyses of molting, metamorphosis, and gonad development have been performed using model insects such as the red flour beetle, *Tribolium castaneum* (Konopova and Jindra 2007; Konopova et al. 2011; Marchal et al. 2014). However, the functions of JH downstream genes in termite soldier differentiation induced by elevated JH levels remain unknown. We investigated the expression level of *Met* during soldier differentiation under natural conditions in the aforementioned early colonies of *Z. nevadensis* and attempted to estimate its function using RNA interference (RNAi) methods (Masuoka et al. 2015). Real-time quantitative polymerase chain reaction (PCR) analysis of *Met* expression in the head revealed a rapid increase in expression levels immediately after differentiation into presoldiers. When this rapid increase in expression was suppressed by RNAi methods, the molt from presoldier to soldier was not particularly affected, but the development of weapon traits (head enlargement and mandibular elongation) was severely impaired (Fig. 13.2; Masuoka et al. 2015). Furthermore, when RNAi of *Met* and *SRC* was performed simultaneously with JH analog (pyriproxyfen) treatment in old workers of this species, the differentiation rate was significantly reduced, and molted individuals displayed a morphology intermediate between workers and presoldiers (Fig. 13.3; Masuoka et al. 2018a). In contrast, *Kr-h1* knockdown did not affect soldier differentiation (Fig. 13.3; Masuoka et al. 2018a). Therefore, there may be a genetic pathway underlying soldier differentiation that is distinct from general JH downstream signaling in insects.

Fig. 13.2 Phenotype of a soldier under the RNAi treatment of *methoprene-tolerant* (*Met*) in *Zootermopsis nevadensis* incipient colony. (Masuoka et al. 2015)

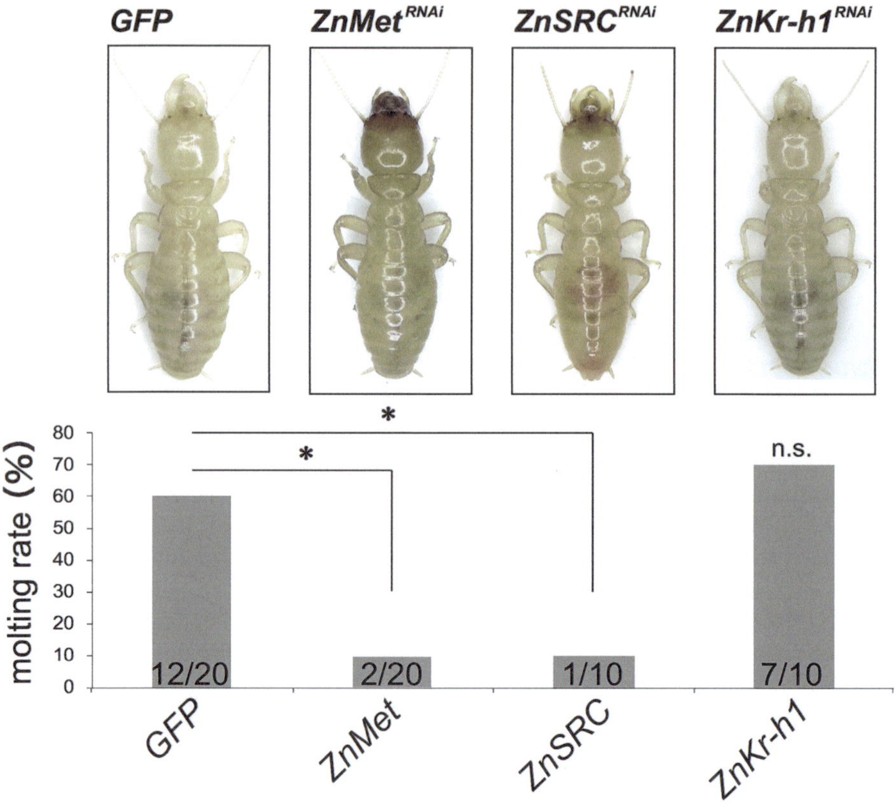

Fig. 13.3 Phenotype and molting rate of a juvenile hormone (JH)-induced presoldier under the RNAi treatment of JH signaling genes in *Zootermopsis nevadensis.* (Masuoka et al. 2018a)

13.5 The Relationship Between JH and 20-Hydroxyecdysone (20E) in Soldier Differentiation

As mentioned above, an increase in the amount of JH in the body induces soldier differentiation, which is accompanied by molting. Therefore, it is possible that JH induces molting via the molting hormone 20-hydroxyecdysone (20E) in termites. Similar to the JH pathway, the 20E signaling pathway is well-understood, especially in model insects such as the silkworm, *Bombyx mori,* and *T. castaneum*. First, heterodimers of the ecdysone receptor (EcR) and its partner ultraspiracle (USP) bind to 20E, and this complex interacts with DNA to promote the expression of early response genes, such as hormone receptor 4 (HR4) and ecdysone-inducible protein E75 (E75) (Thomas et al. 1993; Nakagawa and Henrich 2009). In the tobacco hornworm, *Manduca sexta,* and the honeybee, *Apis mellifera,* the 20E signaling

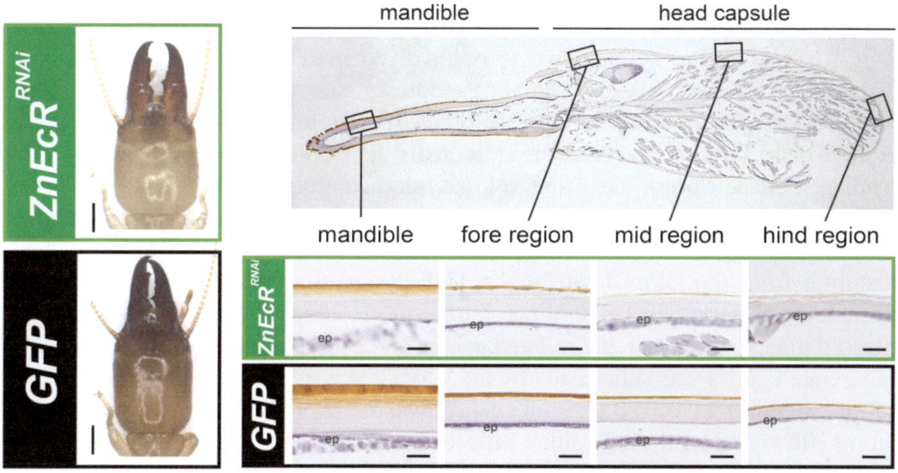

Fig. 13.4 Phenotype and mandibular epidermal tissue of a soldier under the RNAi treatment of *ecdysone receptor* (*EcR*) in *Zootermopsis nevadensis*. (Modified from Masuoka et al. 2016a)

pathway, including HR4 and E75, affects cuticle sclerotization and pigmentation (Hiruma and Riddiford 2009; Elias-Neto et al. 2010). A set of genes involved in 20E synthesis (Halloween genes) has also been identified. Termites also have homologs of these 20E-related genes, and knockdown of some synthesis pathway genes or *EcR* inhibited molting into presoldiers under JHA treatment (Masuoka et al. 2018a). Knockdown of *EcR* late in the soldier differentiation process also inhibited soldier-specific cuticle formation (Fig. 13.4; Masuoka and Maekawa 2016a). The phenotype of these individuals was similar to that of individuals with knockdown of tyrosine pathway genes involved in cuticle formation (Masuoka and Maekawa 2016b). Knockdown of *EcR* inhibited *Lac2* expression, suggesting that 20E is involved in soldier-specific cuticular formation via the tyrosine metabolic pathway genes. Thus, 20E was found to play an important role in soldier differentiation. However, what about its relationship with JH? Knockdown of the JH receptor Met during soldier differentiation suppressed the expression of several 20E-related genes, including *EcR* and 20E synthesis pathway genes (Masuoka et al. 2018a). Therefore, it is suggested that JH triggers the synthesis of 20E and regulates downstream gene expression during soldier differentiation. A comparative transcriptome analysis of soldier differentiation and worker molting processes under natural conditions revealed genes involved in JH and 20E mediation. We searched for a set of genes specifically upregulated during three different molting processes: worker-presoldier, presoldier-soldier, and worker-next-instar worker. The genes specifically upregulated during the presoldier molt included those involved in transforming growth factor-beta (TGF-β) signaling, a signaling pathway that is conserved in animals and has diverse functions, including regulation of cell proliferation, growth differentiation, and osteogenesis. In the two-spotted cricket, *Gryllus bimaculatus*, TGF-β signaling has been shown to be involved in JH synthesis (Ishimaru et al. 2016). Knockdown of TGF-β signaling–related genes, which are highly expressed during presoldier molting, affects presoldier morphology and behavior (Masuoka et al. 2018b). That is, knockdown individuals differentiated into individuals intermediate between presoldiers and soldiers, and molting into soldiers was subsequently inhibited. Normally, presoldiers have a colorless, soft exoskeleton and do not exhibit aggression toward enemies. Interestingly, intermediate individuals resulting from the knockdown of the TGF-β signaling–related gene showed the same level of aggression as soldiers. Furthermore, knockdown of these genes suppressed the expression of 20E-related genes. The expression of these TGF-β signaling–related genes was suppressed by the knockdown of the JH receptor *Met*. These results suggest that TGF-β signaling acts downstream of JH and is involved in regulating 20E signaling during soldier differentiation.

13.6 The Relationship Between JH and Sexual Differentiation of Soldiers

Unlike Hymenoptera, termite colonies are composed of both males and females. In lower groups, the sex ratio of soldiers is often 50:50. However, in higher groups (especially in the family Termitidae), there are species in which only one sex produces soldiers. The nasute termite, *Nasutitermes takasagoensis*, which belongs to the Termitidae family, has four different workers with different body sizes (male minor, male medium, female medium, and female major workers), and only male minor workers differentiate into soldiers. However, artificial JHA treatment can lead to soldiers being differentiated from female workers. Therefore, the ability to differentiate into soldiers is retained in both sexes, and the resulting differences in JH titer may contribute to sex differences in soldier differentiation rates. There was no difference in JH titer between the sexes of workers in the Japanese subterranean termite, *Reticulitermes speratus*, which has a low sex ratio of soldiers, whereas *N. takasagoensis* had a higher JH titer in male miner workers (Toga et al. 2016). Such sex differences in JH titer may be regulated by sex-determining gene pathways. This suggests that JH synthesis genes are under the control of the sex-determining gene pathway. In insects, differences in JH titer are also involved in gonadal development in adults. Indeed, in termites, adult females have been shown to have higher JH titers than males (Cornette et al. 2008). Termite soldiers are not capable of reproduction, but like adults, they are in the terminal stages of development. Therefore, it is possible that the mechanisms responsible for the sex differences in JH titer observed in adults may also be involved in soldier differentiation.

13.7 Soldier Evolution and Changes of JH Downstream Genes

In many species, treatment of workers with JH or JHA causes soldier differentiation with molting, suggesting that in termites, JH may function to activate the molting process via 20E. In contrast, in some species, JH and JHA treatments are known to prolong the larval period and delay molting. However, JH promotes the secretion of prothoracic gland stimulating hormone Prothoracicotropic hormone (PTTH) and induces molting into pupae in the last-instar larvae of *B. mori* (Mizoguchi 2001). In the German cockroach, *Blattella germanica,* JHA treatment inhibits 20E synthesis and larval molting (Hangartner and Masner 1973; Masner et al. 1975). The author has also treated larvae of *P. americana* and wood-feeding cockroaches, *Panesthia angustipennis* and *Salganea esakii*, which are more closely related to termites, with JHA and observed no effect on molting (unpublished data by Y. Masuoka). Interestingly, JHA treatment induced molting to the next-instar larvae in larvae of the sister group of termites, the wood roach, *Cryptocercus punctulatus* (Masuoka et al. 2018a). However, no morphological changes were observed in this species, as seen in termites. JH-induced molting was inhibited by *Met* RNAi. Therefore, the mechanism of molting promotion by JH–Met signaling is thought to be conserved in both wood roaches and in termites. The author predicted that 20E-related genes, which vary in a JH-dependent manner only in termites, are involved in the morphological transformation of soldiers. The expression of 20E-related genes was analyzed in *Met* knockdown termites and wood roaches. Among the genes commonly downregulated by *Met* knockdown in both species, *HR3* knockdown suppressed molting in JH-treated termites. In contrast, *HR39* was downregulated by *Met* knockdown only in termites. Knockdown of *HR39* in JH-treated termites did not suppress molting, but the molted individuals exhibited worker-like morphology and suppressed morphological changes (Fig. 13.5; Masuoka et al. 2018a). This gene is known to be active during pupation in holometabolous insects. It is possible that pupation with dramatic morphological changes and worker differentiation share a common mechanism. These results suggest that the molt-promoting function of JH may have been acquired preadaptively in the common ancestor of termites and wood roaches. In addition, termites may have acquired a morphogenetic mechanism involving molting by recruiting preexisting genes downstream of JH. Although unpublished, *HR39* has also been shown to be involved in the expression of *insulin-like receptor 3 (InR3*), a cell growth factor gene. Insects generally have two *InR*s (*InR1* and *InR2*) resulting from gene duplication; however, an additional *InR3* is acquired as a duplicated gene in the order Blattodea, including termites (Kremer et al. 2018). Termites have three *InR*s; however, *HR39* seems to be involved in the expression of only one of them, InR3. During soldier differentiation, *InR* is involved in mandibular development, which may be regulated by *HR39*.

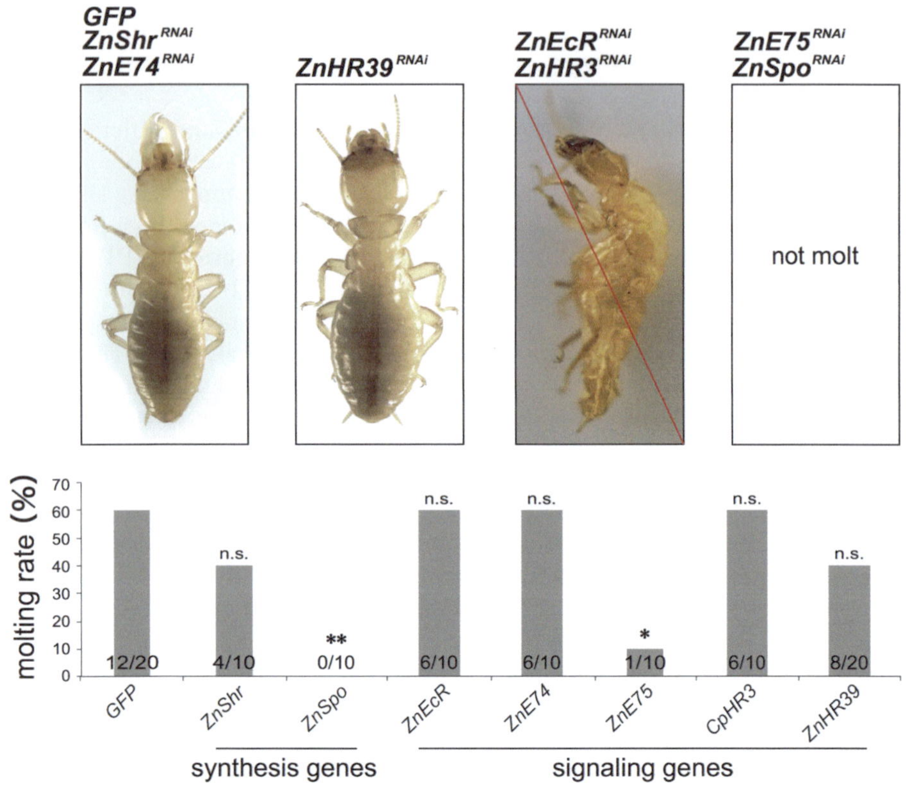

Fig. 13.5 Phenotype and molting rate of a juvenile hormone-induced presoldier under the RNAi treatment of 20-hydroxyecdysone-related genes in *Zootermopsis nevadensis*. (Masuoka et al. 2018a)

13.8 The Role of JH in Reproductive Caste

JHs are expected to play an important role in the differentiation of the reproductive caste. In termites, in addition to primary reproductives (kings and queens) that establish colonies, there are neotenic reproductives (replacement kings and queens) that replace the primary reproductives and take over reproduction (Roisin 2000). Since neotenic reproductives are specialized individuals in which only the gonads are significantly developed among the adult organs, it is possible that JH plays an important role in the development of localized adult traits. *R. speratus* would be a good material to test this possibility since the differentiation of neotenic reproductives can be artificially induced by isolating individuals from their maternal nests. Currently, we do not have any data directly indicating that JH is involved in determining the differentiation of neotenic reproductives. However, it is likely that the activation of the Met-mediated JH signaling pathway is important, at least for the prominent gonadal development of neotenic reproductives (Saiki et al. 2015). Knockdown of Met expression during the differentiation of female neotenic reproductives resulted in a significant decrease in vitellogenin (yolk precursor protein) expression after differentiation. The multiple functions of JH may operate in a heterochronic and site-specific manner to shape the diverse castes of termites.

13.9 The Role of JH in Caste Differentiation of Hymenopteran Social Insects

It has long been known that JH plays an important role in queen differentiation. JH treatment of third-instar honeybee larvae promotes their differentiation into queens. Royalactin, a component of royal jelly, has been associated with increased JH levels via the epidermal growth factor (EGF) receptor (Kamakura 2011). Differentiated workers also show a division of labor between internal and external work, and JH is known to affect work transitions. In workers, the corpora allata gradually enlarge, JH synthesis increases, and the transition from internal to external work occurs as the amount of JH in the body increases (Sasagawa et al. 1989). JHA (methoprene) treatment of workers promoted forager transformation (Robinson 1985). Thus, in hymenopteran social insects, JH is not only a determinant of caste differentiation but also plays an important role in caste modification after differentiation. In a species of wasp, the difference in the amount of JH titer is known to be involved in ranking individuals (Röseler et al. 1984). JH is also known to be involved in soldier differentiation in ants. Individuals with JH levels above a certain threshold have a prolonged larval period and differentiate into workers with larger body sizes. These larger individuals act as soldiers to defend the nest. JH treatment of larvae is also known to induce differentiation into larger-than-normal supersoldiers (Rajakumar et al. 2012).

13.10 The Role of JH in Other Polyphenisms

So far, The author has introduced the role of JH in caste differentiation in social insects, but JH also plays an important role in many other polyphenisms. In one dung beetle species, *Onthophagus taurus*, the horn size of adult males is dimorphic, depending on nutritional conditions during the larval stage. In this case, the nutritional condition during the larval stage is translated into the amount of JH titer in the body, which manifests as horn polyphenism (Emlen and Nijhout 2001). Similarly, male mandibular polyphenism in the stag beetle, *Cyclommatus metallifer*, is known to involve fluctuations in JH titer in response to nutritional conditions during the larval stage (Goto et al. 2011). The polymorphism of such sex-specific traits involves a sex-determining gene (*double sex*: *dsx*), and knockdown of *dsx* reduces susceptibility to JH in males while conversely increasing susceptibility in females (Goto et al. 2014). Variation in JH titer due to nutritional conditions during the larval stage and male-specific JH sensitivity due to sex-determining genes produce mandibular polyphenism in this species. Another well-known example of weapon trait polyphenism is the defensive form of the water flea. In some daphnids, offspring of females sensitive to kairomones released by natural enemies differentiate into individuals with defensive traits. Again, JH acts as a reactor to environmental factors, triggering plastic morphogenesis. In the crustacean *Daphnia magna*, the precursor of JH, methyl farnesoate (MF), acts as a "juvenile hormone." In several *Daphnia* species (*D. pulex* and *D. galeata*), MF treatment has been shown to promote the expression of defense traits (Miyakawa et al. 2013; Oda et al. 2011). Additionally, MF plays a crucial role in *Daphnia* development, as discussed in detail in Chap. 14. This JH-mediated reproductive polymorphism is also known in insects. One aphid species switches from monogenetic to sexual reproduction under short-day cold conditions, and the expression of JH esterase, which is involved in the metabolism of JH, is upregulated during this process (Ishikawa et al. 2012, 2025). JH is also known to be involved in polyphenism, which is associated with plastic changes in the body color. The most famous example is the color change associated with phase polyphenism in the desert locust, *Schistocerca gregaria*. When raised under high-density conditions, individuals develop a gregarious phase, characterized by black body coloration and long wings suited for flight. Conversely, in low-density situations, they differentiate into a solitary phase, exhibiting body colors suited for mimicry. Solitary individuals show green body coloration, and JH is involved in developing this body coloration. Experiments with albino individuals have shown that JH and a pigmentation hormone (His-corazonin) cooperate to determine the body color of solitary-phase individuals (Tanaka 2000). Phase polyphenism is also observed in the common cutworm, *Spodoptera litura*. In this species, rearing under high-density conditions produces a gregarious phase with strong flight ability, while solitary individuals have a longer larval period and heavier pupae than colonial individuals, which is related to JH titer during the final larval instar. When final-instar larvae reared under high-density conditions are subjected to JH treatment, their instar period is extended, and they exhibit pupal weights equivalent to those of solitarily reared individuals (Tojo et al. 1985). In addition, JH is involved in wing polymorphism associated with phase polyphenism in plant hoppers (Iwanaga and Tojo 1986; Xu and Zhang 2017). In the sphingid, *Manduca quinquemaculata*, larval body coloration changes depending on the temperature conditions at hatching. Larvae hatched at low temperatures have black body coloration, whereas those hatched at high temperatures have green body coloration (Hudson 1966). A related species, *M. sexta*, does not exhibit this plasticity, but mutations can produce individuals that respond to heat shock at 42 °C. This body color change is caused by a change in JH titer. Suzuki and Nijhout (2006) used these mutants to experimentally demonstrate genetic accommodation and assimilation. By the 13th generation of crosses, all mutants had changed body color in response to heat shock and had become responsive to heat shock at 28 °C. In other words, genetic accommodation that enhances plasticity in response to heat shock when selection is applied has been reproduced. Conversely, when unresponsive populations were crossed, body color plasticity disappeared, and genetic assimilation occurred. In nature, the temperature-dependent wing patterns of butterflies have been shown to be similar to those of other species in other regions at different temperatures (Shapiro 1976; Nijhout 1984). Such plastic phenotypes may be a driving force for new speciation, and JH may play an important role in the variation of plastic traits.

References

Charles JP, Iwema T, Epa VC, Takaki K, Rynes J, Jindra M (2011) Ligand-binding properties of a juvenile hormone receptor, Methoprene-tolerant. Proc Natl Acad Sci USA 108:21128–21133. https://doi.org/10.1073/pnas.1116123109

Cornette R, Gotoh H, Koshikawa S, Miura T (2008) Juvenile hormone titers and caste differentiation in the damp-wood termite *Hodotermopsis sjostedti*. J Insect Physiol 54:922–930. https://doi.org/10.1016/j.jinsphys.2008.04.017

Deligne J, Quennedey A, Blum MS (1981) The enemies and defense mechanisms of termites. In: Hermann HR (ed) Social insects, vol 2. Academic Press, New York, pp 1–76

Elias-Neto M, Soares MP, Simoes ZL, Hartfelder K, Bitondi MM (2010) Developmental characterization, function and regulation of a Laccase2 encoding gene in the honey bee, *Apis mellifera* (hymenoptera, Apinae). Insect Biochem Mol Biol 40:241–251. https://doi.org/10.1016/j.ibmb.2010.02.004

Emlen DJ, Nijhout HF (2001) Hormonal control of male horn length dimorphism in *Onthophagus taurus* (Coleoptera: Scarabaeidae): a second critical period of sensitivity to juvenile hormone. J Insect Physiol 47:1045–1054. https://doi.org/10.1016/s0022-1910(01)00084-1

Evans PD (1980) Biogenic amines in the insect nervous system. Adv Insect Physiol 15:317–473. https://doi.org/10.1016/S0065-2806(08)60143-5

Gilbert SF, Epel D (2009) Ecological developmental biology: integrating epigenetics, medicine, and evolution. Sinauer Associates, Sunderland, MA

Gotoh H, Cornette R, Koshikawa S, Okada Y, Miura T (2011) Juvenile hormone regulates extreme mandible growth in male stag beetles. PLoS One 6:e21139. https://doi.org/10.1371/journal.pone.0021139

Gotoh H, Miyakawa H, Ishikawa A et al (2014) Developmental link between sex and nutrition; doublesex regulates sex-specific mandible growth via juvenile hormone signaling in stag beetles. PLoS Genet 10:e1004098. https://doi.org/10.1371/journal.pgen.1004098

Hangartner W, Masner P (1973) Juvenile hormone: inhibition of ecdysis in larvae of the German cockroach, *Blattella germanica*. Experientia 9:1358–1359. https://doi.org/10.1007/bf01922816

Hiruma K, Riddiford LM (2009) The molecular mechanisms of cuticular melanization: the ecdysone cascade leading to dopa decarboxylase expression in Manduca sexta. Insect Biochem Mol Biol 39:245–253. https://doi.org/10.1016/j.ibmb.2009.01.008

Howard RW, Haverty MI (1979) Termites and juvenile hormone analogues: a review of methodology and observed effects. Sociobiology 4:269–278

Hrdy I, Krecek J (1972) Development of superfluous soldiers induced by juvenile hormone analogues in the termite, *Reticulitermes lucifugus santonensis*. Insect Soc 14:104–109. https://doi.org/10.1007/BF02224728

Hudson A (1966) Proteins in the haemolymph and other tissues of the developing tomato hornworm, *Protoparce quinquemaculata* Haworth. Can J Zool 44:541–555. https://doi.org/10.1139/z66-058

Ishikawa A, Ogawa K, Gotoh H, Walsh TK, Tagu D, Brisson JA et al (2012) Juvenile hormone titre and related gene expression during the change of reproductive modes in the pea aphid. Insect Mol Biol 21:49–60. https://doi.org/10.1111/j.1365-2583.2011.01111.x

Ishikawa A, Gotoh H, Ogawa K, Kanbe T, Akimoto SI, Miura T (2025) Loss of photoperiodic control of juvenile-hormone signaling pathway underlying the evolution of obligate parthenogenesis in the pea aphid. Zool Sci 42:186–195. https://doi.org/10.2108/zs240075

Ishimaru Y, Tomonari S, Matsuoka Y, Watanabe T, Miyawaki K, Bando T et al (2016) TGF-β signaling in insects regulates metamorphosis via juvenile hormone biosynthesis. Proc Natl Acad Sci USA 113:5634–5639. https://doi.org/10.1073/pnas.1600612113

Iwanaga K, Tojo S (1986) Effects of juvenile hormone and rearing density on wing dimorphism and oöcyte development in the brown planthopper, *Nilaparvata lugens*. J Insect Physiol 32:585–590. https://doi.org/10.1016/0022-1910(86)90076-4

Jindra M, Uhlirova M, Charles JP, Smykal V, Hill RJ (2015) Genetic evidence for function of the bHLH-PAS protein Gce/met as a juvenile hormone receptor. PLoS Genet 11:e1005394. https://doi.org/10.1371/journal.pgen.1005394

Kamakura M (2011) Royalactin induces queen differentiation in honeybees. Nature 473:478–483. https://doi.org/10.1038/nature10093

Kayukawa T, Minakuchi C, Namiki T, Togawa T, Yoshiyama M, Kamimura M et al (2012) Transcriptional regulation of juvenile hormone-mediated induction of Krüppel homolog 1, a repressor of insect metamorphosis. Proc Natl Acad Sci USA 109:11729–11734. https://doi.org/10.1073/pnas.1204951109

Konopova B, Jindra M (2007) Juvenile hormone resistance gene *Methoprene-tolerant* controls entry into metamorphosis in the beetle *Tribolium castaneum*. Proc Natl Acad Sci USA 104:10488–10493. https://doi.org/10.1073/pnas.0703719104

Konopova B, Smykal V, Jindra M (2011) Common and distinct roles of juvenile hormone signaling genes in metamorphosis of holometabolous and hemimetabolous insects. PLoS One 6:e28728. https://doi.org/10.1371/journal.pone.0028728

Korb J (2015) Juvenile hormone: a central regulator of termite caste polyphenism. In: Zayed A, Kent CF (eds) Advances in insect physiology, vol. 48. Academic Press, London, pp 131–161

Kremer LPM, Korb J, Bornberg-Bauer E (2018) Reconstructed evolution of insulin receptors in insects reveals duplications in early insects and cockroaches. J Exp Zool B 330:305–311. https://doi.org/10.1002/jez.b.22809

Lo N, Eggeleton P (2011) Termite phylogenetics and co-cladogenesis with symbionts. In: Bignell DE, Roisin Y, Lo N (eds) Biology of termite: a modern synthesis. Springer, Heidelberg, pp 27–50

Lüscher M (1958) Experimentelle Erzeugung von Soldaten bei der Termite Kalotermes flavicollis (Fabr). Naturwissenschaften 45:69–70. https://doi.org/10.1007/BF00638631

Ma Z, Guo W, Guo X, Wang X, Kang L (2011) Modulation of behavioral phase changes of the migratory locust by the catecholamine metabolic pathway. Proc Natl Acad Sci USA 108:3882–3887. https://doi.org/10.1073/pnas.1015098108

Maekawa K, Nakamura S, Watanabe D (2012) Termite soldier differentiation in incipient colonies is related to parental proctodeal trophallactic behavior. Zool Sci 29:213–217. https://doi.org/10.2108/zsj.29.213

Maekawa K, Hayashi Y, Lee T, Lo N (2014) Presoldier differentiation of Australian termite species induced by juvenile hormone analogues. Austral Entomol 53:138–143. https://doi.org/10.1111/aen.12060

Marchal E, Hult FE, Huang J, Pang Z, Stay B, Tobe SS (2014) Methoprene-tolerant (met) knockdown in the adult female cockroach, *Diploptera punctata* completely inhibits ovarian development. PLoS One 9:e106737. https://doi.org/10.1371/journal.pone.0106737

Masner P, Hangartner W, Suchy M (1975) Reduced titers of ecdysone following juvenile hormone treatment in the German cockroach, Blattella germanica. J Insect Physiol 21:1755–1762. https://doi.org/10.1016/0022-1910(75)90237-1

Masuoka Y, Maekawa K (2016a) Ecdysone signaling regulates soldier-specific cuticular pigmentation in the termite *Zootermopsis nevadensis*. FEBS Lett (12):1694–1703. https://doi.org/10.1002/1873-3468.12219

Masuoka Y, Maekawa K (2016b) Gene expression changes in the tyrosine metabolic pathway regulate caste-specific cuticular pigmentation of termites. Insect Biochem Mol Biol 74:21–31. https://doi.org/10.1016/j.ibmb.2016.04.006

Masuoka Y, Yaguchi H, Suzuki R, Maekawa K (2015) Knockdown of the juvenile hormone receptor gene inhibits soldier-specific morphogenesis in the damp-wood termite *Zootermopsis nevadensis* (Isoptera: Archotermopsidae). Insect Biochem Mol Biol 64:25–31. https://doi.org/10.1016/j.ibmb.2015.07.013

Masuoka Y, Toga K, Nalepa CA, Maekawa K (2018a) A crucial caste regulation gene detected by comparing termites and sister group cockroaches. Genetics 209:1225–1234. https://doi.org/10.1534/genetics.118.301038

Masuoka Y, Yaguchi H, Toga K, Shigenobu S, Maekawa K (2018b) TGF β signaling related genes are involved in hormonal mediation during termite soldier differentiation. PLoS Genet 14:e1007338. https://doi.org/10.1371/journal.pgen.1007338

Matsuyama S, Nagao T, Sasaki K (2015) Consumption of tyrosine in royal jelly increases brain levels of dopamine and tyramine and promotes transition from normal to reproductive workers in queenless honey bee colonies. Gen Comp Endocrinol 211:1–8. https://doi.org/10.1016/j.ygcen.2014.11.005

Minakuchi C, Zhou X, Riddiford LM (2008) Krüppel homolog 1 (Kr-h1) mediates juvenile hormone action during metamorphosis of *Drosophila melanogaster*. Mech Dev 125:91–105. https://doi.org/10.1016/j.mod.2007.10.002

Miura T, Scharf ME (2011) Molecular basis underlying caste differentiation in termites. In: Bignell DE, Roisin Y, Lo N (eds) Biology of termite: a modern synthesis. Springer, Heidelberg, pp 211–253

Miyakawa H, Gotoh H, Sugimoto N, Miura T (2013) Effect of juvenoids on predator-induced polyphenism in the water flea, *Daphnia pulex*. J Exp Zool A 319:440–450. https://doi.org/10.1002/jez.1807

Mizoguchi A (2001) Effects of juvenile hormone on the secretion of prothoracicotropic hormone in the last-and penultimate-instar larvae of the silkworm *Bombyx mori*. J Insect Physiol 47:767–775. https://doi.org/10.1016/s0022-1910(01)00052-x

Nakagawa Y, Henrich VC (2009) Arthropod nuclear receptors and their role in molting. FEBS J 276:6128–6157. https://doi.org/10.1111/j.1742-4658.2009.07347.x

Nijhout HF (1984) Colour pattern modification by coldshock in Lepidoptera. Development 81:287–305. https://doi.org/10.1242/dev.81.1.287

Nijhout HF, Wheeler DE (1982) Juvenile hormone and the physiological basis of insect polyphenism. Q Rev Biol 57:109–133. https://doi.org/10.1086/412671

Oda S, Kato Y, Watanabe H, Tatarazako N, Iguchi T (2011) Morphological changes in *Daphnia galeata* induced by a crustacean terpenoid hormone and its analog. Environ Toxicol Chem 30:232–238. https://doi.org/10.1002/etc.378

Okada Y, Sasaki K, Miyazaki S, Shimoji H, Tsuji K, Miura T (2015) Social dominance and reproductive differentiation mediated by dopaminergic signaling in a queenless ant. J Exp Biol 218:1091–1098. https://doi.org/10.1242/jeb.118414

Rajakumar R, San Mauro D, Dijkstra MB, Huang MH, Wheeler DE, Hiou-Tim F et al (2012) Ancestral developmental potential facilitates parallel evolution in ants. Science 335:79–82. https://doi.org/10.1126/science.1211451

Robinson GE (1985) Effects of a juvenile hormone analogue on honey bee foraging behaviour and alarm pheromone production. J Insect Physiol 31:277–282. https://doi.org/10.1016/0022-1910(85)90003-4

Roisin Y (2000) Diversity and evolution of caste patterns. In: Abe T, Bignell DE, Higashi M (eds) Termites: evolution, sociality, symbioses, ecology. Kluwer Academic Press, Dordrecht, pp 95–119

Röseler PF, Röseler I, Strambi A, Augier R (1984) Influence of insect hormones on the establishment of dominance hierarchies among foundresses of the paper wasp, *Polistes gallicus*. Behav Ecol Sociobiol 15:133–142. https://doi.org/10.1007/BF00299381

Saiki R, Gotoh H, Toga K, Miura T, Maekawa K (2015) High juvenile hormone titer and abdominal activation of the JH signaling may induce reproduction of termite neotenics. Insect Mol Biol 24:432–441. https://doi.org/10.1111/imb.12169

Sasagawa H, Sasaki M, Okada I (1989) Hormonal control of the division of labor in adult honeybees (Apis mellifera L.): I. Effect of methoprene on corpora allata and hypopharyngeal gland, and its α-glucosidase activity. Appl Entomol Zool 24:66–77. https://doi.org/10.1303/aez.24.66

Scharf ME, Ratliff CR, Hoteling JT, Pittendrigh BR, Bennett GW (2003) Caste differentiation responses of two sympatric *Reticulitermes* termite species to juvenile hormone homologs and synthetic juvenoids in two laboratory assays. Insect Soc 50:346–354. https://doi.org/10.1007/s00040-003-0690-4

Scharf ME, Ratliff CR, Wu-Scharf D, Zhou X, Pittendrigh BR, Bennett GW (2005) Effects of juvenile hormone III on Reticulitermes flavipes: changes in hemolymph protein composition and gene expression. Insect Biochem Mol Biol 35:207–215. https://doi.org/10.1016/j.ibmb.2004.12.001

Shapiro AM (1976) Seasonal polyphenism. In: Hecht MK, Steere WC, Wallace B (eds) Evolutionary biology, vol 9. Springer, Boston, pp 259–333

Simpson SJ, Sword GA, Lo N (2011) Polyphenism in insects. Curr Biol 21:R738–R749. https://doi.org/10.1016/j.cub.2011.06.006

Suzuki Y, Nijhout HF (2006) Evolution of a polyphenism by genetic accommodation. Science:650–652. https://doi.org/10.1126/science.1118888

Tanaka S (2000) Induction of darkening by corazonins in several species of Orthoptera and their possible presence in ten insect orders. Appl Entomol Zool 35:509–517. https://doi.org/10.1303/aez.2000.509

Terrapon N, Li C, Robertson HM, Ji L, Meng X, Booth W et al (2014) Molecular traces of alternative social organization in a termite genome. Nat Commun 5:3636. https://doi.org/10.1038/ncomms4636

Thomas HE, Stunnenberg HG, Stewart AF (1993) Heterodimerization of the drosophila ecdysone receptor with retinoid X receptor and ultraspiracle. Nature 362:471–475. https://doi.org/10.1038/362471a0

Thorne BL (1996) Termite terminology. Sociobiology 28:253–263

Tian L, Zhou X (2014) The soldiers in societies: defense, regulation, and evolution. Int J Biol Sci 10:296. https://doi.org/10.7150/ijbs.6847

Toga K, Hojo M, Miura T, Maekawa K (2009) Presoldier induction by a juvenile hormone analog in the nasute termite *Nasutitermes takasagoensis* (Isoptera: Termitidae). Zool Sci 26:382–388. https://doi.org/10.2108/zsj.26.382

Toga K, Hanmoto S, Suzuki R, Watanabe D, Miura T, Maekawa K (2016) Sexual difference in juvenile-hormone titer in workers leads to sex-biased soldier differentiation in termites. J Insect Physiol 87:63–70. https://doi.org/10.1016/j.jinsphys.2016.02.005

Tojo S, Morita M, Hiruma K (1985) Effects of juvenile hormone on some phase characteristics in the common cutworm, *Spodoptera litura*. J Insect Physiol 31:243–249. https://doi.org/10.1016/0022-1910(85)90126-X

Tsuchiya M, Watanabe D, Maekawa K (2008) Effect on mandibular length of juvenile hormones and regulation of soldier differentiation in the termite *Reticulitermes speratus* (Isoptera: Rhinotermitidae). Appl Entomol Zool 43:307–314. https://doi.org/10.1303/aez.2008.307

Watanabe D, Gotoh H, Miura T, Maekawa K (2014) Social interactions affecting caste development through physiological actions in termites. Front Physiol 5:127. https://doi.org/10.3389/fphys.2014.00127

Weisel-Eichler A, Haspel G, Libersat F (1999) Venom of a parasitoid wasp induces prolonged grooming in the cockroach. J Exp Biol 202:957–964. https://doi.org/10.1242/jeb.202.8.957

West-Eberhard MJ (2003) Developmental plasticity and evolution. Oxford University Press, New York

Xu HJ, Zhang CX (2017) Insulin receptors and wing dimorphism in rice planthoppers. Philos Trans R Soc Lond Ser B Biol Sci 372:20150489. https://doi.org/10.1098/rstb.2015.0489

Yaguchi H, Masuoka Y, Inoue T, Maekawa K (2015) Expressions of juvenile hormone biosynthetic genes during presoldier differentiation in the incipient colony of *Zootermopsis nevadensis* (Isoptera: Archotermopsidae). Appl Entomol Zool 50:497–508. https://doi.org/10.1098/rstb.2015.0489

Yaguchi H, Shigenobu S, Hayashi Y, Miyazaki S, Toga K, Masuoka Y et al (2018) A lipocalin protein, Neural Lazarillo, is key to social interactions that promote termite soldier differentiation. Proc Royal Soc B 285:20180707. https://doi.org/10.1098/rspb.2018.0707

Zhou X, Oi FM, Scharf ME (2006) Social exploitation of hexamerin: RNAi reveals a major caste-regulatory factor in termites. Proc Natl Acad Sci USA 103:4499–4504. https://doi.org/10.1073/pnas.0508866103

Chapter 14
Juvenile Hormone in the Environmental Response of Crustacea

Hitoshi Miyakawa and Kenji Toyota

Abstract Juvenile hormones (JHs) are key endocrine factors that regulate various physiological phenomena in crustaceans, similar to insects. In crustaceans, such as *Daphnia*, JHs regulate unique environmental responses, including environmental sex determination (ESD) and inducible defenses. Functional analyses of JH receptors involved in signal transduction are also underway. However, the biosynthetic organs responsible for JH production and the structural details of JH in the body remain unclear. In contrast to *Daphnia*, methyl farnesoate (MF) is synthesized in the mandibular organ of decapod crustaceans, such as shrimp and crabs, and is known to regulate molting and reproduction. However, research on the molecular signaling cascade of JHs in these species has made little progress. When discussing JHs in crustaceans, it is crucial to avoid lumping together the knowledge from these entirely different taxonomic groups under the umbrella of "crustaceans." Instead, a comparative approach should be adopted, considering the phylogenetic relationships within Pancrustacea, which includes both insects and crustaceans.

Keywords Pancrustacean phylogeny · Environmental sex determination · Inducible defense · Methyl farnesoate · *Daphnia* · Decapod

H. Miyakawa (✉)
Center for Bioscience Research and Education, Utsunomiya University, Utsunomiya, Japan
e-mail: h-miya@cc.utsunomiya-u.ac.jp

K. Toyota
Graduate School of Integrated Sciences for Life, Hiroshima University, Higashihiroshima, Japan

T. Shinoda et al. (eds.), *Juvenile Hormone*, Entomology Monographs,
https://doi.org/10.1007/978-981-95-8605-9_14

14.1 Introduction

Juvenile hormones (JHs), similar to those of insects, exist in crustaceans and play a vital role in regulating molting, metamorphosis, reproduction, and various environmental responses (Laufer et al. 1993; Nagaraju 2007; Miyakawa et al. 2014). Findings from studies on representative crustacean species are often assumed to apply universally across all crustaceans. A prime example of this generalization is the recognition of "methyl farnesoate (MF) as a crustacean juvenile hormone." MF, an unepoxidized precursor of JH III and the primary JH in insects, was first identified in the crab *Libinia emarginata* (Laufer et al. 1987). Over the past four decades, MF has been confirmed in many decapod crustaceans (Laufer et al. 1987; Laufer and Biggers 2001). However, despite the vast diversity of crustaceans, direct evidence supporting the use of MF as a JH outside of decapods is scarce. In contrast, the availability of genomic information for various species has suggested that the endocrine signaling pathways centered around JH (the JH signaling pathway) are broadly conserved across crustaceans (Miyakawa et al. 2014, 2018a, b).

This chapter begins by organizing the phylogenetic relationships among crustacean taxa and between crustaceans and insects, followed by an examination of the genes related to the JH signaling pathway conserved in each taxonomic group. Subsequently, it highlights two examples: *Daphnia* species, which are relatively closely related to insects and have recently been the focus of molecular studies on JH mechanisms, and decapods, the most extensively studied group in JH research. The physiological roles of JHs in these species are discussed, including their adaptation to fluctuating natural environments.

14.2 Pancrustacean Phylogeny

The phylogenetic relationships of arthropods, which encompass an extremely diverse array of species, have long been a topic of debate. However, it is now widely accepted that there is a clade known as Pancrustacea, comprising Hexapoda (including insects) and Crustacea. Within this clade, Hexapoda forms a monophyletic group, whereas the remaining crustaceans do not form a monophyletic group. One of the most critical questions in the molecular phylogenetics of Pancrustacea is determining which group of crustaceans is the sister group to the Hexapoda. Oakley et al. (2013) proposed that Hexapoda, Remipedia, Branchiopoda, and Cephalocarida form a monophyletic group named Allotriocarida, with Remipedia being the closest sister group to Hexapoda. According to their findings, Multicrustacea—a group including Malacostraca and Copepoda (as defined by Regier et al. (2010))—is the sister group to Allotriocarida, whereas Oligostraca, which includes Ostracoda, is placed outside these clades. This major phylogenetic framework has been supported

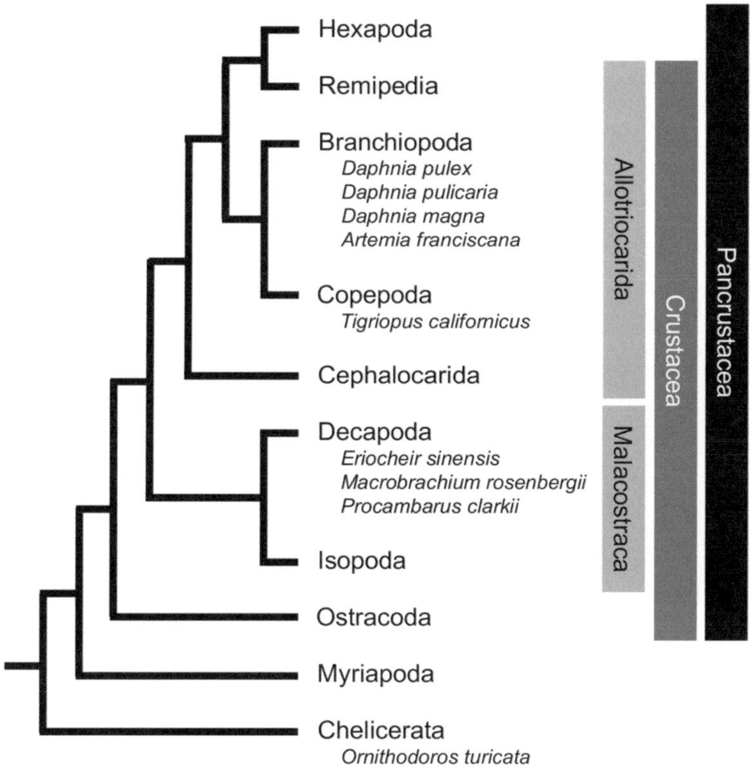

Fig. 14.1 Phylogenetic tree of Arthropoda. The branching pattern is based on Bernot et al. (2023), with some clades omitted for simplicity. Species names used for homolog searches are listed under each taxon

by various studies over the subsequent decade (Schwentner et al. 2017, 2018; Lozano-Fernandez et al. 2019). However, recent findings have surprisingly shown that Copepoda is part of the Allotriocarida clade, suggesting that the monophyletic group Multicrustacea does not exist (Bernot et al. 2023; Yu et al. 2024; Su et al. 2024). As genomic data from more species become available, the phylogenetic relationships within Pancrustacea are likely to be revisited and potentially revised. Therefore, when discussing the commonalities and diversity of JH within Pancrustacea and their evolutionary trajectories, it is essential to stay informed about and reference the latest molecular phylogenetic insights. In this chapter, we base our discussion on the phylogenetic relationships proposed by Bernot et al. (2023), as shown in Fig. 14.1.

Table 14.1 Eight Crustacea and one Chelicerata with chromosome-level assemblies

Species name	Common name	Taxon	Complete by Benchmarking Universal Single-Copy Orthologs (BUSCO)[a]	Accession number
Daphnia pulex	Water flea	Branchiopoda: Cladocera	97.2%	GCF_021134715.1
Daphnia pulicaria	Water flea	Branchiopoda: Cladocera	97.6%	GCF_021234035.1
Daphnia magna	Water flea	Branchiopoda: Cladocera	97.8%	GCF_020631705.1
Artemia franciscana	Brine shrimp	Branchiopoda: Anostraca	94.8%	GCF_032884065.1
Tigriopus californicus	–	Copepoda: Harpacticoida	97.6%	GCF_007210705.1
Eriocheir sinensis	Chinese mitten crab	Malacostraca: Decapoda	97.9%	GCF_024679095.1
Macrobrachium rosenbergii	Giant river prawn	Malacostraca: Decapoda	97.1%	GCF_040412425.1
Procambarus clarkii	Red swamp crayfish	Malacostraca: Decapoda	98.4%	GCF_040958095.1
Ornithodoros	Relapsing fever tick	Chelicerata	97.8%	GCF_037126465.1

[a]Sum of "single-copy" and "duplicated"

14.3 Conservation of Juvenile Hormone Signaling–Related Genes in Crustacea

To determine whether an endocrine mechanism involving JH exists in noninsect arthropods, particularly crustaceans, the primary approach is to identify homologs of known JH signaling–related genes in their genomes. Various homologs of JH signaling factors have been cloned in several crustacean species, and comprehensive searches for these homologs have been conducted whenever draft genomes or transcriptomes were analyzed. In this section, we provide an overview of the conservation of the JH signaling pathway in Pancrustacea by examining the presence of homologs of major JH signaling–related genes in eight crustacean species with highly reliable chromosome-level assemblies enabled by recent advances in long-read sequencing technologies (Table 14.1). Additionally, one chelicerate species that serves as an outgroup to Pancrustacea was included in this analysis (Table 14.1). Table 14.2 lists the National Center for Biotechnology Information (NCBI) RefSeq accession numbers of the homologs identified in each species. Homolog identification was primarily based on reciprocal top hits from Basic Local Alignment Search Tool (BLAST) searches using the amino acid sequences of JH signaling–related genes from the fruit fly, *Drosophila melanogaster*; the red flour beetle, *Tribolium castaneum*; and the silkworm, *Bombyx mori* as references. For genes that produce multiple isoforms, only one representative isoform is listed. If multiple homologs are listed for a single gene, these homologs are encoded in different genomic regions.

14.3.1 Genes Involved in Juvenile Hormone Biosynthesis

The biosynthetic pathway of juvenile hormone (JH), as elucidated from insect studies, is divided into two phases: the "early step," where farnesyl diphosphate (FPP) is synthesized from acetyl-CoA via the mevalonate pathway—a process widely conserved in eukaryotes—and the "late step," where FPP is converted into JH III or other related compounds specific to JH biosynthesis (Bellés et al. 2005; Goodman and Cusson 2011; Noriega 2014; Rivera-Pérez et al. 2020). In the early steps, several enzyme genes play crucial roles in the transformation: *acetoacetyl-CoA thiolase*, *3-hydroxy-3-methylglutaryl-CoA synthase* (*HMG-CoA synthase*), *3-hydroxy-3-methylglutaryl-CoA reductase* (*HMG-CoA reductase*), *mevalonate kinase*, *phosphomevalonate kinase*, *diphosphomevalonate decarboxylase*, *isopentenyl diphosphate isomerase* (*IPP isomerase*), and *farnesyl diphosphate synthase* (*FPP synthase*). These genes are generally conserved as single copies across crustaceans and mites, with some exceptions of duplication in specific species (Table 14.2). However, *IPP isomerase* has not been identified in the genome of *Daphnia magna*. This absence is likely due to the incompleteness of the genome data, as other *Daphnia* species, such as *D. pulex* and *D. pulicaria*, possess this gene.

In the late steps, farnesyl pyrophosphate is converted to farnesoic acid via the actions of *farnesyl pyrophosphatase* (*FPPase*), *farnesol dehydrogenase* (*FDH*), and *farnesal dehydrogenase* (*FaDH*) (Bellés et al. 2005; Goodman and Cusson 2011; Noriega 2014). These enzyme genes exhibit diversity even within insects, suggesting that a high sequence similarity does not necessarily indicate functional conservation. For instance, homologs of insect *FPPase* have not been identified in crustaceans or mites (Table 14.2), implying the involvement of entirely different genes in these taxa. In decapods, *FaDH* homologs are often duplicated, with the Chinese mitten crab, *Eriocheir sinensis,* possessing over ten tandemly duplicated copies (Table 14.2). This raises the possibility that *FaDH* may perform functions distinct from those of its insect counterparts in decapod crustaceans.

In contrast, the enzyme *juvenile hormone acid O-methyltransferase* (*JHAMT*), which converts farnesoic acid into methyl farnesoate (MF) (Shinoda and Itoyama 2003; Minakuchi et al. 2008a), is broadly conserved as a single-copy gene across crustaceans (Table 14.2). Functional studies using recombinant proteins have shown that *D. pulex JHAMT* can convert farnesoic acid to MF (Toyota et al. 2015a). Moreover, because MF exhibits physiological activity in many crustaceans (Laufer et al. 1993; Nagaraju 2007; Miyakawa et al. 2014), it is presumed that *JHAMT* plays a critical role in the final step of JH biosynthesis in crustaceans, similar to its role in insects. On the other hand, in the tick *Ornithodoros turicata*, numerous putative homologs with low similarity to insect *JHAMT* have been identified. Further

Table 14.2 Homologs of juvenile hormone signaling–related genes in eight Crustacea and one Chelicerata genomes

	Daphnia pulex	*Daphnia pulicaria*	*Daphnia magna*	*Artemia franciscana*	*Tigriopus californicus*	*Eriocheir sinensis*	*Macrobrachium rosenbergii*	*Procambarus clarkii*	*Ornithodoros turicata*
Biosynthesis (early step: mevalonate pathway)									
acetoacetyl-CoA thiolase	XM_046582584	XM_046789080	XM_032931338	XM_065712570	XM_059238776	XM_050836945	XM_067106479	XM_069313415	XM_064632041
HMG-CoA synthase	XM_046608162	XM_046795829	XM_045171530	XM_065713204	XM_059225529	XM_050858282	XM_067084337	XM_045769134	XM_064621396
HMG-CoA reductase	XM_046590381	XM_046778933	XM_045168908	XM_065725043	XM_059239638	XM_050842930	XM_067102161	XM_045757547	XM_064605371
mevalonate kinase	XM_046593871	XM_046786571	XM_045177065XM_032934489	XM_065713372	XM_059222814XM_059224763	XM_050841286	XM_067089669	XM_069301437	XM_064604782
phosphome valonate kinase	XM_046600147	XM_046789196	XM_032927319XM_045173081	XM_065724919	XM_059239768	XM_050870793	XM_067132647	XM_045759206	XM_064618333
diphosphome valonate decarboxylase	XM_046598934	XM_046801487	XM_032939731	XM_065718689XM_065718690	XM_059233958	XM_050882712	XM_067125563	XM_045738685	XM_064629460
IPP isomerase	XM_046605449	XM_046778153	No hit	XM_065710820	XM_059237464	XM_050832344	XM_067104521	XM_045764816	XM_064600029
FPP synthase	XM_046586596	XM_046794060	XM_032943805XM_045168226	XM_065701065XM_065700472XM_065700470XM_065713691XM_065713693	XM_059234395	XM_050868035XM_050868072	XM_067126286	XM_069310371	XM_064620946
Biosynthesis (late step)									
FPPase	No hit	No hit	No hit	No hit	No hit	No hit	No hit	No hit	No hit

	FDH	XM_046606932	XM_046799557	XM_045179000	No hit	XM_059223827XM_059235035	XM_050875648XM_050875653XM_050867048XM_050867055XM_050867058XM_050867125XM_050867059XM_050867057XM_050867047XM_050868088XM_050868089XM_050868076XM_050867126XM_050868090	XM_067102110XM_067102103XM_067133794	XM_045757537XM_069310473XM_045757544XM_069310476	No hit
	FaDH	XM_046591007	XM_046794902	XM_032920499	XM_065713527	XM_059228795XM_059227250	XM_050856391XM_050830805XM_050838408	XM_067104159XM_067112032XM_067085484	XM_045766495	XM_064620624
	JHAMT	XM_046603631	XM_046797443	XM_032924760	XM_065718746	XM_059236325	XM_050870644	XM_067082829	XM_045754877	XM_064600027XM_064620102XM_064620099XM_064620098XM_064620101XM_064617797XM_064617801XM_064620104XM_064616513XM_064627288

(continued)

Table 14.2 (continued)

	Daphnia pulex	*Daphnia pulicaria*	*Daphnia magna*	*Artemia franciscana*	*Tigriopus californicus*	*Eriocheir sinensis*	*Macrobrachium rosenbergii*	*Procambarus clarkii*	*Ornithodoros turicata*
MFE (*CYP15*)	XM_046598614XM_046599921XM_046599676XM_046599680XM_046598383XM_046598448XM_046598386XM_046582874XM_046598380XM_046599677XM_046606147XM_046598384XM_046598449	XM_046800432XM_046800424XM_046800406XM_046800442XM_046800391XM_046789081XM_046800426XM_046801743XM_046800430XM_046800496XM_046798734	XM_032940885XM_032940702XM_045179982XM_045179987XM_045179986XM_032941094XM_032941092XM_045179988XM_032941095XM_032939763XM_032939764	XM_065718415XM_065710932	XM_059227753XM_059237887XM_059231998XM_059243298XM_059233012XM_059242723XM_059230214XM_059233363XM_059231852	XM_050856131XM_050856132XM_050856130XM_050834393	XM_067123441XM_067123442XM_067122797(XM_067122797)(XM_067122797)	XM_069304726XM_069300217XM_069305552	No hit
FAMeT	XM_046586437[a]	XM_046786195[a]	XM_032938811[a]	No hit	No hit	XM_050838336	XM_067121502	XM_045739500	No hit
Reception									
Methoprene-tolerant	XM_046599149	XM_046801895XM_046801888	XM_032940759	XM_065700985XM_065702419XM_065728611	XM_059224664	XM_050854850	XM_067095490XM_067096918	XM_045761163	XM_064626174
Steroid receptor coactivator	XM_046582814	XM_046792515	XM_045175470	XM_065710935XM_065709866	No hit	XM_050841910	XM_067132856	XM_069328374	XM_064624859
cycle	XM_046608366	XM_046798058	XM_032925372	XM_065712643	XM_059231951	XM_050835880	XM_067131822	XM_069334459	XM_064626381
hsp83	XM_046601065	XM_046781971	XM_032928151	XM_065722459XM_065722441XM_065721806	XM_059223452	XM_050873094	XM_067092136XM_067092134	XM_069301739XM_045762480	XM_064633528

Degradation									
juvenile hormone esterase	XM_046597282XM_046588263XM_046585591	XM_046783361XM_046775349XM_046792564	XM_045167659XM_032935568XM_045177634	XM_065704633 (KxxHxxD/E)	XM_059241756XM_059241809	XM_050860636	XM_067118575	XM_069339945	XM_064613139XM_064613141XM_064614041XM_064615128XM_064615130XM_064615186XM_064615191XM_064615192XM_064615193XM_064615194XM_064615200XM_064615362XM_064616341XM_064616355
juvenile hormone epoxide hydrolase	XM_046600547	XM_046789940XM_046782842	XM_032927775	XM_065718563	XM_059230420	XM_050873940XM_050873943	XM_067091504XM_067091506	XM_045762455	XM_064632772
juvenile hormone diol kinase	XM_046596014	XM_046783317	XM_045173713	XM_065710868	XM_059228866	XM_050882299	XM_067106285	XM_069323749	XM_064615699
Downstream									
Krüppel--homolog 1	XM_046581836	XM_046776522	XM_045171309	XM_065716420	XM_059233228	XM_050864205	XM_067129038	XM_045768477	No hit

[a]Low similarity to decapod *FAMeT*
[b]This gene lacks GxxHxxD/E motif

research is necessary to determine whether these genes contribute to JH biosynthesis or perform other functions. Interestingly, in some decapods, *farnesoic acid O-methyltransferase* (*FAMeT*), which encodes a protein distinct from *JHAMT*, has been reported to convert farnesoic acid into MF (Gunawardene et al. 2002; Liu et al. 2022). Homologs of *Metapenaeus ensis FAMeT* were identified as single copies in three decapod species but were absent in Allotriocarida and *O. turicata* (Table 14.2). This indicates that even if *FAMeT* serves as the primary biosynthetic enzyme, its role may be restricted to decapods or Malacostraca.

In insects, methyl farnesoate epoxidase (MFE), a CYP15 enzyme, epoxidizes MF to form JH III, the most common active JH (Helvig et al. 2004). While previous studies suggested the absence of *CYP15* genes in *Daphnia* and mites (Daimon and Shinoda 2013), our examination of chromosome-level assemblies revealed multiple *CYP15-like* genes with high similarity to insect *MFE* in all eight crustacean species analyzed, though not in the tick *O. turicata* (Table 14.2). Notably, *Daphnia* exhibits lineage-specific duplications of these genes, raising the possibility that one or more of these paralogs may function as epoxidases in JH biosynthesis.

Overall, a broad overview of the homologs of JH biosynthetic enzymes in crustaceans reveals significant differences in the late steps between insects and crustaceans, as well as among crustacean groups. While MF has been experimentally confirmed as the JH in several decapod species (Laufer et al. 1987; Laufer and Biggers 2001), the structure of JH in other crustaceans remains uncertain. Our liquid chromatography–tandem mass spectrometry (LC-MS/MS) analyses did not detect MF or JH III in *Daphnia* species, suggesting that the diversification of late-step biosynthetic genes may result in the production of distinct JH structures in different taxonomic groups.

14.3.2 Genes Involved in Juvenile Hormone Degradation

The timely degradation of JH is crucial for suppressing its physiological effects and regulating various life processes mediated by JH. The key genes involved in the degradation and metabolism of active JH include *juvenile hormone esterase* (*JHE*), *juvenile hormone epoxide hydrolase* (*JHEH*), and *juvenile hormone diol kinase* (*JHDK*) (Rivera-Pérez et al. 2020). *JHE* converts JHs into JH acids. However, insect genomes contain numerous *JHE*-like sequences that cannot be distinguished from *JHE* based solely on sequence similarity (Ward et al. 1992; Kamita and Hammock 2010). Similarly, in crustaceans, many genes exhibit high similarity to the insect *JHE*. By focusing on sequences containing conserved motifs, that is, the RF, DQ, GQSAG, and GxxHxxD/E motifs (Ward et al. 1992; Kamita and Hammock 2010), one to three candidate *JHE* homologs were identified in each crustacean species analyzed (Table 14.2). For *juvenile hormone epoxide hydrolase* (*JHEH*), which converts JH into JH diol (Debernard et al. 1998; Tsubota et al. 2010), and *juvenile hormone diol kinase* (*JHDK*), which converts JH diol into JH diol phosphate (Fu et al. 2015; Zeng et al. 2016), one or two clear homologs were identified in each of the crustacean species (Table 14.2). These findings suggest that, unlike the JH biosynthetic pathway, the degradation pathway exhibits minimal variation among taxonomic groups. In the tick *O. turicata*, while only one homolog was found for each of *JHEH* and *JHDK*, 14 genes containing the conserved motifs characteristic of *JHE* were identified (Table 14.2). Whether these genes actually use JH as a substrate remains to be clarified in future studies, for example, using recombinant proteins.

14.3.3 Genes Involved in Juvenile Hormone Reception

In insects, Methoprene-tolerant (Met), a member of the basic helix–loop–helix/Per-ARNT-SIM (bHLH-PAS) family, generally functions as a receptor for JH. Met acts as a transcription factor by forming a JH-dependent heterodimer with another bHLH-PAS protein, steroid receptor coactivator (*SRC*), also known as Taiman or FISC, to transmit downstream signals (Charles et al. 2011; Li et al. 2011; Zhang et al. 2011; Jindra et al. 2013, 2015). Homologs of *Met* were widely found in the genomes of all eight crustacean species and the one tick species examined, and *SRC* homologs were identified in all species except the harpacticoid copepod *Tigriopus californicus* (Table 14.2). In *Daphnia*, it has been demonstrated via reporter assays using cultured cells that Met and SRC form a JH-dependent heterodimer and exhibit transcriptional activity through the same specific genomic sequence (JH response element (JHRE)) as in insects (Miyakawa et al. 2013a, 2018a, b; Tanaka et al. 2019). These findings suggest that even if the JH ligand used in vivo differs among taxonomic groups, the JH reception mechanism involving Met is conserved between *Daphnia* and insects and may even be preserved across all arthropods. In addition to SRC, Met forms JH-dependent heterodimers with Cycle, a bHLH-PAS protein and clock gene (Shin et al. 2012). Furthermore, Met interacts with Hsp83 (Jindra et al. 2024). Both the *cycle* and *hsp83* genes are widely conserved across arthropods.

14.3.4 Krüppel Homolog 1 in Crustacea

In insects, the JH-bound Met–SRC complex directly activates the transcription of *Krüppel homolog 1* (*Kr-h1*) (Minakuchi et al. 2008b, 2009). The suppression of *Broad-Complex* and *E93* expression by Kr-h1 constitutes the core mechanism of JH-mediated metamorphosis regulation (Kayukawa et al. 2016, 2017), making *Kr-h1* the most critical target of Met. Although no homolog of *Kr-h1* has been observed in ticks, it is widely conserved among crustaceans (Table 14.2). In decapods, *Kr-h1* expression increases following MF injection (Li et al. 2021a). On the other hand, in *Daphnia*, *Kr-h1* expression is minimally induced by JH, as shown by gene expression analyses and reporter assays (Miyakawa et al. 2018a, b). Furthermore, the JH response element (JHRE) sequence (CCTCCACGTG), commonly found near *Kr-h1* in insects (Kayukawa et al. 2012, 2013) was absent in the vicinity of *Kr-h1* genes across the eight crustacean genomes examined in this chapter (Table 14.2). A single-nucleotide shorter sequence (CTCCACGTG) exists in the genomes of *D. pulex* and related species (Miyakawa et al. 2018a, b). Thus, whether *Kr-h1* serves as a downstream gene of Met across all crustaceans requires careful investigation through further experiments. At the very least, it is predicted that *Kr-h1* contributes minimally to the physiological roles of JH in *Daphnia* and its relatives.

14.4 Juvenile Hormone in the Environmental Response of *Daphnia*

14.4.1 Regulation of Environmental Sex Determination

Daphnia, commonly known as water fleas, exhibit a reproductive strategy called cyclical parthenogenesis, in which they switch between asexual and sexual reproduction depending on environmental conditions (Decaestecker et al. 2009; Fig. 14.2). The parthenogenetic process under favorable conditions is clonal reproduction, where the first meiotic division is skipped, allowing the mother to pass an identical set of chromosomes to the offspring (Hiruta et al. 2010). In contrast, sexual

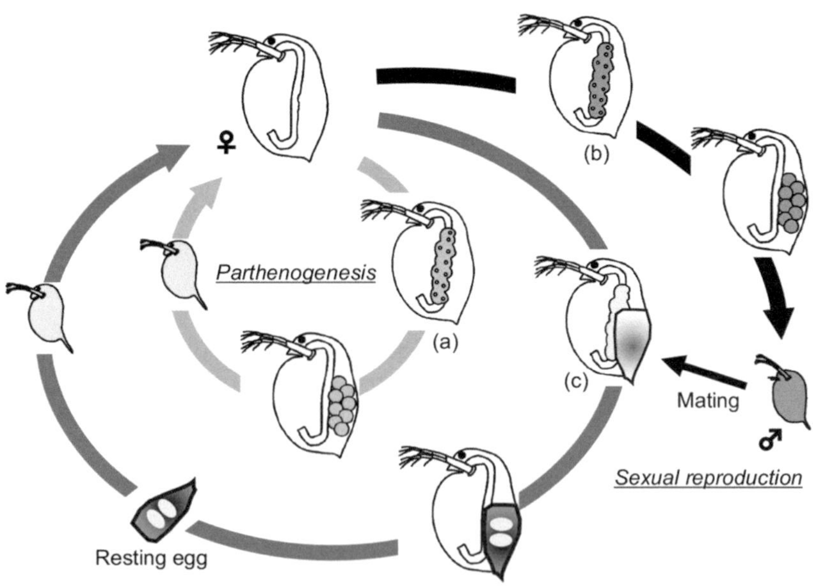

Fig. 14.2 Alteration of reproductive strategies in *Daphnia*. Under favorable conditions, *Daphnia* produces only females via parthenogenesis (**a**). Under unfavorable conditions, some individuals produce males parthenogenetically (**b**), while others produce meiotic oocytes, which become resting eggs (**c**)

reproduction is triggered by environmental stimuli, such as low temperatures, short day lengths, and poor nutritional conditions (Kleiven et al. 1992). Upon sensing these environmental cues, the mother alters the offspring's sex from female to male (environmental sex determination (ESD)). The eggs destined to develop into males inherit the same chromosomal set as their mothers through clonal reproduction. However, their sex is determined within the mother's ovary before being oviposited, and they autonomously develop into males by expressing male-specific differentiation genes, such as doublesex (Kato and Watanabe 2022).

Insights into how mothers regulate the sex of oocytes in their ovaries were obtained accidentally during toxicological studies on the biological effects of various chemicals using *D. magna*. It was discovered that exposing *D. magna* to various JHs or JH analogs used as insect growth regulators could induce male-biased offspring production, regardless of environmental conditions (Olmstead and LeBlanc 2002; Tatarazako et al. 2003). Subsequent studies confirmed that this induction of male production by JH exposure is widespread among cyclical parthenogenetic cladocerans, including *D. pulex*, *D. galeata*, *Ceriodaphnia dubia*, and *Moina macrocopa* (Oda et al. 2005; Toyota et al. 2013a). These findings suggest that elevated JH titers mediate the production of males. Furthermore, the differences in the male-inducing activity of various JHs closely matched the ligand selectivity of the JH receptor Met, indicating that JH induces male production through Met-mediated signaling (Miyakawa et al. 2013a; Miyakawa and Iguchi 2017). The discovery of the WTN6 strain, capable of stable male induction solely through short-day signals without JH exposure, enabled the exploration of the upstream signals that activate JH signaling during male production. Toyota et al. (2015a) suggested that the increase in JH levels during male production is driven by the expression of the JH biosynthetic enzyme, *JHAMT*. Transcriptomic and metabolomic analyses have implicated several factors upstream of JH signaling during male production, including N-methyl-D-aspartate (NMDA)-type glutamate receptors (Toyota et al. 2015b), pantothenic acid (Toyota et al. 2016), and protein kinase C (Toyota et al. 2017). More recently, knockout of the clock gene *period* by genome editing in the WTN6 strain revealed that *Daphnia* lacking *period* continue to produce females even under short-day conditions. This finding demonstrates that the reception of short-day signals that induce male production requires a photoperiod recognition mechanism that involves the circadian clock (Abe et al. 2024). Furthermore, recent findings regarding the downstream signaling of JH have emerged. Transcriptional analyses, including in vitro reporter assays and in vivo analyses using genome-edited *Daphnia* lacking the JHRE, have demonstrated that *Daphnia* Met directly activates the expression of *vrille* in a JH-dependent manner, thereby promoting male determination (Takahata et al. 2026). However, the mechanisms by which *Daphnia* perceive low temperatures or poor food conditions to activate JH signaling remain unclear. Further research is ongoing to unravel the full regulatory mechanism of environment-dependent sex determination in *Daphnia*.

14.4.2 Regulation of Resting Egg Production

Mature male *Daphnia* produce the next generation with a new chromosomal set by mating with females carrying meiotic eggs. Similar to male production, the induction of these meiotic eggs is triggered when the mother detects environmental deterioration (Takahata et al. 2022; Fig. 14.2). However, the regulatory mechanisms underlying this process are less understood than those underlying male production. Unlike parthenogenetic eggs, which are produced in large numbers (> 20) during each oviposition, only one meiotic egg is produced from each ovary at a time. Meiotic eggs contain fine yellowish yolk granules, and once fertilized, they become resting eggs (also referred to as dormant or ephippial eggs) capable of withstanding low temperatures and desiccation (Schults 1977; Hiruta and Tochinai 2014). Several studies have shown that JH exposure influences the formation of meiotic (resting) eggs in *Daphnia* species; however, the effects vary among species. In *D. magna*, JH exposure promotes the formation of resting eggs (Olmstead and LeBlanc 2001; Navis et al. 2018), whereas in *D. pulex*, it suppresses resting egg formation (Suppa et al. 2021). These differences may reflect variations in dormant strategies, such as differences in the threshold of environmental deterioration needed to trigger dormancy, which are thought to play a key role in enabling the coexistence of *Daphnia* species within the same ecological niche (Maruoka et al. 2024). The genetic diversity in JH responsiveness may underlie these differences in dormant strategies at the molecular level.

14.4.3 Regulation of Inducible Defense

Along with environmentally dependent sex determination, another well-known environmental response in *Daphnia* is the formation of defensive morphologies in response to the presence of predators (inducible defenses). When *Daphnia* species detect predator-released chemicals (kairomones), they alter their morphology (such as body shape and size) and life history traits (such as fecundity and behavior) (Tollrian and Dodson 1999). The molecular basis of inducible defense has been studied most extensively in *D. pulex*. When *D. pulex* receives kairomones released by the predator, the phantom midge *Chaoborus* larvae during embryonic development, it forms a spiny defensive structure called "neckteeth" on the head of the juvenile after hatching (Tollrian and Dodson 1999). Individuals with neckteeth are less likely to be consumed by *Chaoborus*, making this defense adaptive to predation pressure (Miyakawa et al. 2015a). To date, several genes whose expression fluctuates in response to *Chaoborus* kairomones have been identified (Miyakawa et al. 2010, 2015b), including the JH biosynthetic enzyme gene *JHAMT* and the JH receptor gene *Met*. As the expression of these genes increases before neckteeth formation, it is predicted that in *D. pulex*, signaling through an increase in internal JH titer is involved when kairomones are sensed (Miyakawa et al. 2010). Indeed, exposing *Daphnia* embryos to JH in combination with *Chaoborus* kairomones significantly increases neckteeth formation (Miyakawa et al. 2013b). However, as exposure to JH alone does not induce neckteeth formation, this suggests that additional essential factors, besides JH, are involved in the inducible defense of *D. pulex*.

On the other hand, in species other than *D. pulex*, JH has been suggested to play a role in inducible defense in *D. galeata*. When *D. galeata* detects kairomones released by various predators, such as *Chaoborus*, *Notonecta*, and fish, it elongates its head to form a helmet, and similar head elongation is observed when exposed to JH (Oda et al. 2011). Although the shapes and locations of neckteeth and helmets differ, these structures are thought to have evolved in response to the different predators present in each species' habitat (Colbourne et al. 1997; Tollrian and Dodson 1999). It is intriguing that JH is involved in controlling both structures. However, the fact that helmet elongation in *D. galeata* can be induced by JH alone, unlike in *D. pulex*, highlights some differences between the species. Future detailed comparative analyses are expected to shed light on how *Daphnia* species have evolved a remarkable inducible defense system.

14.4.4 Regulation of Hemoglobin Synthesis

The final topic regarding *Daphnia* is the involvement of JH in the hypoxic response of *Daphnia*. When exposed to chronic low oxygen levels, *Daphnia* species increase the expression of *hemoglobin* genes to utilize oxygen efficiently (Zeis et al. 2003; Eads et al. 2008). This increase in *hemoglobin* gene expression can be induced by JH exposure, even in the absence of hypoxic conditions (Gorr et al. 2006; Toyota et al. 2013b). Additionally, a report indicated the presence of JH receptor–binding sequences in the regulatory regions of the *hemoglobin* gene in *D. magna*, suggesting that hemoglobin expression in response to hypoxia may be controlled by JH signaling (Gorr et al. 2006). However, this study was conducted before the JH receptor mechanisms were fully elucidated in insects, and the analysis of the binding sequences was not targeted at the Met-binding sequences, which have recently been shown to be widely conserved in insects. While the ecological significance of hemoglobin production in response to hypoxia has long been investigated (Kring and O'Brien 1976; Heisey and Porter 1977), whether JH is involved in its regulation needs to be reassessed based on recent findings.

14.5 Methyl Farnesoate in Decapods

14.5.1 Mandibular Organ as a Methyl Farnesoate Synthesis and Secretory Organ in Decapods

As described in the introduction of this chapter, MF is a crustacean juvenile hormone. Laufer et al. (1987) first identified MF as an endogenous juvenoid molecule in decapod crustaceans using the spider crab, *Libinia emarginata*. Since then, MF has been recognized as the primary juvenoid substance present in malacostracan crustaceans (Tobe et al. 1989; Laufer and Biggers 2001; Nagaraju 2007). In malacostracan crustaceans, the mandibular organ (MO) located near the mandible is considered functionally analogous to the insect corpus allatum, which synthesizes JH. Laufer et al. (1987) and Homola et al. (1991) demonstrated that the MO is responsible for synthesizing and secreting MF. To date, anatomically and functionally homologous organs to the MO have not been identified in non-malacostracan crustaceans, such as daphnids and ostracods. The regulation of MF synthesis was further elucidated by identifying a sinus gland–derived peptide hormone, termed mandibular organ–inhibiting hormone (MOIH). The MOIH negatively regulates MF synthesis and secretion in *L. emarginata* (Liu and Laufer 1996) and the brown crab, *Cancer pagurus* (Wainwright et al. 1996). Recently, using the black tiger shrimp, *Penaeus monodon*, the anatomical location of the MO at the base of the posterior mandibular tendon was demonstrated using histological sections (Semchuchot et al. 2023). Furthermore, *PmJHAMT* was highly expressed in the MO, and its expression was significantly upregulated following eyestalk ablation, suggesting the involvement of MOIH in regulating MF biosynthesis in *P. monodon*. This study provides a clear and detailed demonstration of the anatomical location of the MO, hoping that this work will serve as a catalyst for advancing MF research, particularly focusing on the MO of decapod crustaceans.

14.5.2 The Role of MF in Reproduction of Decapods

In insects, the physiological roles of JH extend beyond metamorphosis and molting, contributing to a wide range of phenotypic plasticity phenomena, such as caste differentiation in social insects and the development of weapon traits in stag beetles (Miura 2019). Similarly, in daphnids, JH serves as a mediator of environmental signals, such as seasonal changes and predator presence, converting various types of environmental information into endocrine information. This allows daphnids to rapidly adapt to environmental changes through mechanisms such as environmental sex determination, production of diapause eggs, and inducible defense. In contrast, despite a long history of research on MF, its known physiological functions in malacostracan crustaceans remain unclear.

Numerous studies have supported the pivotal role of MF in promoting reproductive maturation in female decapod crustaceans. Notably, enhanced reproductive indices, such as increased gonadosomatic indices and ovarian growth, have been observed following MF treatment in various species, such as the freshwater crab, *Oziotelphusa senex senex,* and the ridgeback prawn, *Sicyonia ingentis* (Reddy et al. 2004; Paran et al. 2010). Furthermore, a strong positive correlation has been reported between circulating MF titers and the stage of ovarian development in various decapod species, such as the freshwater crab, *Travancoriana schirnerae*; the Chinese mitten crab, *Eriocheir sinensis*; and the swimming crab, *Portunus trituberculatus*, indicating that MF functions as a critical endocrine signal regulating reproductive processes (Ayanath and Raghavan 2020; Li et al. 2021b; Xie et al. 2015). These findings highlight the essential contribution of MF to reproductive physiology, acting in conjunction with other hormonal factors. The biosynthesis of MF, which is essential for ovarian maturation, has also been found to be negatively regulated by MOIH. Studies on the American lobster *Homarus americanus* have shown that eyestalk ablation leads to a substantial increase in MF levels due to the removal of the MOIH's inhibitory effect. However, upon MOIH injection, both circulating MF titers and the enzymatic activity of FAMeT in the MO were dramatically suppressed (Li et al. 2010). This evidence underscores the antagonistic relationship between MOIH and MF biosynthesis, with MOIH playing a key role in regulating ovarian development. Although future studies using economically important species, such as the kuruma prawn, *Marsupenaeus japonicus*, are essential to further elucidate the role of MF, preliminary findings in some decapod species suggest that MF administration could serve as an alternative to the conventional eyestalk ablation method for inducing ovarian maturation and spawning. This approach has the potential to advance aquaculture technologies by promoting better animal welfare and developing more ethical and sustainable practices.

14.5.3 Larval Metamorphosis in Decapods

In insects, the growth and metamorphosis from the larval stage to the adult stage are known to be cooperatively regulated by JH and 20-hydroxyecdysone (20E), with JH-responsive genes, such as *Kr-h1,* playing a central role (Truman and Riddiford 2019). In contrast, marine and brackish water decapod species spend their larval stage as plankton and transition to a benthic lifestyle through molting and metamorphosis. However, the mechanisms underlying larval metamorphosis in these species remain poorly understood. To address this issue, advancing our understanding of the endocrine regulation of larval metamorphosis and associated gene regulatory networks across decapod species could enable interspecific comparative analyses, similar to the progress achieved in insect research. Despite numerous investigations into the roles of MF and 20E in larval metamorphosis, the findings remain inconsistent. For instance, our studies demonstrated that the administration of MF and 20E disrupted molting-associated metamorphosis in the kuruma prawn, *M. japonicus*, leading to high mortality rates (Toyota et al. 2020). Similarly, MF treatment inhibits metamorphosis in the freshwater prawn, *Macrobrachium rosenbergii* (Abdu et al. 1998), and mud crab, *Scylla paramamosain* (Zhao et al. 2021). In contrast, other studies have reported that MF accelerates metamorphosis in late-stage shrimp and barnacle larvae (Laufer and Biggers 2001; McKenney et al. 1993; Smith et al. 2000; Yamamoto et al. 1997). These conflicting results suggest that the role of MF in decapod metamorphosis is not yet fully understood. An alternative approach to studying this phenomenon involves eyestalk ablation in intermediate larval stages. This procedure impairs or completely inhibits metamorphosis to the megalopa stage in several crab species, including the panopeid crab, *Rhithropanopeus harrisii* (Freeman and Costlow 1980), and swimming crab, *P. trituberculatus* (Dan et al. 2014), as well as *H. americanus* (Synder and Chang 1986). Subsequent studies revealed that eyestalk ablation caused a rise in circulating MF levels, indicating that the observed inhibitory effects on metamorphosis were likely due to MF's regulatory effects (Laufer et al. 1997). Interestingly, our previous research highlighted stage-specific differences in endocrine responses during metamorphosis. For example, nauplius larvae of *M. japonicus* demonstrated higher tolerance to the lethal effects of MF and 20E than later stages, such as zoea and mysids. This finding suggests the involvement of distinct endocrine regulatory mechanisms during the transition from nauplius to zoea and subsequent metamorphic stages (Toyota et al. 2020). Similarly, transcriptomic analyses of larval metamorphosis in the spiny lobster, *Sagmariasus verreauxi,* suggested that puerulus (megalopa)–juvenile metamorphosis is regulated by the MF signaling pathway. In contrast, phyllosoma (zoea)–puerulus (megalopa) metamorphosis is associated with sustained expression of FAMeT, the enzyme responsible for MF biosynthesis (Ventura et al. 2015). These observations imply that the conventional inhibitory role of the MF may not apply to early-stage metamorphosis, such as the nauplius stage in *M. japonicus* or the phyllosoma stage in *S. verreauxi.*

14.5.4 Behavioral Modifications Following the Terminal Molt

Beyond its roles in reproduction and larval metamorphosis, the physiological function of MF in decapods also includes its involvement in terminal molt. Molting is a vital biological process that underpins arthropod growth, morphological development, and transitions through life stages defined by distinct physical characteristics. While insects typically halt molting after reaching a specific developmental stage, many decapod crustaceans continue to molt throughout their lives. In contrast, some brachyuran species, such as spider crabs (*Libinia* spp.) and snow crabs (*Chionoecetes* spp.), undergo a predetermined number of molts, usually concluding with sexual maturation. The terminal molt is particularly important, as it triggers significant changes, including the allometric growth of chelae and notable shifts in male behavior. Following this molt, males tend to exhibit increased aggression, a behavioral adaptation that is believed to enhance reproductive success (Alunno-Bruscia and Sainte-Marie 1998; Conan and Comeau 1986). Although ecdysteroids, especially 20E, are well-recognized as the key endocrine factors driving the molting process in arthropods (Song et al. 2017), they alone cannot fully account for the physical and behavioral changes observed after terminal molt. It has been hypothesized that the MF may play a critical role in mediating these changes. Evidence supporting this hypothesis includes observations in males of *L. emarginata* and the snow crab, *Chionoecetes opilio*, where individuals with larger carapaces and chelae—attributes associated with terminal molt—demonstrate heightened reproductive activity. These males also show elevated levels of circulating MF in their hemolymph (Rotllant et al. 2000; Sagi et al. 1994; Toyota et al. 2023). Additionally, earlier studies on *L. emarginata* have revealed a positive correlation between high MF titers and prominent reproductive behavior (Laufer et al. 1996; Sagi et al. 1994).

14.6 Outlook

As discussed in this chapter, JH is involved in controlling metamorphosis, reproduction, and novel life phenomena that have evolved specifically within different taxa, not only in insects but also in crustaceans. Understanding how JH exerts its various functions and how these functions have evolved is crucial for understanding the evolutionary processes in diverse arthropods. However, research on JH in non-model insects and noninsect arthropods, which is essential for understanding this process, is still lacking. Questions such as "Where are the biosynthetic organs?" and "What structures of JH are synthesized and used?" may seem like classic problems that were resolved decades ago for insect researchers, but these questions have only been clarified in the very limited groups of hexapods and decapods within the broader arthropod phylum. Comparative physiological analyses using the canonical insect JH signaling pathway as a model to investigate how JH signaling pathways differ in species such as *Daphnia*, which have acquired different functions of JH, are expected to become increasingly important in the future. Furthermore, from the perspective that JH exposure triggers physiological changes, research on tardigrades and nematodes may also become a potential area of study (Yoshida et al. 2019; Hoss et al. 2007). Studies using these taxa may shed light on the still-mysterious evolutionary origins of JH.

References

Abdu U, Takac P, Laufer H, Sagi A (1998) Effect of methyl farnesoate on late larval development and metamorphosis in the prawn *Macrobrachium rosenbergii* (Decapoda, Palaemonidae): a juvenoid-like effect? Biol Bull 195:112–119. https://doi.org/10.2307/1542818

Abe S, Takahata Y, Miyakawa H (2024) *Daphnia* uses its circadian clock for short-day recognition in environmental sex determination. Curr Biol 34:2002–2010. https://doi.org/10.1016/j.cub.2024.03.027

Alunno-Bruscia M, Sainte-Marie B (1998) Abdomen allometry, ovary development, and growth of female snow crab, *Chionoecetes opilio* (Brachyura, Majidae), in the northwestern Gulf of St. Lawrence. Can J Fish Aquat Sci 55:459–477. https://doi.org/10.1139/f97-241

Ayanath A, Raghavan S (2020) Effect of exogenously administered methyl farnesoate and 20-OH ecdysone on polypeptide profile of vitellogenin and vitellin in the freshwater crab *Travancoriana schirnerae*. Asian J Life Sci 8:55–63. https://doi.org/10.17582/journal.sajls/2020/8.2.55.63

Bellés X, Martín D, Piulachs MD (2005) The mevalonate pathway and the synthesis of juvenile hormone in insects. Annu Rev Entomol 50:181–199. https://doi.org/10.1146/annurev.ento.50.071803.130356

Bernot JP, Owen CL, Wolfe JM, Meland K, Olesen J, Crandall KA (2023) Major revisions in pancrustacean phylogeny and evidence of sensitivity to taxon sampling. Mol Biol Evol 40(8):msad175. https://doi.org/10.1093/molbev/msad175

Charles JP, Iwema T, Epa VC, Takaki K, Rynes J, Jindra M (2011) Ligand-binding properties of a juvenile hormone receptor, methoprene-tolerant. Proc Natl Acad Sci USA 108:21128–21133. https://doi.org/10.1073/pnas.1116123109

Colbourne JK, Hebert PDN, Taylor DJ (1997) Evolutionary origins of phenotypic diversity in *Daphnia*. In: Givnish TJ, Sytsma KJ (eds) Molecular evolution and adaptive radiation. Cambridge University Press, New York, pp 163–188

Conan GY, Comeau M (1986) Functional maturity and terminal molt of male snow crab, *Chionoecetes opilio*. Can J Fish Aquat Sci 43:1710–1719. https://doi.org/10.1139/f86-214

Daimon T, Shinoda T (2013) Function, diversity, and application of insect juvenile hormone epoxidases (CYP15). Biotechnol Appl Biochem 60:82–91. https://doi.org/10.1139/f86-214

Dan S, Kaneko T, Takeshima S, Ashidate M, Hamasaki K (2014) Eyestalk ablation affects larval morphogenesis in the swimming crab *Portunus trituberculatus* during metamorphosis into megalopae. Sex Early Dev Aquat Org 1:57–73. https://doi.org/10.3354/sedao00007

Debernard S, Morisseau C, Severson TF, Feng L, Wojtasek H, Prestwich GD, Hammock BD (1998) Expression and characterization of the recombinant juvenile hormone epoxide hydrolase (JHEH) from *Manduca sexta*. Insect Biochem Mol Biol 28:409–419. https://doi.org/10.1016/S0965-1748(98)00014-9

Decaestecker E, De Meester L, Mergeay J (2009) Cyclical parthenogenesis in *Daphnia*: sexual versus asexual reproduction. In: Schön I, Martens K, van Dijk P (eds) Lost sex: the evolutionary biology of parthenogenesis. Springer, Dordrecht, pp 295–316

Eads BD, Andrews J, Colbourne JK (2008) Ecological genomics in *Daphnia*: stress responses and environmental sex determination. Heredity 100:184–190. https://doi.org/10.1038/sj.hdy.6800999

Freeman JA, Costlow JD (1980) The molt cycle and its hormonal control in *Rhithropanopeus harrisii* larvae. Dev Biol 74:479–485. https://doi.org/10.1016/0012-1606(80)90447-9

Fu KY, Lu FG, Guo WC, Li GQ (2015) Characterization and functional study of a putative juvenile hormone diol kinase in the colorado potato beetle *Leptinotarsa decemlineata* (Say). Arch Insect Biochem Physiol 90:154–167. https://doi.org/10.1002/arch.21251

Goodman WG, Cusson M (2011) The juvenile hormones. In: Gilbert LI (ed) Insect endocrinology. Elsevier, Amsterdam, pp 310–365

Gorr TA, Rider CV, Wang HY, Olmstead AW, LeBlanc GA (2006) A candidate juvenoid hormone receptor *cis*-element in the *Daphnia magna* hb2 hemoglobin gene promoter. Mol Cell Endocrinol 247:91–102. https://doi.org/10.1016/j.mce.2005.11.022

Gunawardene YINS, Tobe SS, Bendena WG, Chow BKC, Yagi KJ, Chan SM (2002) Function and cellular localization of farnesoic acid *O*-methyltransferase (FAMeT) in the shrimp, *Metapenaeus ensis*. Eur J Biochem 269:3587–3595. https://doi.org/10.1046/j.1432-1033.2002.03048.x

Heisey D, Porter KG (1977) The effect of ambient oxygen concentration on filtering and respiration rates of *Daphnia galeata mendotae* and *Daphnia magna*. Limnol Oceanogr 22:839–845. https://doi.org/10.4319/lo.1977.22.5.0839

Helvig C, Koener JF, Unnithan GC, Feyereisen R (2004) CYP15A1, the cytochrome P450 that catalyzes epoxidation of methyl farnesoate to juvenile hormone III in cockroach corpora allata. Proc Natl Acad Sci USA 101:4024–4029. https://doi.org/10.1073/pnas.0306980101

Hiruta C, Tochinai S (2014) Formation and structure of the ephippium (resting egg case) in relation to molting and egg laying in the water flea *Daphnia pulex* De Geer (Cladocera: Daphniidae). J Morphol 275:760–767. https://doi.org/10.1002/jmor.20255

Hiruta C, Nishida C, Tochinai S (2010) Abortive meiosis in the oogenesis of parthenogenetic *Daphnia pulex*. Chromosom Res 18:833–840. https://doi.org/10.1007/s10577-010-9159-2

Homola E, Sagi A, Laufer H (1991) Relationship of claw form and exoskeleton condition to reproductive system size and methyl farnesoate in the male spider crab, *Libinia emarginata*. Invertebr Reprod Dev 20:219–225. https://doi.org/10.1080/07924259.1991.9672202

Hoss S, Weltje L (2007) Endocrine disruption in nematodes: effects and mechanisms. Ecotoxicology 16:15–28. https://doi.org/10.1007/s10646-006-0108-y

Jindra M, Palli SR, Riddiford LM (2013) The juvenile hormone signaling pathway in insect development. Annu Rev Entomol 58:181–204. https://doi.org/10.1146/annurev-ento-120811-153700

Jindra M, Bellés X, Shinoda T (2015) Molecular basis of juvenile hormone signaling. Curr Opin Insect Sci 11:39–46. https://doi.org/10.1016/j.cois.2015.08.004

Jindra M, Tumova S, Bittova L, Tuma R, Sedlak D (2024) Agonist-dependent action of the juvenile hormone receptor. Curr Opin Insect Sci 65:101234. https://doi.org/10.1016/j.cois.2024.101234

Kamita SG, Hammock BD (2010) Juvenile hormone esterase: biochemistry and structure. J Pestic Sci 35:265–274. https://doi.org/10.1584/jpestics.R10-09

Kato Y, Watanabe H (2022) Regulation of doublesex1 expression for environmental sex determination in the cladoceran crustacean *Daphnia*. Front Cell Dev Biol 10:881255. https://doi.org/10.3389/fcell.2022.881255

Kayukawa T, Minakuchi C, Namiki T, Togawa T, Yoshiyama M, Kamimura M et al (2012) Transcriptional regulation of juvenile hormone-mediated induction of Krüppel homolog 1, a repressor of insect metamorphosis. Proc Natl Acad Sci USA 109:11729–11734. https://doi.org/10.1073/pnas.1204951109

Kayukawa T, Tateishi K, Shinoda T (2013) Establishment of a versatile cell line for juvenile hormone signaling analysis in *Tribolium castaneum*. Sci Rep 3:1570. https://doi.org/10.1038/srep01570

Kayukawa T, Nagamine K, Ito Y, Nishita Y, Ishikawa Y, Shinoda T (2016) Krüppel homolog 1 inhibits insect metamorphosis via direct transcriptional repression of *Broad-Complex*, a pupal specifier gene. J Biol Chem 291:1751–1762. https://doi.org/10.1074/jbc.M115.686121

Kayukawa T, Jouraku A, Ito Y, Shinoda T (2017) Molecular mechanism underlying juvenile hormone-mediated repression of precocious larval-adult metamorphosis. Proc Natl Acad Sci USA 114:1057–1062. https://doi.org/10.1073/pnas.1615423114

Kleiven OT, Larsson P, Hobaek A (1992) Sexual reproduction in *Daphnia magna* requires three stimuli. Oikos 65:197–206. https://doi.org/10.2307/3545010

Kring RL, O'Brien WJ (1976) Effect of varying oxygen concentrations on the filtering rate of *Daphnia pulex*. Ecology 57:808–814. https://doi.org/10.2307/1936194

Laufer H, Biggers WJ (2001) Unifying concepts learned from methyl farnesoate for invertebrate reproduction and post-embryonic development. Am Zool 41:442–457. https://doi.org/10.1093/icb/41.3.442

Laufer H, Borst D, Baker FC, Reuter CC, Tsai LW, Schooley DA et al (1987) Identification of a juvenile hormone-like compound in a crustacean. Science 235:202–205. https://doi.org/10.1126/science.235.4785.2

Laufer H, Ahl JSB, Sagi A (1993) The role of juvenile hormones in crustacean reproduction. Am Zool 33:365–374. https://doi.org/10.1093/icb/33.3.365

Laufer H, Ahl JSB, Takac P, Scott G, Stevens B, Munk JE et al (1996) Hormone and reproductive strategies in spider crabs with emphasis on commercially important species. In: Keller S, Carol K (eds) High latitude crabs: biology, management, and economics. University of Alaska Fairbanks, School of Fisheries and Ocean Sciences, Fairbanks, pp 383–387

Laufer H, Takac P, Ahl J, Laufer M (1997) Methyl farnesoate and the effect of eyestalk ablation on the morphogenesis of the juvenile female spider crab *Libinia emarginata*. Invertebr Reprod Dev 31:63–68. https://doi.org/10.1080/07924259.1997.9672564

Li S, Friesen JA, Holford KC, Borst DW (2010) Methyl farnesoate synthesis in the lobster mandibular organ: the roles of HMG-CoA reductase and farnesoic acid O-methyltransferase. Comp Biochem Physiol A 155:49–55. https://doi.org/10.1016/j.cbpa.2009.09.016

Li M, Mead EA, Zhu JS (2011) Heterodimer of two bHLH-PAS proteins mediates juvenile hormone-induced gene expression. Proc Natl Acad Sci USA 108:638–643. https://doi.org/10.1073/pnas.1013914108

Li X, Chen T, Han Y, Huang M, Jiang H, Huang J et al (2021a) Potential role of Methoprene-tolerant (Met) in methyl farnesoate-mediated vitellogenesis in the Chinese mitten crab (*Eriocheir sinensis*). Comp Biochem Physiol B 252:110524. https://doi.org/10.1016/j.cbpb.2020.110524

Li X, Chen T, Jiang H, Huang J, Huang M, Xu R et al (2021b) Effects of methyl farnesoate on Krüppel homolog 1 (Kr-h1) during vitellogenesis in the Chinese mitten crab (*Eriocheir sinensis*). Anim Reprod Sci 224:106653. https://doi.org/10.1016/j.anireprosci.2020.106653

Liu L, Laufer H (1996) Isolation and characterization of sinus gland neuropeptides with both mandibular organ inhibiting and hyperglycemic effects from the spider crab *Libinia emarginata*. Arch Insect Biochem Physiol 32:375–385. https://doi.org/10.1002/(SICI)1520-6327(1996)32:3/4<375::AID-ARCH10>3.0.CO;2-9

Liu M, Yan C, Liu Y, Wu Z, Zhang J, Sun Y (2022) Cloning, expression analysis and RNAi of farnesoic acid *O*-methylransferase gene from *Neocaridina denticulata sinensis*. Comp Biochem Physiol B 259:110719. https://doi.org/10.1016/j.cbpb.2022.110719
Lozano-Fernandez J, Giacomelli M, Fleming JF, Chen A, Vinther J, Thomsen PF et al (2019) Pancrustacean evolution illuminated by taxon-rich genomic-scale data sets with an expanded remipede sampling. Genome Biol Evol 11:2055–2070. https://doi.org/10.1093/gbe/evz097
Maruoka N, Yamaki K, Makino W, Urabe J (2024) To lose is to win: long-term co-occurrence of two asexual populations realized by a dormant strategy of the inferior competitor. Funct Ecol 38:1121–1133. https://doi.org/10.1111/1365-2435.14551
McKenney CL Jr, Celestial DM (1993) Variation in larval growth and metabolism of an estuarine shrimp *Palaemonetes pugio* during toxicosis by an insect growth regulator. Comp Biochem Physiol C 105:239–245. https://doi.org/10.1016/0742-8413(93)90201-U
Minakuchi C, Namiki T, Yoshiyama M, Shinoda T (2008a) RNAi-mediated knockdown of *juvenile hormone acid O-methyltransferase* gene causes precocious metamorphosis in the red flour beetle *Tribolium castaneum*. FEBS J 275:2919–2931. https://doi.org/10.1111/j.1742-4658.2008.06428.x
Minakuchi C, Zhou X, Riddiford LM (2008b) *Krüppel homolog 1* (*Kr-h1*) mediates juvenile hormone action during metamorphosis of *Drosophila melanogaster*. Mech Dev 125:91–105. https://doi.org/10.1016/j.mod.2007.10.002
Minakuchi C, Namiki T, Shinoda T (2009) *Krüppel homolog 1*, an early juvenile hormone-response gene downstream of *Methoprene-tolerant*, mediates its anti-metamorphic action in the red flour beetle *Tribolium castaneum*. Dev Biol 325:341–350. https://doi.org/10.1016/j.ydbio.2008.10.016
Miura T (2019) Juvenile hormone as a physiological regulator mediating phenotypic plasticity in pancrustaceans. Develop Growth Differ 61:85–96. https://doi.org/10.1111/dgd.12572
Miyakawa H, Iguchi T (2017) Comparative luciferase assay for establishing reliable *in vitro* screening system of juvenile hormone agonists. J Appl Toxicol 37:1082–1090. https://doi.org/10.1002/jat.3459
Miyakawa H, Imai M, Sugimoto N, Ishikawa Y, Ishikawa A, Ishigaki H et al (2010) Gene up-regulation in response to predator kairomones in the water flea, *Daphnia pulex*. BMC Dev Biol 10:45. https://doi.org/10.1186/1471-213X-10-45
Miyakawa H, Gotoh H, Sugimoto N, Miura T (2013a) Effect of juvenoids on predator-induced polyphenism in the water flea, *Daphnia pulex*. J Exp Zool A 319:440–450. https://doi.org/10.1002/jez.1807
Miyakawa H, Toyota K, Hirakawa I, Ogino Y, Miyagawa S, Oda S et al (2013b) A mutation in the receptor Methoprene-tolerant alters juvenile hormone response in insects and crustaceans. Nat Commun 4:1856. https://doi.org/10.1038/ncomms2868
Miyakawa H, Toyota K, Sumiya E, Iguchi T (2014) Comparison of JH signaling in insects and crustaceans. Curr Opin Insect Sci 1:81–87. https://doi.org/10.1016/j.cois.2014.04.006
Miyakawa H, Sato M, Colbourne JK, Iguchi T (2015a) Ionotropic glutamate receptors mediate inducible defense in the water flea *Daphnia pulex*. PLoS One 10:e0121324. https://doi.org/10.1371/journal.pone.0121324
Miyakawa H, Sugimoto N, Kohyama TI, Iguchi T, Miura T (2015b) Intra-specific variations in reaction norms of predator-induced polyphenism in the water flea *Daphnia pulex*. Ecol Res 30:705–713. https://doi.org/10.1007/s11284-015-1272-4
Miyakawa H, Sato T, Song Y, Tollefsen KE, Iguchi T (2018a) Ecdysteroid and juvenile hormone biosynthesis, receptors and their signaling in the freshwater microcrustacean *Daphnia*. J Steroid Biochem Mol Biol 184:62–68. https://doi.org/10.1016/j.jsbmb.2017.12.006
Miyakawa H, Watanabe M, Araki M, Ogino Y, Miyagawa S, Iguchi T (2018b) Juvenile hormone-independent function of *Krüppel homolog 1* in early development of water flea *Daphnia pulex*. Insect Biochem Mol Biol 93:12–18. https://doi.org/10.1016/j.ibmb.2017.12.007
Nagaraju GPC (2007) Is methyl farnesoate a crustacean hormone? Aquaculture 272:39–54. https://doi.org/10.1016/j.aquaculture.2007.05.014

Navis S, Waterkeyn A, De Meester L, Brendonck L (2018) Acute and chronic effects of exposure to the juvenile hormone analog fenoxycarb during sexual reproduction in *Daphnia magna*. Ecotoxicology 27:627–634. https://doi.org/10.1007/s10646-018-1935-3

Noriega FG (2014) Juvenile hormone biosynthesis in insects: what is new, what do we know, and what questions remain? Int Sch Res Notices 2014:967361. https://doi.org/10.1155/2014/967361

Oakley TH, Wolfe JM, Lindgren AR, Zaharoff AK (2013) Phylotranscriptomics to bring the under-studied into the fold: monophyletic ostracoda, fossil placement, and pancrustacean phylogeny. Mol Biol Evol 30:215–233. https://doi.org/10.1093/molbev/mss216

Oda S, Tatarazako N, Watanabe H, Morita M, Iguchi T (2005) Production of male neonates in four cladoceran species exposed to a juvenile hormone analog, fenoxycarb. Chemosphere 60:74–78. https://doi.org/10.1016/j.chemosphere.2004.12.080

Oda S, Kato Y, Watanabe H, Tatarazako N, Iguchi T (2011) Morphological changes in *Daphnia galeata* induced by a crustacean terpenoid hormone and its analog. Environ Toxicol Chem 30:232–238. https://doi.org/10.1002/etc.378

Olmstead AW, LeBlanc GA (2001) Temporal and quantitative changes in sexual reproductive cycling of the cladoceran *Daphnia magna* by a juvenile hormone analog. J Exp Zool 290:148–155. https://doi.org/10.1002/jez.1044

Olmstead AW, Leblanc GA (2002) Juvenoid hormone methyl farnesoate is a sex determinant in the crustacean *Daphnia magna*. J Exp Zool 293:736–739. https://doi.org/10.1002/jez.10162

Paran BC, Fierro IJ, Tsukimura B (2010) Stimulation of ovarian growth by methyl farnesoate and eyestalk ablation in penaeoidean model shrimp, *Sicyonia ingentis* Burkenroad, 1938. Aquac Res 41:1887–1897. https://doi.org/10.1111/j.1365-2109.2010.02612.x

Reddy PR, Nagaraju GPC, Reddy PS (2004) Involvement of methyl farnesoate in the regulation of molting and reproduction in the freshwater crab *Oziotelphusa senex senex*. J Crustac Biol 24:511–515. https://doi.org/10.1651/C-2478

Regier JC, Shultz JW, Zwick A, Hussey A, Ball B, Wetzer R et al (2010) Arthropod relationships revealed by phylogenomic analysis of nuclear protein-coding sequences. Nature 463:1079–1183. https://doi.org/10.1038/nature08742

Rivera-Pérez C, Clifton M, Noriega F, Jindra M (2020) Juvenile hormone regulation and action. In: Saleuddin S, Lange A, Orchard I (eds) Advances in invertebrate (neuro) endocrinology. Apple Academic Press, Burlington, pp 1–76

Rotllant G, Takac P, Liu L, Scott GL, Laufer H (2000) Role of ecdysteroids and methyl farnesoate in morphogenesis and terminal moult in polymorphic males of the spider crab *Libinia emarginata*. Aquaculture 190:103–118. https://doi.org/10.1016/S0044-8486(00)00388-4

Sagi A, Ahl JS, Danaee H, Laufer H (1994) Methyl farnesoate levels in male spider crabs exhibiting active reproductive behavior. Horm Behav 28:261–272. https://doi.org/10.1006/hbeh.1994.1022

Schultz TW (1977) Fine structure of the ephippium of *Daphnia pulex* (Crustacea: Cladocera). Trans Am Microsc Soc 96:313–321. https://doi.org/10.2307/3225860

Schwentner M, Combosch DJ, Pakes Nelson J, Giribet G (2017) A phylogenomic solution to the origin of insects by resolving crustacean-hexapod relationships. Curr Biol 27:1818–1824. https://doi.org/10.1016/j.cub.2017.05.040

Schwentner M, Richter S, Rogers DC, Giribet G (2018) Tetraconatan phylogeny with special focus on Malacostraca and Branchiopoda: highlighting the strength of taxon-specific matrices in phylogenomics. Proc Biol Sci 285:20181524. https://doi.org/10.1098/rspb.2018.1524

Semchuchot W, Chotwiwatthanakun C, Santimanawong W, Kruangkum T, Thaijongrak P, Withyachumnarnkul B et al (2023) Sesquiterpenoid pathway in the mandibular organ of *Penaeus monodon*: cloning, expression, characterization of *PmJHAMT* and its alteration response to eyestalk ablation. Gen Comp Endocrinol 331:114176. https://doi.org/10.1016/j.ygcen.2022.114176

Shin SW, Zou Z, Saha TT, Raikhel AS (2012) bHLH-PAS heterodimer of methoprene-tolerant and cycle mediates circadian expression of juvenile hormone-induced mosquito genes. Proc Natl Acad Sci USA 109:16576–16581. https://doi.org/10.1073/pnas.1214209109

Shinoda T, Itoyama K (2003) Juvenile hormone acid methyltransferase: a key regulatory enzyme for insect metamorphosis. Proc Natl Acad Sci USA 100:11986–11991. https://doi.org/10.1073/pnas.2134232100

Smith PA, Clare AS, Rees HH, Prescott MC, Wainwright G, Thorndyke MC (2000) Identification of methyl farnesoate in the cypris larva of the barnacle, *Balanus amphitrite*, and its role as a juvenile hormone. Insect Biochem Mol Biol 30:885–890. https://doi.org/10.1016/S0965-1748(00)00062-X

Snyder MJ, Chang ES (1986) Effects of eyestalk ablation on larval molting rates and morphological development of the American lobster, *Homarus americanus*. Biol Bull 170:232–243. https://doi.org/10.2307/1541805

Song Y, Villeneuve DL, Toyota K, Iguchi T, Tollefsen KE (2017) Ecdysone receptor agonism leading to lethal molting disruption in arthropods: review and adverse outcome pathway development. Environ Sci Technol 51:4142–4157. https://doi.org/10.1021/acs.est.7b00480

Su ZH, Sasaki A, Minami H, Ozaki K (2024) Arthropod phylotranscriptomics with a special focus on the basal phylogeny of the myriapoda. Genome Biol Evol 16. https://doi.org/10.1093/gbe/evae189

Suppa A, Gorbi G, Marková S, Buschini A, Rossi V (2021) Transgenerational effects of methyl farnesoate on *Daphnia pulex* clones: male and ephippia production and expression of genes involved in sex determination. Freshw Biol 66:374–390. https://doi.org/10.1111/fwb.13644

Takahata Y, Miyakawa H (2022) Developmental staging of sexual egg formation in *Daphnia pulex*: unmated females resorb meiotic oocytes to resist starvation. Zool Sci 39:407–412. https://doi.org/10.2108/zs220010

Takahata Y, Kusajima M, Abe S, Miyakawa MO, Suzuki T, Dohra H et al. (2026) Evolution of environmental sex determination via juvenile hormone-induced gene co-option in *Daphnia*. Proc Natl Acad Sci USA 123:e2525480123. https://doi.org/10.1073/pnas.2525480123

Tanaka T, Iguchi T, Miyakawa H (2019) Establishment of a high-sensitivity reporter system in mammalian cells for detecting juvenoids using juvenile hormone receptors of *Daphnia pulex*. J Appl Toxicol 39:241–246. https://doi.org/10.1002/jat.3713

Tatarazako N, Oda S, Watanabe H, Morita M, Iguchi T (2003) Juvenile hormone agonists affect the occurrence of male *Daphnia*. Chemosphere 53:827–833. https://doi.org/10.1016/S0045-6535(03)00761-6

Tobe SS, Young DA, Khoo HW (1989) Production of methyl farnesoate by the mandibular organs of the mud crab, *Scylla serrata*: validation of a radiochemical assay. Gen Comp Endocrinol 73:342–353. https://doi.org/10.1016/0016-6480(89)90190-1

Tollrian R, Dodson SI (1999) Inducible defenses in Cladocera: constraints, costs, and multipredator environments. In: Tollrian R, Harvell CD (eds) The ecology and evolution of inducible defenses. Princeton University Press, Princeton, pp 177–202

Toyota K, Kato Y, Miyakawa H, Yatsu R, Mizutani T, Ogino Y et al (2013a) Molecular impact of juvenile hormone agonists on neonatal *Daphnia magna*. J Appl Toxicol:537–544. https://doi.org/10.1002/jat.2922

Toyota K, Kato Y, Sato M, Sugiura N, Miyagawa S, Miyakawa H et al (2013b) Molecular cloning of doublesex genes of four cladocera (water flea) species. BMC Genomics 14:239. https://doi.org/10.1186/1471-2164-14-239

Toyota K, Miyakawa H, Hiruta C, Furuta K, Ogino Y, Shinoda T et al (2015a) Methyl farnesoate synthesis is necessary for the environmental sex determination in the water flea *Daphnia pulex*. J Insect Physiol 80:22–30. https://doi.org/10.1016/j.jinsphys.2015.02.002

Toyota K, Miyakawa H, Yamaguchi K, Shigenobu S, Ogino Y, Tatarazako N et al (2015b) NMDA receptor activation upstream of methyl farnesoate signaling for short day-induced male offspring production in the water flea, *Daphnia pulex*. BMC Genomics 16:186. https://doi.org/10.1186/s12864-015-1392-9

Toyota K, Gavin A, Miyagawa S, Viant MR, Iguchi T (2016) Metabolomics reveals an involvement of pantothenate for male production responding to the short-day stimulus in the water flea, *Daphnia pulex*. Sci Rep 6:25125. https://doi.org/10.1038/srep25125

Toyota K, Sato T, Tatarazako N, Iguchi T (2017) Protein kinase C is involved with upstream signaling of methyl farnesoate for photoperiod-dependent sex determination in the water flea *Daphnia pulex*. Biol Open 6:161–164. https://doi.org/10.1242/bio.021857

Toyota K, Yamane F, Ohira T (2020) Impacts of methyl farnesoate and 20-hydroxyecdysone on larval mortality and metamorphosis in the kuruma prawn *Marsupenaeus japonicus*. Front Endocrinol 11:475. https://doi.org/10.3389/fendo.2020.00475

Toyota K, Yamamoto T, Mori T, Mekuchi M, Miyagawa S, Ihara M et al (2023) Eyestalk transcriptome and methyl farnesoate titers provide insight into the physiological changes in the male snow crab, *Chionoecetes opilio*, after its terminal molt. Sci Rep 13:7204. https://doi.org/10.1038/s41598-023-34159-y

Truman JW, Riddiford LM (2019) The evolution of insect metamorphosis: a developmental and endocrine view. Philos Trans R Soc B 374:20190070. https://doi.org/10.1098/rstb.2019.0070

Tsubota T, Nakakura T, Shiotsuki T (2010) Molecular characterization and enzymatic analysis of juvenile hormone epoxide hydrolase genes in the red flour beetle *Tribolium castaneum*. Insect Mol Biol 19:399–408. https://doi.org/10.1111/j.1365-2583.2010.01001.x

Ventura T, Fitzgibbon QP, Battaglene SC, Elizur A (2015) Redefining metamorphosis in spiny lobsters: molecular analysis of the phyllosoma to puerulus transition in *Sagmariasus verreauxi*. Sci Rep 5:13537. https://doi.org/10.1038/srep13537

Wainwright G, Webster SG, Wilkinson MC, Chung JS, Rees HH (1996) Structure and significance of mandibular organ-inhibiting hormone in the crab, *Cancer pagurus*. Involvement in multihormonal regulation of growth and reproduction. J Biol Chem 271:12749–12754. https://doi.org/10.1074/jbc.271.22.12749

Ward VK, Bonning BC, Huang T, Shiotsuki T, Griffeth VN, Hammock BD (1992) Analysis of the catalytic mechanism of juvenile hormone esterase by site-directed mutagenesis. Int J Biochem 24:1933–1941. https://doi.org/10.1016/0020-711X(92)90289-D

Xie X, Zhu D, Li Y, Qiu X, Cui X, Tang J (2015) Hemolymph levels of methyl farnesoate during ovarian development of the swimming crab *Portunus trituberculatus*, and its relation to transcript levels of HMG-CoA reductase and farnesoic acid *O*-methyltransferase. Biol Bull 228:118–124. https://doi.org/10.1086/BBLv228n2p118

Yamamoto H, Okino T, Yoshimura E, Tachibana A, Shimizu K, Fusetani N (1997) Methyl farnesoate induces larval metamorphosis of the barnacle, *Balanus amphitrite* via protein kinase C activation. J Exp Zool 278:349–355. https://doi.org/10.1002/(SICI)1097-010X(19970815)278:6<349::AID-JEZ2>3.0.CO;2-O

Yoshida Y, Sugiura K, Tomita M, Matsumoto M, Arakawa K (2019) Comparison of the transcriptomes of two tardigrades with different hatching coordination. BMC Dev Biol 19:24. https://doi.org/10.1186/s12861-019-0205-9

Yu HY, Chu KH, Tsang LM, Ma KY (2024) Incomplete lineage sorting and long-branch attraction confound phylogenomic inference of Pancrustacea. Front Ecol Evol 12:1243221. https://doi.org/10.3389/fevo.2024.1243221

Zeis B, Becher B, Goldmann T, Clark R, Vollmer E, Bolke B et al (2003) Differential haemoglobin gene expression in the crustacean *Daphnia magna* exposed to different oxygen partial pressures. Biol Chem 384:1133–1145. https://doi.org/10.1515/BC.2003.126

Zeng BJ, Lv Y, Zhang LL, Huang LH, Feng QL (2016) Cloning and structural characterization of juvenile hormone diol kinase in *Spodoptera litura*. Insect Sci 23:819–828. https://doi.org/10.1111/1744-7917.12236

Zhang ZL, Xu JJ, Sheng ZT, Sui YP, Palli SR (2011) Steroid receptor co-activator is required for juvenile hormone signal transduction through a bHLH-PAS transcription factor, Methoprene tolerant. J Biol Chem 286:8437–8447. https://doi.org/10.1074/jbc.M110.191684

Zhao M, Zhang F, Wang W, Liu Z, Ma L (2021) Effect of methyl farnesoate and farnesoic acid during 5th zoea and megalopa metamorphosis in the mud crab *Scylla paramamosain* Estampador, 1950 (Decapoda, Brachyura, Portunidae). Crustaceana 94:855–863. https://doi.org/10.1163/15685403-bja10133

Part IV
Application

Chapter 15
Regulation of Body Size in Edible Crickets Through Myoglianin–Juvenile Hormone Signaling

Yoshiyasu Ishimaru, Takahito Watanabe, Sumihare Noji, and Taro Mito

Abstract The two-spotted cricket, *Gryllus bimaculatus*, a hemimetabolous insect, is a valuable model for investigating the hormonal and molecular mechanisms that regulate growth and metamorphosis. Recent studies have shown that the transforming growth factor-β (TGF-β) family ligand myoglianin (Myo) modulates juvenile hormone (JH) biosynthesis in the corpus allatum, thereby helping coordinate somatic growth with endocrine regulation. Functional analyses using RNA interference (RNAi) and CRISPR/Cas9-mediated gene knockout demonstrated that Myo is required for normal nymphal growth and suppression of JH, which enables the onset of metamorphosis. Based on these findings, we propose a threshold-based regulatory model in which Myo levels, potentially reflecting the progression of somatic growth, reach a critical physiological level at which JH synthesis is downregulated, and additional growth is terminated. This Myo–JH regulatory system provides a homeostatic mechanism that links growth progression to the determination of the final body size. Comparative considerations based on findings from insects and vertebrates further suggest that vertebrate homologs of Myo, such as myostatin (growth differentiation factor 8 (GDF8)) and GDF11, may retain evolutionarily conserved functions related to body size regulation. Collectively, these insights support an integrative framework in which Myo-mediated signaling contributes to the coordination of somatic growth, developmental endocrinology, and organism size determination. Manipulation of this pathway may offer a promising strategy for engineering crickets with optimized body sizes for sustainable food production and biomaterial applications.

Y. Ishimaru
Division of Bioscience and Bioindustry, Graduate School of Technology, Industrial and Social Sciences, Tokushima University, Tokushima, Japan

T. Watanabe · T. Mito (✉)
Bio-Innovation Research Center, Tokushima University, Tokushima, Japan
e-mail: mito.taro@tokushima-u.ac.jp

S. Noji
EvoDvo Tech Co., Ltd., Tokushima, Japan

T. Shinoda et al. (eds.), *Juvenile Hormone*, Entomology Monographs,
https://doi.org/10.1007/978-981-95-8605-9_15

Keywords *Gryllus bimaculatus* · Hemimetabolous insects · Holometabolous insects · Juvenile hormone · Decapentaplegic (Dpp) · Myoglianin (Myo) · Metamorphosis · Body size

15.1 Introduction: Crickets as Model Organisms and Food Resources

15.1.1 **Gryllus bimaculatus** *as a Model Organism for Research*

We selected the two-spotted cricket, *Gryllus bimaculatus*, a hemimetabolous insect, as a model organism to study development, regeneration, and evolutionary processes (Noji 2022). This choice represents a shift from the traditional focus on holometabolous insects, such as the fruit fly, *Drosophila melanogaster,* and the silkworm, *Bombyx mori*. Achieving a deeper understanding of insect biology, including the mechanisms underlying insect evolution, requires the investigation of hemimetabolous species alongside well-established holometabolous models.

15.1.2 Global Food Crisis and Insect-Based Food

The utilization of crickets as a food source offers a promising opportunity to address urgent global challenges while contributing to societal development, particularly in the context of Sustainable Development Goals (SDGs) (Sogari et al. 2023). This approach not only leverages scientific findings but also provides tangible societal benefits, thereby fulfilling the dual role of advancing scientific knowledge and supporting sustainable development.

Edible crickets hold considerable potential for mitigating global food shortages, particularly the growing protein deficit driven by the rapid increase in the world's population (Hanboonsong et al. 2013; Mito et al. 2022). The global population is projected to reach 9.7 billion by 2050, a substantial increase from 7.8 billion in 2020 (United Nations 2024). This demographic expansion is expected to exacerbate multiple resource-related challenges, including rising energy demand and, most critically, food scarcity (Godfray et al. 2010). Despite technological advances in agriculture and biotechnology, many countries continue to experience food insecurity. Furthermore, a global deficit of approximately 100 million tons of animal--derived protein is anticipated as population growth outpaces current production capacity.

The Food and Agriculture Organization (FAO) of the United Nations has proposed entomophagy, the consumption of insects, as a viable and sustainable solution to emerging global food challenges (Aiello et al. 2023). Notably, insects are already part of the diet in approximately 130 countries, with an estimated two billion people consuming more than 2000 species. Insects are highly valued not only for their substantial protein content but also for their rich composition of vitamins, minerals, and unsaturated fatty acids, making them promising functional food ingredients for humans. The environmental impact of insect consumption is remarkably low. The greenhouse gas emissions and water use associated with insect farming are significantly lower than those associated with conventional livestock production. Moreover, insects exhibit feed conversion efficiency several times higher than that of traditional livestock, further highlighting their potential as sustainable protein sources.

15.1.3 Advantages of Crickets as Edible Insects

Among the various edible insects, crickets are particularly well-suited for human consumption (Kemsawasd et al. 2022). Table 15.1 compares the characteristics of insects currently used for food and feed. Notably, among the six criteria listed in the table, crickets are the only species that meet five criteria.

Historically, large-scale locust outbreaks, which belong to the same order (Orthoptera) as crickets, have caused substantial agricultural damage. Although locusts can undergo explosive population increases under favorable natural conditions, achieving stable mass production in controlled rearing environments remains challenging. In contrast, crickets demonstrate rapid and consistent population growth under artificial rearing conditions, making them highly accessible and practical food resources.

Table 15.1 Comparison of the characteristics of the insects currently used for food and feed

	Cricket	Locust	Mealworm	Silkworm	Housefly	Soldier fly	Termite
Breeding ease	Easy	Difficulty	Easy	Easy	Easy	Easy	Difficulty
Generation time	1–2 months	2 weeks	Several months	1 month	1–2 weeks	3 weeks	Several months
Body size	Large	Large	Middle	Large	Small	Middle	Small
Feeding habitat	Ominivorous	Herbivorous	Ominivorous	Herbivorous (Mulberry)	Corrosion	Corrosion	Wood food
Usage	Food and feed	Edible	Food and feed	Food and feed	Feed	Feed	Food and feed
Feed efficiency	High	Low	High	Low	Low	Low	Low

Building on these research themes, this chapter introduces studies on insect body size determination regulated by juvenile hormones (JHs) and growth factors, as well as their potential applications.

15.2 Mechanisms Regulating Body Size and Metamorphosis

15.2.1 Determination of Insect Body Size and the Threshold Concept

The diversity of insects is immense, with more than one million species estimated to inhabit Earth. Compared with mammals, insects are considerably smaller, and no insect approaches the body size of an elephant. Even within the class Insecta, body size varies remarkably, ranging from tiny fairyflies (body length: 0.14 mm) to giant wētā (body length: 10 cm). Importantly, the final adult body size is species-specific and heritable, indicating that the genetic blueprint determining size is encoded in the genome of the organism. Many fundamental and complex biological traits exhibit simple scaling relationships with body size. Allometric scaling of metabolic rate from molecules and mitochondria to cells and mammals (West and Brown 2005). Despite these patterns, the nature of the information that determines species-specific body size remains unresolved. The developmental and genetic control of insect body size, including its diversity and genetic basis, has been discussed in detail by Nijhout et al. (2014) and Grunert et al. (2015). However, the precise mechanisms that determine the final adult body size remain unknown (Truman et al. 2024). Our recent research using crickets has provided promising clues for understanding this mechanism. In this section, we introduce this finding and summarize the current knowledge on the mechanisms underlying insect body size determination.

Insects can be classified based on their developmental strategies. The most primitive insects, such as silverfish (Ametabola), do not undergo metamorphosis and continue to molt and grow after reaching sexual maturity. In contrast, most insects undergo metamorphosis, which is broadly divided into complete (Holometabola) and incomplete (Hemimetabola) metamorphoses. In holometabolous insects, morphologically distinct larvae transform into pupae and subsequently into adults through dramatic morphological remodeling. In contrast, hemimetabolous insects maintain a more uniform morphology throughout development, with nymphs resembling miniature adults. In both developmental strategies, growth ceases at the adult molt, and the exoskeleton fixes the final body size. In hemimetabolous insects, in which body shape changes only a little, body size can be readily evaluated. In contrast, in holometabolous insects, in which body shape changes significantly, evaluating changes in body size is more difficult. This metamorphic transition is triggered by the cessation of JH secretion. Metamorphosis occurs only after the larva or nymph reaches a particular body size, suggesting the evolution of an internal size assessment mechanism (Nijhout and Williams 1974).

Extensive research on *D. melanogaster* has illuminated many of the molecular, genetic, and physiological pathways that regulate growth and metamorphosis (Tyson et al. 2023). Nevertheless, how animals evaluate their body size and determine when to terminate growth remains one of the central unsolved questions in developmental biology. In holometabolous insects, feeding and most somatic growth occur during the larval stages. At the end of the final instar, larvae enter the pupal stage, during which profound morphological changes occur. After metamorphosis, the adult emerges with a fixed body size, as the rigid exoskeleton prevents further expansion of the body. Thus, the size of the larva at the end of the final instar is directly correlated with the final adult size.

Studies have shown that once larvae reach a species-specific "threshold size," they terminate further growth, initiate pupation, and proceed to metamorphosis. The threshold size was first identified in the tobacco hornworm, *Manduca sexta* (Nijhout 1975). Under laboratory conditions, *M. sexta* typically undergoes five molts. However, under nutrient-poor conditions, larval growth slows, and pupation may occur after one to three additional instars. Previous studies have defined threshold checkpoints at the onset of each instar, which determine whether larvae proceed to the next molt or remain in the same larval stage and continue feeding (Kingsolve 2007; Nijhout 1975) (Fig. 15.1). Importantly, among these checkpoints, attainment of the metamorphic threshold size during the penultimate instar represents a distinct and decisive event that commits the larva to enter the final instar and initiate the developmental program for pupation at the subsequent molt. Thus, while size-dependent checkpoints operate at multiple larval stages, the metamorphic threshold size functions as an absolute determinant of a species' characteristic body size. A central question, therefore, is what biological factors define this critical threshold.

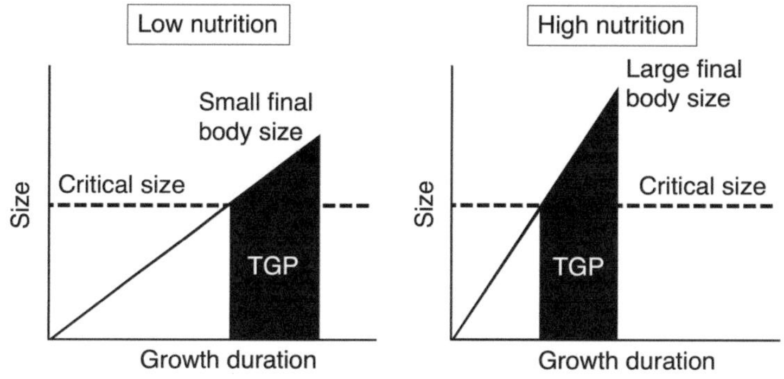

Fig. 15.1 Relationship between body growth, nutrition status, and critical size (modified from Texada et al. 2020, Fig. 6). Schematic representation of how nutritional conditions influence growth rate, attainment of critical size, and final body size. Under low-nutrition conditions (left), the growth rate decreases, delaying the attainment of critical size and ultimately resulting in a smaller final body size. Under high-nutrition conditions (right), an increased growth rate enables earlier achievement of the critical size and supports the development of a larger final body size. The shaded region indicates the terminal growth period (TGP), a phase of feeding and growth that is largely insensitive to nutritional status and occurs after the critical size has been reached

In previous studies by Kingsolver (2007) and Nijhout (1975), body weight or head capsule width was used as an external metric to determine whether larvae had reached the fifth instar and entered the final larval stage. However, these metrics are operational measures used by researchers and do not necessarily reflect the actual parameters assessed by the larvae themselves. Although the critical weight is externally measurable and represents an absolute quantity, the mechanism by which larvae internally evaluate this "threshold size" likely relies on intrinsic physiological cues rather than relative comparisons with earlier states or environmental conditions.

15.2.2 JH and Regulation of Metamorphosis in Holometabolous Insects

Larvae of holometabolous insects, such as *M. sexta*; the red flour beetle, *Tribolium castaneum*; and *B. mori*, undergo precocious metamorphosis when JH is deficient (Daimon et al. 2012; Minakuchi et al. 2008; Suzuki et al. 2013). Experimental suppression of JH synthesis similarly induces precocious metamorphosis (Cheng et al. 2014). It has long been known that the removal of the corpus allatum, the endocrine gland responsible for JH secretion, results in premature metamorphosis and the formation of small adult bodies. JH functions, in part, by modulating the activity of the molting hormone 20-hydroxyecdysone (20E), thereby maintaining juvenile larval characteristics and preventing metamorphosis (Riddiford 1996). Figure 15.2 illustrates the molecular mechanisms of JH and 20E action in holometabolous insects, along with the corresponding mechanisms in hemimetabolous species for comparison.

JH exerts its action by binding to its receptor, Methoprene-tolerant (Met), a member of the basic helix–loop–helix-Per-ARNT-Sim (bHLH-PAS) transcription factor family (Jindra et al. 2015). In *T. castaneum*, RNA interference (RNAi)–mediated knockdown of *Met* reduces larval sensitivity to JH and induces precocious metamorphosis, producing small pupae with adult-like characteristics (Konopova and Jindra 2007). Similarly, a decrease in the expression of the JH-response gene *Krüppel homolog 1* (*Kr-h1*) results in an early onset of metamorphosis in this species (Minakuchi et al. 2008).

Subsequent studies identified the ecdysone response gene *Ecdysone-induced protein 93F* (*E93*) as a downstream target of Kr-h1, which functions as a master regulator of metamorphosis initiation. Knockdown of *E93* in *T. castaneum* causes severe hyper-ecdysis, demonstrating its essential role (Chafino et al. 2019). Chafino et al. (2019) further proposed that the threshold size represents the minimum body weight that a larva must attain during the previous instar to commit to entering the final instar and initiating metamorphosis. They also demonstrated that this developmental commitment was accompanied by an increase in *E93* expression. Importantly, the decision to pupate is determined within 24 h of entering the final instar.

In summary, once the larvae reach the threshold size, JH titers decline, leading to reduced *Kr-h1* expression. This reduction permits an increase in *E93* expression, thereby initiating the metamorphic program (Bellés and Santos 2014). Although the decline in JH is essential for metamorphosis, it likely represents a proximate regulatory event that enables metamorphosis to occur rather than a mechanism by which larvae directly assess the body size.

15.2.3 Comparative Mechanisms of Metamorphic Regulation in Hemimetabolous Insects

The mechanisms underlying metamorphosis in hemimetabolous insects are less well-characterized than those in holometabolous insects, although recent studies have begun to clarify the hormonal and molecular frameworks involved, as summarized in Fig. 15.2 (Ishimaru et al. 2019). Although the overall metamorphic processes differ markedly between the two developmental modes, the fundamental endocrine regulation governing these transitions is evolutionarily conserved. The key distinction lies in the developmental trajectories after hatching: Holometabolous insects pass through a morphologically distinct larval stage before forming a pupa, whereas hemimetabolous insects progress through nymphal stages that closely resemble the adult form. In both groups, sequential molting and metamorphosis are orchestrated by the interplay between JH and 20E, the relative concentrations of which determine the developmental fate at each stage.

In response to 20E, transcription of early response genes, such as *Broad-Complex* (*Br-C*), *E74*, *E75*, and *E93*, is activated. This activation occurs when 20E binds to the ultraspiracle protein/ecdysone receptor (USP/EcR) complex, which subsequently interacts with ecdysone response elements (EcREs) located in the promoter regions of the early 20E-response genes. These factors constitute a transcriptional cascade that mediates molting and metamorphic transitions (Truman 2019). In holometabolous insects, Br-C plays a pivotal role in specifying the pupal stage, and its transient induction during the prepupal period is required for proper adult formation. Conversely, suppression of *Br-C* results in defective pupae that fail to complete adult development.

The activity of 20E signaling is tightly modulated by the presence of JH. High JH titers during early larval or nymphal instars maintain juvenile characteristics by preventing premature activation of the metamorphic program. JH exerts this regulatory effect by activating the JH receptor Met, which induces the expression of *Kr-h1*. Kr-h1 suppresses metamorphosis by repressing *E93* expression. As development proceeds toward the final instar, JH levels decline, leading to a reduction in *Kr-h1* expression and a concomitant induction of *E93*, which promotes adult differentiation (Truman 2019; Ishimaru et al. 2019). Depletion of *E93* in the final-instar nymphs of the German cockroach, *Blattella germanica,* or in the pupae of *T. castaneum* and *D. melanogaster* results in the failure of adult formation (Bellés and Santos 2014; Chafino et al. 2019; Lam et al. 2022), demonstrating that Kr-h1 and E93 function as mutual antagonists. Collectively, these findings led to the formulation of the conserved "Met-Kr-h1-E93 (MEKRE93) pathway" (Bellés and Santos

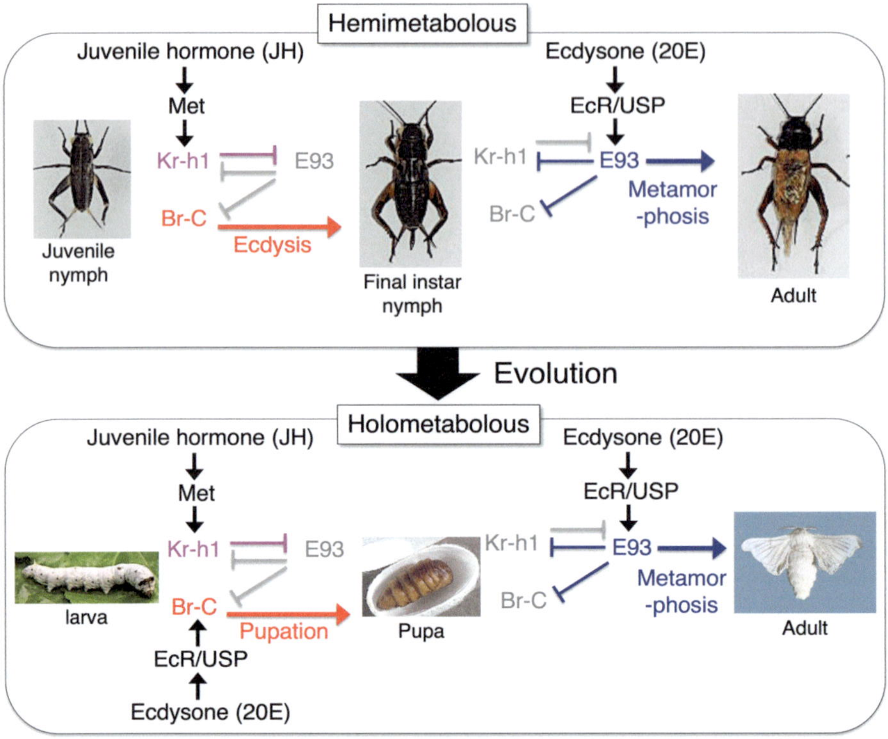

Fig. 15.2 Mechanism of metamorphosis in hemimetabolous cricket and holometabolous silkworm. Schematic comparison of the hormonal regulatory networks controlling metamorphosis in hemimetabolous insects (upper panel: *Gryllus bimaculatus*) and holometabolous insects (lower panel: *Bombyx mori*). In both groups, the juvenile hormone (JH) maintains juvenile characteristics by inducing the transcription factor *Krüppel homolog 1* (*Kr-h1*), which suppresses the expression of the metamorphosis factor *Ecdysone-induced protein 93F* (*E93*). In holometabolous insects, ecdysone (20-hydroxyecdysone; 20E) induces the expression of *Broad-Complex* (*Br-C*), which specifies pupal identity and drives pupal molting. When JH levels decline during the final nymphal instar in hemimetabolous insects or during the pupal stage in holometabolous insects, 20E signal upregulates *E93*, thereby initiating metamorphosis into the adult form

2014), which integrates JH and 20E signaling to regulate metamorphosis in diverse insect taxa.

The functional roles of Kr-h1, Br-C, and E93 have been elucidated in *G. bimaculatus* (Ishimaru et al. 2019). After hatching, crickets progress through seven nymphal molts and then molt a final (eighth) time to undergo metamorphosis into adults. The expression levels of *Kr-h1* and *Br-C* peaked at the sixth instar, after which they markedly decreased in the penultimate (seventh) instar. RNAi-mediated knockdown of either gene in fifth-instar nymphs induced precocious metamorphosis at the seventh instar, producing small-sized adults, indicating that both genes act to prevent premature metamorphosis. In contrast, *E93* expression peaked in the seventh instar. Knockdown of *E93* suppresses adult differentiation, resulting in repeated molting beyond the normal developmental cycle, up to the 11th instar. Gene expression analyses of RNAi-treated nymphs revealed reciprocal regulatory relationships among these three factors. Knockdown of *Kr-h1* or *Br-C* leads to precocious induction of *E93* in the sixth instar, whereas knockdown of *E93* maintains high levels of *Kr-h1* and *Br-C* in the seventh instar, when their expression normally declines. These results demonstrate a mutually inhibitory network in which Kr-h1 and Br-C suppress *E93* to maintain juvenile molting, whereas E93 represses *Kr-h1* and *Br-C* to trigger adult metamorphosis. Thus, the coordinated shift from Kr-h1/Br-C dominance to E93 dominance operates as a molecular switch that governs the timing of metamorphosis in hemimetabolous insects.

Functional analyses in *G. bimaculatus* further indicated that this two-stage, mutually inhibitory regulatory system is evolutionarily conserved between hemimetabolous and holometabolous insects. Based on the temporal dynamics of these regulatory interactions, Ishimaru et al. (2019) proposed that the penultimate (seventh) and final (eighth) nymphal instars of hemimetabolous insects correspond functionally to the pupal stage of the holometabolous insects. Moreover, they identified lineage-specific differences in *Br-C* expression patterns and isoform diversity: Holometabolous insects typically possess four or five *Br-C* isoforms, whereas hemimetabolous insects such as crickets and cockroaches possess six or more. This diversification of *Br-C* isoforms may have played a crucial role in the evolutionary origin of the pupal stage and the emergence of complete metamorphosis. Thus, the functional divergence of *Br-C* isoforms provides important insights into the evolution of the molecular framework underlying complete metamorphosis in insects.

15.3 The Role of TGF-β Signaling in the Regulation of JH

15.3.1 Decapentaplegic Signaling Promotes JH Biosynthesis

Since insect body size is influenced by changes in JH levels, it has been hypothesized that the mechanism by which insects assess body size is linked to the regulation of JH biosynthesis. Research on cricket metamorphosis has provided important insights into this regulatory mechanism. In the final step of JH biosynthesis, juvenile hormone acid methyltransferase (JHAMT) and cytochrome P450 (CYP15) were identified as the key enzymes, first in *B. mori* (Shinoda and Itoyama 2003) and the cockroach, *Diploptera punctata* (Helvig et al. 2004), respectively. In *B. mori*, *jhamt* expression in the corpora allata persists during the early larval stages but declines during the late larval stage, coinciding with the cessation of JH biosynthesis and the onset of metamorphosis. These findings suggest that the suppression of *jhamt* expression serves as a critical molecular switch that triggers metamorphosis.

Recent studies in *D. melanogaster* have revealed that the decapentaplegic (Dpp) signaling pathway, a member of the transforming growth factor-β (TGF-β) family, plays a crucial role in the transcriptional induction of *jhamt* (Huang et al. 2011). In

G. bimaculatus, *dpp* is expressed continuously in the corpus allatum. To investigate the functional role of Dpp signaling in regulating *jhamt*, we performed RNAi targeting *Mothers against dpp* (*mad*), a key intracellular transducer of Dpp signaling. Third-instar nymphs treated with *mad* RNAi underwent precocious metamorphosis and became adults at the seventh instar, resulting in a significantly reduced body size. This phenotype was accompanied by a marked decrease in *jhamt* expression and JH titers, suggesting that the reduction of JH underlies the premature metamorphosis observed. Similar results were obtained when RNAi was applied to the ligands *dpp* and *glass bottom boat* (*gbb*), as well as the type I receptor *thickveins* (*tkv*), all components of the Dpp/Gbb signaling pathway. Inhibition of Dpp/Gbb signaling led to decreased *jhamt* expression, reduced JH levels, and early metamorphic onset. These findings indicate that Dpp/Gbb signaling functions as a positive regulator of *jhamt* expression during early nymphal stages, thereby promoting JH biosynthesis and preventing precocious metamorphosis in insects (Fig. 15.3).

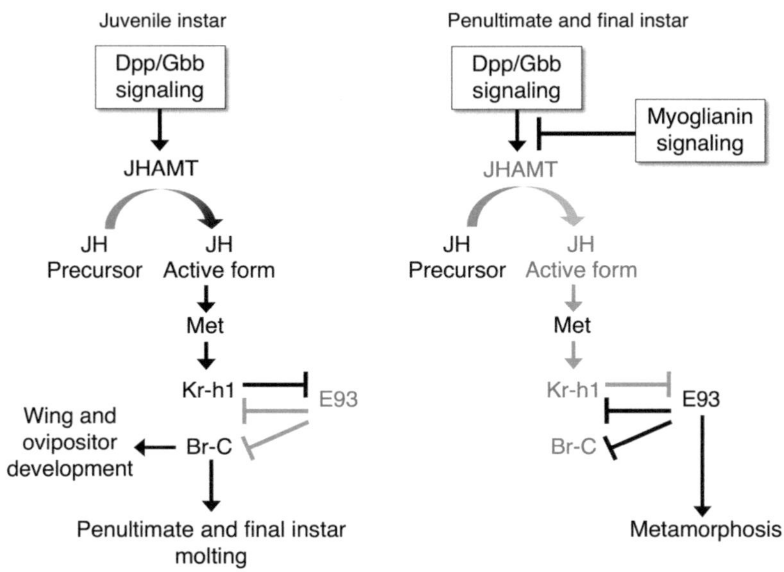

Fig. 15.3 Transforming growth fartor-β (TGF-β) signaling controls the production of juvenile hormone (JH). The schematic representation illustrates how the TGF-β signaling pathways, including Myoglianin (Myo) and decapentaplegic (Dpp)/glass bottom boat (Gbb), regulate JH biosynthesis by modulating *juvenile hormone acid O-methyltransferase* (*jhamt*) expression. During the juvenile (sixth) instar, Dpp/Gbb signaling promotes the production of the JH-active form by inducing *jhamt*, thereby maintaining high levels of *Krüppel homolog 1* (*Kr-h1*) and *Broad-Complex* (*Br-C*) that support nymphal molting and the growth of structures such as wing or the ovipositor. In contrast, during the penultimate (seventh) and final (eighth) instars, Myo signaling suppresses *jhamt* expression, leading to a reduction in the JH titer. This decrease permits *Ecdysone-induced protein 93F* (*E93*) induction by 20-hydroxyecdysone (20E) signaling, which triggers adult metamorphosis

15.3.2 Myoglianin Negatively Regulates JH Biosynthesis and Controls Body Size

Our research has revealed, for the first time, that myoglianin (Myo), another member of the TGF-β superfamily, plays a critical role in regulating metamorphosis in *G. bimaculatus* (Ishimaru et al. 2016). Myo is homologous to vertebrate myostatin (growth differentiation factor 8 (GDF8)) and GDF11, and its expression in the corpus allatum correlates with the dynamics of *jhamt* expression. RNAi-mediated inhibition of the Myo signaling pathway, including the ligand *myo*, the type I receptor *baboon* (*babo*), and the downstream effector *smox* (the ortholog of vertebrate Smad2), in early-instar nymphs (e.g., third-instar) induced supernumerary molts, resulting in prolonged maintenance of juvenile nymphal stages. During this period, *jhamt* expression substantially increased, accompanied by elevated JH titers. The *myo* RNAi nymphs continued to molt while retaining characteristics of third- and fourth-instar nymphs, with their body size remaining relatively small but gradually increasing. Once the RNAi effect subsided, the nymphs attained a size larger than that of normal fourth-instar nymphs and resumed the regular developmental sequence. They subsequently underwent molts corresponding to the fifth through eighth instars, with each stage exhibiting a body size larger than that of normal nymphs at the same instar, ultimately emerging as adults with a significantly increased body size (Fig. 15.4). In contrast, when *myo* RNAi was performed in fifth-instar nymphs, molting ceased at the sixth instar, and individuals remained in the nymphal stage and eventually died. In these cases, both *jhamt* expression and JH

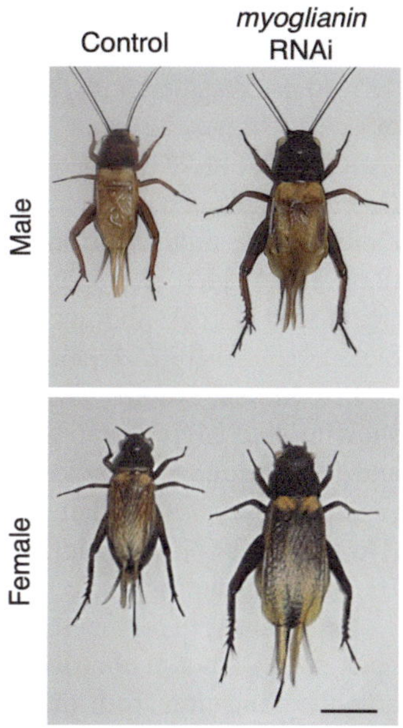

Fig. 15.4 Effects of *myoglianin* (*myo*) RNAi on cricket body size. Representative images of adult crickets subjected to RNAi-mediated knockdown of *myo* (upper panel: males; lower panel: females). Reduction of *myo* expression results in an additional molt and produces adults that are noticeably larger (right) than the controls (left). This finding demonstrates that Myo plays a crucial role in determining the final body size during postembryonic development. Scale bar: 1 cm

titers were markedly elevated. These observations indicate that Myo normally functions as a negative regulator of JH biosynthesis, serving as a systemic signal that establishes the growth termination threshold. By suppressing Myo activity in early instars, the signal to halt growth and initiate metamorphosis is delayed, allowing the nymphs to continue growing beyond the normal species-specific size before JH levels decline and metamorphosis occurs.

Collectively, these findings demonstrate that the two TGF-β family members, Dpp and Myo, regulate JH production, thereby controlling both the timing of metamorphosis and the final adult body size of *G. bimaculatus*. Our observations in *G. bimaculatus* are consistent with the recent results obtained for other hemimetabolous insects. Specifically, RNAi-mediated knockdown of *myo* induces additional nymphal molts in both *G. bimaculatus* and *B. germanica* (Kamsoi and Bellés 2019). In *B. germanica*, *myo* is expressed at much higher levels in the prothoracic gland than in the corpus allatum, where it stimulates the expression of ecdysone biosynthesis–related genes. This indicates that, in addition to its role in regulating JH synthesis, Myo may also positively influence ecdysteroid production (Kamsoi and Bellés 2019).

15.3.3 Evolutionary Conservation of Myoglianin Function and the Threshold Size Hypothesis

In *D. melanogaster*, *myo* is expressed in embryonic muscles and glial cells (Lo and Frasch 1999). During the larval stage, Myo functions similarly to vertebrate myostatin at the neuromuscular junction (NMJ), impeding synaptic transmission, growth, and branching of NMJ (Augustin et al. 2017). Knockdown of muscle-derived *myo* results in increased larval body weight and expansion of the body-wall muscles, suggesting that Myo acts as a negative regulator of muscle growth (Augustin et al. 2017). However, unlike many other insects, *myo* mutant larvae of *D. melanogaster* do not exhibit additional larval instars (Augustin et al. 2017). Additionally, final-instar larvae display a weak response to JH, as topical application of JH does not substantially delay metamorphosis or induce extra molts (Riddiford and Ashburner 1991). Consequently, defining the threshold size for metamorphosis in *D. melanogaster* remains challenging.

In contrast, studies in other holometabolous insects provide clearer insights, showing that Myo signaling functions in a conserved manner to regulate body size and the timing of metamorphosis. In the 28-spotted lady beetle, *Henosepilachna vigintioctopunctata* (Chen et al. 2022); *M. sexta* (He et al. 2020); and *T. castaneum* (He et al. 2020), Myo signaling suppresses *jhamt* expression, thereby modulating JH biosynthesis.

Importantly, in holometabolous insects, Myo does not directly determine the adult body size. Rather, it contributes to the definition of a critical size threshold that governs the transition from the penultimate to the final larval instar, a developmental checkpoint that precedes irreversible commitment to metamorphosis. He et al. (2020) investigated how *M. sexta* larvae assess body size at this checkpoint under two growth-restricting conditions, nutrient deprivation and hypoxia. These treatments altered larval growth rates, resulting in variation in body size at the end of the fourth instar, which serves as the threshold checkpoint for entry into the final instar. Nutrient deprivation, the standard approach for assessing threshold size in *M. sexta*, shifted the threshold depending on the timing and duration of food restriction. In contrast, hypoxia, another condition known to suppress larval growth in *M. sexta* and other insects (Callier and Nijhout 2011), consistently reduced the threshold size regardless of the timing of exposure. Across these conditions, threshold size was strongly correlated with the relative dry mass of muscle and integument, suggesting that these tissues act as key indicators of larval growth status. Consistent with this interpretation, expression of *myo* in muscle increased markedly as larvae approached the threshold size for the fourth to fifth instar transition. Because in vivo gene knockdown is not feasible in *M. sexta*, the functional role of Myo was examined in *T. castaneum*, which is amenable to systemic RNAi (He et al. 2020). Knockdown of *myo* in *T. castaneum* resulted in larvae that continued to molt beyond the normal threshold size and failed to enter the metamorphic program, with some individuals remaining in the larval state indefinitely. Together, these findings support the idea that increasing levels of *myo* produced by growing tissues act as a systemic signal that enables larvae to assess body size and trigger developmental progression once the threshold size is reached.

Thus, although the developmental stage at which size is assessed differs between hemimetabolous and holometabolous insects, Myo-dependent size sensing may represent an evolutionarily conserved component of size regulation. In holometabolous insects, this mechanism constrains adult body size indirectly by regulating entry into the final larval instar and subsequent pupal development, whereas in hemimetabolous insects, it more directly contributes to the determination of the final adult size.

Recent studies have revealed that Myo plays a critical role in determining species-specific body size in both holometabolous and hemimetabolous insects (He et al. 2020; Ishimaru et al. 2016). Based on these findings, we hypothesized that the threshold size may be closely related to muscle mass (M), a relationship that can be conceptually illustrated using a simple mathematical model. Specifically, if Myo production increases proportionally with muscle mass (M) and suppresses further muscle growth at a constant rate λ, the dynamics of muscle growth can be described by the following differential equation:

$$\frac{dM}{dt} = \alpha - \lambda M$$

where α represents the rate of muscle growth.

Solving this equation yields:

$$M(t) = M_\infty + (M_0 - M_\infty)e^{-\lambda t}$$

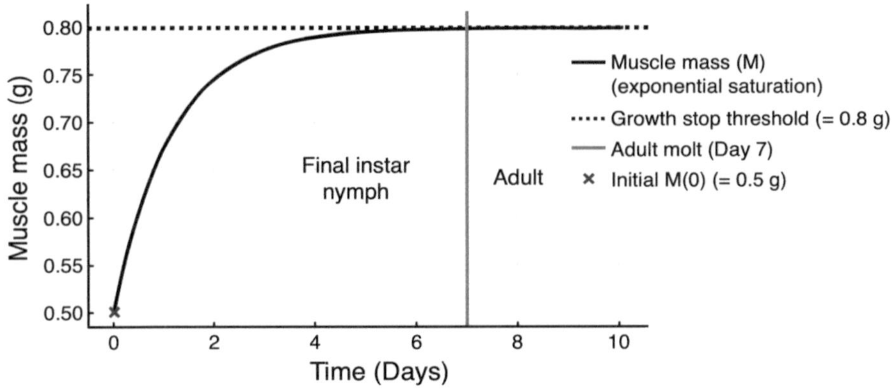

Fig. 15.5 Exponential saturation model of muscle growth in the final-instar cricket. Exponential saturation model illustrating muscle growth dynamics during the final (eighth) instar of *Gryllus bimaculatus*. Muscle mass (M) increases from 0.5 g to an asymptotic maximum of 0.8 g over the 7-day period preceding the final molt, as indicated by the gray dotted line. The saturation point (black dotted line) represents the physiological threshold at which Myoglianin (Myo) signaling is sufficient to suppress juvenile hormone (JH) synthesis, thereby initiating adult differentiation. This model highlights how endocrine feedback establishes a size-dependent limit for growth

Here, M_0 denotes the initial muscle mass, and M_∞ represents the threshold muscle mass. Assuming that body weight scales proportionally with muscle mass, a final-instar cricket with an initial body weight of 0.5 g is estimated to reach approximately 0.8 g by day 7, as illustrated in Fig. 15.5.

Although the precise mechanism establishing the Myo threshold remains unknown, it likely represents a functional rather than an absolute boundary. In this context, the effective threshold may emerge from the balance between Myo secretion from muscle tissues and the responsiveness of corpus allatum cells that synthesize JH. When Myo levels increase sufficiently to saturate Activin-type receptors and fully activate Sma and Mad (Smad) on X (Smox) signaling, the downstream transcriptional repression of JH biosynthetic enzymes such as *jhamt* is triggered. This feedback mechanism results in a rapid decline in circulating JH, thereby initiating the final molt to adulthood. In this framework, the "Myo threshold" is dynamically defined by the interplay among ligand concentration, receptor responsiveness, and Smad-mediated transcriptional regulation, representing a homeostatic tipping point within the Myo–JH regulatory loop that terminates the growth.

In *Drosophila,* Dpp is produced by a restricted subset of imaginal disc cells and acts as a classical morphogen that diffuses across the disc to regulate tissue patterning and growth. Similarly, TGF-β/Activin-like ligands, including Myo, function as systemic or interorgan signals that coordinate the organ size. Upadhyay et al. (2020) demonstrated that muscle-derived Myo regulates wing size through the canonical Smad2-mediated signaling pathway. This pathway links Myo activity to transcriptional programs that restrict wing disc growth, thereby mediating the heterometric relationship between muscle tissues and associated organs. These findings suggest that the level of Myo production may act as a cue for establishing the developmental size threshold, a mechanism that may be conserved across diverse insect lineages.

Similar phenomena have been observed in vertebrates. In humans, longitudinal height growth ceases in adulthood when epiphyseal plates ossify. Bone elongation is initially promoted by growth hormone and insulin-like growth factor I (IGF-1), and sex hormones induce the pubertal growth spurt. Ultimately, increased sex hormone levels drive the complete closure of the epiphyseal plates, thereby terminating further growth. Genetic predisposition largely determines the upper limit of adult height, whereas environmental factors, such as nutrition, sleep, and general health, modulate the realized phenotypes (Nilsson et al. 2005). Evolutionarily, insect Myo appears to have diverged into vertebrate myostatin (GDF8) and GDF11 (Suh and Lee 2020). The mature domain of Myo shares approximately 74% amino acid identity with myostatin and 72% with GDF11 (Fig. 15.6). In vertebrates, myostatin was first identified in mice as a negative regulator of skeletal muscle growth (McPherron et al. 1997). Myostatin is produced by muscle fibers, circulates systemically, and functions as a feedback inhibitor of muscle growth, consistent with the classical Chalone hypothesis of circulating tissue-specific growth inhibitors (Lee 2023). Although direct evidence linking these signaling pathways to human height determination is currently lacking, the evolutionary conservation of Myo-related signaling raises the possibility that muscle-derived TGF-β ligands participate in conserved growth-limiting mechanisms across animals. From this perspective, muscle-derived growth regulators can be viewed as integrative signals that couple tissue growth to species-specific size thresholds and developmental timing.

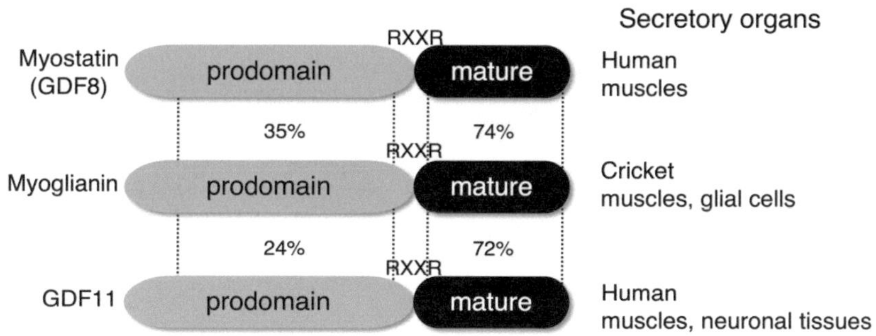

Fig. 15.6 Evolutionary conservation of structure and function among Myoglianin (Myo), myostatin (GDF8), and GDF11. Comparative domain architecture of insect Myo and vertebrate myostatin (GDF8) and GDF11. Each protein contains an N-terminal prodomain and a C-terminal mature ligand domain separated by a conserved proteolytic cleavage motif (RXXR). The mature domain of Myo shares 74% amino acid identity with myostatin (GDF8) and 72% with GDF11, whereas the prodomains exhibit lower conservation (35% and 24%, respectively). Functionally, myostatin (GDF8) and GDF11 are secreted mainly from vertebrate skeletal muscles and neuronal tissues and act as systemic regulators of growth. Likewise, Myo, which is produced in insect muscles and glial cells, is suggested to function as a muscle-derived Transforming growth fartor-β (TGF-β) ligand that may coordinate systemic growth and metamorphosis. These structural and functional correspondences imply that Myo/myostatin (GDF8)/GDF11 may constitute an evolutionarily conserved muscle-based endocrine system that links muscle status to whole-body growth regulation

15.3.4 Functional Analysis of Myoglianin Using the CRISPR/Cas9 System and Integration with RNAi Findings

Using CRISPR/Cas9-mediated gene knockout, we recently examined the role of Myo during early development in *G. bimaculatus* (Kawamoto et al. 2025). *myo* knockout (KO) mutants exhibit delayed embryonic development, severe defects in nymphal growth, and abnormal molting patterns. Although *myo* KO nymphs underwent the normal number of seven molts, their body size remained consistently smaller than that of wild-type (WT) individuals, and they ultimately failed to complete metamorphosis, dying as arrested nymphs (left panel of Fig. 15.7). Notably, leg muscles were significantly enlarged in KO nymphs (right panel of Fig. 15.7), and wing primordia failed to develop properly, suggesting a disruption of JH-dependent developmental processes, likely due to persistently high JH titers.

Contrary to our initial expectations, the *myo* KO nymphs were extremely small, reaching only approximately one-third of the length of WT individuals (Fig. 15.8). To interpret this phenotype, we considered the developmental framework proposed by Truman et al. (2024), who described three phases of nymphal morphogenesis in hemimetabolous insects: JH-independent, JH-dependent, and JH-free phases. In *G. bimaculatus*, the first to third instars constitute the JH-independent phase, during which wing primordia are formed. From the fourth instar, nymphs enter the JH-dependent phase, characterized by wing pad elongation and rapid body growth. In *myo* KO nymphs, wing pads never formed, indicating that development failed to progress beyond the equivalent of the third instar. This developmental arrest provides a mechanistic explanation for the drastically reduced body size.

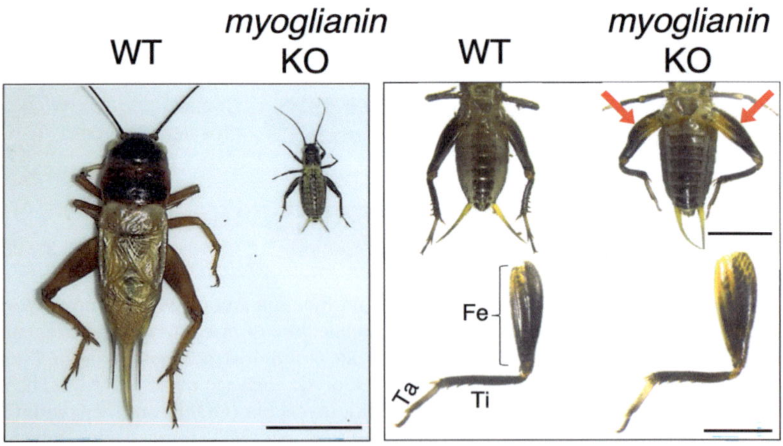

Fig. 15.7 Leg muscle hypertrophy in *myo* KO nymphs. Representative images comparing wild--type (WT) and *myoglianin* (*myo*) knockout (KO) crickets. Left panel: Comparison of overall body size between WT (left) and *myo* KO (right) individuals, showing that *myo* KO crickets exhibited reduced body size. Scale bar: 1 cm. Right panel: Comparison of legs from body-size-matched individuals. The legs of the *myo* KO nymph (right; red arrows) were markedly thicker than those of the WT nymph (left), indicating muscle hypertrophy resulting from the loss of Myo function. Fe femur, Ti tibia, and Ta tarsus. Scale bar: upper: 3 mm and lower: 2 mm

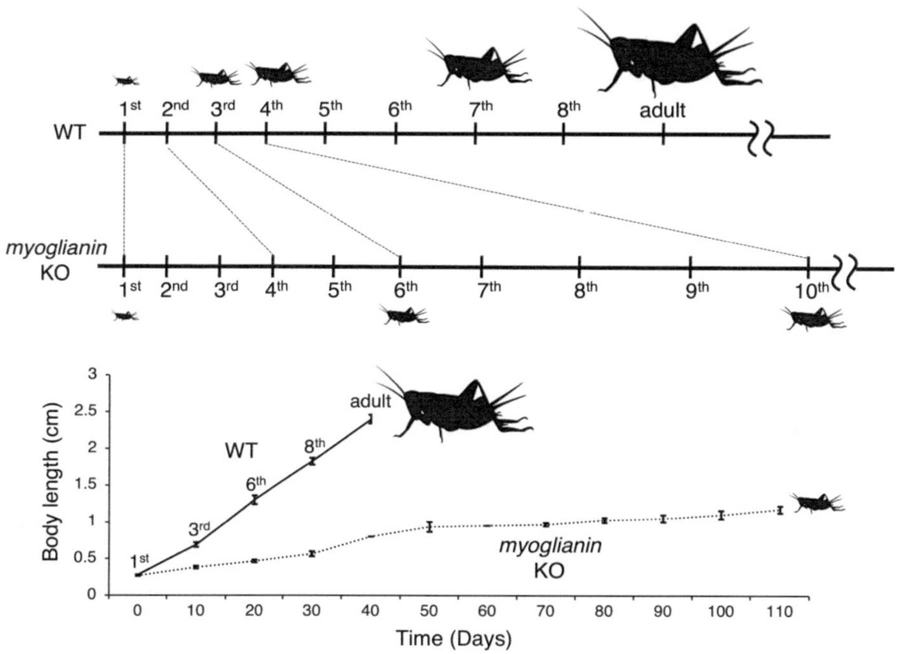

Fig. 15.8 Myoglianin is required for the nymphal growth. Comparison of developmental progression and growth dynamics between wild-type (WT) and *myoglianin* (*myo*) knockout (KO) crickets. Top: Developmental timeline illustrating the number of molts from the first instar to adulthood. WT crickets normally undergo seven molts before reaching the adult stage, whereas *myo* KO crickets continue to molt repeatedly without completing normal development. Bottom: Growth curve showing the body length over time. In *myo* KO nymphs, growth markedly slowed after the third instar, and the final body length remained approximately one-third of that of WT adults. Despite their smaller size, *myo* KO nymphs displayed an extended lifespan, up to twice that of WT individuals, indicating that the loss of *myo* uncouples somatic growth from developmental timing

This phenotype stands in stark contrast to our previous *myo* RNAi results, in which partial suppression of Myo signaling caused an increase in body size due to the prolonged duration of the JH-dependent phase. The difference between RNAi and KO phenotypes suggests that Myo performs dual functions: (1) as a systemic signal that terminates JH-dependent growth and (2) as an essential factor required for the basic somatic growth machinery during the JH-independent phase. In the absence of *myo* (KO), the fundamental capacity for body growth appears to collapse, regardless of the JH status. In contrast, partial reduction of Myo signaling (RNAi) preserves its essential growth-supporting function while delaying the growth termination cue, resulting in a larger body size.

These findings indicate that Myo is a critical regulator of nymphal development, coordinating both the transition from the JH-independent to the JH-dependent phase, which we designate as "nymphal metamorphosis," and the subsequent transition from the JH-dependent to the JH-free phase, leading to adult metamorphosis. Thus, Myo regulates not only the timing of metamorphosis but also the capacity for somatic growth and the proper development of structures such as wing primordia. It is necessary to introduce a new concept to explain the functions of Myo in growth and metamorphosis in crickets.

15.4 Applications: Producing Crickets of Desired Sizes and Functions

15.4.1 Genome Editing and Size Engineering

Research on the production of edible crickets of varying sizes is currently underway. In the present study, the adult body length of *G. bimaculatus* was approximately 3 cm. Artificially altering the body length has the potential to generate crickets of different adult sizes, thereby meeting diverse utilization needs. For example, when crickets are used as feed for various livestock, it is important to determine the optimal size that maximizes the overall cost efficiency and yield by streamlining collection, cleaning, and primary processing. To address this challenge, we are developing a strategy that employs genome editing technology (Watanabe et al. 2012, 2014, 2017; Matsuoka et al. 2024) to produce crickets of varying sizes, guided by the available genomic information of *G. bimaculatus* (Ylla et al. 2021).

15.4.2 Edible Cricket Products and Food Recycling Systems

Many insects are herbivorous, such as silkworms, which feed exclusively on mulberry. In contrast, crickets are omnivorous, and their ability to utilize food scraps as nutrient sources represents a significant advantage. For instance, in Japan, a substantial amount of food is imported annually (approximately 32.41 million tons), and the management of food waste (i.e., edible food that is discarded) remains a major social concern. In the course of our research, we obtained data supporting the feasibility of replacing a portion of the feed used for rearing crickets with food waste (Watanabe 2024). Utilizing crickets as a novel protein production system within a recycling framework has the potential to contribute to food security in the future.

The challenge posed by food shortages lies in the inability to respond promptly. For instance, in the event of supply disruptions caused by climate change–related disasters, agricultural product shortages, or war, a significant number of people may face starvation. Even considering self-sufficiency, immediate solutions are not feasible. To ensure food security, it is crucial to maintain a stable supply of crickets, which are excellent sources of protein. Consequently, we are conducting research on mass-rearing techniques for the two-spotted

Fig. 15.9 Edible cricket–based food product: cricket crackers. Representative examples of food products produced using powder derived from *Gryllus bimaculatus*. This photo shows "cricket crackers," which is a representative example of food products incorporating cricket powder. It is said that the taste resembles shrimp crackers. With their high protein content, they are likely to become a food source that addresses future protein shortages

cricket and exploring its use as a food source, for example, in products such as cricket-based rice crackers (Fig. 15.9).

To promote the utilization of insects as food, we are addressing various challenges, including but not limited to improving productivity, assessing functionality and safety, and reducing psychological resistance to consuming insects.

15.4.3 Prospects for Functional Food and Pharmaceutical Applications

Furthermore, through the application of genome editing and other advanced methodologies, crickets can serve as a biological platform for the sustainable production of high-value biomaterials, including pharmaceuticals and vaccines. This highlights their potential not only as a sustainable food source but also as a versatile bioresource platform that contributes to human health and global sustainability.

Our findings demonstrate that the modulation of *myo* expression directly influences the final body size of crickets. By controlling the balance between muscle-derived Myo and JH activity, Myo regulates the timing of the transition from growth to metamorphosis in the nymphal stage. These results suggest that Myo functions as a systemic signal, conveying the developmental state of muscle tissues to the endocrine system, thereby determining the appropriate point at which somatic growth should cease. This mechanism may represent a fundamental biological principle underlying the size determination in animals.

From an evolutionary perspective, Myo-mediated regulation of body size appears to be conserved across different animal lineages. Myostatin and GDF11 share high sequence similarity in their mature domains and perform analogous functions in restricting the growth of the body and organs. Accordingly, the Myo–myostatin/GDF11 pathway likely represents an ancient, conserved regulatory system that coordinates tissue growth and overall body size via TGF-β/Smad-mediated signaling.

Understanding the interplay between muscle-derived signals and hormonal control provides insight not only into the fundamental mechanisms of growth regulation but also into potential practical applications. Fine-tuning growth through *myo* expression could enable the production of crickets optimized for diverse purposes, ranging from food to experimental model systems. Furthermore, this principle may extend to other agricultural species, such as livestock and aquaculture animals, contributing to sustainable production strategies and improved environmental efficiencies.

In summary, Myo functions as a key molecular mediator linking muscle development, endocrine signaling, and the final body size. Studies on hemimetabolous insects, such as *G. bimaculatus*, provide critical insights into the evolutionary conservation and diversification of the mechanisms controlling growth termination and size homeostasis across the animal kingdom.

Acknowledgments We thank Dr. Sayuri Tomonari, Dr. Shintaro Inoue, and Dr. Yuki Nakamura of Tokushima University for their valuable contributions to this study through insightful discussions.

References

Aiello D, Barbera M, Bongiorno D, Cammarata M, Censi V, Indelicato S et al (2023) Edible insects an alternative nutritional source of bioactive compounds: a review. Molecules 28:699. https://doi.org/10.3390/molecules28020699

Augustin H, McGourty K, Steinert JR, Cochemé HM, Adcott J, Cabecinha M et al (2017) Myostatin-like proteins regulate synaptic function and neuronal morphology. Development 144:2445–2455. https://doi.org/10.1242/dev.152975

Bellés X, Santos CG (2014) The MEKRE93 (Methoprene tolerant-Krüppel homolog 1-E93) pathway in the regulation of insect metamorphosis, and the homology of the pupal stage. Insect Biochem Mol Biol 52:60–68. https://doi.org/10.1016/j.ibmb.2014.06.009

Callier V, Nijhout HF (2011) Control of body size by oxygen supply reveals size-dependent and size-independent mechanisms of molting and metamorphosis. Proc Natl Acad Sci USA 108:14664–14669. https://doi.org/10.1073/pnas.1106556108

Chafino S, Ureña E, Casanova J, Casacuberta E, Franch-Marro X, Martín D (2019) Upregulation of *E93* gene expression acts as the trigger for metamorphosis independently of the threshold size in the beetle *Tribolium castaneum*. Cell Rep 27:1039–1049. https://doi.org/10.1016/j.celrep.2019.03.094

Cheng C, Ko A, Chaieb L, Koyama T, Sarwar P, Mirth CK et al (2014) The POU factor ventral veins lacking/Drifter directs the timing of metamorphosis through ecdysteroid and juvenile hormone signaling. PLoS Genet 10(6):e1004425. https://doi.org/10.1371/journal.pgen.1004425

Chen F, Zhang XQ, Wu JJ, Jin L, Li GQ (2022) Requirement of Myoglianin for metamorphosis in the beetle *Henosepilachna vigintioctopunctata*. Insect Mol Biol 31:671–685. https://doi.org/10.1111/imb.12795

Daimon T, Kozaki T, Niwa R, Kobayashi I, Furuta K, Namiki T et al (2012) Precocious metamorphosis in the juvenile hormone-deficient mutant of the silkworm, *Bombyx mori*. PLoS Genet 8:e1002486. https://doi.org/10.1371/journal.pgen.1002486

Godfray HC, Beddington JR, Crute IR, Haddad L, Lawrence D, Muir JF et al (2010) Food security: the challenge of feeding 9 billion people. Science 327:812–818. https://doi.org/10.1126/science.1185383

Grunert LW, Clarke JW, Ahuja C, Eswaran H, Nijhout HF (2015) A quantitative analysis of growth and size regulation in *Manduca sexta*: the physiological basis of variation in size and age at metamorphosis. PLoS One 10:e0127988. https://doi.org/10.1371/journal.pone.0127988

Hanboonsong Y, Jamjanya T, Durst PB (2013) Six-legged livestock: edible insect farming, collection and marketing in Thailand. Agric Food Sci 3:8–21

He LL, Shin SH, Wang Z, Yuan I, Weschler R, Chiou A et al (2020) Mechanism of threshold size assessment: metamorphosis is triggered by the TGF-beta/Activin ligand Myoglianin. Insect Biochem Mol Biol 126:103452. https://doi.org/10.1016/j.ibmb.2020.103452

Helvig C, Koener JF, Unnithan GC, Feyereisen R (2004) CYP15A1, the cytochrome P450 that catalyzes epoxidation of methyl farnesoate to juvenile hormone III in cockroach corpora allata. Proc Natl Acad Sci USA 101:4024–4029. https://doi.org/10.1073/pnas.0306980101

Huang J, Tian L, Peng C, Abdou M, Wen D, Wang Y et al (2011) DPP-mediated TGFβ signaling regulates juvenile hormone biosynthesis by activating the expression of juvenile hormone acid methyltransferase. Development 138:2283–2291. https://doi.org/10.1242/dev.057687

Ishimaru Y, Tomonari S, Matsuoka Y, Watanabe T, Miyawaki K, Bando T et al (2016) TGF-β signaling in insects regulates metamorphosis via juvenile hormone biosynthesis. Proc Natl Acad Sci USA 113:5634–5639. https://doi.org/10.1073/pnas.1600612113

Ishimaru Y, Tomonari S, Watanabe T, Noji S, Mito T (2019) Regulatory mechanisms underlying the specification of the pupal-homologous stage in a hemimetabolous insect. Philos Trans R Soc B 374:20190225. https://doi.org/10.1098/rstb.2019.0225

Jindra M, Uhlirova M, Charles JP, Smykal CV, Hill RJ (2015) Genetic evidence for function of the bHLH-PAS protein Gce/Met as a juvenile hormone receptor. PLoS Genet 11:e1005394. https://doi.org/10.1371/journal.pgen.1005394

Kamsoi O, Bellés X (2019) Myoglianin triggers the premetamorphosis stage in hemimetabolan insects. FASEB J 33:3659–3669. https://doi.org/10.1096/fj.201801511R

Kawamoto K, Ishimaru Y, Tomonari S, Watanabe T, Noji S, Mito T (2025) Myoglianin is a crucial factor for the transition to the juvenile hormone-dependent phase during hemimetabolous nymphal development. Insect Biochem Mol Biol 178:104274. https://doi.org/10.1016/j.ibmb.2025.104274

Kemsawasd V, Inthachat W, Suttisansanee U, Temviriyanukul P (2022) Road to the red carpet of edible crickets through integration into the human food chain with biofunctions and sustainability: a review. Int J Mol Sci 23:1801. https://doi.org/10.3390/ijms23031801

Kingsolver JG (2007) Variation in growth and instar number in field and laboratory *Manduca sexta*. Proc R Soc B 274:977–981. https://doi.org/10.1098/rspb.2006.0036

Konopova B, Jindra M (2007) Juvenile hormone resistance gene *Methoprene-tolerant* controls entry into metamorphosis in the beetle *Tribolium castaneum*. Proc Natl Acad Sci USA 104:10488–10493. https://doi.org/10.1073/pnas.0703719104

Lam G, Nam HJ, Velentzas PD, Baehrecke EH, Thummel CS (2022) *Drosophila E93* promotes adult development and suppresses larval responses to ecdysone during metamorphosis. Dev Biol 481:104–115. https://doi.org/10.1016/j.ydbio.2021.10.001

Lee S-J (2023) Myostatin: a skeletal muscle Chalone. Annu Rev Physiol 85:269–291. https://doi.org/10.1146/annurev-physiol-012422-112116

Lo PC, Frasch M (1999) Sequence and expression of *myoglianin*, a novel *Drosophila* gene of the TGF-beta superfamily. Mech Dev 86:171–175. https://doi.org/10.1016/s0925-4773(99)00108-2

McPherron AC, Lawler AM, Lee S-J (1997) Regulation of skeletal muscle mass in mice by a new TGF-β superfamily member. Nature 387:83–90. https://doi.org/10.1038/387083a0

Matsuoka Y, Nakamura T, Watanabe T, Barnett AA, Tomonari S, Ylla G et al (2024) Establishment of CRISPR/Cas9-based knock-in in a hemimetabolous insect: targeted gene tagging in the cricket *Gryllus bimaculatus*. Development 152:dev199746. https://doi.org/10.1242/dev.199746

Minakuchi C, Namiki T, Yoshiyama M, Shinoda T (2008) RNAi-mediated knockdown of *juvenile hormone acid O-methyltransferase* gene causes precocious metamorphosis in the red flour beetle *Tribolium castaneum*. FEBS J 275:2919–2931. https://doi.org/10.1111/j.1742-4658.2008.06428.x

Mito T, Ishimaru Y, Watanabe T, Nakamura T, Ylla G, Noji S et al (2022) Cricket: the third domesticated insect. Curr Top Dev Biol 147:291–306. https://doi.org/10.1016/bs.ctdb.2022.02.003

Nijhout HF, Williams CM (1974) Control of moulting and metamorphosis in the tobacco hornworm, *Manduca sexta* (L.): growth of the last-instar larva and the critical size for pupation. J Exp Biol 61:493–501. https://doi.org/10.1242/jeb.61.2.493

Nijhout HF (1975) A threshold size for metamorphosis in the tobacco hornworm, *Manduca sexta* (L.). Biol Bull 149:214–225. https://doi.org/10.2307/1540491

Nijhout HF, Riddiford LM, Mirth C, Shingleton AW, Suzuki Y, Callier V (2014) The developmental control of size in insects. WIREs Dev Biol 3:113–134. https://doi.org/10.1002/wdev.124

Nilsson O, Marino R, De Luca F, Phillip M, Baron J (2005) Endocrine regulation of the growth plate. Horm Res 64:157–165. https://doi.org/10.1159/000088791

Noji S (ed) (2022) The latest cricketology. Hokuryukan, Tokyo. (in Japanese)

Riddiford LM, Ashburner M (1991) Effects of juvenile hormone mimics on larval development and metamorphosis of *Drosophila melanogaster*. Gen Comp Endocrinol 82:172–183. https://doi.org/10.1016/0016-6480(91)90181-5

Riddiford LM (1996) Juvenile hormone: the status of its "status quo" action. Arch Insect Biochem Physiol 32:271–286. https://doi.org/10.1002/(SICI)1520-6327(1996)32:3/4<271::AID-ARCH2>3.0.CO;2-W

Shinoda T, Itoyama K (2003) Juvenile hormone acid methyltransferase: a key regulatory enzyme for insect metamorphosis. Proc Natl Acad Sci USA 100:11986–11991. https://doi.org/10.1073/pnas.2134232100

Sogari G, Amato M, Palmieri R, Hadj Saadoun J, Formici G, Verneau F et al (2023) The future is crawling: evaluating the potential of insects for food and feed security. Curr Res Food Sci 6:100504. https://doi.org/10.1016/j.crfs.2023.100504

Suh J, Lee Y-S (2020) Similar sequences but dissimilar biological functions of GDF11 and myostatin. Exp Mol Med 52:1673–1693. https://doi.org/10.1038/s12276-020-00516-4

Suzuki Y, Koyama T, Hiruma K, Riddiford LM, Truman JW (2013) A molt timer is involved in the metamorphic molt in *Manduca sexta* larvae. Proc Natl Acad Sci USA 110:12518–12525. https://doi.org/10.1073/pnas.1311405110

Texada MJ, Koyama T, Rewitz K (2020) Regulation of body size and growth control. Genetics 216:269–313. https://doi.org/10.1534/genetics.120.303095

Truman JW (2019) The evolution of insect metamorphosis. Curr Biol 29:R1252–R1268. https://doi.org/10.1016/j.cub.2019.10.009

Truman JW, Riddiford LM, Konopova B, Nouzova M, Noriega FG, Herko M (2024) The embryonic role of juvenile hormone in the firebrat, *Thermobia domestica*, reveals its function before its involvement in metamorphosis. eLife 12:RP92643. https://doi.org/10.7554/eLife.92643.3

Tyson JJ, Monshizadeh A, Shvartsman SY, Shingleton AW (2023) A dynamical model of growth and maturation in *Drosophila*. Proc Natl Acad Sci USA 120:e2313224120. https://doi.org/10.1073/pnas.2313224120

Upadhyay A, Peterson AJ, Kim MJ, O'Connor MB (2020) Muscle-derived Myoglianin regulates *Drosophila* imaginal disc growth. eLife 9:e51710. https://doi.org/10.7554/eLife.51710

United Nations (2024) World population prospects. United Nations Department of Economic and Social Affairs, Population Division, New York. https://population.un.org/wpp/

Watanabe T, Noji S, Mito T (2017) Genome editing in the cricket, *Gryllus bimaculatus*. Methods Mol Biol 1630:219–233. https://doi.org/10.1007/978-1-4939-7128-2_18

Watanabe T, Ochiai H, Sakuma T, Horch HW, Hamaguchi N, Nakamura T et al (2012) Non-transgenic genome modifications in a hemimetabolous insect using zinc-finger and TAL effector nucleases. Nat Commun 3:1017. https://doi.org/10.1038/ncomms2020

Watanabe T, Noji S, Mito T (2014) Gene knockout by targeted mutagenesis in a hemimetabolous insect, the two-spotted cricket *Gryllus bimaculatus*, using TALENs. Methods 69:17–21. https://doi.org/10.1016/j.ymeth.2014.05.006

Watanabe T (2024) Edible crickets as a sustainable food source, alternative protein technology and market. In: Technology and market of alternative proteins. CMC Publishing, Tokyo, pp 96–103. (in Japanese)

West GB, Brown JH (2005) The origin of allometric scaling laws in biology from genomes to ecosystems: towards a quantitative unifying theory of biological structure and organization. J Exp Biol 208:1575–1592. https://doi.org/10.1242/jeb.01589

Ylla G, Nakamura T, Itoh T, Kajitani R, Toyoda A, Tomonari S et al (2021) Insights into the genomic evolution of insects from cricket genomes. Commun Biol 4:733. https://doi.org/10.1038/s42003-021-02197-9

Chapter 16
Juvenile Hormone Analogs (JHAs) and Their Application to Pest Management

Hitoshi Kawada

Abstract To apply juvenile hormone (JH) actions to insecticide development, many studies have been conducted on the synthesis of JH-like substances (JH analog (JHA) or JH mimic (JHM)). Many compounds synthesized in these studies have not reached practical use due to low environmental stability, high production costs, and inadequate insecticide efficacy. Methoprene, hydroprene, fenoxycarb, and pyriproxyfen are representative JHAs that have been used in practice. Terpene structural analogs of natural JHs, such as methoprene and hydroprene, have been applied exclusively in the field of public health pest control because of their low stability in the field. JHAs with non-terpene structures, such as fenoxycarb and pyriproxyfen, are examples of JHAs applied in agriculture.

Keywords Autodissemination · Formulation · Juvenile hormone analog · Mosquito · Pyriproxyfen · Sterilization

16.1 Juvenile Hormone Analogs as Insecticides

Juvenile hormone (JH), which is secreted from the corpus allatum of insects, acts together with ecdysone, a molting hormone secreted from the prothoracic gland, to regulate insect metamorphosis. Insects are most sensitive to JH from the last larval instar to the prepupal stage, when they undergo a dramatic metamorphosis from larva to pupa or nymph to adult. If excessive JH is administered during this period, maladaptive phenomena, such as pupal death, abnormal metamorphosis, and excessive molting, are observed. The "paper factor" discovered by Sláma and Williams (1965) was the first event in the development of JH analogs (JHAs). With the discovery of juvabione isolated from "the paper factor" and the first naturally occurring JHA (Bowers et al. 1966), Williams (1967) predicted that JHAs would become ideal insecticides that act specifically on insects with low toxicity. To make use of JH

H. Kawada (✉)
Institute of Tropical Medicine, Nagasaki University, Nagasaki, Japan

T. Shinoda et al. (eds.), *Juvenile Hormone*, Entomology Monographs,
https://doi.org/10.1007/978-981-95-8605-9_16

action for insecticidal applications, many studies have been conducted on the synthesis of juvenile hormone–like substances (JHAs) and their application to insecticides (Bowers et al. 1965, 1966; Henrick et al. 1973, 1976; Bowers and Nishida 1980; Dorn et al. 1981). Many of the compounds synthesized in these studies have not reached the stage of practical use due to their low stability in the environment, high production costs, and insufficient efficacy as insecticides. Methoprene, hydroprene (Henrick et al. 1973, 1976), kinoprene (Hrdy 1974), fenoxycarb (Dorn et al. 1981), and pyriproxyfen (Ohsumi et al. 1985; Hatakoshi et al. 1986; Hatakoshi 1992) are representative JHAs that have been practically used (Fig. 16.1). Many of the JHAs' structures deviate considerably from those of natural sesquiterpenoid JHs, raising questions about whether their molecular actions are identical to those of natural JHs. However, the recent discovery of JH signaling through the intracellular JH receptor, methoprene-tolerant (Met), has demonstrated that chemically distinct JHAs, such as methoprene, fenoxycarb, and pyriproxyfen, act as agonist ligands for the JH receptor (Jindra and Bittova 2020).

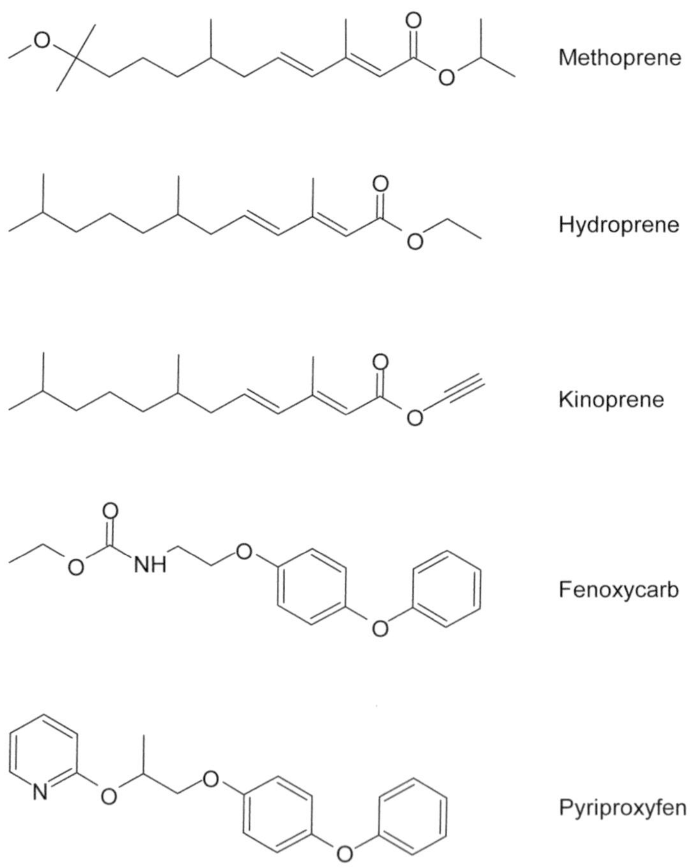

Fig. 16.1 Chemical structures of juvenile hormone analogs

These groups of compounds with JH agonist activity are collectively referred to as insect growth regulators (IGRs), along with other non-insecticide groups, such as ecdysone agonists (tebufenozide) (Ghoneim et al. 1991) and chitin synthesis inhibitors (CSIs), represented by benzoylphenylurea-type compounds such as diflubenzuron (Wellinga et al. 1973). CSI is characterized by its ability to inhibit the synthesis of a new cuticle during insect molting, leading to death, and has a completely different mode of action from JHAs. Cyromazine (Hart et al. 1982), an inhibitor of molting in the housefly, *Musca domestica*, and tebufenozide (Smagghe and Degheele 1994), an agonist of the molting hormone, are also known as IGRs. These non-insecticide compounds with new modes of action have been attracting attention as "third-generation" insecticides, as proposed by Williams (1967), and are still being developed by many researchers in the present day.

Among the numerous sesquiterpenoid JHA compounds screened, the acyclic isoprenoid compound methoprene exhibited outstandingly high JH activity and was the first commercial JHA synthesized by Zoëcon, USA, in 1973. Methoprene is effective against dipteran insects, such as flies and mosquitoes, stored grain pests, and fleas. Methoprene has been marketed under the brand names "Altosid®" (for flies and mosquitoes), "Diacon®" (for stored grain pests), and "Precor®" (for fleas). Hydroprene has been commercialized for cockroach control under the brand name "Gentrol®." Kinoprene has been used in greenhouses as an aphid and whitefly control agent under the brand name "Enstar®." The main reason why these JHAs have been applied exclusively in the field of public health and greenhouse pest control is that these structural mimics of natural JHs (terpene compounds) have low stability in the outside environment and cannot be applied in the farm field. To protect this unstable nature from the outside environment, special formulations in which the active ingredient was adsorbed on fine-grained activated carbon (Altosid® 10F) or microencapsulated with polyacrylamide resin (Altosid® SR10) were developed for fly and mosquito control. A unique formulation using methoprene is Manta® (Kamada 1992), developed by Otsuka Pharmaceutical in Japan in 1978. Silkworm moth, *Bombyx mori*, larvae treated with Manta® increased cocoon weight by more than 10% by extending the fifth instar period by half a day to one day.

Fenoxycarb was the first JHA compound to be commercialized for controlling agricultural pests by Maag, Switzerland, in 1977. Fenoxycarb acts as a JH mimic in the physiological processes of insects, such as metamorphosis and reproduction. Fenoxycarb also exhibits larvicidal and ovicidal activities against numerous insect species (Masner et al. 1987). This new type of JHA was marketed as a selective insecticide. Its characteristics include disruption of insect metamorphosis and effectiveness against various Lepidoptera and scale insect pests of fruits, citrus, olives, and vines. Fenoxycarb is also effective against mosquitoes, cockroaches, grain pests, ants, chironomids, and whiteflies and is commercialized under the trade names Insegar®, Precision®, Award®, Logic®, Polyon®, Torus®, Pictyl®, Varikill®, Dicare®, and Helgar® (Grenier and Grenier 1993).

Ohsumi et al. (1985) found that an oxime ether compound with a significantly different structure from methoprene or natural JHs exhibited JH activity. In a subsequent screening, Hatakoshi et al. (1986, 1987a, b) found that a pyridyl ether compound with a 4-phenoxyphenoxy structure, 4-phenoxyphenyl (RS)-2-(2-pyridyloxy) propyl ether (pyriproxyfen), showed extremely high inhibitory activity against the metamorphosis of dipteran and lepidopteran larvae. Pyriproxyfen is extremely active compared to terpene compounds such as methoprene and is characterized by its high stability in the environment (Kawada et al. 1987). These compounds are also highly active against agricultural insect pests, such as aphids, scale insects, and whiteflies (Hirano et al. 1998). Pyriproxyfen was first developed as a mosquito and housefly larvicide in 1989 in Japan and was marketed as "Sumilarv®" (Fig. 16.2). Pyriproxyfen was registered with the USEPA (U.S. Environmental Protection Agency) in 1995 and marketed under the brand name "Nylar®." In the agricultural field, it has been marketed for the control of whiteflies, thrips, and scale insects.

Conventional insecticides, such as organophosphates, carbamates, and pyrethroids, which act on the nervous system of insects, exhibit the same insecticidal

Fig. 16.2 Treatment with pyriproxyfen granules (Sumilarv® 0.5G) at the garbage treatment plant in Tokyo, Japan

activity regardless of the developmental stage of the pest. In contrast, JHAs are characterized by the expression of various activities according to the developmental stage of the pests. Therefore, to apply JHAs for pest control, it is necessary to consider the most effective timing and treatment methods.

16.2 Characteristics of Pyriproxyfen Action

Pyriproxyfen is characterized by its high JH-like activity. It has no effect when endogenous JH is present in the body of the insect pest. Insects develop from eggs to adults through repeated molting and metamorphosis, and there is a very short period during which JH disappears from the insect body. In the case of lepidopteran insects, JH is not detected in the early stages of egg, middle and late stages of the last-instar larvae, and pupal stage. Therefore, the time at which pyriproxyfen acts depends on the developmental stage of the target pest. The effects of pyriproxyfen include (I) hatching inhibition (egg-killing) in eggs, (II) metamorphosis inhibition in larvae, (III) reproductive inhibition in pupae, and (IV) reproductive inhibition in adults (a decrease in the number of eggs laid and the hatching rate of laid eggs) (Fig. 16.3).

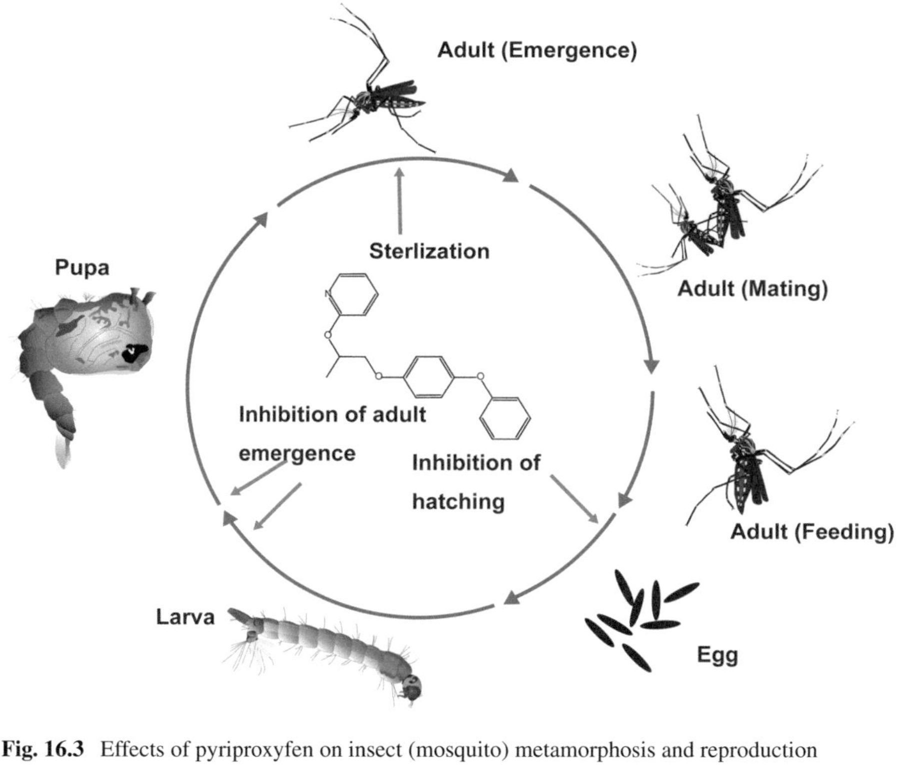

Fig. 16.3 Effects of pyriproxyfen on insect (mosquito) metamorphosis and reproduction

Because of its ability to inhibit hatching in whitefly eggs and sterilize adult whiteflies (Ishaaya and Horowitz 1992), pyriproxyfen can be used as a highly prophylactic pesticide when applied early in the pest infestation, unlike conventional pesticides that are applied after the pest infestation has increased. In the case of scale insects, only the limited direct application of conventional pesticides to the first-instar nymphs immediately after hatching (crawlers) was effective since this stage is the only defenseless stage of the scale insects, during which they are highly locomotive and have no protective scale. Pyriproxyfen enabled ambush tactics by application before the emergence of crawlers because of its high stability in the field. Adults are not directly affected by pyriproxyfen, but the first-instar nymphs that hatch after spraying stop growing from the first to the second instar stage and die (Isayama and Tsuda 2008). A yellow plastic tape device containing pyriproxyfen has been commercialized for the control of sweet potato whitefly, *Bemisia tabaci*, infestations in greenhouses, taking advantage of the attraction of adult insects to the color yellow. The tape device is stretched in a banner or hung in strips near the crop to attract and sterilize adult female whiteflies (Nakamura et al. 1994) (Fig. 16.4). A very small amount of the active ingredient attached to the adult whitefly also inhibits the hatching of healthy eggs when they come into contact with it (Nakamura et al. 2007).

Pyriproxyfen causes the destruction and structural failure of organelles, such as mitochondria, resulting in vacuolation and stunting of the imaginal disc of larvae in the yellow fever mosquito *Aedes aegypti* (Syafruddin et al. 1990). In addition, adults of the malaria-vector mosquito *Anopheles balabacensis* immersed for 48 h in

Fig. 16.4 Treatment with yellow tapes containing pyriproxyfen (Lano® Tape) in the eggplant farm. (Photo provided by Sumitomo Chemical Co., Ltd.)

0.005 ppb pyriproxyfen (one-eighth of a 50% lethal concentration (LC_{50})) during the last-instar larval stage showed greatly reduced sperm and egg production, as well as reduced blood-sucking and mating abilities (Iwanaga and Kanda 1988). Through the exposure of *Ae. aegypti* larvae to sublethal doses of pyriproxyfen, adult emergence was reduced by 48.7%, the hatchability of eggs laid by survivors was 36.8% lower than that of normal, and 39.9% of eggs were unfertilized (Loh and Yap 1989). Several laboratory studies have been conducted on the larvicidal effects of pyriproxyfen against infectious-disease-vector mosquitoes (Table 16.1). The concentration for 50% emergence inhibition (IC_{50}) of pyriproxyfen varies with mosquito species (0.00042–0.11 ppb) (Schaefer et al. 1988; Kawada et al. 1993b). Pyriproxyfen appears to be highly active against anopheline mosquitoes, which are important human malaria vectors.

Table 16.1 Emergence inhibition effect of pyriproxyfen on mosquito larvae

Species[a]	IC_{50} (concentration for 50% inhibition of emergence)	Reference
Aedes taeniorhynchus	0.010 ppb	Schaefer et al. (1988)
Ae. aegypti	0.023 ppb	Hatakoshi et al. (1986)
Ae. aegypti	0.0039 ppb	Henrick (1994)
Ae. aegypti	0.056 ppb	Itoh et al. (1994)
Ae. aegypti	0.011 ppb	Itoh et al. (1994)
Ae. argypti	0.01–0.048 ppb	Seccacini et al. (2008)
Ae. albopictus	0.11 ppb	Ali et al. (1995)
Ae. albopictus	0.009 ppb	Ohashi (2017)
Anopheles albimanus	0.016 ppb	Kawada et al. (1993a)
An. albimanus (OP-R)	0.00042 ppb	Kawada et al. (1993a)
An. balabacensis	0.04 ppb	Iwanaga and Kanda (1988)
An. gambiae	0.025 ppb	Kawada et al. (1993a)
An. gambiae (ICIPE strain)	0.00050 ppb	Kawada et al. (unpublished)
An. gambiae (OC-R)	0.0098 ppb	Kawada et al. (1993a)
An. gambiae (OC-R)	0.0040 ppb	Kawada et al. (1993a)
An. gambiae (PY-R)	0.0098–0.037 ppb	Kawada et al. (unpublished)
An. arabiensis (ICIPE strain)	0.0060ppb	Kawada et al. (unpublished)
An. arabiensis (PY-R)	0.027 ppb	Kawada et al. (unpublished)
An. funestus s.s. (PY-R)	0.014 ppb	Kawada et al. (unpublished)
An. farauti	0.0017 ppb	Kawada et al. (1993a)
An. stephensi	0.043 ppb	Hatakoshi et al. (1987a)
An. stephensi (OP-R)	0.025ppb	Kawada et al. (1993a)
Culex pipiens pallens	0.0046 ppb	Hatakoshi et al. (1987a)

(continued)

Table 16.1 (continued)

Species[a]	IC_{50} (concentration for 50% inhibition of emergence)	Reference
C. pipiens pallens (OP-R)	0.016 ppb	Kawada et al. (1987)
C. pipiens molestus (OP-R)	0.029 ppb	Kawada et al. (1994)
C. quinquefasciatus	0.04 ppb	Mulla et al. (1986)
C. quinquefasciatus	0.018 ppb	Schaefer et al. (1988)
C. quinquefasciatus (OP-R)	0.022 ppb	Schaefer et al. (1988)
C. tarsalis	0.021 ppb	Schaefer et al. (1988)
C. tarsalis (OP-R)	0.052 ppb	Schaefer et al. (1988)

[a]*OP-R* Organophosphate-resistant, *OC-R* Organochlorine-resistant, *PY-R* Pyrethroid-resistant

Table 16.2 summarizes the field efficacy of pyriproxyfen formulations against infectious-disease-vector mosquito larvae. When a juvenile hormone analog (JHA) is applied in the field where larvae of various stages exist, it must remain active until they develop to the last larval stage or early pupal stage, in which mosquitoes are most susceptible. Therefore, in addition to the chemical stability of JHA itself, a controlled-release formulation that releases a small but sufficient amount of the active ingredient into the environment is required to enhance residual efficacy. A granular or block-type formulation of pyriproxyfen showed the longest and most stable efficacy among several candidate formulations (Mulla et al. 1986; Kawada et al. 1987, 2006). The granular formulation of pyriproxyfen was treated at an initial concentration of 0.1 ppm in the water of a small pool in a breeding area of *Anopheles punctulatus* larvae (a small water pool that repeatedly appears and disappears due to rainfall and drying) in a hilly area of the Solomon Islands, and this treatment showed activity even after the source remained dry for 50 days (Fig. 16.5). Pyriproxyfen in a slow-release plastic formulation remained active in water jars for a long period of time against *Ae. aegypti* larvae, even when the water was frequently changed (Itoh 1993; Ohashi and Shono 2015) (Fig. 16.6).

Granules and slow-release formulations are suitable for relatively small-scale applications where the applicator is accessible and can be applied easily without the need for special equipment. In running water, they have been widely used for the control of chironomid larvae in urban rivers because they can be applied in batches, allowing the active ingredients to be released slowly into the water and have an effect downstream. However, granules are not suitable for spraying in breeding areas that are out of reach of sprayers, such as large wetlands or marshes. Emulsions and water-soluble granules that can be used with power sprayers are suitable for such sites (Hemingway et al. 1988; Suzuki et al. 1989; Chavasse et al. 1995). Water

Table 16.2 Field efficacy of pyriproxyfen formulations against mosquitoes

Species	Formulation[a]	Dosage as active ingredient	Efficacy (residual duration)	Reference
Aedes melanimon	0.5% granule	0.005–0.01 lb./acre	80–81% for 4 days	Mulla et al. (1986)
Ae. nigromaculis	0.5% granule	0.0025–0.005 lb/acre	66–79% for 4 days	Mulla et al. (1986)
Ae. nigromaculis *Ae. melanimon*	10% EC	0.0011–0.0056 kg/ha	73–100% at 48 hr after treatment	Schaefer et al. (1988)
Ae. aegypti	0.5% granule	0.025–0.05 ppm	82–100%	Adames and Rovira (1993)
Ae. aegypti	0.5% granule	0.02–0.05 ppm	> 6 weeks	Nayar et al. (2002)
Ae. aegypti	0.5% granule	0.01–0.02 ppm	4 months	Vythilingam et al. (2005)
Ae. aegypti	0.5% granule	0.02 ppm	5 months	Darriet et al. (2010)
Ae. aegypti	2% MR	1 disc/40 L	> 6 months	Ohashi and Shono (2015)
Ae. aegypti, C. quinquefasciatus	0.5% granule	2.5 ppm	> 26 weeks	Su (2023)
Ae. aegypti	2% MR	1 disc/250–1000 L	224–420 days	Müller et al. (2024)
C. pipiens pallens, Ae. albopictus	2% MR	1 disc/basin	> 6 months	Ohashi (2017)
Ae. togoi	0.5% granule	0.01–0.5 ppm	> 40 days	Lee (2001)
Anopheles minimus *An. maculatus*	0.5% granule	1 ppm to the water volume of 24 hr	> 4 weeks	Kerdpibule (1989)
An. farauti	1% EC	0.1 ppm	> 3 months	Suzuki et al. (1989)
An. punctulatus	0.5% granule	0.01–0.1 ppm	> 5 months at 0.1 ppm, > 3 months at 0.02–0.05 ppm	Okazawa et al. (1991)
An. albimanus	0.5% granule	0.025–0.05 ppm	82–100%	Adames and Rovira (1993)
An. subpictus *An. nigerrimus*	10% EC	0.1 ppm	> 71 days	Hemingway et al. (1988)
An. stephensi	2% MR	1 disc/100–500 L	> 31 weeks	Yehwalaw et al. (2025)
Culex tarsalis	0.5% granule	0.005–0.025 lb./acre	85–100% for 7 days	Mulla et al. (1986)
C. pipiens pallens	0.5% tablet	0.03–0.1 ppm	100% for 10 days	Ishii et al. (1990)

(continued)

Table 16.2 (continued)

Species	Formulation[a]	Dosage as active ingredient	Efficacy (residual duration)	Reference
C. pipiens pallens	0.5% granule	0.05–0.1 ppm	> 5–6 weeks	Kamimura and Arakawa (1991)
C. pipiens molestus	0.5% water-dispensable granule	0.01 ppm	> 1 month	Kawada et al. (1994)
C. tritaeniorhynchus	0.5% granule	0.01 ppm	> 3 months	Kamimura and Arakawa (1991)
C. tritaeniorhynchus	0.5% granule	–	> 90%	Thongrungiat and Kanda (1991)
C. peus	0.5% granule	0.025–0.05 lb./acre	63–66% for 2 days	Mulla and Darwazeh (1988)
Culex spp.	0.5% granule	0.1 kg/ha	28–66 days	Mulligan and Schaefer (1990)
C. quinquefasciatus	0.5% granule	0.025–0.05 ppm	82–100%	Adames and Rovira (1993)
C. quinquefasciatus	10% EC	0.1 ppm	4 weeks in the rainy season, 11 weeks in the dry season	Chavasse et al. (1995)

[a]*EC* Emulsifiable concentrate, *MR* Matrix release formulation

depth, vegetation, flow velocity, and other habitat parameters also affect the in situ effectiveness of JHAs. At greater depths, the application of pyriproxyfen granules had less effect on larvae near the water surface. This is because pyriproxyfen is easily adsorbed by bottom soils, in addition to the slow-release effect of the active ingredient from the granules (Kawada et al. 1987). Contaminated water requires pyriproxyfen doses up to 20 times higher than those required for clean water (Mulla 1995). Schaefer et al. (1991) reported a half-life of 7.5 days for pyriproxyfen in organic matter. In contrast, Mulligan and Schaefer (1990) reported that long-term residual effects of pyriproxyfen were also found in contaminated sources. Ohashi (2017) suggested that the high residual efficacy of pyriproxyfen in water may be due to (1) the adsorption of pyriproxyfen on organic matter, such as fallen leaves; (2) its subsequent gradual release from organic matter into the water; and (3) the larval ingestion of pyriproxyfen-adsorbed organic matter.

Fig. 16.5 Treatment with granular formulation of pyriproxyfen at the breeding sites of *Anopheles punctulatus* (Papua New Guinea)

16.3 New Potential of Juvenile Hormone Analogs

16.3.1 Autodissemination of Pyriproxyfen for the Control of Aedes aegypti

The slow action of JHAs due to the lack of insecticidal activity requires technical skills in pest control. However, their unique characteristics also provide new ideas for realizing the potential of JHAs. A trial of the new control technology was initiated as a countermeasure against *Ae. aegypti*, an important dengue fever vector in Southeast Asia and Central America. *Ae. aegypti* is anthropophilic and breeds in and around human habitation. People who live in areas without a well-developed water supply collect water from wells and rivers and store the water in bottles for drinking and other purposes, and these bottles in the vicinity of houses are an important breeding source of *Ae. aegypti*. The idea of autodissemination technology is to dredge or adhere pyriproxyfen to the body surface of adult females and have them

Fig. 16.6 Treatment with plastic formulation of pyriproxyfen (Sumilarv® 2MR) at the breeding site of *Aedes aegypti.* (Photo provided by Sumitomo Chemical Co., Ltd.)

carry it to the oviposition sites where larvae are breeding, thereby inhibiting the emergence of adults (Fig. 16.7).

The outline of this new concept was first presented in the 1990s (Kawada et al. 1993a; Itoh et al. 1994). A small-scale laboratory experiment was conducted, and a simple technique for attaching pyriproxyfen to the body surface of female mosquitoes was demonstrated. The idea was also floated that adult females of *Toxorhynchites* mosquitoes, natural predators of *Aedes* mosquitoes, could transmit pyriproxyfen to the breeding sites when they deposit their eggs. This idea might be realistic because the sensitivity of *Toxorhynchites* larvae to pyriproxyfen is much lower than that of *Ae. aegypti* (Kawada et al. 1993a). In a semi–field trial in Bangkok, Thailand, significant emergence inhibition of *Ae. aegypti* larvae was successfully obtained using an indoor-placed pyriproxyfen trap. Pyriproxyfen dissolved in rapeseed oil was applied to a PET (polyethylene terephthalate) film inside a black box. *Aedes aegypti* females trapped in this box picked up pyriproxyfen and transferred it onto the water surface of the breeding container when they contacted the water surface for oviposition (Itoh 1995).

The term "autodissemination"has been used in the literature since 1978. The original idea was to kill agricultural pests, such as lepidopteran insects, by having them carry pathogens or *Bacillus thuringiensis* (BT). This term was probably first used for mosquito control by Scholte et al. (2004). Although the term "autodissemination" was not used, the experiment of Itoh et al. (1994) was replicated in the laboratory by Chism and Apperson (2003) and Sihuincha et al. (2005). Chism and Apperson (2003) reported a significant linear response in the inhibition of the emergence of two mosquitoes, *Ochlerotatus triseriatus* and *Aedes albopictus*, by the tarsal contact exposure of gravid females to pyriproxyfen-treated surfaces. Sihuincha

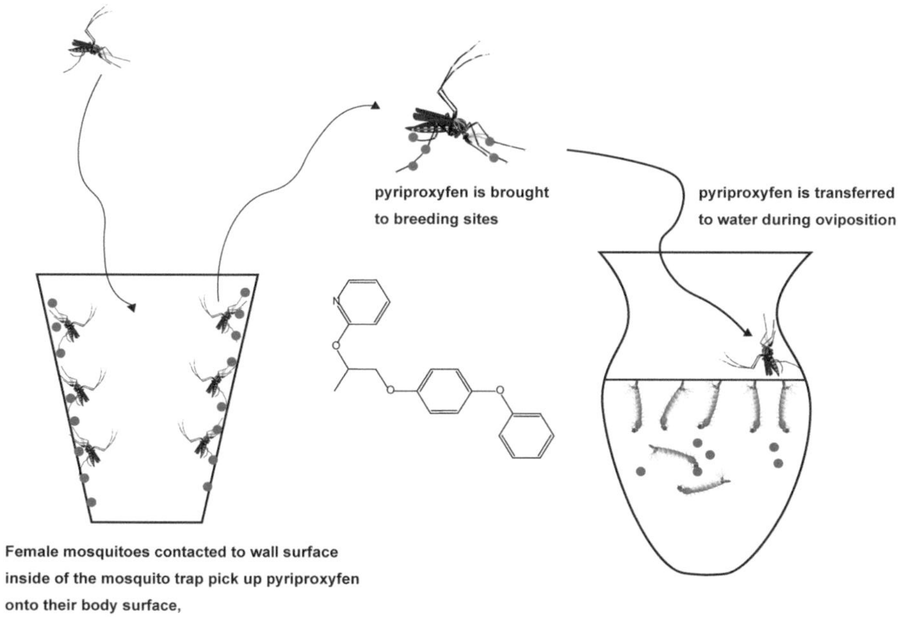

Fig. 16.7 Schematic diagram of autodissemination of pyriproxyfen by *Aedes aegypti* female

et al. (2005) reported that exposure of female *Ae. aegypti* to 0.003 g/m^2 pyriproxyfen showed a reduction of approximately 80% in the hatching of larvae at the egg-laying site and in the hatching rate of laid eggs by 70–90%. The authors' group further obtained 42–98% emergence inhibition of *Ae. aegypti* in field tests in Iquitos, Peru (Devine et al. 2009).

A Google Scholar search revealed that the number of articles on pyriproxyfen autodissemination, which was zero until 2008, has rapidly increased, starting with the report by Devine et al. (2009), and now exceeds 300 hits. This is a major breakthrough in the control of infectious-disease-vector mosquitoes. However, most studies have focused on *Ae. aegypti* control, and some have expanded the use of *Ae. albopictus* (Ohba et al. 2013; Chandel et al. 2016; Unlu et al. 2020) and anopheline mosquitoes (Lwetoijera et al. 2019; Kunambi et al. 2023). Autodissemination with pyriproxyfen has attracted much attention because (1) it is not insecticidal, (2) it does not affect adult egg-laying behavior, and (3) it is effective in emergence inhibition at very low doses. Although several products using the autodissemination of pyriproxyfen are now available, field trials based on reductions in dengue cases and dengue antibody titers are necessary to determine whether this new method will become the mainstay of dengue control in the future.

16.3.2 Pyriproxyfen-Impregnated Mosquito Nets for Sterilizing Anopheline Mosquitoes

Kawada et al. (1992) reported that the minimum dose of pyriproxyfen that caused infertility in female houseflies was 5 μg per female. An aliquot of 0.05–0.49 μg of pyriproxyfen per female was thought to be enough for the reduction of fertility in *Ae. aegypti* by contact with pyriproxyfen-treated PET films (Itoh et al. 1994). Several reports have described attempts to sterilize *Anopheles gambiae* mosquitoes by impregnating mosquito nets with pyriproxyfen (Aiku et al. 2006; Ohashi et al. 2012; Ohashi and Shono 2015). Ohashi et al. (2012) reported that exposure of adult female *An. gambiae* to pyriproxyfen-treated nets (3.5 mg/m^2) before and after blood feeding resulted in 100% sterility, and exposure to pyriproxyfen resulted in a dose-dependent decrease in adult longevity. Similarly, Harris et al. (2013) reported that *Anopheles arabiensis* fed blood 1 d before exposure to pyriproxyfen (3 mg/m^2) was sterilized, whereas other treatments (exposure 3 days and 1 d before and 3 days after blood feeding) had no significant effect. The pyriproxyfen-based mosquito sterilization technique may be one of the most effective means of controlling pyrethroid-resistant malaria-vector mosquitoes in the future.

Kawada et al. (2014a) investigated the effects of pyriproxyfen on pyrethroid-resistant wild populations of malaria-vector mosquitoes in western Kenya, where the predominant *An. gambiae* s.s. was resistant to pyrethroids, with more than 95% of the knockdown-resistant mutation (L1014S) in the voltage-sensitive sodium channel (VSSC) (Kawada et al. 2011). Mosquitoes were collected from houses with three types of bed net interventions: (1) 1% pyriproxyfen–impregnated, (2) 1% pyriproxyfen +2% permethrin–impregnated, and (3) 2% permethrin–impregnated (Olyset® nets). The deaths of blood-fed female mosquitoes within 3 days after collection were considered to be due to the effect of pyriproxyfen because the mean longevity of adult *An. gambiae* s.s. females exposed to pyriproxyfen was 5.6 days or less (Ohashi et al. 2012). Although there was no significant difference in the number of adult *An. gambiae* s. s. females collected from the houses of three types of intervention, the number of blood-fed mosquitoes killed within three days significantly increased in the houses with pyriproxyfen-impregnated mosquito nets (Fig. 16.8). Additionally, a significant decrease in the average number of eggs laid was observed in the houses with pyriproxyfen-impregnated mosquito nets. The number of progeny, calculated from the number of eggs laid and hatchability, also significantly decreased in houses with pyriproxyfen-impregnated mosquito nets (Fig. 16.9).

The Anopheles gambiae s.s. population in the study area is highly anthropophilic and has high blood-sucking activity, mainly at midnight, independent of the use of insecticide-impregnated mosquito nets (LLINs) (Kawada et al. 2014c). In this study, the bedtime of the residents was not restricted, so it is likely that mosquitoes had a chance to bite the residents who spent the night outside the nets before sleeping or slept with incomplete netting. Most of the *An. gambiae* s.s. females were assumed

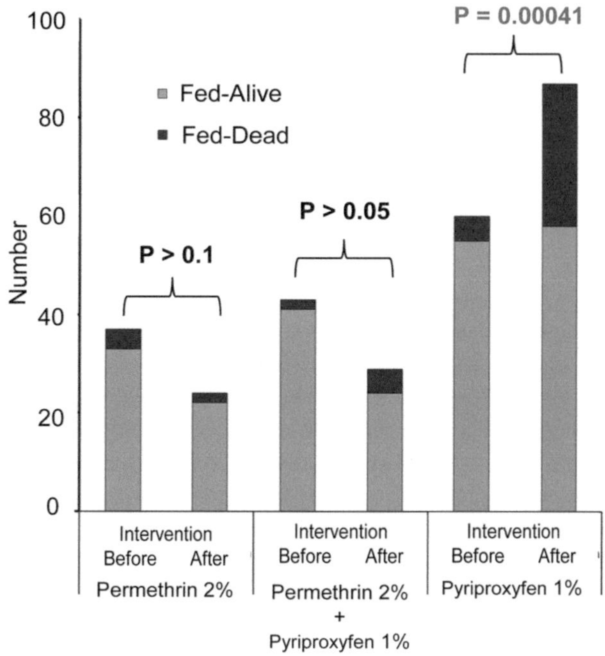

Fig. 16.8 Total number of dead (Fed-Dead) and alive (Fed-Alive) blood-fed females of *Anopheles gambiae* s.s. before and after intervention with three types of bed nets with permethrin and pyriproxyfen. *P*-values by χ^2 test are shown above the bars

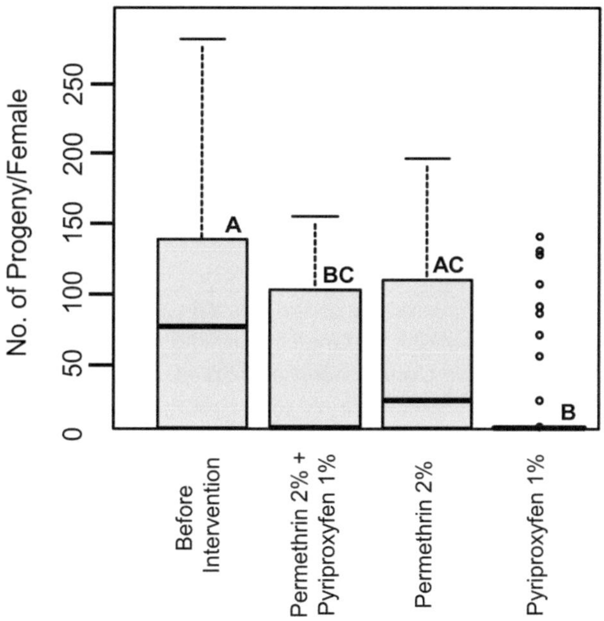

Fig. 16.9 Box-and-whisker plots (medians: thick lines, with interquartile ranges) of the number of progenies produced by *Anopheles gambiae* s.s. females collected from houses with three types of bed nets with permethrin and pyriproxyfen. Different letters indicate a significant difference by Kruskal–Wallis one-way analysis of variance

to be in contact with the nets before and after blood sucking, although we did not measure the frequency or total contact time of mosquitoes with the nets in this study.

This study also revealed the effects of a combination of permethrin and pyriproxyfen. Permethrin plays a role in reducing the number of mosquitoes that come into contact with bed nets or enter the nets. When pyriproxyfen is used without permethrin in mosquito nets, the risk of humans being bitten by mosquitoes may increase, even if mosquitoes that come into contact with pyriproxyfen die after a few days. However, when permethrin was sufficiently active as a repellent, the chance of contact with the nets might be reduced, resulting in a reduction in the intake of pyriproxyfen. Therefore, a contradiction exists in the combination of permethrin and pyriproxyfen treatments. In this study, however, the combination of permethrin and pyriproxyfen was demonstrated to be effective in reducing the number of females that laid eggs and the number of eggs laid, resulting in a reduction in the next generation. The effectiveness of the combination of permethrin as a repellent and pyriproxyfen in this study may be due, in part, to the reduced repellency of the target species, *An. gambiae* s.s., to permethrin (Kawada et al. 2014b). The effectiveness of the combination of permethrin and pyriproxyfen was also demonstrated by Ngufor et al. (2014), who confirmed that the combination of permethrin and pyriproxyfen was highly effective in sterilizing *An. gambiae* s.s. with a knockdown resistance (*kdr*) factor (probably by the mutation at the VSSC, L1014F) of >90% but was less effective against pyrethroid-resistant *Culex quinquefasciatus* (resistance factor not mentioned but resistant to pyrethroids, carbamates, and organophosphates). The relationship between mosquito repellency and the efficacy of pyriproxyfen when combined with pyrethroids and the relationship between resistance factors and the efficacy of pyriproxyfen in different mosquito species should be pursued in depth in the future.

16.4 Resistance to Juvenile Hormone Analogs

Pyriproxyfen resistance in agricultural pests was first discovered in the sweet potato whitefly, *B. tabaci*, in Israel (Horowitz and Ishaaya 1994) and was subsequently reported in the United States and Spain. Several biotypes of this species exist, and biotype Q is less susceptible to pyriproxyfen than biotype B (Horowitz et al. 2003). Low susceptibility to pyriproxyfen has also been reported in the greenhouse whitefly, *Trialeurodes vaporariorum*, and detoxification metabolism by cytochrome P450 (CYP4G61) was thought to be involved as a factor (Karatolos et al. 2012). However, susceptibility can be restored or replaced by other susceptible biotypes when pyriproxyfen use is discontinued (Horowitz et al. 2005). The melon thrips, *Thrips palmi*, has also become less susceptible to pyriproxyfen (Nishi et al. 2013), and two- to 128-fold resistance has been reported in the western flower thrips, *Frankliniella occidentalis* (Wang et al. 2016).

Only a few reports on resistance to JHA in pests of public health importance have been published. *Ochlerotatus nigromaculis*, a mosquito species in California, USA, provides one of the few examples where high resistance to methoprene has developed, resulting in the failure of control (Cornel et al. 2002). Field populations with reduced susceptibility to methoprene have also been found in houseflies in Denmark and *Ae. aegypti* in Malaysia. Interestingly, cross-resistance to pyriproxyfen has not been observed in *Ae. aegypti* that developed high resistance to methoprene (Lau et al. 2015).

The problem of insecticide resistance in malaria- and dengue-vector mosquitoes is serious. When selecting pyriproxyfen as the active ingredient for controlling them, it is important that there is no cross-resistance with other insecticides. Pyriproxyfen has been reported to be effective as a larvicide against anopheline mosquitoes resistant to DDT, dieldrin, organophosphates, and carbamates, and it has been suggested that there is no cross-resistance to these insecticides (Kawada et al. 1993b). Kasai et al. (2007) reported the efficacy of insect growth regulators (IGRs), such as diflubenzuron and pyriproxyfen, in Japanese mosquito populations and suggested the existence of cross-resistance between IGRs and pyrethroids (etofenprox) in the mosquito *Culex pipiens* form *molestus*. Although there are only a few reports on resistance to pyriproxyfen in mosquitoes, there seems to be some causal relationship regarding cross-resistance since P450, a metabolic factor involved in the oxidative metabolism of pyrethroids (Yunta et al. 2016), is also involved in the metabolism of pyriproxyfen. Further studies are needed to confirm this hypothesis.

16.5 Conclusion

In this chapter, the author evaluated the physiological effects of JHAs on agricultural and public health pests and their basic performance as insecticides. The author also clarified the problems associated with the formulation and practical application of JHAs in the field. The fact that pyriproxyfen is effective against mosquitoes with knockdown resistance (*kdr*) factors caused by point mutations in the voltage-sensitive sodium channel suggests that this agent might be used as a countermeasure against pyrethroid resistance. Williams (1967) predicted that JHAs would be the ideal insecticides, *which are not only more specific but also proof against the evolution of resistance*. Although it has been realized that the latter prediction was too optimistic since the publication of experimental results by Vinson and Plapp (1974), which contradicted this prediction, JHA's low toxicity, low environmental impact, and action are different from those of second-generation insecticides (pyrethroids, carbamates, organophosphates, etc.) and should still be highly valued. Therefore, the remaining question is how to incorporate these ideal insecticides into pest control programs in a way that is effective and does not shorten their own lifespan. For example, more research is needed to develop treatments and formulations that minimize the development of resistance, search for effective synergists, and mix or rotate the use of these insecticides with other active insecticides.

References

Adames E, Rovira J (1993) Evaluation of the juvenile growth regulator pyriproxyfen (S-31183) against three species of mosquitoes in Panama. J Am Mosq Control Assoc 9:452–453

Aiku AO, Yates A, Rowland M (2006) Laboratory evaluation of pyriproxyfen treated bed nets on mosquito fertility and fecundity. A preliminary study. West Afr J Med 25:22–26. https://doi.org/10.4314/wajm.v25i1.28240

Ali A, Nayar JK, Xue R-D (1995) Comparative toxicity of selected larvicides and insect growth regulators to a Florida laboratory population of *Aedes albopictus*. J Am Mosq Control Assoc 11:72–76. PMID: 7616194

Bowers WS, Nishida R (1980) Potent juvenile hormone mimics from sweet basil. Science 209:1030–1032. https://doi.org/10.1126/science.154.3752.1020

Bowers WS, Thompson MJ, Uebel EC (1965) Juvenile and gonadotropic activity of 10, 11 epoxy farnesoic acid methyl ester. Life Sci 4:2323–2331. https://doi.org/10.1016/0024-3205(65)90256-0

Bowers WS, Fales HM, Thompson MJ, Uebel EC (1966) Juvenile hormone: identification of an active compound from Balsam fir. Science 154:1020–1021. https://doi.org/10.1126/science.154.3752.1020

Chandel K, Suman DS, Wang Y, Unlu I, Williges E, Williams GM et al (2016) Targeting a hidden enemy: Pyriproxyfen autodissemination strategy for the control of the container mosquito *Aedes albopictus* in cryptic habitats. PLoS Negl Trop Dis 10:e0005235. https://doi.org/10.1371/journal.pntd.0005235

Chavasse DC, Lines JD, Ichimori K, Majara AR, Minjas JN, Marijani J (1995) Mosquito control in Dar Es Salaam. II. Impact of expanded polystyrene beads and pyriproxyfen treatment of breeding sites on *Culex quinquefasciatus* densities. Med Vet Entomol 9:147–154. https://doi.org/10.1111/j.1365-2915.1995.tb00171.x

Chism BD, Apperson CS (2003) Horizontal transfer of the insect growth regulator pyriproxyfen to larval microcosms by gravid *Aedes albopictus* and *Ochlerotatus triseriatus* mosquitoes in the laboratory. Med Vet Entomol 17:211–220. https://doi.org/10.1046/j.1365-2915.2003.00433.x

Cornel AJ, Stanich MA, McAbee RD, Mulligan FS III (2002) High level methoprene resistance in the mosquito *Ochlerotatus nigromaculis* in Central California. Pest Manag Sci 58:791–798. https://doi.org/10.1002/ps.521

Darriet F, Marcombe S, Etienne M, Yébakima A, Agnew P, Yp-Tcha M-M et al (2010) Field evaluation of pyriproxyfen and spinosad mixture for the control of insecticide resistant *Aedes aegypti* in Martinique (French West Indies). Parasit Vect 3:88. https://doi.org/10.1186/1756-3305-3-88

Devine DJ, Pereab EZ, Killeen GF, Stancile JD, Clark SJ, Morrison AC (2009) Using adult mosquitoes to transfer insecticides to *Aedes aegypti* larval habitats. Proc Natl Acad Sci USA 106:11530–11534. https://doi.org/10.1073/pnas.0901369106

Dorn S, Frischkeeht MS, Martinez V, Zurfluch R, Fischer R (1981) A novel non-neurotoxic insecticide with a broad activity spectrum. Z Pflanzenkrankh Pflanzenschutz 88:269–275. https://www.jstor.org/stable/43214853

Ghoneim KS, Fouda MA, Bream AS (1991) Effectiveness of the non-steroidal ecdysone mimic, RH-5849 for the control of *Musca domestica vicina*. J Egypt Soc Parasitol 21:723–733

Grenier S, Grenier A-M (1993) Fenoxycarb, a fairly new insect growth regulator: review of its effects on insects. Ann App Biol 122:369–403. https://doi.org/10.1111/j.1744-7348.1993.tb04042.x

Harris C, Lwetoijera DW, Dongus S, Matowo NS, Lorenz LM, Devine GJ et al (2013) Sterilizing effects of pyriproxyfen on *Anopheles arabiensis* and its potential use in malaria control. Parasit Vect 6:144. https://doi.org/10.1186/1756-3305-6-144

Hart RJ, Cavey WA, Ryan KJ, Strong MB, Moore B, Thomas PL et al (1982) CGA-72662—a new sheep blowfly insecticide. Aust Vet J 59:104–109. https://doi.org/10.1111/j.1751-0813.1982.tb02741.x

Hatakoshi M (1992) An inhibitory mechanism over oviposition in the tobacco cutworm, *Spodoptera litura*, by juvenile hormone analogue pyriproxyfen. J Insect Physiol 38:793–801. https://doi.org/10.1016/0022-1910(92)90032-9

Hatakoshi M, Agui N, Nakayama I (1986) 2-[1-Methyl-2(4-phenoxyphen-oxy)ethoxy] pyridine as a new insect juvenile hormone analogue: induction of supernumerary larvae in *Spodoptera litura* (Lepidoptera: Noctuidae). Appl Entomol Zool 21:351–353. https://doi.org/10.1303/aez.21.351

Hatakoshi M, Kawada H, Nishida S, Kishida H, Nakayama I (1987a) Laboratory evaluation of 2-[1-methyl-2-(4-phenoxy-phenoxy)ethoxy]pyridine against larvae of mosquitoes and housefly. Jpn J Sanit Zool 38:271–274. https://doi.org/10.7601/mez.38.271

Hatakoshi M, Nakayama I, Riddiford LM (1987b) Penetration and stability of juvenile hormone analogues in *Manduca sexta* L. (Lepidoptera: Sphingidae). Appl Entomol Zool 22:641–644. https://doi.org/10.1303/aez.22.641

Hemingway J, Bonning BC, Jayawardena KGI, Weerasinghe IS, Herath PRJ, Ouchi H (1988) Possible selective advantage of *anopheles* spp. (Diptera: Culicidae) with the oxidase- and acetylcholinesterase- based insecticide resistance genes after exposure to organophosphates or an insect growth regulator in Sri Lankan rice fields. Bull Entomol Res 78:471–478. https://doi.org/10.1017/S0007485300013225

Henrick CA (1994) Juvenoids (Conf. Proc.). In: Godfrey CRA (ed) Agrochemicals from natural products. Marcel Dekker, New York, pp 147–213

Henrick CA, Staal GB, Sidaal JB (1973) Alkyl-3,7,11-trimethyl-2,4-dodecadienoates, a new class of potent insect growth regulators with juvenile hormone activity. J Agric Food Chem 21:354–359. https://doi.org/10.1021/jf60187a043

Henrick CA, Willy WE, Staal GB (1976) Insect juvenile hormone activity of alkyl (2E,4E)-3,7,11-trimethyl-2,4-dodecadienoates. Variations in the ester function and the carbon chain. J Agric Food Chem 24:207–218. https://doi.org/10.1021/jf60204a018

Hirano M, Hatakoshi M, Kawada H, Takimoto Y (1998) Pyriproxyfen and other juvenile hormone analogues. Rev Toxicol 2:357–394

Horowitz AR, Ishaaya I (1994) Managing resistance to insect growth regulators in the sweetpotato whitefly (Homoptera: Aleyrodidae). J Econ Entomol 87:866–871. https://doi.org/10.1093/jee/87.4.866

Horowitz AR, Denholm I, Gorman K, Cenis JL, Kontsedalov S, Ishaaya I (2003) Biotype Q of *Bemisia tabaci* identified in Israel. Phytoparasitica 31:94–98. https://doi.org/10.1007/BF02979772

Horowitz AR, Kontsedalov S, Khasdan V, Ishaaya I (2005) Biotypes B and Q of *Bemisia tabaci* and their relevance to neonicotinoid and pyriproxyfen resistance. Arch Insect Biochem Physiol 58:216–225. https://doi.org/10.1002/arch.20044

Hrdy I (1974) Effects of juvenoids on insecticide susceptible and resistant aphids (*Myzus persicae*, *Aphis fabae* and *Therioaphis maculata*; Homoptera, Aphidiae). Acta Entomol Bohemoslov 71:367–381

Isayama S, Tsuda N (2008) Development of a novel formulation: Pluto® MC to control mulberry scale on tea. Sumitomo Kagaku 2008-II:4–13

Ishaaya I, Horowitz AR (1992) Novel phenoxy juvenile hormone analog (pyriproxyfen) suppresses embryogenesis and adult emergence of sweetpotato whitefly (Homoptera: Aleyrodidae). J Econ Entomol 85:2113–2117. https://doi.org/10.1093/jee/85.6.2113

Ishii T, Utsumi Y, Kamata A, Kamei S (1990) Field trials of BCP-8702 against mosquito larvae in ditches. J Sci Univ Tokushima 23:9–19. (in Japanese)

Itoh T (1993) Control of DF/DHF vector, *Aedes* mosquito, with insecticides. Trop Med 35:259–267. https://archive.org/details/tropmed100694651

Itoh T (1995) Utilization of blood fed females of *Aedes aegypti* as a vehicle for the transfer of the insect growth regulator, pyriproxyfen, to larval habitats. Trop Med 36:243–248

Itoh T, Kawada H, Abe Y, Eshita Y, Rongsriyam Y, Igarashi A (1994) Utilization of blood-fed females of *Aedes aegypti* as a vehicle for the transfer of the insect growth regulator pyriproxyfen to larval habitats. J Am Mosq Control Assoc 10:344–347. PMID: 7807075

Iwanaga K, Kanda T (1988) The effects of a juvenile hormone active oxime ether compound on the metamorphosis and reproduction of an anopheline vector, *Anopheles balabacensis*. Appl Entomol Zool 23:186–193. https://doi.org/10.1303/aez.23.186

Jindra M, Bittova L (2020) The juvenile hormone receptor as a target of juvenoid "insect growth regulators". Arc Insect Biochem Physiol 103:e21615. https://doi.org/10.1002/arch.21615

Kamada A (1992) Effects of methoprene (Manta®) application on growth and cocoon production in the silkworm, *Bombyx mori*, of different rearing temperatures. J Seric Sci Jpn 61:116–122. https://doi.org/10.11416/kontyushigen1930.61.116. (in Japanese with English abstract)

Kamimura K, Arakawa R (1991) Field evaluation of an insect growth regulator, pyriproxyfen, against *Culex pipiens pallens* and *Culex tritaeniorhynchus*. Jpn J Sanit Zool 42:249–254. https://doi.org/10.7601/mez.42.249

Karatolos N, Williamson MS, Denholm I, Gorman K, ffrench-Constant RH, Bass C (2012) Overexpression of a cytochrome P450 is associated with resistance to pyriproxyfen in the greenhouse whitefly *Trialeurodes vaporariorum*. PLoS One:e31077. https://doi.org/10.1371/journal.pone.0031077

Kasai S, Shono T, Komagata O, Tsuda Y, Kobayashi M, Motoki M et al (2007) Insecticide resistance in potential vector mosquitoes for West Nile virus in Japan. J Med Entomol 44:822–829. https://doi.org/10.1093/jmedent/44.5.822

Kawada H, Dohara K, Shinjo G (1987) Laboratory and field evaluation of an insect growth regulator, 4-phenoxyphenyl (RS)-2-(2-pyridyloxy) propyl ether as a mosquito larvicide. Jpn J Sanit Zool 39:339–346. https://doi.org/10.7601/mez.39.339

Kawada H, Senbo S, Abe Y (1992) Effects of pyriproxyfen on the reproduction of the housefly, *Musca domestica*, and the German cockroach, *Blattella germanica*. Jpn J Sanit Zool 43:169–175. https://doi.org/10.7601/mez.43.169

Kawada H, Itoh T, Abe Y, Horio M (1993a) Can mosquito be a carrier of larvicides? In: Wildey KB, Robinson WH (eds) Proceedings of the 1st international congress on insect pests in the urban environment. UK, Cambridge, p 497

Kawada H, Shono Y, Itoh T, Abe Y (1993b) Laboratory evaluation of insect growth regulators against several species of anopheline mosquitoes. Jpn J Sanit Zool 44:349–353. https://doi.org/10.7601/mez.44.349

Kawada H, Kohama T, Abe Y (1994) Larvicidal activity of a water soluble granular formulation of the insect growth regulator, pyriproxyfen, against *Culex* mosquito. Jpn J Environ Entomol Zool 6:68–77. (in Japanese). https://doi.org/10.11257/jjeez.6.68

Kawada H, Saita S, Shimabukuro K, Hirano M, Koga M, Iwashita T et al (2006) Mosquito larvicidal effectiveness of EcoBio-block S: a novel integrated water-purifying concrete block formulation containing insect growth regulator pyriproxyfen. J Am Mosq Control Assoc 22:451–456. https://doi.org/10.2987/8756-971X(2006)22[451:MLEOES]2.0.CO;2

Kawada H, Dida GO, Ohashi K, Komagata O, Kasai S, Tomita T et al (2011) Multimodal pyrethroid resistance in malaria vectors, *Anopheles gambiae* s.s., *Anopheles arabiensis*, and *Anopheles funestus* s.s. in western Kenya. PLoS One 6:e22574. https://doi.org/10.1371/journal.pone.0022574

Kawada H, Dida GO, Ohashi K, Kawashima E, Sonye G, Njenga SM et al (2014a) A small-scale field trial of pyriproxyfen-impregnated bed nets against pyrethroid-resistant *Anopheles gambiae* s.s. in western Kenya. PLoS One 9:e111195. https://doi.org/10.1371/journal.pone.0111195

Kawada H, Ohashi K, Dida GO, Sonye G, Njenga SM, Mwandawiro C et al (2014b) Insecticidal and repellent activities of pyrethroids to the three major pyrethroid-resistant malaria vectors in western Kenya. Parasit Vect 7:208. https://doi.org/10.1186/1756-3305-7-208

Kawada H, Ohashi K, Dida GO, Sonye G, Njenga SM, Mwandawiro C et al (2014c) Preventive effect of permethrin-impregnated long-lasting insecticidal nets on the blood feeding of three major pyrethroid-resistant malaria vectors in western Kenya. Parasit Vect 7:383. https://doi.org/10.1186/1756-3305-7-383

Kerdpibule V (1989) A field test of 2-[1-methyl-2-(4-phenoxyphenoxy)ethoxy] pyridine against principal vectors of malaria in a foot-hill area in Thailand. Jpn J Trop Med Hyg 17:175–183. https://doi.org/10.11501/11393823

Kunambi HJ, Ngowo H, Ali A, Urio N, Ngonzi AJ, Mwalugelo YA et al (2023) Sterilized *Anopheles funestus* can autodisseminate sufficient pyriproxyfen to the breeding habitat under semi-field settings. Mal J 22:280. https://doi.org/10.1186/s12936-023-04699-9

Lau KW, Chen CD, Lee HL, Norma-Rashid Y, Sofian-Azirun M (2015) Evaluation of insect growth regulators against field-collected *Aedes aegypti* and *Aedes albopictus* (Diptera: Culicidae) from Malaysia. J Med Entomol 52:199–206. https://doi.org/10.1093/jme/tju019

Lee D-K (2001) Field evaluation of an insect growth regulator, pyriproxyfen, against *Aedes togoi* larvae in brackish water in South Korea. J Vect Ecol 26:3942. PMID: 11469183

Loh PY, Yap HH (1989) Laboratory studies on the efficacy and sublethal effects of an insect growth regulator, pyriproxyfen (S-31183) against *Aedes aegypti*. Trop Biomed 6:7–12

Lwetoijera D, Kiware S, Okumu F, Devine GJ, Majambere S (2019) Autodissemination of pyriproxyfen suppresses stable populations of *anopheles arabiensis* under semi-controlled settings. Mal J 18:166. https://doi.org/10.1186/s12936-019-2803-1

Masner P, Angst M, Dorn S (1987) Fenoxycarb, an insect growth regulator with juvenile hormone activity: a candidate for *Heliothis virescens* (F.) control on cotton. Pestic Sci 18:89–94. https://doi.org/10.1002/ps.2780180203

Mulla MS (1995) The future of insect growth regulators in vector control. J Am Mosq Control Assoc 11:269–273. PMID: 7595460

Mulla MS, Darwazeh HA (1988) Efficacy of new insect growth regulators against mosquito larvae in dairy waste water lagoons. J Am Mosq Control Assoc 4:322–325. PMID: 3199121

Mulla MS, Darwazeh HM, Kennedy B, Dawson DM (1986) Evaluation of new insect growth regulators against mosquitoes with notes on nontarget organisms. J Am Mosq Control Assoc 2:314–320. PMID: 2906977

Müller JN, Galardo AKR, Corrêa APS, Macoris ML, Melo-Santos MAV, Nakazawa MM et al (2024) Impact of SumiLarv® 2MR on *Aedes aegypti* larvae: a multicenter study in Brazil. Parasit Vect 18:88. https://doi.org/10.1186/s13071-023-06064-w

Mulligan FS III, Schaefer CH (1990) Efficacy of a juvenile hormone mimic, pyriproxyfen (S-31183), for mosquito control in dairy wastewater lagoons. J Am Mosq Control Assoc 6:89–92. PMID: 2324729

Nakamura S, Inoue M, Fujimoto H, Kasamatsu K (1994) A new application method of pyriproxyfen for controlling sweet potato whitefly, *Bemisia tabaci* (Gennadius) (Homoptera: Aleyrodidae). Appl Entomol Zool 29:454–456. https://doi.org/10.1303/aez.29.454

Nakamura S, Inoue M, Fujimoto H, Kasamatsu K (2007) Ovicidal effect of pyriproxyfen transported from a tape formulation by adult sweet potato whitefly, *Bemisia tabaci* (Homoptera: Aleyrodidae). Jpn J Appl Entomol Zool 51:58–59. (in Japanese with English abstract). https://doi.org/10.1303/jjaez.2007.58

Nayar JK, Ali A, Zaim M (2002) Effectiveness and residual activity comparison of granular formulations of insect growth regulators pyriproxyfen and s-methoprene against Florida mosquitoes in laboratory and outdoor conditions. J Am Mosq Control Assoc 18:196–201. PMID: 12322941

Ngufor C, N'Guessan R, Fagbohoun J, Odjo A, Malone D, Akogbeto M et al (2014) Olyset Duo® (a pyriproxyfen and permethrin mixture net): an experimental hut trial against pyrethroid resistant *Anopheles gambiae* and *Culex quinquefasciatus* in Southern Benin. PLoS One 9:e93603. https://doi.org/10.1371/journal.pone.0093603

Nishi Y, Sano T, Koma H, Tanaka R, Nagamori S (2013) Effect of several commercial pesticides on the survival of *Thrips palmi* Karny collected from eggplant greenhouses in Okayama prefecture, Japan. Kyushu Pl Prot Res 59:81–85. (in Japanese). https://doi.org/10.4241/kyubyochu.59.81

Ohashi K (2017) Control of mosquito larvae in catch basins using pyriproxyfen and the mechanism underlying residual efficacy. Med Entomol Zool 68:127–135. https://doi.org/10.7601/mez.68.127

Ohashi K, Shono Y (2015) Recent progress in the research and development of new products for malaria and dengue vector control. Sumitomo Kagaku 2015:4–14

Ohashi K, Nakada K, Ishiwatari T, Miyaguchi J, Shono Y, Lucas JR et al (2012) Efficacy of pyriproxyfen-treated nets in sterilizing and shortening the longevity of *Anopheles gambiae* (Diptera: Culicidae). J Med Entomol 49:1052–1058. https://doi.org/10.1603/me12006

Ohba S, Ohashi K, Pujiyati E, Higa Y, Kawada H, Mito N et al (2013) The effect of pyriproxyfen as a "population growth regulator" against *Aedes albopictus* under semi-field conditions. PLoS One 8:e67045. https://doi.org/10.1371/journal.pone.0067045

Ohsumi T, Hatakoshi M, Kishida H, Matsuo N, Nakayama I, Itaya N (1985) Oxime ethers: new potent insect growth regulators. Agric Biol Chem 49:3197–3202. https://doi.org/10.1271/bbb1961.49.3197

Okazawa T, Bakote'e B, Suzuki H, Kawada H, Kere N (1991) Field evaluation of an insect growth regulator, pyriproxyfen, against *Anopheles punctulatus* on North Guadalcanal, Solomon Islands. J Am Mosq Control Assoc 7:604–607. PMID: 1787406

Schaefer CH, Miura T, Dupras EF Jr, Mulligan FS III, Wilder WH (1988) Efficacy, nontarget effects, and chemical persistence of S-31183, a promising mosquito (Diptera: Culicidae) control agent. J Econ Entomol 81:1648–1655. https://doi.org/10.1093/jee/81.6.1648

Schaefer CH, Dupras EF Jr, Mulligan FS III (1991) Studies on the environmental persistence of S-31183 (pyriproxyfen): adsorption onto organic matter and potential for leaching through soil. Ecotoxicol Environ Saf 21:207–214. https://doi.org/10.1016/0147-6513(91)90022-h

Scholte E-J, Knols BG, Takken W (2004) Autodissemination of the entomopathogenic fungus *Metarhizium anisopliae* amongst adults of the malaria vector *Anopheles gambiae* s.s. Malar J 3:45. https://doi.org/10.1186/1475-2875-3-45

Seccacini E, Lucia A, Harburguer L, Zerba E, Licastro S, Masuh H (2008) Effectiveness of pyriproxyfen and diflubenzuron formulations as larvicides against *Aedes aegypti*. J Am Mosq Control Assoc 24:398–403. https://doi.org/10.2987/5697.1

Sihuincha MS, Zamora-Perea E, Orellana-Rios W, Stancil JD, Pez-Sifuentes VL, Vidal-Ore C et al (2005) Potential use of pyriproxyfen for control of *Aedes aegypti* (Diptera: Culicidae) in Iquitos, Perú. J Med Entomol 42:620–630. https://doi.org/10.1093/jmedent/42.4.620

Sláma K, Williams CM (1965) Juvenile hormone activity for the bug *Pyrrhocoris apterus*. Proc Natl Acad Sci USA 54:411–414. https://doi.org/10.1073/pnas.54.2.411

Smagghe G, Degheele D (1994) Action of a novel nonsteroidal ecdysteroid mimic, tebufenozide (RH-5992), on insects of different orders. Pestic Sci 42:85–92. https://doi.org/10.1002/ps.2780420204

Su T (2023) Comparative activity and efficacy of Sumilarv 0.5G and Altosid Xr Briquet against *Culex quinquefasciatus* and *Aedes aegypti* in simulated catch basins. J Am Mosq Control Assoc 39:173–182. https://doi.org/10.2987/23-7134

Suzuki H, Okazawa T, Kere N, Kawada H (1989) Field evaluation of a new insect growth regulator, pyriproxyfen, against *Anopheles farauti*, the main vector of malaria in the Solomon Islands. Jpn J Sanit Zool 40:253–257. https://doi.org/10.7601/mez.40.253

Syafruddin AR, Kamimura K, Kawamoto F (1990) Histopathological effects of an insect growth regulator, 4-phenoxyphenyl (RS)-2-(2-pyridyloxy)propyl ether (pyriproxyfen), on the larvae of *Aedes aegypti*. Jpn J Sanit Zool 41:15–22. https://doi.org/10.7601/mez.41.15

Thongrungiat S, Kanda T (1991) Efficacy of pyriproxyfen for the control of *Culex tritaeniorhynchus* at an open rice field in Bang Len, Nakhon Pathom province, Thailand. Trop Biomed 8:113–116

Unlu I, Rochlin I, Suman DS, Wang Y, Chandel K, Gaugler R (2020) Large-scale operational pyriproxyfen autodissemination deployment to suppress the immature Asian tiger mosquito (Diptera: Culicidae) populations. J Med Entomol 57:1120–1130. https://doi.org/10.1093/jme/tjaa011

Vinson SB, Plapp FW (1974) Third generation pesticides. Potential for the development of resistance by insects. J Agric Food Chem 22:356–360. https://doi.org/10.1021/jf60193a026

Vythilingam I, Luz BM, Hanni R, Beng TS, Huat TC (2005) Laboratory and field evaluation of the insect growth regulator pyriproxyfen (Sumilarv® 0.5G) against dengue vectors. J Am Mosq Control Assoc 21:296–300. https://doi.org/10.2987/8756-971X(2005)21[296:LAFEOT]2.0.CO;2

Wang Z-H, Gong Y-J, Jin G-H, Li B-Y, Chen J-C, Kang Z-J et al (2016) Field-evolved resistance to insecticides in the invasive western flower thrips *Frankliniella occidentalis* (Pergande) (Thysanoptera: Thripidae) in China. Pest Manag Sci 72:1440–1444. https://doi.org/10.1002/ps.4200

Wellinga K, Mulder R, van Daalen JJ (1973) Synthesis and laboratory evaluation of 1-(2,6-disubstituted benzoyl)-3-phenylureas, a new class of insecticides. II. Influence of the acyl moiety on insecticidal activity. J Agric Food Chem 21:993–998. https://doi.org/10.1021/jf60190a052

Williams CM (1967) Third-generation pesticides. Sci Am 217:13–17. https://doi.org/10.1038/scientificamerican0767-13

Yehwalaw D, Erena E, Degefa T, Zemene E, Simma EA, Kifle YG (2025) Biological and residual activity of candidate larvicide formulation, SumiLarv® TM 2MR, against an exotic invasive mosquito *Anopheles stephensi* Liston in Ethiopia. Sci Rep 15:291. https://doi.org/10.1038/s41598-024-80019-8

Yunta C, Grisales N, Nász S, Hemmings K, Pignatelli P, Voice M, Ranson H, Paine MJI (2016) Pyriproxyfen is metabolized by P450s associated with pyrethroid resistance in *An. gambiae*. Insect Biochem Mol Biol 78:50–57. https://doi.org/10.1016/j.ibmb.2016.09.001

Chapter 17
Development of Juvenile Hormone Inhibitors

Kenjiro Furuta

Abstract Juvenile hormone (JH) plays a crucial role in insect development by regulating molting and metamorphosis. Due to its insect-specific functions, JH biosynthesis and signaling have long been attractive targets for insect growth regulators (IGRs). JH analogs, such as methoprene, pyriproxyfen, and fenoxycarb, have been successfully commercialized; however, they often prolong the larval stage in Lepidoptera, which can exacerbate crop damage. In contrast, anti-JH agents, classified as either JH biosynthesis inhibitors or JH antagonists, induce precocious metamorphosis, offering a promising approach for controlling pests in the larval stage. Several JH biosynthesis inhibitors, such as precocene and compactin, have been reported; however, their limited target spectrum and toxicity to mammals hinder their practical applications. Recent advances have clarified the JH signaling pathway, including the JH receptor, methoprene-tolerant (Met). Potential JH antagonists, including *N*-[2-(4-tert-butyl-2-chlorophenoxy)ethyl]-1*H*-1,2,4-triazole (JHSI48) and kingidiol (SS5A), were identified using JH receptor–based in vitro screening. Ethyl (*E*)-3-(4-{[7-(4-methoxycarbonylbenzyloxy)-1,4-benzodioxan-6-yl]methyl}phenyl)-2-propenoate (EMBP) has demonstrated the most potent JH antagonistic activity in vitro and in vivo by specifically suppressing JH-responsive gene expression, thereby inducing precocious metamorphosis.

Keywords Biosynthesis inhibitors · Antagonist · Biological evaluation · Mode of action · Molecular design

K. Furuta (✉)
Institute of Agrobiological Sciences, National Agriculture and Food Research Organization, Tsukuba, Japan
e-mail: furuta.kenjiro067@naro.go.jp

T. Shinoda et al. (eds.), *Juvenile Hormone*, Entomology Monographs,
https://doi.org/10.1007/978-981-95-8605-9_17

17.1 Introduction

Juvenile hormone (JH) is involved in the physiological regulatory mechanisms in insects, such as molting and metamorphosis. Therefore, JH has long been explored as a target for insect growth regulators (IGRs). Terpenoid compounds, such as methoprene and kinoprene, have been developed by modifying the structure of JH and are used as insecticides against sanitary pests, such as mosquitoes and flies (Henrick et al. 1973). Subsequently, JH analogs with a 4-phenoxyphenoxy structure, such as pyriproxyfen and fenoxycarb, have been commercialized (Kawada et al. 1989; Grenier and Grenier 1993) (see Chap. 16 for details). These JH analogs are not only used as pesticides but are also employed in basic research as substitutes for JH because of their relative stability within insect bodies (Fig. 17.1). Natural substances with biological activities similar to those of JH have also been identified. These include juvabione, identified as the "paper factor" and juvocimene, which was isolated from the sweet basil, *Ocimum basilicum,* and exhibits ovicidal activity against eggs of the firebug, *Pyrrhocoris apterus* (Bowers et al. 1966; Bowers and Nishida 1980). Among the JH analogs, it has been confirmed that methoprene, pyriproxyfen, and fenoxycarb bind to the JH receptor, methoprene-tolerant (Met) (Jindra et al. 2015; Bittova et al. 2019).

JH analogs are effective against pests that cause damage in the adult stage, such as mosquitoes and flies. However, there is a concern that they may prolong the larval period, thereby exacerbating the damage caused by pests, such as lepidopteran

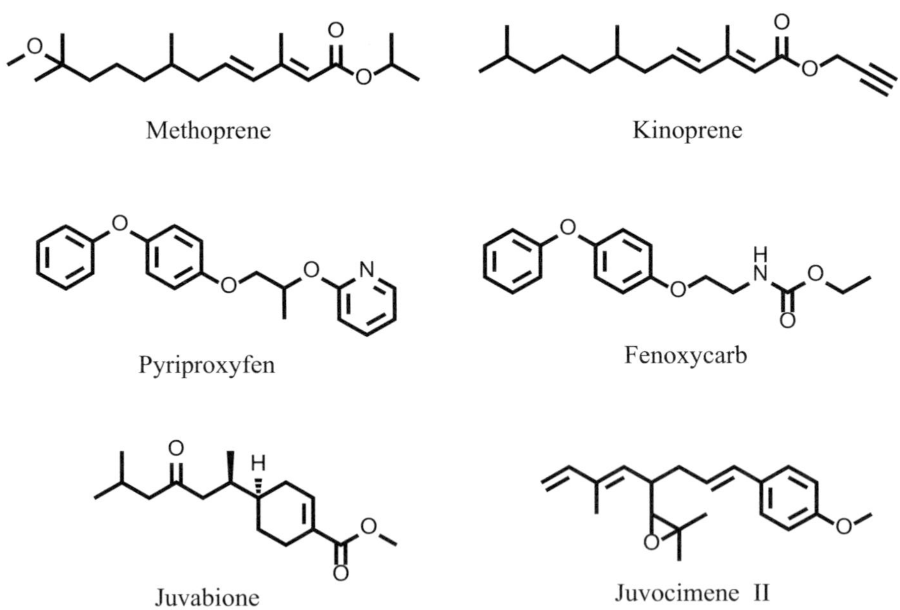

Fig. 17.1 The chemical structures of juvenile hormone analogs

insects, during the larval stage. Conversely, anti-JH agents can induce precocious metamorphosis by inhibiting the action of endogenous JH. This could reduce feeding damage during the larval stage of lepidopteran insects. Therefore, anti-JH agents are expected to be used as IGRs. Anti-JH agents can be broadly categorized as either JH biosynthesis inhibitors, which block the action of JH biosynthetic enzymes in the corpora allata (CA), or JH antagonists, which prevent JH from binding to the JH receptors. However, none of these have been commercialized to date. This is largely because the mechanisms of JH biosynthesis and signaling, including the identification of the JH receptor, have remained unclear for many years. Recently, however, the JH receptor has been identified (Konopova et al. 2007; Li et al. 2011), and the JH signaling pathway has been progressively elucidated. Additionally, key enzymes involved in JH biosynthesis have been identified in the silkworm, *Bombyx mori*, including JH epoxidase (CYP15C1) (Helvig et al. 2004; Daimon et al. 2012), which converts farnesoic acid to JH acid, and juvenile hormone acid *O*-methyltransferase (JHAMT) (Shinoda et al. 2003), which converts JH acid to JH.

Given these recent advances in JH research, the development of more effective anti-JH agents is highly anticipated. This chapter presents the structures and mechanisms of action of the anti-JH agents reported to date, with particular emphasis on the results of our study on *B. mori*.

17.2 JH Biosynthesis Inhibitors

The JH biosynthesis pathway involves both the generally conserved mevalonate pathway, which produces mevalonate in both insects and vertebrates, and insect-specific enzymatic reactions that generate the final active hormones. Inhibitors targeting this pathway can be classified into two major types: (1) cytotoxic compounds that damage CA and (2) enzyme inhibitors that interfere with one or more biochemical steps in the biosynthetic pathway. The former group includes well-characterized precocenes, originally isolated from plants and known to induce precocious metamorphosis in certain insects by selectively damaging the CA (Brooks et al. 1979). The latter group includes structurally diverse compounds, such as fluoromevalonate (FMeV), compactin, and imidazole derivatives, which inhibit key enzymes involved in JH biosynthesis.

In the following sections, representative examples of each group are described, detailing their mechanisms of action, target specificity, and biological effects.

17.2.1 Precocenes

Precocene is an anti-JH agent originally isolated from the flossflower, *Ageratum houstonianum,* and exists in two natural forms: 7-methoxy-2,2-dimethylchromene (precocene I) and 6,7-dimethoxy-2,2-dimethylchromene (precocene II) (Fig. 17.2).

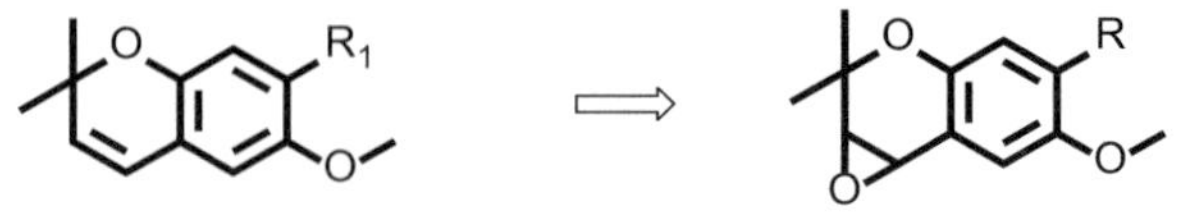

Precocene I : R = H
Precocene II : R = OMe
Precocene III : R = OEt

Fig. 17.2 The chemical structures of precocenes and their metabolites

These compounds induce precocious metamorphosis in the large milkweed bug, *Oncopeltus fasciatus,* by causing atrophy of the CA, thereby inhibiting JH biosynthesis and secretion (Brooks et al. 1979). However, the biological activity of natural precocenes is limited to hemipterous and orthopterous insects only. Therefore, various analogs were synthesized, among which 6-methoxy-7-ethoxy-2,2-dimethylchromene, known as precocene III, exhibited ten times higher anti-JH activity than precocene II. Precocene III induced precocious metamorphosis in the green peach aphid, *Myzus persicae,* by destroying its CA (Hales et al. 1981). Precocenes are not intrinsically cytotoxic. It has been proposed that the double bond between positions 3 and 4 on the chromene ring is oxidized in vivo to form a reactive epoxide (Soderlund et al. 1981). This metabolite can alkylate proteins and nucleic acids within the CA, leading to glandular degeneration. Despite their promising anti-JH effects in insects, precocenes exhibit hepatotoxicity and nephrotoxicity in mammals, limiting their practical application as insecticides (Hsia et al. 1981; Schranke et al. 1982).

17.2.2 JH Biosynthetic Enzyme Inhibitors

17.2.2.1 Fluoromevalonate and Compactin

Well-known inhibitors of the JH biosynthetic pathway include fluoromevalonate (FMeV) (Quistad et al. 1981) and compactin (Monger et al. 1982) (Fig. 17.3). Originally developed as inhibitors of cholesterol synthesis in mammals, these compounds exert their effects by inhibiting 3-hydroxy-3-methylglutaryl-CoA reductase (HMG-CoA reductase), a key enzyme in the mevalonate pathway. In insects, this inhibition affects the production of prenyl intermediates, such as farnesol and farnesoic acid, which are essential for JH synthesis. Compactin has been reported to exhibit anti-JH activity in the tobacco hornworm, *Manduca sexta*, resulting in disrupted development and altered metamorphosis. However, as these compounds target a pathway that is also found in mammals, they lack selectivity for insect-specific processes and could pose a risk of off-target toxicity to nontarget organisms, including humans. This restricts their practical use as IGRs despite their research usefulness.

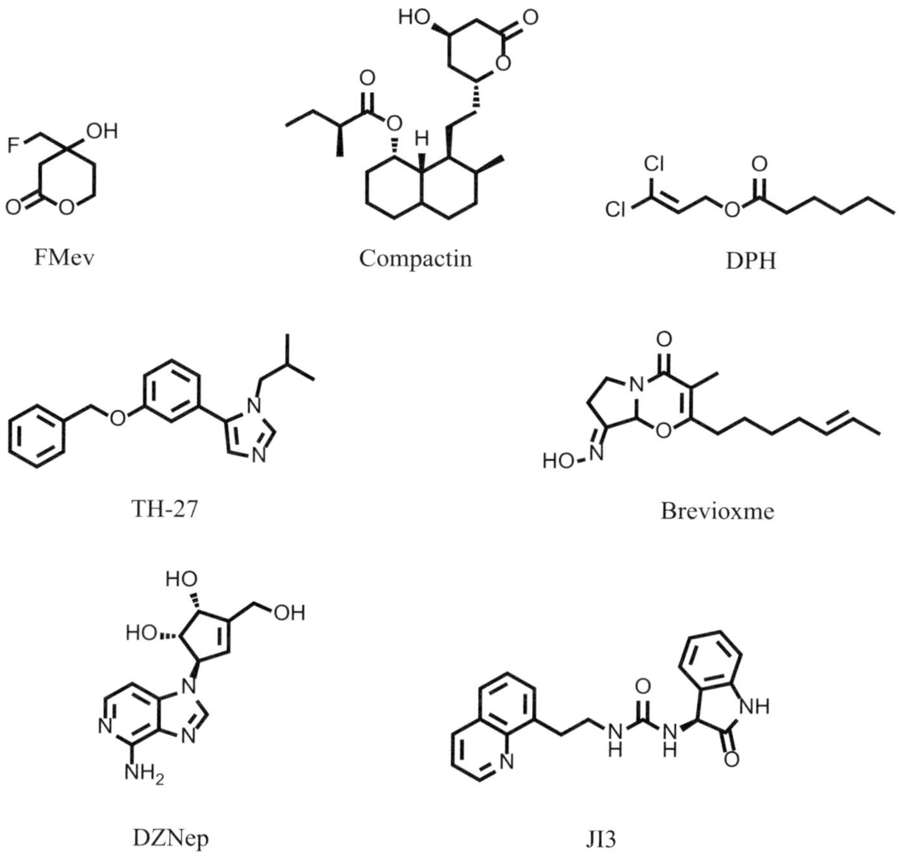

Fig. 17.3 The chemical structures of juvenile hormone biosynthesis inhibitors

17.2.2.2 3,3-Dichloro-2-Pentenyl Hexanoate

3,3-dichloro-2-pentenyl hexanoate (DPH) has also been identified as a potent inhibitor of JH biosynthesis (Quistad et al. 1985) (Fig. 17.3). In vitro studies using CA have demonstrated that DPH substantially inhibits the production of JH. Interestingly, this inhibitory effect was reversed by the addition of farnesoic acid, which is produced in the later stages of the JH biosynthetic pathway. This suggests that DPH acts upstream of farnesoic acid, possibly by targeting the enzymes that convert mevalonate into isoprenoid precursors early in this pathway. However, given the overlap of these early steps with the vertebrate mevalonate pathway, DPH, like FMeV and compactin, suffers from a lack of specificity. Its potential toxicity to vertebrates raises concerns about its use as an insecticide and necessitates the development of more selective inhibitors that target insect-specific enzymes in the downstream portion of the JH pathway.

17.2.2.3 1,5-Disubstituted Imidazole Derivative

To develop more selective inhibitors, Unnithan et al. (1995) examined a series of 1,5-disubstituted imidazole derivatives to determine their inhibitory effects on JH biosynthesis. Using a radiochemical in vitro assay with CA from the Pacific beetle cockroach, *Diploptera punctata*, they evaluated the activities of 17 different compounds. Among these, 1-isobutyl-5-(3-benzyloxyphenyl)imidazole (TH-27) exhibited the most potent activity, with a half-maximal inhibitory concentration (IC_{50}) value of 36 nM (Fig. 17.3).

In addition to its in vitro potency, TH-27 caused the accumulation of methyl farnesoate, a direct JH precursor, in larvae. This accumulation suggests that TH-27 inhibits JH epoxidase, the enzyme responsible for converting methyl farnesoate into active JH. Furthermore, when administered to *B. mori*, TH-27 induced precocious metamorphosis, supporting its in vivo efficacy and physiological relevance as an anti-JH agent.

17.2.2.4 Brevioxime: A Fungal Metabolite

Brevioxime, a naturally occurring compound isolated from the fungus *Penicillium brevicompactum* is an interesting JH biosynthesis inhibitor that targets the final steps of the pathway (Fig. 17.3). In vitro assays using CA from the migratory locust, *Locusta migratoria,* showed that brevioxime effectively suppressed JH production. Significantly, the addition of farnesol or farnesoic acid did not rescue JH biosynthesis in brevioxime-treated glands, indicating that brevioxime inhibits either the epoxidation or methylation step necessary for the final conversion to active JH (Moya et al. 1997).

Brevioxime also induced precocious metamorphosis in *O. fasciatus*, demonstrating its interference with JH biosynthesis in vivo. This unique mode of action, which targets late-stage biosynthetic enzymes that are more specific to insects, makes brevioxime a promising candidate for selective JH inhibition with potentially lower toxicity risks.

17.2.2.5 JHAMT Inhibitors

3-deazaneplanocin A (DZNep) was initially recognized for its ability to suppress JH synthesis, effectively lowering JH titers in the yellow fever mosquito, *Aedes aegypti* (Fig. 17.3) (Nouzová et al. 2019). However, DZNep does not directly inhibit the activity of JHAMT. Instead, it primarily targets *S*-adenosyl-*L*-homocysteine (SAH) hydrolase, leading to the accumulation of SAH, a potent competitive inhibitor of methylation. Consequently, DZNep indirectly suppresses the activity of a broad range of *S*-adenosylmethionine (SAM)-dependent methyltransferases, including JHAMT. Although DZNep validated the concept of reducing JH biosynthesis via

JHAMT modulation, its low selectivity and nonspecific toxicity limit its practical application as a lead compound for insecticide development.

(*S*)-1-(2-oxoindolin-3-yl)-3-[2-(quinoline-8-yl)ethyl]urea (JI3), identified through structure-based screening targeting TcJHAMT3 from the red flour beetle, *Tribolium castaneum*, exhibits a distinct mechanism of action and potent biological activity compared to nonselective inhibitors such as DZNep (Fig. 17.3) (Yang et al. 2025). In vitro analyses demonstrated that JI3 is a direct inhibitor of JHAMT, showing marked inhibitory activity against recombinant TcJHAMT3 with an IC_{50} value of 65.04 ± 1.98 μM. In vivo experiments further demonstrated that oral administration of *T. castaneum* larvae induced high mortality. Transcriptional analysis confirmed this mechanism: JI3 treatment significantly downregulated the expression of JH signaling components in *T. castaneum*, including *TcJHAMT3*, *Met*, and *Krüppel homolog 1* (*Kr-h1*), while upregulating *Broad-Complex* (*BR-C*). These shifts indicate that JI3 effectively removes the JH-mediated "brake" on metamorphosis, allowing the pupation signals to predominate. These findings validate JI3 as a lead compound for the development of selective JHAMT inhibitors.

17.3 JH Antagonists

For many years, progress in the discovery of JH antagonists has been limited because JH receptors have not yet been clearly identified. Consequently, early studies predominantly relied on in vivo assays using insect models, making it difficult to characterize the precise mode of action or target specificity of candidate compounds.

However, the subsequent identification of the JH receptor, Met, along with its transcriptional co-regulators, has led to substantial advances in our understanding of JH signaling pathways (Konopova et al. 2007; Li et al. 2011). This has also enabled the development of reliable in vitro screening systems, such as reporter gene assays and receptor-based interaction models. These tools now allow for the high-throughput mechanistic evaluation of both JH agonists and antagonists.

This section provides an overview of JH antagonists, ranging from early synthetic compounds such as ethyl 4-[2-(tert-butylcarbonyloxy)butoxy]benzoate (ETB) and its derivative, ethyl 4-(2-benzylbutyloxy)benzoate (KF-13), to more potent agents such as the retinoid-inspired ethyl 4-[(6-methoxy-2,2-dimethyl-2H-chromen-7-yl)methoxy]benzoate (KF-38). These three compounds exhibit dual activities, functioning primarily as antagonists while retaining weak agonistic properties.

In contrast, the overview then extends to highly potent benzodioxane derivatives, including ethyl 4-([7-benzyloxy-1,4-benzodioxan-6-yl]methyl)benzoate (NY-04) and ethyl (*E*)-3-(4-{[7-(4-methoxycarbonylbenzyloxy)-1,4-benzodioxan-6-yl]methyl} phenyl)-2-propenate (EMBP), both of which act as full antagonists with no detectable agonistic activity against *B. mori*. The structure–activity relationships (SARs) and biological effects on insect development are also discussed.

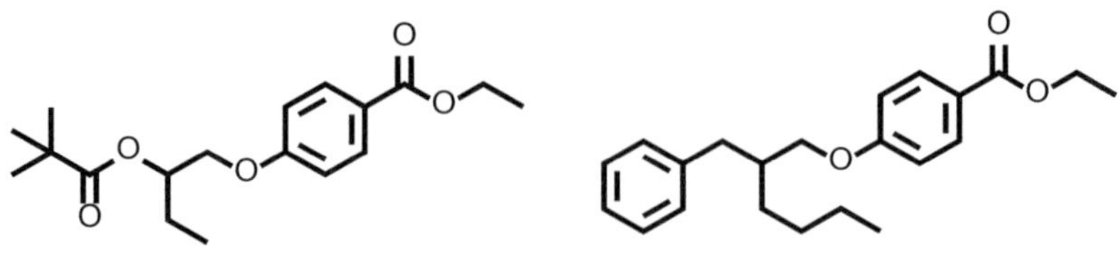

ETB KF-13

Fig. 17.4 The chemical structures of ETB and KF-13

17.3.1 ETB and Its Derivatives

Before the identification of the JH receptor, the biological evaluation of JH antagonists primarily relied on in vivo phenotypic assays using model insects, such as *B. mori* and *M. sexta*. One of the most extensively studied JH antagonists is ETB, which disrupts molting and metamorphosis (Fig. 17.4). ETB exhibits dose-dependent dual effects: At low doses, it induces precocious metamorphosis, indicating JH antagonism, whereas at higher doses, it inhibits metamorphosis, indicating JH-mimicking activity. This biphasic response suggests that ETB acts as a partial JH antagonist (Staal 1986).

Furuta et al. (2007) conducted SAR studies to develop more potent ETB derivatives. Among them, KF-13 exhibited significantly higher precocious metamorphosis–inducing activity than ETB against *B. mori*. Interestingly, co-treatment with methoprene completely counteracted these antagonistic effects, whereas 20-hydroxyecdysone did not. This pharmacological profile confirms that KF-13 specifically interferes with JH function. The SAR analysis of KF-13 revealed several important findings. First, the ethyl ester moiety was essential for activity. Second, the optimal alkyl side chain was identified as *n*-butyl or isobutyl. Third, the *S*-enantiomer (KF-13S) was identified as a biologically active stereoisomer. KF-13S exhibited anti-JH activity at low doses, but its activity was reduced at higher doses. When *B. mori* whose CA had been surgically removed—thereby eliminating endogenous JH synthesis—was treated with KF-13S, the larvae molted into final instar larvae. Since the removal of the CA normally induces precocious metamorphosis due to the absence of JH, these results suggest that KF-13S possesses both anti-JH and JH agonistic activities, functioning as a partial JH antagonist similar to ETB (Fujita et al. 2008).

Kaneko et al. (2011) further investigated the effects of KF-13S on CA using an organ culture system. They reported that KF-13S inhibited JH biosynthesis in the CA. Subsequent analysis of the expression of JH biosynthetic enzymes revealed that KF-13S suppressed the expression of several enzyme genes, including *HMG-CoA synthase*, *HMG-CoA reductase*, and *JHAMT*. These findings demonstrate that KF-13S reduces JH production by decreasing the levels of key enzymes involved in its synthesis.

17.3.2 Retinoid Derivatives

Palli et al. (1991) reported that two retinoic acid analogs, SRI 5942-64 and Ro13-6298, exhibited modest JH agonistic activity in black mutant larvae of *M. sexta* compared to natural JH. Since retinoic acid and JH are both terpenoids with similar structural backbones, Furuta et al. explored novel anti-JH agents based on retinoid analogs. Building on these findings, they designed and synthesized KF-38 (Fig. 17.5), in which the six-membered ring of retinoic acid was replaced by a 2,2-dimethyl-2*H*-chromene ring containing an oxygen atom at the corresponding position (Furuta et al. 2010).

When evaluated for anti-JH activity against *B. mori* larvae, KF-38 induced appreciable precocious metamorphosis. In contrast to KF-13S, KF-38 exhibited a clear dose-dependent increase in activity. When applied to fourth-instar *B. mori* larvae with the CA surgically removed, KF-38 showed only weak JH agonistic activity compared to that of KF-13S. SAR studies revealed that extending the alkyl chain of the methoxy group at position 6 on the chromene ring substantially reduced anti-JH activity, highlighting the importance of this substituent in the chromene ring. Similarly, the ester group at the position 4 was essential for activity, as observed for KF-13S.

Using liquid chromatography–mass spectrometry (LC-MS), the concentration of JH I in the hemolymph of *B. mori* treated with KF-38 was measured to clarify the mode of action of KF-38. The JH I titer declined sharply within 24 h and remained consistently low until the onset of pupation. Since KF-38 contains the 2,2-dimethyl-2*H*-chromene ring found in precocenes, it was proposed that KF-38 might inhibit JH production through a mechanism similar to that of precocenes. To investigate this, a derivative with a double bond in the chromene ring reduced to a chromane ring was synthesized. Although its anti-JH activity slightly decreased, it remained considerable, suggesting that KF-38 acts via a mechanism distinct from that of precocene.

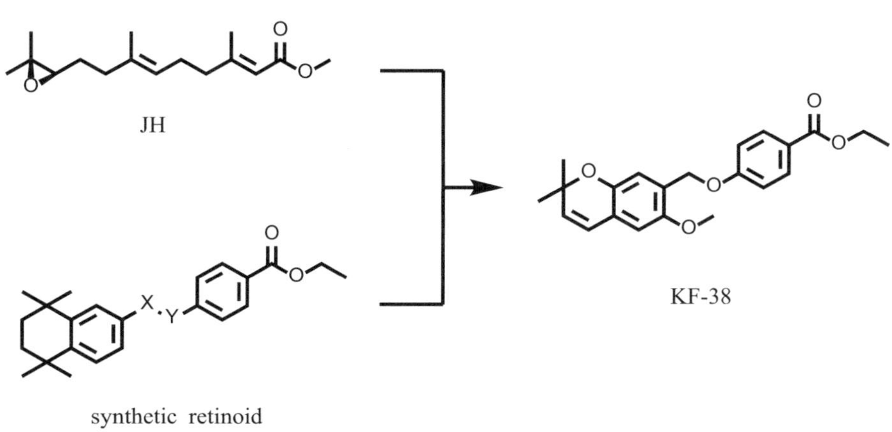

Fig. 17.5 The design of the chromene derivatives based on juvenile hormone and synthetic retinoid

17.3.3 JH Antagonists Identified by In Vitro Assay Systems

Kayukawa et al. (2020) developed an in vitro screening system based on a reporter assay designed to enable rapid, sensitive, and straightforward evaluation of the JH agonist and antagonistic activities of various test compounds. This system employs a *B. mori*–derived cell line, BmN_JF&AR, which stably expresses two plasmids. One plasmid, BmN_JHRE-Fluc, contains the JH response element (JHRE) cloned upstream of a luc2P firefly luciferase gene, which incorporates a PEST degradation sequence to reduce the background signal. The second plasmid carried a Renilla luciferase gene (hRlucP) driven by the BmA3 promoter, serving as a reference reporter to normalize the monitoring of cytotoxicity.

The JH agonistic activity of a test compound was assessed by measuring firefly luciferase activity. Upon treatment with JH I, BmN_JF&AR cells exhibited a clear, concentration-dependent increase in luminescence, with a median effective concentration (EC_{50}) of 0.37 nM, demonstrating the high sensitivity of this system (Fig. 17.6a). To evaluate JH antagonistic activity, cells were co-treated with JH I and the test compound, and the antagonistic effects were quantified by the degree of suppression of JH-induced luminescence (Fig. 17.6b). Furthermore, measuring Renilla luciferase activity after treatment with the test compound alone allows for the exclusion of false positives caused by cytotoxicity, thereby improving assay reliability.

Kayukawa et al. performed high-throughput screening using a chemical library containing 9600 diverse compounds to identify potential JH antagonists. This

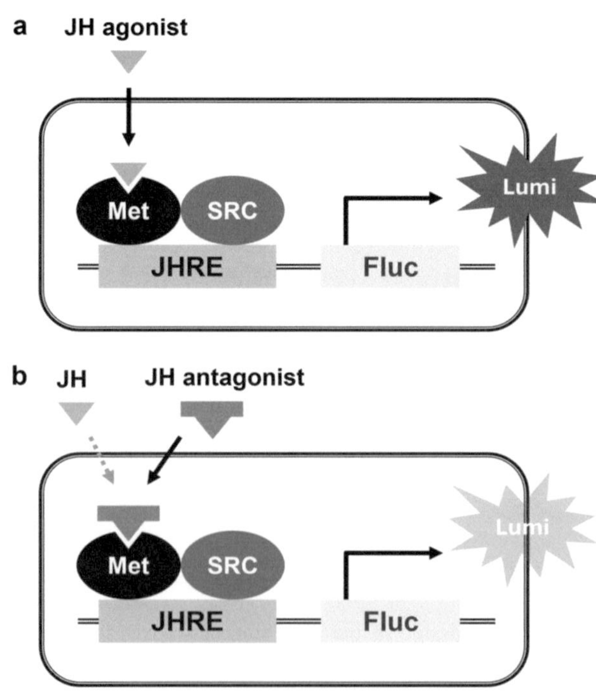

Fig. 17.6 Schematic diagram of the reporter assays using BmN_JF&AR cells. (**a**) juvenile hormone (JH) agonist test and (**b**) JH antagonist test. MET Methoprene-tolerant, SRC Steroid receptor coactivator, JHRE JH response element, Fluc Firefly luciferase gene, Lumi Luminescence

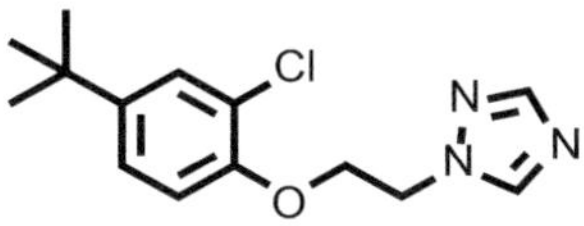

JHSI48

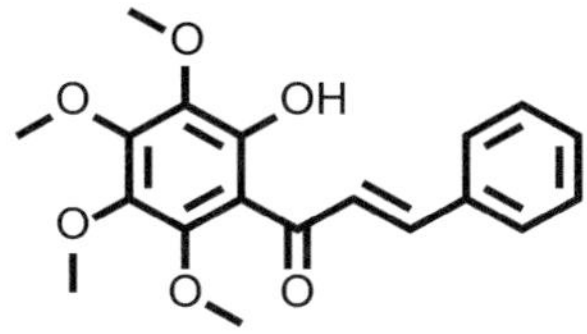

LE3B

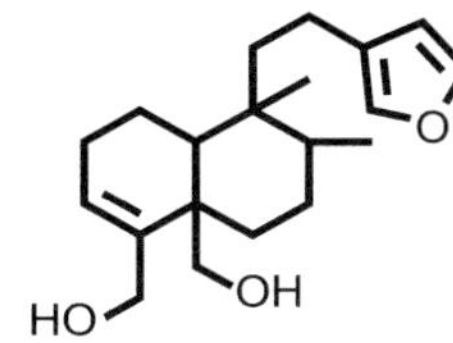

SS5A

Fig. 17.7 The chemical structures of juvenile hormone antagonists identified through in vitro screening

screening yielded 69 compounds with inhibitory activity. Notably, *N*-[2-(4-tert-butyl-2-chlorophenoxy)ethyl]-1*H*-1,2,4-triazole (JHSI48) induced precocious metamorphosis in *B. mori* larvae in vivo (Fig. 17.7). Further investigation of the mechanism of JHSI48 revealed that it suppresses *Kr-h1* expression in *B. mori* larval epidermis in a concentration-dependent manner. Additionally, a competitive inhibition assay using a reporter system demonstrated that the JH dose–response curve shifted rightward in the presence of JHSI48, indicating that JHSI48 acts as a competitive antagonist by inhibiting JH binding to its receptors.

Lee et al. (2015) developed a yeast two-hybrid assay targeting the interaction between *A. aegypti* JH receptor, Met, and its cofactors Cycle (CYC) or βFtz-F1-interacting steroid receptor coactivator (FISC). The addition of pyriproxyfen triggers Met–CYC complex formation and activates β-galactosidase. This assay provides an efficient *in vitro* method for evaluating JH antagonists by measuring their potency in preventing the interaction between the JH receptor and its cofactors. They used this assay to screen 1651 plant extracts for inhibitors of this interaction. This led to the identification of 83 extracts that disrupted the Met–CYC binding. Of these, 53 compounds exhibited significant larval lethality in bioassays. Five diterpene compounds were isolated from two of these plants: *Lindera erythrocarpa* and *Solidago serotina*. Of these, kanakugiol (LE3B) and kingidiol (SS5A) inhibited the Met–CYC interaction in vitro; suppressed the expression of two Met downstream target genes, Hairy and Kr-h1; and delayed ovarian development in adult mosquitoes (Fig. 17.7).

Recently, Tumova and Jindra (2023) developed a real-time NanoBiT assay that allows dynamic monitoring of JH receptor activation via a ligand-dependent interaction between Met and taiman (Tai). When exposed to JH, Met undergoes a conformational change and rapidly forms a heterodimer with Tai, resulting in a quantifiable luminescent signal within living cells. Owing to its sensitivity, speed, and ability to capture receptor-level interactions in a physiologically relevant context, this assay system is highly promising for screening potential JH antagonists. In principle, compounds that interfere with JH binding or Met–Tai dimerization can be identified by their ability to reduce or inhibit luminescent responses. Further validation of these assays is needed; however, this approach could be a valuable alternative to conventional transcription-based assays for discovering novel JH antagonists.

17.3.4 1,4-Benzodioxane Derivatives

The author conducted in vitro screening for JH agonists and antagonists using a reporter gene assay based on transient expression, as described by Kayukawa et al. (2012), as follows. ETB, KF-13, and KF-38, which act as partial antagonists in vivo, suppressed the JH response induced by methoprene in vitro (unpublished data). When tested individually, each compound exhibited weak JH agonistic activity, indicating that they also function as partial antagonists in vitro, consistent with their in vivo effects. Through this process, methyl 4-(1,4-benzodioxan-6-yl)carbonylbenzoate (DM-10) was identified as a JH antagonist, exhibiting only antagonistic activity and no detectable agonistic activity; however, its activity was extremely weak, and it showed no observable biological effect in *B. mori* larvae (Fig. 17.8).

In response, Yamada et al. (2016) synthesized a variety of DM-10 derivatives and evaluated their precocious metamorphosis–inducing activity against *B. mori* to investigate the SAR in detail. The findings were as follows: (1) The 1,4-benzodioxane ring is essential for activity, (2) the introduction of an alkoxy substituent at the position 7 of the benzodioxane ring markedly increases the activity, and (3) an ester moiety at the para-position of the terminal benzene ring is necessary for biological activity. Based on these findings, ethyl 4-([7-benzyloxy-1,4-benzodioxan-6-yl]methyl) benzoate (NY-04) was synthesized (Fig. 17.8). NY-04 showed significant precocious metamorphosis–inducing activity against *B. mori* (effective dose for 50% of individuals (ED_{50}) = 41 ng/larva). In vitro, NY-04 also displayed potent JH-antagonistic activity (half-maximal inhibitory concentration, IC_{50} = 123 nM) (Fig. 17.9a). Unlike ETB, however, NY-04 exhibited almost no JH agonistic activity, suggesting that it acts as a complete JH antagonist.

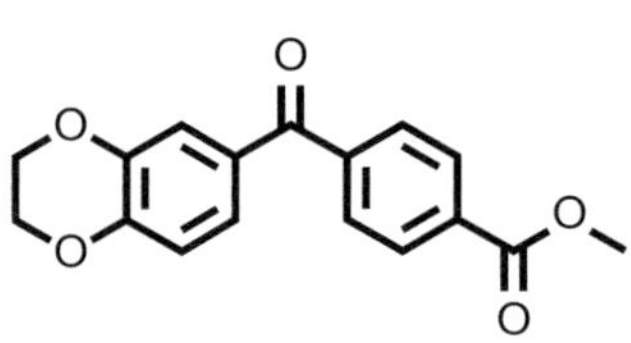

DM-10

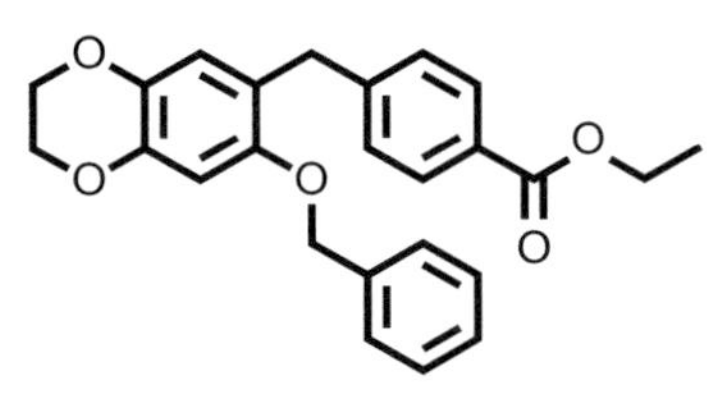

NY-04

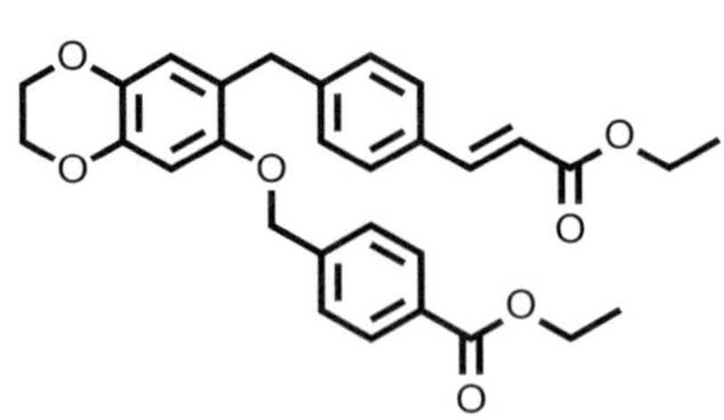

EMBP

Fig. 17.8 The chemical structures of 1,4-benzodioxane derivatives

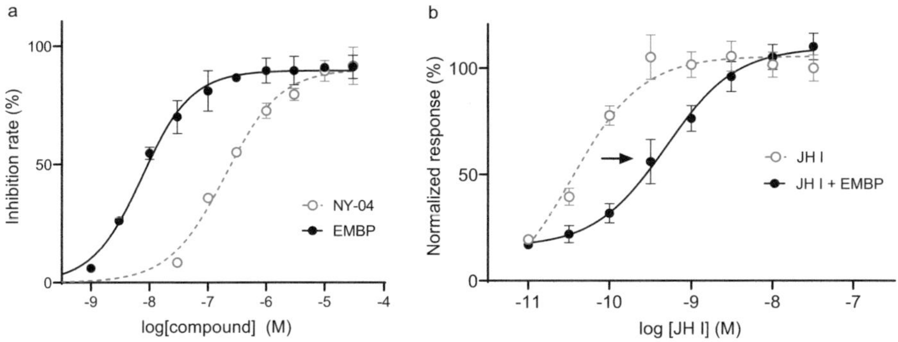

Fig. 17.9 (**a**) Juvenile hormone I (JH I) signaling inhibitory activity of NY-04 and EMBP in the reporter assay. (**b**) Concentration–response curves of JH I in the presence and absence of EMBP. Data represent the mean ± SD ($n = 3$). (From Furuta et al. (2023), John Wiley & Sons. Used with permission)

Subsequently, Furuta et al. modified NY-04 to further enhance its biological activity by introducing a 4-methoxycarbonylbenzyl group into the benzyloxy moiety and inserting a vinylene group between the ethoxycarbonyl group and the benzene ring. This modification produced ethyl (E)-3-(4-{[7-(4-methoxycarbonylbenzyloxy)-1,4-benzodioxan-6-yl]methyl}phenyl)-2-propenate (EMBP) (Fig. 17.8) (Furuta et al. 2023). The precocious metamorphosis–inducing activity of EMBP was approximately seven times higher than that of NY-04 in *B. mori* larvae (ED_{50} = 6.2 ng/larva). EMBP exhibited extremely high JH-antagonistic activity in vitro (IC_{50} = 7.5 nM), which was over 15 times more potent than that of NY-04 (Fig. 17.9a). Furthermore, in a competitive inhibition assay, the JH dose–response curve shifted to the right in the presence of EMBP, suggesting that it also acts as a competitive antagonist, similar to JHSI48 (Fig. 17.9b). To elucidate the mechanism of EMBP in vivo, the effect of EMBP on the expression levels of JH- and ecdysone-responsive genes in the epidermis of *B. mori* larvae was examined. Significant decreases in the expression levels of the JH-responsive genes *BmKr-h1α* and *BmKr-h1β* were observed within 12 h of EMBP treatment (Fig. 17.10a). In contrast, the expression of the ecdysone-responsive gene *BmBR-C* was unaffected (Fig. 17.10b). These results suggest that EMBP specifically inhibits the JH signaling pathway to induce precocious metamorphosis.

17.4 Future Perspectives

Recent advances in our understanding of the biosynthesis and signaling pathways of JH have laid a solid foundation for the rational design of potent anti-JH agents. Several compounds targeting mevalonate pathway enzymes have been identified as inhibitors of JH biosynthesis. However, specific inhibitors of key enzymes, such as

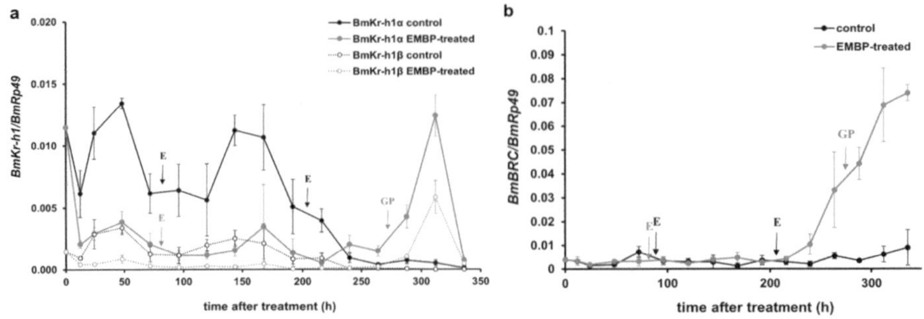

Fig. 17.10 Developmental changes in *Krüppel homolog 1* (*BmKr-h1*) and *Broad-Complex* (*BmBR-C*) in the epidermis of *Bombyx mori* larvae treated with EMBP. (**a**) Expression levels of *BmKr-h1* and (**b**) all isoforms of *BmBR-C*. E: ecdysis and GP: gut purge. Data represent the mean ± SD ($n = 3$). (From Furuta et al. (2023), John Wiley & Sons. Used with permission)

JH epoxidase and JHAMT, which play central roles in JH biosynthesis, remain largely unexplored. In contrast, several partial JH antagonists, including ETB, KF-13, and KF-38, have been reported. Notably, EMBP was recently identified as a full JH antagonist that binds with high affinity to the JH receptor Met and effectively suppresses JH-induced signaling. This discovery represents a significant breakthrough in the field.

Despite their promise, EMBP and other currently known JH antagonists face challenges in terms of potency, selectivity, species specificity, and metabolic stability. Further structural optimization is essential for practical applications.

Notably, the increasing availability of structural information on JH receptors and biosynthetic enzymes has paved the way for in silico screening and structure-based drug design (SBDD). In particular, high-precision modeling of ligand-binding domains enables virtual screening and AI-assisted molecular design, facilitating the efficient identification of novel lead compounds. Recent efforts using pharmacophore modeling and structural similarity–based screening have yielded promising candidates. High-throughput screening technologies enable the predictive identification of anti-JH agents in large chemical libraries. These methods markedly accelerate early-stage discoveries and reduce the need for costly in vivo testing. Additionally, targeting non-model insects could allow the customization of anti-JH agents for specific pest species, thereby enhancing both efficacy and safety.

Moving forward, the integration of computational drug discovery techniques with new insights into JH is expected to significantly accelerate the development of high-affinity anti-JH compounds. These efforts could lead to the development of next-generation insect control agents that are safer, more specific, and more effective.

References

Bittova L, Jedlicka P, Dracinsky M, Kirubakaran P, Vondrasek J, Hanus R et al (2019) Exquisite ligand stereoselectivity of a *Drosophila* juvenile hormone receptor contrasts with its broad agonist repertoire. J Biol Chem 294:410–423. https://doi.org/10.1074/jbc.RA118.005992

Bowers WS, Nishida R (1980) Juvocimenes: potent juvenile hormone mimics from sweet basil. Science 209:1030–1032. https://doi.org/10.1126/science.209.4460.1030

Bowers WS, Fales HM, Thompson MJ, Uebel EC (1966) Juvenile hormone: identification of an active compound from balsam fir. Science 154:1020–1021. https://doi.org/10.1126/science.154.3752.1020

Brooks GT, Pratt GE, Jennings RC (1979) The action of precocenes in milkweed bugs (*Oncopeltus fasciatus*) and locusts (*Locusta migratoria*). Nature 281:570–572. https://doi.org/10.1038/281570a0

Daimon T, Kozaki T, Niwa R, Kobayashi I, Furuta K, Namiki T et al (2012) Precocious metamorphosis in the juvenile hormone–deficient mutant of the silkworm *Bombyx mori*. PLoS Genet 8:e1002486. https://doi.org/10.1371/journal.pgen.1002486

Fujita N, Furuta K, Ashibe K, Yoshida S, Yamada N, Shiotsuki T et al (2008) Juvenile hormone activity of optically active ethyl 4-(2-benzylalkyloxy)benzoates inducing precocious metamorphosis. J Pestic Sci 33:383–386. https://doi.org/10.1584/jpestics.G08-22

Furuta K, Ashibe K, Shirahashi H, Fujita N, Yamashita H, Yamada N et al (2007) Synthesis and anti-juvenile hormone activity of ethyl 4-(2-benzylalkyloxy)benzoates and their enantiomers. J Pestic Sci 32:99–105. https://doi.org/10.1584/jpestics.G06-46

Furuta K, Fujita N, Ibushi T, Shiotsuki T, Yamada N, Kuwano E (2010) Synthesis and anti-juvenile hormone activity of ethyl 4-[(6-substituted 2,2-dimethyl-2*H*-chromen-7-yl)methoxy]benzoates. J Pestic Sci 35:405–411. https://doi.org/10.1584/jpestics.G09-71

Furuta K, Yamada N, Kayukawa T (2023) Synthesis of 1,4-benzodioxan derivatives and the evaluation of their biological activity as a novel juvenile hormone signaling inhibitor. Pest Manag Sci 79:5341–5348. https://doi.org/10.1002/ps.7744

Hales DF, Mittler TE (1981) Precocious metamorphosis of the aphid *Myzus persicae* induced by the precocene analogue 6-methoxy-7-ethoxy-2,2-dimethylchromene. J Insect Physiol 27:333–341. https://doi.org/10.1016/0022-1910(81)90079-2

Helvig C, Koener JF, Unnithan GC, Feyereisen R (2004) CYP15A1, the cytochrome P450 that catalyzes epoxidation of methyl farnesoate to juvenile hormone III in cockroach corpora allata. Proc Natl Acad Sci USA 101:4024–4029. https://doi.org/10.1073/pnas.0306980101

Jindra M, Uhlirova M, Charles JP, Smykal V, Hill RJ (2015) Genetic evidence for function of the bHLH-PAS protein Gce/Met as a juvenile hormone receptor. PLoS Genet 11:e1005394. https://doi.org/10.1371/journal.pgen.1005394

Kaneko Y, Furuta K, Kuwano E, Hiruma K (2011) An anti-juvenile hormone agent, ethyl 4-(2-benzylhexyloxy)benzoate, inhibits juvenile hormone synthesis through the suppression of transcription of juvenile hormone biosynthetic enzymes in the corpora allata in *Bombyx mori*. Insect Biochem Mol Biol 41:788–794. https://doi.org/10.1016/j.ibmb.2011.05.009

Kawada H, Kojima I, Shinjo G (1989) Laboratory evaluation of a new insect growth regulator pyriproxyfen, as a cockroach control agent. Jpn J Sanit Zool 40:195–201. https://doi.org/10.7601/mez.40.195

Kayukawa T, Minakuchi C, Namiki T, Togawa T, Yoshiyama M, Kamimura M et al (2012) Transcriptional regulation of juvenile hormone-mediated induction of Krüppel homolog 1, a repressor of insect metamorphosis. Proc Natl Acad Sci USA 109:11729–11734. https://doi.org/10.1073/pnas.1204951109

Kayukawa T, Furuta K, Nagamine K, Shinoda T, Yonesu K, Okabe T (2020) Identification of a juvenile-hormone signaling inhibitor via high-throughput screening of a chemical library. Sci Rep 10:18413. https://doi.org/10.1038/s41598-020-75386-x

Konopova B, Jindra M (2007) Juvenile hormone resistance gene Methoprene-tolerant controls entry into metamorphosis in the beetle *Tribolium castaneum*. Proc Natl Acad Sci USA 104:1488–1493. https://doi.org/10.1073/pnas.0703719104

Lee SH, Oh HW, Fang Y, An SB, Park DS, Song HH et al (2015) Identification of plant compounds that disrupt the insect juvenile hormone receptor complex. Proc Natl Acad Sci USA 112:1733–1738. https://doi.org/10.1073/pnas.1424386112

Li M, Mead EA, Zhu J (2011) Heterodimer of two bHLH-PAS proteins mediates juvenile hormone-induced gene expression. Proc Natl Acad Sci USA 18:638–643. https://doi.org/10.1073/pnas.1013914108

Monger DJ, Lim WA, Kézdy FJ, Law JH (1982) Compactin inhibits insect HMG-CoA reductase and juvenile hormone biosynthesis. Biochem Biophys Res Commun 105:1374–1380. https://doi.org/10.1016/0006-291X(82)90939-1

Moya P, Castillo M, Primo-Yúfera E, Couillaud F, Martínez-Mañez R, Garcerá MD et al (1997) Brevioxime: a new juvenile hormone biosynthesis inhibitor isolated from *Penicillium brevicompactum*. J Org Chem 62:8544–8545. https://doi.org/10.1021/jo970397y

Nouzová M, Ethen C, Noriega FG (2019) Inhibition of juvenile hormone synthesis in mosquitoes by the methylation inhibitor 3-deazaneplanocin A (DZNep). Insect Biochem Mol Biol 113:103183. https://doi.org/10.1016/j.ibmb.2019.103183

Palli SR, Riddiford LM, Hiruma K (1991) Juvenile hormone and "retinoic acid" receptors in Manduca epidermis. Insect Biochem 21:7–15. https://doi.org/10.1016/0020-1790(91)90059-N

Quistad GB, Cerf DC, Schooley DA, Staal GB (1981) Fluoromevalonate acts as an inhibitor of insect juvenile hormone biosynthesis. Nature 289:176–177. https://doi.org/10.1038/289176a0

Quistad GB, Cerf DC, Kramer SJ, Bergot JB, Schooley DA (1985) Design of novel anti-juvenile hormones: allylic alcohol derivatives. J Agric Food Chem 33:47–50. https://doi.org/10.1021/jf00061a014

Shinoda T, Itoyama K (2003) Juvenile hormone acid methyltransferase: a key regulatory enzyme for insect metamorphosis. Proc Natl Acad Sci USA 100:11986–11991. https://doi.org/10.1073/pnas.2134232100

Soderlund DM, Messeguer A, Bowers WS (1981) Precocene II metabolism in insects: synthesis of potential metabolites and identification of initial in vitro biotransformation products. Pestic Biochem Physiol 15:281–292. https://doi.org/10.1021/jf60230a053

Staal GB (1986) Anti-juvenile agents. Annu Rev Entomol 31:391–429. https://doi.org/10.1146/annurev.en.31.010186.002135

Tumova S, Jindra M (2023) Ligand-dependent protein interactions of the juvenile hormone receptor captured in real time. FEBS J 290:2881–2894. https://doi.org/10.1111/febs.16719

Unnithan GC, Andersen JF, Hisano T, Kuwano E, Feyereisen R (1995) Inhibition of juvenile hormone biosynthesis and methyl farnesoate epoxidase activity by 1,5-disubstituted imidazoles in the cockroach *Diploptera punctata*. Pestic Sci 43:13–19. https://doi.org/10.1002/ps.2780430103

Yamada N, Maeda K, Masumoto M, Inagaki Y, Furuta K (2016) Anti-juvenile hormone activity of ethyl 4-[(7-substituted 1,4-benzodioxan-6-yl)methyl]benzoates and their effect on the juvenile hormone titer in the hemolymph of the silkworm *Bombyx mori*. J Pestic Sci 41:38–43. https://doi.org/10.1584/jpestics.D15-072

Yang Y, Li Q, Zheng H, Shi J, Hu Y, Xia Y et al (2025) Unveiling the molecular architecture of TcJHAMT: structural basis for catalytic efficiency and targeted inhibitor screening in the red flour beetle, Tribolium castaneum. Pest Manag Sci. https://doi.org/10.1002/ps.70267

Index

T. Shinoda et al. (eds.), *Juvenile Hormone*, Entomology Monographs,
https://doi.org/10.1007/978-981-95-8605-9

C

D

E

F

G

H

FSC
www.fsc.org
MIX
Papier aus verantwortungsvollen Quellen
Paper from responsible sources
FSC® C105338

If you have any concerns about our products, you can contact us on
ProductSafety@springernature.com

In case Publisher is established outside the EU, the EU authorized representative is:
Springer Nature Customer Service Center GmbH
Europaplatz 3, 69115 Heidelberg, Germany

Printed by Libri Plureos GmbH
in Hamburg, Germany